Thi ... hown
ed by

Teaching Geographic Information Science and Technology in Higher Education

Teaching Geographic Information Science and Technology in Higher Education

Editors

David J. Unwin

Birkbeck College London, UK

Kenneth E. Foote

University of Colorado, US

Nicholas J. Tate

University of Leicester, UK

David DiBiase

Pennsylvania State University, US

WILEY-BLACKWELL

A John Wiley & Sons, Ltd., Publication

This edition first published 2012

Wiley-Blackwell is an imprint of John Wiley & Sons, formed by the merger of Wiley's global Scientific, Technical and Medical business with Blackwell Publishing.

Registered office: John Wiley & Sons, Ltd, The Atrium, Southern Gate, Chichester, West Sussex, PO19 8SQ, UK
Editorial offices: 600 Garsington Road, Oxford, OX4 2DQ, UK
The Atrium, Southern Gate, Chichester, West Sussex, PO19 8SQ, UK
111 River Street, Hoboken, NJ 07030-5774, USA

For details of our global editorial offices, for customer services and for information about how to apply for permission to reuse the copyright material in this book please see our website at www.wiley.com/wiley-blackwell.

Library of Congress Cataloging-in-Publication Data

Teaching geographic information science and technology in higher education / David Unwin . . . [et al.].
p. cm.
Includes bibliographical references and index.
ISBN 978-0-470-74856-5 (cloth)
1. Geographic information systems–Study and teaching (Higher) I. Unwin, D. (David John)
G70.212.T36 2011
910.71′1–dc23

2011022718

A catalogue record for this book is available from the British Library.

This book is published in the following electronic formats: ePDF 9781119950585;
Wiley Online Library 9781119950592; ePub 9781119962434; Mobi 9781119962441

Set in 10/12.5pt Minion by Aptara Inc., New Delhi, India.
Printed in Singapore by Ho Printing Singapore Pte Ltd

First Impression 2012

Contents

About the editors

David J. Unwin

David, formerly professor of geography at Birkbeck College in the University of London is now retired. He was one of the team that established the *Journal of Geography in Higher Education* and in 2006 received recognition from the US *University Consortium for Geographic Information Science* as *Educator of the Year*. In 1999 his work in establishing its continuing professional development scheme led to the award of the UK *Association for Geographic Information's Past President's Prize*. He has experience of teaching GIS&T in universities in the UK, Canada, New Zealand and the USA. With David O'Sullivan he is co-author of *Geographic Information Analysis*, also published by Wiley.

Kenneth E. Foote

Ken is a professor of geography and former department chair at the University of Colorado at Boulder. His interests focus on improving geography in higher education, learning technologies, and GIScience, particularly multimedia cartography. He has served as president of both the *Association of American Geographers* (AAG) (2010–11) and *National Council for Geographic Education* (NCGE) (2006). He edited the *NCGE Pathways* series for geography educators (2000–05) and was the North American editor of the *Journal of Geography in Higher Education* (1999–2003). He received the AAG's Honors in Geographic Education in 2005.

Nicholas J. Tate

Nick is senior lecturer at the Department of Geography, University of Leicester. He was PI and Director of the Spatial Literacy in Teaching (SPLINT) CETL from 2005–10, where he directed the pedagogic activities across the consortium (University of Leicester, University of Nottingham and UCL) towards the development of geospatial technologies and spatial thinking for a predominantly graduate audience. He is currently GIS section editor for *Geography Compass* and has been on the editorial board of three other journals. He has also served on a variety of committees for the AGI, RGS-IBG and RSPSoc. With Peter Atkinson he is editor of *Advances in Remote Sensing and GIS* and *Modelling Scale in Geographical Information Science*, both published by Wiley.

David DiBiase

David led the *Penn State Online* certificate and Master of GIS degree programs in GIS from their inception in 1998 until 2011. The certificate program earned ESRI's *Special Achievement in GIS* award in 2004, and the masters program earned the *Sloan Consortium's* 2009 award for *Most Outstanding Online Teaching and Learning Program*. David led the editorial teams that completed both the UCGIS *GIS&T Body of Knowledge* and the US Department of Labor's *Geospatial Technology Competency Model*. He participated in the URISA Certification Committee that helped established the GIS Certification Institute, and served as GISCI President from 2010–11. In August 2011, David moves to Esri as Director of Education, Industry Solutions.

List of contributors

Matthew Bampton
Department of Geography-Anthropology
University of Southern Maine
37 College Avenue
Gorham ME 04038 USA

Robert S. Bednarz
Department of Geography
Texas A&M University
College Station TX 77843-3147 USA

Sarah W. Bednarz
Department of Geography
Texas A&M University
College Station TX 77843-3147 USA

Susanne Bleisch
FHNW University of Applied Sciences
Northwestern Switzerland
Institute of Geomatics Engineering
4132 Muttenz SWITZERLAND

David J. Bodenhamer
The Polis Center
1200 Waterway Boulevard
Suite 100
Indianapolis IN 46202-2157 USA

Andy Burton
Computing and Technology Team
Nottingham Trent University
Room 302 Computing and Informatics Building
Clifton Campus Nottingham NG11 8NS UK

William Cartwright
School of Mathematical and Geospatial Sciences
RMIT University
GPO Box 2476
Melbourne 3001 AUSTRALIA

Paula Curvelo
Institute for Statistics and Information Management
New University of Lisbon
Campus de Campolide
1070-312 Lisbon PORTUGAL

David DiBiase
John A. Dutton e-Education Institute
418 Earth-Engineering Sciences Building
Penn State University
University Park PA 16802 USA

Kenneth E. Foote
Department of Geography
Campus Box 260
University of Colorado at Boulder
Boulder CO 80309 USA

Nick Frunzi
Esri
380 New York Street
Redlands, CA 92373 USA

Michael F. Goodchild
Center for Spatial Studies and Department of Geography
University of California
Santa Barbara, CA 93106-4060 USA

Christopher Goranson
GIS Center
Department of Health and Mental Hygiene
City of New York
125 Worth Street New York NY 10013 USA

Ian N. Gregory
Department of History
Lancaster University
Lancaster LA1 4YG UK

Thomas Grossmann
ETH Zurich
Institute of Terrestrial Ecosystems
8092 Zurich SWITZERLAND

Francis Harvey
Department of Geography
University of Minnesota
267 19th Ave S
Minneapolis MN 55455 USA

Claire Jarvis
Department of Geography
University of Leicester
Leicester LE1 7RH UK

Injeong Jo
Department of Geography
Texas A&M University
College Station TX 77843-3147 USA

Ann Johnson
National Geospatial Technology Center
Del Mar College
Corpus Christi, TX 78404 USA

Karen K. Kemp
Spatial Sciences Institute
University of Southern California
3616 Trousdale Parkway
Los Angeles CA, 90089-0374 USA

Andrew Klein
Department of Geography
Texas A&M University
College Station TX 77843-3147 USA

Patrick Lüscher
Department of Geography
University of Zurich
Winterthurerstraβe 190
8057 Zurich SWITZERLAND

Susanna A. McMaster
Department of Geography
University of Minnesota
267 19th Ave S
Minneapolis MN 55455 USA

Robert B. McMaster
Department of Geography
University of Minnesota
267 19th Ave S
Minneapolis MN 55455 USA

Nick J. Mount
School of Geography
University of Nottingham
University Park
Nottingham NG7 2RD UK

David M. Mountain
School of Informatics
City University London
Northampton Square
London, EC1V 0HB UK

Monika Niederhuber
ETH Zurich
Institute of Terrestrial Ecosystems
8092 Zurich SWITZERLAND

Marco Painho
Institute for Statistics and Information Management
New University of Lisbon
Campus de Campolide
1070-312 Lisbon PORTUGAL

Michael Phoenix
Esri
380 New York Street
Redlands, CA 92373 USA

Steven D. Prager
Department of Geography
University of Wyoming
1000 E. University Ave
Laramie WY 82071 USA

Gary Priestnall
School of Geography
University of Nottingham
University Park
Nottingham NG7 2RD UK

Richard B. Schultz
Department of Geography and Geosciences
Elmhurst College
190 Prospect Avenue
Elmhurst IL 60126-3296 USA

Diana S. Sinton
University of Redlands
1200 E. Colton Avenue
Redlands California 92374 USA

Martin J. Smith
Department of Civil Engineering
University of Nottingham
The Nottingham Geospatial Building
Triumph Road
Nottingham NG7 2TU UK

Lynn Songer
Department of Social Science
Lane Community College
Eugene Oregon 97405 USA

Josef Strobl
University of Salzburg
Center for Geoinformatics (Z_GIS)
Hellbrunnerstraße 30
A-5020 Salzburg AUSTRIA

Nicholas J. Tate
Department of Geography
University of Leicester
Leicester LE1 7RH UK

David J. Unwin
School of Geography
Birkbeck College, University of London
Malet Street
London WC1E 7HX UK

Robert Weibel
Department of Geography
University of Zurich
Winterthurerstraβe 190
8057 Zurich SWITZERLAND

Eric West
Department of Geography
Southern Connecticut State University
118 Morrill Hall
501 Crescent Street
New Haven CT 06515-1355 USA

Dawn Wright
Department of Geosciences
Oregon State University
Corvallis OR 97331-5506 USA

Foreword

Information about the Earth's surface, about the nature of places and the routes that connect them, is vital to almost all aspects of life today. For centuries such information has been captured and disseminated in the form of maps, but in recent decades a suite of new tools and technologies has become available that has vastly increased the range of what can be captured and how it is applied. Today we make constant use of the Global Positioning System, online mapping services such as Google Earth, imagery captured by Earth-orbiting satellites, and the analytic capabilities of geographic information systems. Moreover the need to solve problems that arise in developing and using these geographic information technologies, and the need to discover general principles that can be used to improve them, are of sufficient significance and difficulty as to constitute a research field of their own, a field known as geographic information science (GIScience).

One of the most pressing of the problems of geographic information science and technology (GIS&T) concerns representation: how to design an effective and efficient way of capturing the infinite complexity of the geographic world in the absurdly limited space and two-character alphabet of a digital computer. We have learned over the past four decades that such designs involve a host of choices: what to capture and what to leave out, which of innumerable coding schemes to use to convert geographic reality into a binary sequence, and how to make the result understandable by any application system. GIS&T is not a simple matter of a few rules, but a complex world of nuanced alternatives that requires an understanding not only of the technology, but also of the geographic world that the technology is attempting to represent. The fundamental principles of GIScience include some that reflect the nature of computational systems, and some that concern the ways in which the geographic world itself is organized.

Just as there are numerous choices in GIS&T, so also are there numerous choices in how GIS&T is taught. How should we balance training in the technical details of today's technology, with education in the principles, that will still be true when today's technology is a memory? Who are we teaching: the researchers of tomorrow or the next generation of practitioners? How should we balance open-source and commercial software products, and how should students be exposed to them? What is the appropriate mix of lecture, practical exercises, and individual or group projects?

When I started a course in GIS&T over thirty-five years ago I had little doubt of who my audience was: university students majoring in geography who would go on to careers in the fields traditionally staffed by professional geographers, as teachers, environmental consultants, or location analysts. Even then, knowledge of the rapidly expanding field of GIS&T would give them a valuable edge in competing for such jobs. Courses like this proliferated, and GIS&T slowly evolved into a recognized professional qualification. Yet today the situation we face could not be more different. In addition to an ever-increasing demand for professionals, universal access to at least a minimal set of geographic information services has raised a different set of questions: in addition to asking what the professional needs to know, we also need to be asking what *every* well-educated citizen needs to know. While online mapping tools may appear to make working with digital geographic information easy and straightforward, in reality it is all too easy to make mistakes and false inferences, to endanger personal privacy, and to engage with many other ethical issues. We teach mathematics and language skills to everyone – should we not also be teaching some subset of GIS&T to everyone?

This question is becoming more and more important as the phenomenon of *neogeography* takes hold and makes everyone both a consumer and a producer of geographic information. The costs of entry into map-making have declined effectively to zero, and services such as Google's MapMaker now allow anyone not only to make their own maps, but also to contribute geographic information to central repositories where it can be accessed by anyone. Unlike the maps of the past, these new maps are personal, up to date, cheap to produce, and readily distributed. Moreover the people making them, needing a basic understanding of parts of GIS&T, are in many cases long past their period of formal education.

This book provides a very welcome review of the issues surrounding the teaching of GIS&T in higher education. Some of them are longstanding, while others have arisen only recently, and all are being impacted by the rapid evolution of the technologies, the abundance of new research results, and the changing social role of GIS&T. The community of practice that has assembled the book includes many of the world's leading thinkers about GIS&T pedagogy, and its leading innovators. Together its chapters present an intriguing range of options and choices, and much food for thought.

Higher education finds itself today in a state of transition. The traditional notion of public higher education is under threat in numerous parts of the world because of budget pressures; today's students have grown up with advanced technologies and have adopted very different approaches to learning; online and student-centered learning are on the rise; and undergraduates are expected to acquire substantial levels of personal debt. GIS&T, with its strong employment prospects, high-tech appeal, and engagement with many of the major issues facing society, may be better able than many fields to withstand contemporary pressures and better able to adapt to the evolving academic environment.

I have always derived a great deal of satisfaction from the privilege of being able to teach GIS&T to generations of students. If this book achieves nothing else, I hope it helps others to think creatively about their own teaching, and adds an increment to their own satisfaction.

Michael F. Goodchild
Center for Spatial Studies and Department of Geography,
University of California, Santa Barbara, USA

Editors' preface

This book is the outcome of a series of meetings beginning with sessions at the Association of American Geographers annual meeting in San Francisco in 2007 originally organized by David DiBiase, and conversations amongst the editors over a number of years. In the face of an increasing body of work on the subject of GIS&T pedagogy we felt a volume that attempted to assess where we have come from, where we are now, and where to go in the future, was overdue. Specific impetus came from education initiatives such as the SPLINT CETL in the UK, publication of the *GIS&T Body of Knowledge* and the *Geospatial Technology Competency Model*, and numerous workshops and papers sessions at the meetings of University Consortium for GIScience, Association of American Geographers, and National Council for Geographic Education in the US, and the European GIS in Education Seminar and GIS Research UK meetings in Europe in which we were involved, and which evidence the wealth of activity in this area. This book is timely given the recent strides that GIS&T has made onto the web, onto the mobile/cell phone and via *neogeography* into the broader consciousness. Higher education has also been subject to considerable change and in part as a response to the demands of learners the place of formal face-to-face traditional education is now contested as never before.

The contributors to this book are drawn primarily from the USA and UK with additional contributions from elsewhere in Europe and Australia, and the twenty nine chapters are organized into five sections and a conclusion. We have taken the slightly unusual step of including a commentary in which we variously provide a synthesis and forward look for each main section.

As always, the process of getting an edited volume together relies on a great number of people in addition to the editors and contributors. In particular we would like to acknowledge the efforts of all colleagues who provided review comments for the contributed chapters (several for more than one). These include:

Sarah Battersby	Michael Goodchild	Duane Marble
Robert Bednarz	Richard Harris	Ulrik Mårtensson
Allan Brimicombe	David Howard	Duane Nellis
George Chaplin	Andrew Hudson-Smith	Yelena Olgneva-Himmelberger
Peter Dana	Alan Jenkins	Brandon Plewe
Mike DeMers	Ann Johnson	Steve Prager
Michael deSmith	Joseph Kerski	Mike Rudibaugh
Eric Fournier	Khusro Kidwai	Diana Sinton
Mark Gahegan	Andrew Lovett	Richard Schultz
Mark Gillings	Robert Maher	André Skupin

Jan Smith
Lynn Songer
Josef Strobl
Su-Yin Tan
Ming-Hsiang Tsou
Wei Tu
Nigel Waters
John Wilson

In addition we would like to thank Alex Szumski (Leicester) for administrative help as well as the Wiley editorial team (Fiona Woods, Izzy Canning and Rachael Ballard in particular). The usual thanks for forbearance go to Polly, Isobel, Sophie and Cindy.

David J. Unwin, Kenneth E. Foote, Nicholas J. Tate, David DiBiase
London, Boulder, Leicester and University Park, 1 April 2011

Section I

GIS&T in the academic curriculum – introduction

1

GIS&T in higher education: challenges for educators, opportunities for education

Kenneth E. Foote[1], David J. Unwin[2], Nicholas J. Tate[3], and David DiBiase[4]

[1] *Department of Geography, University of Colorado at Boulder, Boulder Colorado, USA*
[2] *School of Geography, Birkbeck College, University of London, London, UK*
[3] *Department of Geography, University of Leicester, Leicester, UK*
[4] *John A. Dutton e-Education Institute, Penn State University, University Park, Pennsylvania, USA*

1.1 Overview and historical context

This book is an effort to document three decades of innovation in geographic information science and technology (GIS&T) education, to take stock of lessons learned, to identify new developments and to flag directions for future advances. These issues will be of interest to those directly involved in GIS&T education as well as a wider audience. This is because GIS&T education has benefited from various innovative developments and many of the issues, techniques and lessons learned are perhaps of wider value to other disciplines and to professions that are beginning to use GIS&T. Innovations in e-learning, open source software, and open educational resources all received a substantial push from GIS&T educators. A more important hallmark of the field is the way GIS&T educators have worked cooperatively across disciplinary and national boundaries to innovate and improve practice. We see such collaboration – what we might now term a type of *community of practice* – as a defining quality of GIS&T and as a model that might be emulated more widely in geography and elsewhere. Our hope is that by documenting features of this community, this will not only be of interest for its own sake, but will encourage others to follow similar pathways.

To understand how we reached this point, it is useful to set the development of GIS&T in brief historical perspective. Geographical information systems (GIS) are computer systems developed for the collection, storage and processing of information referenced to some form of location coordinates, with this location information usually being a key element of any analysis. Histories, such as that edited by Foresman (1998), usually cite the *Canada Geographic Information System* (CGIS) of the mid-1960s as the first such system. Essentially

Teaching Geographic Information Science and Technology in Higher Education, First Edition.
Edited by David J. Unwin, Kenneth E. Foote, Nicholas J. Tate and David DiBiase.

CGIS was an attempt to create in a computer a digital geography of the country using as its input scanned copies of conventional maps. In spirit this was not unlike the pre-computer *Land Systems* inventories conducted in Australia but the entire enterprise was constrained by the available technology. At the same time a number of people began to experiment with methods for creating maps using the computer, with a major development being initiated by Howard Fisher at Harvard University in the creation of the SYMAP mapping program. In retrospect, SYMAP was primitive, making use of a standard line printer for its output and coding its 'geography' by means of a simple raster of location coordinates, but it opened many people's eyes to the potential and rapidly led to systems making use of simple pen plotters and, eventually, light pen and cathode ray tube technology that allowed user interaction with the mapping process. A third input into this development during the same period was that of dedicated image processing hardware and software systems to facilitate the analysis of remotely sensed imagery from a rapidly increasing number of earth orbiting satellites. It was easy to see the potential of combining these technologies, even if their integration was some years ahead.

In fact, the term 'GIS' was not much used until the mid-1970s, by which time it had started to appear more frequently particularly in the context of academic meetings. By the late 1980s and early 1990s GIS had clearly gained a foothold in various academic programs at both undergraduate and postgraduate level and this in turn led to the explicit development of what Goodchild (1992) termed 'geographic information science' (GISc or GIScience). As noted by Tate and Unwin (2009) the history of GIS (or GISc) education can be related to the complex and dynamic interaction between technology, the GIS industry and the academy. Table 1.1 is a summary of Tate and Unwin's (2009) brief discussion of technology and trends in GIS education over the period of the last 30 years.

Goodchild (1998) similarly reflected on the historical development of digital computing/GIS (albeit not with an education focus) and noted that GIS technology was then (1998) at the 'middle of the growth curve' somewhere between 'the computer as an information system' (stage 2) and fully 'digital worlds' (stage 3) with a more pervasive role in 'new societies' (stage 4) envisaged, but not yet realized. Arguably the 'typical technology' identified

Table 1.1 Technology and trends in GIS education

Date	Typical technology	GIS education typified by
1980s	Main frame/workstation	Niche programs (often at taught postgraduate level) in small number of geography departments; teaching *about* GIS&T with emphases on teaching in depth and training to use the available systems
1990s	Desktop/PC	Broadening and deepening to other geography departments/disciplines together with the emergence of teaching *with* GIS&T, and the notion of a 'GIScience'. Learning outcomes associated with breadth of vision introduced and the collaborative development and diffusion of teaching resources such as the NCGIA 'Core Curriculum'
2000+	Mobile device/web 2.0/virtual globe	GIS&T became mainstream, with the 'location' variable used ubiquitously and the consequent emergence of 'neogeography' and a loosening of links to academic geography. GIS is 'everywhere and no-where'

as characteristic of the date period 2000+ in Table 1.1 (such as Web 2.0, virtual globes and the ubiquity of the location variable in various mobile devices) that have enabled user-driven neogeography/VGI are hallmarks of this much more pervasive role. There would seem to be growing evidence that we have indeed reached stage 4 that Goodchild subsequently described as the full 'democratization of GIS' (Butler, 2006). Notwithstanding the complex relationship between GIS technology and people (Harvey and Chrisman, 1998) there appears to be little doubt that technological developments have, on the one hand, allowed more people to access GIS and to 'do GIS' as well as on the other hand enabled new learning opportunities and modes of learning such as e-learning and active learning to facilitate teaching (or learning) both *with* and *about* GIS. In relation to the former we have adopted the term 'geographic information science and technology' (GIS&T) in this book in deliberate reference to the specific technologies which both constitute and are specially shaping GIS and GISc. At the time of writing these encompass the web; internet; mobile and cellular technologies; GNSS such as the US Global Positioning System (GPS) and European GALILEO; satellite-borne sensing, ranging and communication systems; and pervasive and cloud computing technologies. This constellation of technologies still involves the collection, storage and processing of geospatial data, but in very different software and hardware configurations than were used even a few years ago. Critically, the 'democratization of GIS' with GIS being 'everywhere and no-where' has profound educational implications not only for who is doing the learning and what needs to be learned (Goodchild, 2011) but also for who is doing the educating and how. Not only could GIS&T education proceed without much involvement of academic geography, but this could take place without much formal involvement of the academy at all.

1.2 Why GIS&T has challenged educators

The rapid pace of the technological transformation of GIS&T as depicted in Table 1.1 has been matched by rapid innovation in education (Foote *et al.*, 2010) often in response to distinct challenges. From Table 1.1 we can see that in less than two decades GIS&T education has moved from a few niche courses in a small number of academic departments to being a major element of almost all geography and environmental studies programs and a growing presence in other disciplines as well. This expansion responds in part to the dramatic growth in demand for high-quality education and training as the GIS&T industry has spread into new commercial markets, and into more government agencies and NGOs (Gaudet *et al.*, 2003; Phoenix, 2000). Equally important in spurring innovation has been the diffusion into disciplines across the social, natural and engineering sciences. These efforts have presented formidable challenges to educators with some (such as how to fund and maintain needed hardware and software) more concrete and practical, but others more theoretical and conceptual (such as how best to reorganize and rethink traditional and sometimes hidebound disciplinary curricula and adopt new teaching methods in the context of this rapidly evolving field). Among the many challenges faced in teaching GIS&T are:

1. Its very recent evolution as a distinct branch of science, which meant that there was little past 'received wisdom' on which educators could rely and, for those faculty just beginning to teach, little guidance learning and teaching materials and curriculum

plans. Comprehensive textbooks did not appear until the late 1980s and, even then, the very first (Burrough, 1986) actually focused on land resources assessment, rather than GIS&T alone. This is a general issue: the absence of such resources is a problem that will be encountered in many fields new to the academy.

2. Its cross-disciplinary nature, which generated issues of pedagogic transfer across disciplinary boundaries. This also raised the issue of 'ownership' of the entire GIS&T enterprise. In the UK for example, the Royal Institute of Chartered Surveyors (RICS) initially tried to capture the GIS&T field by funding the development of an early curriculum (Unwin *et al*., 1990). In USA there were similar moves from lobbyists and trade groups representing the surveying and photogrammetry professions to bring the GIS enterprise under its wings by suggesting that practitioners would have to be qualified as professional land surveyors before being allowed to drive their GIS.
3. Its heavy emphasis on technology, which generated issues of delivery, especially of hands-on practical work involving considerable investment in hardware and software. Times have changed, and the costs of computation have dropped dramatically, but similar problems are likely to occur in almost any field that is reliant on some relatively expensive technology to which students need exposure.
4. An initial lack of qualified people to instruct, which generated a problem in course provision. The dangers here are those of hiring staff only marginally well-qualified to teach and lacking in the experience necessary to build appropriate courses.
5. Its international character, which led to numerous attempts to internationalize teaching through distance learning (for example Birkbeck London's GIScOnLine, the UNIGIS consortium and Esri's Virtual Campus). These pioneering efforts reveal important issues about the comparability of nomenclature, standards and expectations used in different nations and higher education systems (Harris, 2003; Elsner, 2005; Phoenix 2004). As other disciplines travelling the same way will discover, it is one thing to develop internet teaching resources but quite another to develop appropriate course management, teaching styles, quality assurance and business models that make the offering sustainable.
6. Its role as an integrating or enabling technology with the broad domains of navigation, surveying, positioning, remote sensing and mobile infrastructure have meant that it is becoming a ubiquitous technology, but one not always well understood by users. It can be seen by many as something that is somehow 'obvious' but where failure to understand fundamentals could lead to uncritical use of what greater understanding would have shown to be very sharp tools (Openshaw, 1993). An obvious example of this lies in the very many maps now seen that have been easily drafted using modern tools but which disobey even quite basic cartographic principles (Unwin, 2005).
7. There was, and to an extent this remains today, a very strong 'professional' interest necessitating the development of professional education in systems not initially designed to provide it. Again, this is an example of what is rapidly becoming a more general issue for educators as the public rightly demands a greater and greater level of accountable professionalism in almost all walks of life.
8. The central challenge is that GIS&T is changing so rapidly. Preparing effective courses and curricula is like aiming at a moving target and requiring, among teachers especially, a special commitment to stay abreast of constantly changing concepts, techniques and tools.

GIS&T educators have responded effectively to these challenges and have, over the past three decades, led a substantial number of improvements in Higher Education (HE). Problem-based learning, active pedagogy, open educational resources, web-based instructional materials, e-learning, professional training and certification, and other innovations have all received a push from GIS&T educators (Carver *et al.*, 2004; Clark *et al.*, 2007; DiBiase, 1996). Repeatedly, GIS&T educators have been among the first to take advantage of new developments (Benhart, 2000; Deadman *et al.*, 2000; Giordano *et al.*, 2007; Keller *et al.*, 1996; Wentz and Trapido-Laurie, 2001; Zerger *et al.*, 2002). More recently a new challenge has been how to make best use of web-based mapping including virtual globes, mash-ups and VGI, which have allowed these GIS&T to be used more widely in non-specialist learning and teaching settings, and helping to spur the neogeography movement under the banner the important truism that 'geography is everywhere'.

1.3 Creative responses: a record of innovation in GIS&T education

Perhaps as a consequence of the magnitude of the various educational challenges posed by GIS&T, what is unusual in HE Geography (see Jenkins, 1992), is that its practitioners took pedagogy seriously and widespread (often international) collaboration became the norm. The result was a series of educational meetings and projects, and the emergence of shared teaching resources of which perhaps the most well-known was the original NCGIA *Core Curriculum in GIS* (Kemp and Goodchild, 1992), discussed below. Other early education projects in the UK included *GISTutor*, a pioneer computer tutorial system (Raper and Green, 1992), which, although not used by many, developed a variety of important concepts. Similarly, the ASSIST (Academic Support for Spatial Information Systems) project to develop resources for training GIS-users was funded by UK Universities' Joint Information Systems Committee (JISC) and reflected the relative ease of obtaining support for software and teaching resource development associated with almost any new technology. That not much of the substantive materials developed by these projects remain shouldn't surprise, nor, necessarily should be of concern. Technology was evolving more rapidly than the ability of the education system to produce quality materials that were both academically and technologically 'portable' between institutions, disciplines and systems.

At first some of the key issues under discussion were about what to teach, when and how to teach it. In terms of intended learning outcomes (ILOs), many instructors focused (often by necessity) on relatively low-level 'hands on', outcomes that in Bloom's (1956) taxonomy of learning behaviours in the cognitive domain encompassed *knowing*, *comprehending* and *applying* their knowledge. Through time, it has been possible for most instructors to address higher-order objectives so that students faced with problems which ask them to analyse, synthesize and evaluate possible solutions. At the same time, this has meant that some of this hands-on training has largely disappeared from the curriculum. Relatively few students are now introduced to programming in Visual Basic, C, C+, Java or even Python, but such skills and abilities can help them to better analyse, synthesize and evaluate solutions to practical and theoretical problems. So, tension remains as to how best to focus GIS&T curricula in particular educational settings. GIS&T educators have responded to such challenges in creative, innovative ways. The sections below outline some on these advances as well as our rationale for the organization of this book.

GIS&T and the academic curriculum and issues in course design

In Sections 1 and 2 of this book the focus is on one of the greatest challenges faced in GIS&T education which was to establish its place in existing university and college curricula (Chen, 1998; Gilmartin and Cowen, 1991; Jenkins, 1992; Johnson, 1996; Lloyd, 2001; Nyerges and Chrisman, 1989; Painho *et al.*, 2007; Poiker, 1985; Sui, 1995; Unwin, 1997; Unwin and Dale, 1990). This has raised practical issues developing new courses, as well as theoretical concerns about how GIS&T should be situated within undergraduate and graduate/post-graduate curricula and the rigor of this education (Marble, 1998; 1999). This situation meant that GIS&T educators have tended to be open to new ideas that would help them get started. They welcomed initiatives like the US-based NCGIA and UCGIS and in UK Regional Research Laboratories to education. Although many of the issues faced by the first innovators were different to those of today, the question of how best to fit GIS&T into the academic curriculum remains a moving target and the reason we highlight it so prominently in this book. It is an issue likely to be confronted by almost any recently developed, but reasonably distinct branch, of the academy. One of the key innovations in the GIS world was the development of prototype curriculum materials like the *Core Curriculum in GIS* published by NCGIA in 1990 (Goodchild and Kemp, 1992). As Kemp notes in her chapter, these materials helped educators develop courses in many countries (Coulson and Waters, 1991). Other projects have been aimed at two-year community colleges, such as the GISAccess project, the iGETT project and NCGIA's *Core Curriculum in GIS for Technical Programs* (Allen *et al.*, 2006).

The most recent and most externally significant effort in this direction was the publication of the *Geographic Information Science and Technology Body of Knowledge* (DiBiase *et al.*, 2006). More than a replacement for the earlier *Core Curriculum*, the *Body of Knowledge (BoK)* expands and updates the range of topics included and provides a framework for building and assessing GIS&T curricula (DiBiase *et al.*, 2006, 23–25). There are exceptions, but this is one of the very few attempts that we know of in which a discipline has attempted to formalize and publicize the knowledge that its practitioners might be expected to have, specified in terms of intended learning outcomes. The authors of the *BoK* do point out two areas where more work is needed (Dibiase *et al.*, 2006, 34–36).

First, few departments have the staff and resources to address the full scope of the *BoK*. They must make choices about the core concepts and optional topics they will cover in their curricula. Although the *BoK* suggests developing 'multiple pathways to diverse outcomes', none were developed for the first edition. Second, institutions of HE have widely different educational missions and goals and the *BoK* is not necessarily easily adapted to all of these settings. That is, justifications for GIS&T in the curriculum can vary greatly say between a small, private liberal arts BA program in the US, in which GIS&T may be stressed as a means of cultivating critical thinking and reasoning (Sinton and Lund, 2007), and a two-year college in which the employability of GIS&T graduates may be the key reason for developing GIS&T courses and curricula. In research-intensive universities (such as can be found in the UK), far different rationales are needed particularly those relating to cutting-edge scientific research. It may well be that articulation in the language of intended learning outcomes is a key step in making these transfers between sectors.

One of the most important curriculum debates revolves around establishing programs and standards for professional education and certificate programs. Both the American Society for Photogrammetric Engineering and Remote Sensing (ASPRS) and the GIS Certification Institute (GISCI) now offer successful certification programs for GIS&T professionals, with the latter leading to recognition as a Certified GIS Professional (GISP). In UK during the 1990s the Education and Research Committee of the Association for Geographic Information (AGI) introduced a formal continuing professional development scheme (Unwin *et al.,* 1995), which still runs as a voluntary service to members of the Association (see AGI, undated). This did not lead to any formal recognition, but in 2002 the Royal Geographical Society-Institute of British Geographers (RGS-IBG) and AGI collaborated to introduce a formal 'chartered' geographer qualification with a specialization in GIS&T 'CGEOG (GIS)' for which applicants had to demonstrate a past track record of work involving geography, sign up to a formal code of conduct, and commit to a program of continuing professional development (CPD). The schemes established by GISCI and AGI/RGS-IBG have been running for about the same length of time but at the time of writing in USA (pop: around 310 million) some 4,668 people are registered GISPs, whereas around a quarter of the 350 Chartered Geographers in UK (pop: 68 million) are GIS&T practitioners. Although these schemes go some way towards fulfilling the professional need, it is clear that more discussion at the national or international levels is needed to reach agreement on what a certificate in GIS&T should include. It may well be that such certification is of more value in some areas of GIS&T such as surveying, land-record and cadastral mapping, and photogrammetry, than in others, such as town planning, management and ecology, where there is pre-existing professional framework. In the UK for example, RICS maintain a certification program for courses which include various master's level programs in GIS.

Academic certificate programs are also growing rapidly in both undergraduate and graduate/postgraduate curricula (UCGIS, 2008). For example, Esri's (2009) online database lists 316 such programs internationally. The precise meaning of such certification is not always clear (Obermeyer, 1993). Wikle (1999, 54) notes that these programs are 'different from degree programs mostly in terms of their focus and duration. In contrast to degree programs that include general education courses, certificates are narrowly focused and require less time to complete'. Certificates may, however, differ little from what majors or minors would earn in a traditional degree program by concentrating some of their optional components in GIS&T, though these certificates can also be helpful in documenting a students' in-depth training as they enter the workforce or advance their careers.

Perspectives on teaching GIS&T

GIS&T educators have also been at the forefront of education innovation in other areas, and this is the theme of the third section of this volume. Perhaps the most notable is their embrace of active-learning (Carlson, 2007; Drennon, 2005; Lo *et al.,* 2002; Summerby-Murray, 2001). Active pedagogy is the umbrella term for a variety of related interrelated techniques such as problem-based learning, inquiry-based learning, discovery learning and experiential learning, all rooted in constructivist learning theory. By shifting the focus of the learning experience from the teacher to the student, the aim is to engage students as active – not passive – participants in the learning process. Active pedagogy is not the only area of innovation. Ethics education has been the focus of much recent attention as, for example, in the Ethics Education for Future Geospatial Technology Professionals project (Wright *et al.*, 2009). GIS&T is raising a number of important ethical issues such as privacy when GIS&T is used for surveillance (Fisher and Dobson, 2003) or when data collated by location is used to create profiles such as those used in geodemographics (Crampton, 1995). The use of GIS&T in decision making may lead to harm to people, places and the environment if, for example, data are misused or if erroneous data find their way into use. The widespread use of costly and complex GIS&T can also accentuate the digital divide by limiting access nations, organizations, or individuals who lack the resources to acquire GIS&T. It is likely that these issues will gradually become more prominent in curricula in future years.

Of increasing interest is how GIS&T is being integrated into curricula outside geography and the environmental sciences. Sinton and Lund (2007) overview a range of such examples in the social and natural sciences, but more attention should be devoted to helping educators in these disciplines get started with GIS&T. The Center for Spatially Integrated Social Science (CSISS) in the US and the *Spatial Literacy in Teaching* (SPLINT) CETL in the UK are examples of initiatives which adopted strategies to aid such transfer to other disciplines but much remains to be done.

Digital worlds and teaching GIS&T

The fourth section of this book focuses on how recent innovations such as virtual globes, *Second Life*, and mobile technologies are enriching GIS&T and how educators can make use of such developments. Virtual globes like *Google Earth* and NASA's *World Wind* are providing new methods for the delivery of GIS&T to a wider audience which includes a broader range of academic disciplines and courses. Although map server technologies have advanced very quickly, recent systems like Google Earth, Virtual Earth and ArcGIS Explorer provide online excellent visualization tools and intuitive interfaces which are easier for new users to navigate. Furthermore, the open application programming interfaces (API) of recent systems like Google Earth and Google Maps have made it much easier for users to create custom maps, opening up a world of 'mashups' in which users can overlay their own data on existing maps. They do not offer all of the analytic capabilities of GIS or visualization capabilities of CAVEs and similar high-end expensive VR systems, but have instead helped spur the rise of a neogeography movement reflecting Goodchild's 'democratization of GIS': the use of geographic and spatial data by non-expert users, the rise of user-generated geospatial content, and efforts to use 'crowd sourced' information effectively. All of these developments suggest new directions in which GIS&T education can move so that mashups and virtual globes can support learning both inside and outside geography. Again GIS&T educators have taken the lead in exploring, at least tentatively, the use of virtual worlds and other new internet and virtual reality techniques (Hudson-Smith and Crooks, 2008) in education. Even *Facebook* and *Second Life* sites have been used to promote interactions between teachers and learners (DeMers, 2010; 2011; in press).

E-Learning

The fifth section of this book focuses on e-learning, in enhanced, blended or completely online/distance forms (Garrison and Kanuka, 2004), areas in which GIS&T educators have been leaders (Breetzke 2007; Elsner, 2005; Harris 2003; Onsrud 2005; Rees *et al.* 2009; Wright and DiBiase, 2005). The goals of these projects are varied, but among the top reasons were to expand the potential audience for GIS&T education and to use the multimedia features of the web to create more effective learning materials. Although early experiments in e-learning instruction offered little more than online text and graphics, GIS&T materials began quickly to take advantage of the interactive, digital/hypermedia qualities of the web. The more advanced models are now usually 'asynchronous' (or self-paced) and use 'blended' or 'enhanced' modes of learning aided by increased use of Web 2.0 social networking technologies including online text, discussion boards, blogs, wikis, chat-rooms, help desks, virtual seminars and tutorials.

The popularity of e-learning among students provides evidence of its potential both to attract new students and supplant traditional classroom and laboratory instruction. Companies like Esri have seen enrolment in their e-learning programs skyrocket (Johnson and Boyd, 2005). Professionals and adult learners find these courses attractive for many practical reasons that suit their schedules and budgets. Traditional students also find these classes appealing for the same reasons. Successful examples of what is possible in this area include the UNIGIS program, an international collaboration of universities, offering an MSc in GIScience as well as the master's programs available through Birkbeck London, Penn State and the University of Denver.

Other innovations may be just over the horizon. The trend toward open, flexible and individualized curricular paths and greater reliance on blended educational resources means that, in the future, both non-profit and commercial educational institutions may compete to attract learners from those at the start of their careers to those in senior positions. Despite a desire to promote 'interoperability' in GIS&T education, relatively few programs involve meaningful collaborations. Yet the rise of collaborative organizations such as the Worldwide Universities Network (WUN) may mean that frameworks are emerging for new innovations.

The creation of open educational materials is also an area in which GIS&T educators have been leaders. Starting with projects like The *Geographer's Craft* in the early 1990s (Foote, 1997), GIS&T educators have continued this push toward open resources with online versions of the Core Curricula in GIScience and remote sensing, DiBiase's (2009) online *Nature of Geographic Information* textbook at Penn State, as well as a large number of other high-quality wiki and reference materials. For many instructors, teaching courses exclusively from open-source materials in the web is a viable alternative to using a textbook.

1.4 Refrain and prospect

In the categorization/selection of topics above we have attempted to lay out the significant elements of the landscape of GIS&T education. We have focused on curricular issues initially, followed by other areas of significant contribution by GIS&T educators. Inevitably there are both omissions and partiality displayed in our choices, some of which might be expected (with hopefully some that are not). One element that we will return to in the concluding chapter of this book is the issue of *collaboration.* Perhaps one of the most distinctive features of GIS&T education is the way educators have worked collaboratively (often across disciplinary and national boundaries) to innovate and improve practice. Often this collaboration has been in the guise of formal consortia or formal projects to create specific educational resources or to deliver a specific taught program. However, this does not do justice to the wealth of collaboration which has taken place in a wholly informal context, in what are now termed 'communities of practice'.

References

Association for Geographic Information (AGI) (undated) Education and CPD. http://www.agi.org.uk/education-and-c pd/ (accessed August 15, 2011).

Allen, J., O. Brand, R. Beck, A. B. Johnson, and A. Johnson (2006) *Integrating geographic information systems and remote sensing for technical workforce training at two-year colleges.* http://gistech.delmar.edu/geospatial_ws1.htm (accessed April 16, 2009).

Benhart, J., Jr. (2000) An approach to teaching applied GIS: Implementation for local organizations. *Journal of Geography,* 99(6): 245–252.

Bloom, B. S. (1956) *Taxonomy of Educational Objectives. Volume 1: Cognitive Domain.* New York: Longman.

Breetzke, G. (2007) A critique of distance learning as an educational tool for GIS in South Africa. *Journal of Geography in Higher Education,* 31(1): 197–209.

Butler, D. (2006) The web-wide world. *Nature,* 439: 776–778.

Burrough, P. A. (1986) *Principles of Geographical Information Systems for Land Resource Assessment.* Oxford: Clarendon Press.

Carlson, T. (2007) A field-based learning experience for introductory level GIS students. *Journal of Geography,* 106(5): 193–198.

Carver, S., A. Evans, and R. Kingston (2004) Developing and testing an online tool for teaching GIS concepts applied to spatial decision-making. *Journal of Geography in Higher Education,* 28(3): 425–438.

Chen, X. M. (1998) Integrating GIS education with training: A project-oriented approach. *Journal of Geography,* 97(6): 261–268.

Clark, A. M., J. J. Monk, and S. R. Yool (2007) GIS pedagogy, web-based learning and student achievement. *Journal of Geography in Higher Education,* 31(2): 225–239.

Coulson, M. R. C., and N. M. Waters (1991) Teaching the NCGIA curriculum in practice: Assessment and evaluation. *Cartographica,* 28(3): 94–102.

Cox, M. D., and L. Richlin (eds) (2004) *Building Faculty Learning Communities.* San Francisco: Jossey-Bass.

Crampton, J. (1995) The ethics of GIS. *Cartography and Geographic Information Systems,* 22(1): 84–89.

Deadman, P., G. B. Hall, T. Bain, L., Elliot, and D. Dudycha (2000) Interactive GIS instruction using a multimedia classroom. *Journal of Geography in Higher Education,* 24(3): 365–380.

DeMers, M. N. (2010) Second Life as a surrogate for experiential learning. *International Journal of Virtual and Personal Learning Environments,* 1(2): 17–31.

DeMers, M. N. (2011) Subject matter content creation for Second Life delivery: Teaching GIS in Second Life, IN: G. Vincenti and J. Braman (eds) *Teaching through Multi-User Virtual Environments: Applying Dynamic Elements to the Modern Classroom.* Hershey, PA: Information Science Reference/IGI Global, pages 414–432.

DeMers, M. N. (in press) Linking MUVE education and best educational practices, IN: G. Vincenti and J. Braman (eds), *Multi-User Virtual Environments for the Classroom: Practical Approaches to Teaching in Virtual Worlds.* Hershey, PA: Information Science Reference/IGI Global.

DiBiase, D. (1996) Rethinking laboratory education for an introductory course on geographic information. *Cartographica,* 33(4): 61–72.

DiBiase, D. (2009) *Nature of Geographic Information: An Open Geospatial Textbook.* http://natureofgeoinfo.org (accessed April 16, 2009).

DiBiase, D., M. DeMers, A. Johnson, K. Kemp, A. Luck, B. Plewe, and E. Wentz (eds) (2006) *The Geographic Information Science and Technology Body of Knowledge.* Washington, DC: Association of American Geographers and University Consortium for Geographic Information Science.

Drennon, C. (2005) Teaching geographic information systems in a problem-based learning environment. *Journal of Geography in Higher Education,* 29(3): 385–402.

Elsner, P. (2005) GIS teaching via distance learning experiences and lessons learned. *Planet,* (14): 28–30.

Esri (Environmental Systems Research Institute) (2009) *Esri's online database of academic GIS programs.* http://edcommunity.esri.com/universityprograms/results.cfm (accessed April 16, 2009).

Fisher, P. F., and J. E. Dobson (2003) Who knows where you are, and who should in the era of mobile geographies? *Geography,* 88(4): 331–337.

Foote, K. E. (1997) The geographer's craft: Teaching GIS in the web. *Transactions in Geographic Information Systems,* 2: 137–150.

Foote, K. E., M. Bambacus, Q. Miao, H. Wu, W. Li, J. Li, J. Tan, Y. Tang, M. Sun, and H. Yu (2010) Vision for geoinformation science, IN: C. Yang, D. Wong, Q. Miao, and Y. Ruixin (eds), *Advanced GeoInformation Science.* Boca Raton, FL: CRC Press/Taylor and Francis Group, pages 415–451.

Foresman, T. W. (ed.) (1998) *The History of Geographic Information Systems: Perspectives from the Pioneers.* Upper Saddle River, NJ: Prentice Hall.

Garrison, R. D., and H. Kanuka (2004) Blended learning: uncovering its transformative potential in higher education. *Internet and Higher Education,* 7: 95–105.

Gaudet, C., H. Annulis, and J. Carr (2003) Building the geospatial workforce. *Journal of the Urban and Regional Information Systems Association,* 15(1): 21–30.

Gilmartin, P., and D. Cowen (1991) Educational essentials for today's and tomorrow's jobs in cartography and geographic information systems. *Cartography and Geographic Information Systems,* 18(4): 262–267.

Giordano, A., Y. Lu, S. Anderson, and M. Fonstad (2007) Wireless mapping, GIS, and learning about the digital divide: A classroom experience. *Journal of Geography,* 106(6): 285–295.

Goodchild, M. F. (1992) Geographical information science. *International Journal of Geographical Information Systems,* 6(1): 31–45.

Goodchild, M. F. (1998) What next? Reflections from the middle of the growth curve, IN: T. J. Foresman (ed), *The History of Geographic Information Systems: Perspectives from the Pioneers.* Upper Saddle River, NJ: Prentice Hall, pages 369–381.

Goodchild, M. F. (2011) Information technology as megaengineering: The impact of GIS, IN: S. Brunn (ed), *Engineering Earth: The Impacts of Megaengineering Projects.* Berlin: Springer, pages 37–47.

Goodchild, M. F., and K. K. Kemp (1992) NCGIA education activities: The core curriculum and beyond. *International Journal of Geographical Information Systems,* 6(4): 309–320.

Harris, R. (2003) Building a GIScience community in cyberspace: Reflections on GIScOnline. *Journal of Geography in Higher Education,* 7(3): 279–295.

Harvey, F., and N. Chrisman (1998) Boundary objects and the social construction of GIS technology. *Environment and Planning A,* 30(9): 1683–1694.

Hudson-Smith, A., and A. Crooks (2008) The renaissance of geographic information: Neogeography gaming and Second Life. London: University College London, Centre for Advanced Spatial

Analysis. http://www.casa.ucl.ac.uk/publications/workingPaperDetail.asp?ID=142 (accessed August 15, 2011).

Jenkins, A. (1992) Through a model darkly: An educational postscript. *Cartographica*, 28(3): 103–108.

Johnson, A. B., and J. M. Boyd (2005) Content, community, and collaboration at Esri Virtual Campus: a GIS company's perspective on creating an online learning resource. *Journal of Geography in Higher Education*, 29(1): 115–121.

Johnson, M. L. (1996) GIS in business: Issues to consider in curriculum decision-making. *Journal of Geography*, 95(3): 98–105.

Keller, C. P., T. J. Davis, and R. Canessa (1996) A hypertext tutor for teaching principles and techniques of GIS. *Journal of Geography in Higher Education*, 20(2): 193–207.

Kemp, K. K., and M. F. Goodchild (1992) Evaluating a major innovation in higher education: The NCGIA core curriculum in GIS. *Journal of Geography in Higher Education*, 16(1): 21–36.

Lloyd, W. J. (2001) Integrating GIS into the undergraduate learning environment. *Journal of Geography*, 100(5): 158–163.

Lo, C. P., J. M. Affolter, and T. C. Reeves (2002) Building environmental literacy through participation in GIS and multimedia assisted field research. *Journal of Geography*, 101(1): 10–19.

Marble, D. F. (1998) Rebuilding the top of the pyramid. *ArcNews*, 20(1): 28–29.

Marble, D. F. (1999) Developing a model, multipath curriculum for GIScience. *ArcNews*, 21(2): 31.

Nyerges, T., and N. R. Chrisman (1989) A framework for model curricula development in cartography and geographic information systems. *Professional Geographer*, 41(3): 283–293.

Obermeyer, N. O. (1993) Certifying GIS professionals: Challenges and alternatives. *Journal of the Urban and Regional Information Systems Association*, 5(1): 67–75.

Onsrud, H. J. (2005) Web-casting of geographic information science graduate courses. *Journal of Geography in Higher Education*, 29(1): 123–137.

Openshaw, S. (1993) Commentary: GIS 'crime' and GIS 'criminality.' *Environment and Planning A*, 25: 451–458.

Painho, M., P. Curvelo, and I. Jovani (2007) An ontological-based approach to geographic information science curricula design, IN: S. I. Fabrikant and M. Wachowicz (eds) *The European information Society: Leading the Way with Geo-information*. Berlin: Springer, pages 15–34.

Phoenix, M. (2000) Geography and the demand for GIS education. *Association of American Geographers Newsletter*, June: 13.

Phoenix, M. (2004) Global GIS education. *GIM International*, February: 35–37.

Poiker, T. K. (1985) Geographic information systems in the geographic curriculum. *The Operational Geographer*, 8: 38–41.

Raper, J., and N. Green (1992) Teaching the principles. Lessons from the GIS Tutor project. *International Journal of Geographical Information Systems*, 6: 279–290.

Rees, P., L. MacKay, D. Martin, and H. Durham (eds) (2009) *E-Learning for Geographers: Online Materials, Resources, and Repositories*. Hershey, PA: IGI Information Science Reference.

Sinton, D. S., and J. J. Lund (eds) (2007) *Understanding Place: GIS and Mapping across the Curriculum*. Redlands, CA: Esri Press.

Sui, D. Z. (1995) A pedagogic framework to link GIS to the intellectual core of geography. *Journal of Geography*, 94(6): 578–591.

Summerby-Murray, R. (2001) Analysing heritage landscapes with historical GIS: contributions from problem-based inquiry and constructivist pedagogy. *Journal of Geography in Higher Education*, 25(1): 37–52.

Tate, N. J., and D. Unwin (2009) Teaching GIS&T. *Journal of Geography in Higher Education*, 33(Supplement 1): S1–S6.

UCGIS (University Consortium for GIScience) (2008) GIS certificate and master's programs. http://www.ucgis.org/priorities/education/GIS_Cert+Masters_Prog/certificates.htm (accessed April 16, 2009).

Unwin, D. J. (1997) Unit 159: *Curriculum design for GIS*. NCGIA Core Curriculum in GIScience. http://www.ncgia.ucsb.edu/giscc/units/u159/u159.html (accessed April 16, 2009).

Unwin, D. J. (2005) Fiddling on a different planet. *Geoforum*, 36(5): 681–684.

Unwin, D. J., F. Bennion, and B. Capper (1995) *Professional Development for the Geographic Information Industry*. London: Association for Geographic Information.

Unwin, D. J., and P. Dale (1990) An educationalist's view of GIS: Some educational and sociological concerns. *Journal of Geography in Higher Education*, 14(2): 166–169.

Unwin, D. J., M. J. Blakemore, P. Dale, R. G. Healey, M. Jackson, D. J. Maguire, D. Martin, H. Mounsey, and J. Willis (1990) A syllabus for teaching geographical information systems, *International Journal of Geographical Information Systems*, 4(4): 457–465.

Wentz, E. A., and B. Trapido-Lurie (2001) Structured internships in geographic information science education. *Journal of Geography*, 100(4): 140–144.

West, B. A. (2003) Student attitudes and the impact of GIS on thinking skills and motivation. *Journal of Geography*, 102(6): 267–274.

Wikle, T. A. (1999) GIS education through certification programs. *Journal of the Urban and Regional Information Systems Association*, 11(2): 53–60.

Wright, D., and D. DiBiase (2005) Distance education in geographic information science: Arena symposium and an informal survey. *Journal of Geography in Higher Education*, 29(1): 91–100.

Wright, D., D. DiBiase, and F. Harvey (2009) Ethics education for future geospatial technology professionals. http://gisprofessionalethics.org (accessed April 16, 2009).

Zerger, A., I. D. Bishop, F. Escobar, and G. J. Hunter (2002) A self-learning multimedia approach for enriching GIS education. *Journal of Geography in Higher Education*, 26(1): 67–80.

2

Making the case for GIS&T in higher education

Diana S. Sinton

University of Redlands, Redlands, California, USA

2.1 Introduction

We stand today at the intersection of several trajectories that all point to a broader and more distributed role for GIS&T in higher education. First, as a discipline, as a technology, and as a product, GIS&T is rapidly evolving to meet the needs of an ever larger user community, numbering in the millions worldwide. Second, the proliferation of this and related location and visualization technologies and data in our culture are having a transforming effect on educational approaches and content. Third, there is mounting evidence from scientific and educational research that spatial reasoning and analysis are core components of critical thinking. GIS&T can operate as an effective learning environment, and it supports pedagogical methods like problem-based learning, student-directed inquiry and service-learning. Fourth, market and industry analyses show that competence and confidence with GIS&T are becoming requisite skills for the future workforce across disciplines and business sectors, though current global economic constraints are inhibiting anticipated growth.

Thus it is likely that the future of GIS&T in higher education will be more complex, more integrated and more engaging than ever before, while its versatility contributes to diverse and dynamic roles. Even within one institution, the stakeholders (students, faculty, deans and provosts, career counselors, IT staff, others) might identify different reasons for why and how GIS&T is being used on campus. Commitments to teaching and learning GIS&T cannot be done casually; a significant investment is necessary to create a successful program. Behind every new or renovated GIS lab stand rationales that a department and institution must have found compelling. These rationales are not mutually exclusive, they are constantly evolving, and they vary tremendously by academic context. Moreover, the structure and expectations of higher education look very different from one country to another and this is reflected in the driving factors behind GIS&T education worldwide.

In this chapter I review these issues and suggest a list of dynamic and evolving reasons behind GIS&T education as it is pursued in traditional and emerging academic contexts.

Teaching Geographic Information Science and Technology in Higher Education, First Edition.
Edited by David J. Unwin, Kenneth E. Foote, Nicholas J. Tate and David DiBiase.

2.2 Dominant reasons

These tend to be widely recognized as the reasons why many institutions support, and maybe even embrace, the teaching and learning of GIS&T in one or more departments and programs.

The marketplace

A student's motivation for employment, and an institution's motivation to place its graduates into jobs and careers, continues to drive interest in GIS&T worldwide (Sui, 1995; Dunn *et al.*, 1999; Qin, 2003; Li, Kong, and Peng, 2007; Gibson, 2007; Murphy, 2007; Aina, 2009; Whyatt, Clark, and Davies, 2011). Student choices are predominantly driven by 'career opportunities, employability and potential income' (Trend, 2009, 261). Predictors of labour markets in the US say that the demand for 'geospatial technologists' is on the rise (Gewin, 2004; Reiser, 2009; Estaville, 2010), and GIS&T skills are perceived as adding high value. Workforce studies suggest that the world will need surveyors, experts in logistics, precision agriculture technicians, computer programmers with spatial database expertise, specialists in marketing and tracking systems, and the like (Estaville, 2010; Rudibaugh and Ferguson, 2010). At the same time, the educational demands of ancillary users and incumbent workers who seek only a limited amount of GIS&T training are on the rise (Johnson *et al.*, 2010).

In 2010, the US Department of Labor added 'geospatial technology' to its collection of industry competency models, documenting the skills and educational preparation necessary to become a successful and competitive employee in this field (US Department of Labor, 2010). This US effort to identify common characteristics of the geospatial 'industry' may eventually serve as a model in Europe and elsewhere, where similar efforts have not yet been completed (Davis, 2010). Developing and lower-income countries are still at relatively early stages of defining and supporting their GIS&T workforce needs with educational opportunities (Hall, 1999), but when a country has both income and applied, industry-specific needs for GIS&T, such as the United Arab Emirates, growth of an educational market can develop quickly (Yagoub and Engel, 2009).

Whether a student's employment outcome is a highly technical GIS&T position or a professional career track, the wider labour market needs individuals who embed their GIS&T skills within a capacity for applied spatial reasoning, in both the public and private sector (Solem, Cheung and Schlemper, 2008; Wikle, 2010). Hence, GIS&T as a specialization is emerging within schools of business, health and medicine, law enforcement, and natural resources, as well as other academic disciplines, described in greater detail later in this chapter.

In all situations, designing GIS&T curricula so that it aligns with workforce needs, among other content-based learning outcomes, is an ongoing and critical challenge (Marble, 2006; DiBiase, 2007; Sullivan, Brase and Johnson, 2008; Estaville, 2010; Wikle, 2010). The depth and breadth of what *could* be covered in a GIS&T course has grown tremendously over the last two decades as application areas have expanded, as anticipated (Goodchild and Kemp, 1992). Thus while diverse labour markets may have a demand for workers with GIS&T expertise, there is a counter struggle in higher education to prepare students with the combination of skills and knowledge necessary to apply GIS&T wisely. 'Unless it is taught well – no easy task – the vocational value of GIS in practice will be diminished,' (Whyatt, Clark and Davies, 2011, 4).

Conducting research

What GIS&T brings to research first is its capacity for visualization and manipulation of location-based data, through geographical *and* spatial analysis. Variables, conditions and parameters with spatial characteristics may be combined or manipulated to yield new insights, for example with questions involving distances, densities, or distributions. Second, geospatial technologies facilitate the management and interpretation of large, complex and diverse data sets in a way that goes beyond other database tools such as Access or MySQL. Geo-databases operating in a GIS allow researchers to represent systems dynamics and to explore important connections and patterns through space and time – for example, in modeling the path of a hurricane, or the spread of disease. Furthermore, GIS&T can be combined with other information technologies (whether for visual display or quantitative, statistical analysis), and data derived through GIS&T processes exported or integrated into other analytical models or applications to extend research questions.

The importance of GIS&T in contemporary science is highlighted by a National Research Council (2010) report titled *Understanding the Changing Planet: Strategic Directions for the Geographical Sciences.* The study articulates eleven 'grand challenges' for geography under four major headings: 1) How to understand and respond to environmental change; 2) How to promote sustainability; 3) How to recognize and cope with the rapid spatial reorganization of economy and society; and 4) How to leverage technological change for the benefit of society and environment. All eleven of these challenges will almost of necessity involve using GIS&T, even reflected by the fact that the examples within the publication itself are largely generated through and illustrated by GIS&T.

At the same time, GIS&T is being applied to equally important research challenges within many disciplines, for their support of data management and visualization, spatial analyses, modeling, and other computational processes. Research involving GIS&T extends into geographical questions about the oceans (Wright, 1999), the land (Kheir *et al.*, 2007), the atmosphere (Wong, Nichol and Lee, 2009), animals (Harper, Westervelt and Schapiro, 2002), and people (Leitner *et al.*, 2002), among many possible examples. GIS-informed research exists within essentially every higher education discipline of the natural sciences, social sciences, humanities and professional studies (Goodchild and Janelle, 2004; Church and Murray, 2008; Parker and Asencio, 2008; Scholten, van de Velde and van Manen, 2009; Bodenhamer, Corrigan and Harris, 2010; Oberle, Joseph and May, 2010).

Investigations into GIScience itself are also active, as new ways to model, represent, analyse, and think about spatial information are realized (Goodchild, 2004; Montello, 2009; Devillers *et al.*, 2010). Though geography and GIScience as research fields are evident at only a fraction of institutions that offer GIS&T instruction, their contributions to GIS&T development are essential (Unwin, 2005). The diffusion of GIS&T across many disciplines means that new technical solutions are being created in innovative theoretical contexts, and geographers and GIScientists bring needed insights and perspectives on issues such as scale, spatial dependence and data uncertainty (Goodchild and Janelle, 2010).

Competition for students and between programs

Competition for students within an institution, and between programs at different institutions, are strong rationales for GIS&T investment. Incoming students are recruited by the appeal of gaining 'hands-on' experience working on 'real world' problems while using 'cutting edge' technologies. Graduates expect to be competitive to secure desirable employment or gain acceptance to graduate programs. The competition rationale is particularly powerful with stakeholders who are most likely to measure an educational investment in financial terms, such as parents of traditional-age university students, adult working students, and university administrators (Trend, 2009; Boehm and Mohan, 2010).

Furthermore, in the fields of geography, earth and environmental sciences, public health, planning, and natural resources management, many argue that familiarity with GIS&T has become a disciplinary necessity, essentially the 'cost of doing business' in such programs (Kistemann, Dangendorf and Schweikart, 2002; Richardson and Solís, 2004; Drummond, 2008; Kozak, Graham and Wiens, 2008; LeGates, Tate and Kingston, 2009; Ramamurthy, 2009). Graduates may not be GIS&T experts, but as the technologies become concomitant with the professional work flows and processes in those disciplines, preparatory educational programs must find a way to include the technologies in their curricula, or risk their competitive edge.

Managing the business of the university

Some institutions are leveraging their investment in GIS&T by serving the operational needs of their campuses alongside their curricular activities. In addition to supporting campus 'greening,' geospatial technologies are managing resources of the library, the athletics department and the offices of alumni, admissions, enrollment management, university relations, disability services, logistics and facilities (Teodorescu, 2004; Zhou and Wu, 2005; Valcic, 2007; Florance, 2009; Shepherd, 2009; Bishop and Mandel, 2010; Huang *et al.*, 2010). As for any business enterprise, GIS&T provides university administrators with the ability to leverage information from existing systems (such as enrollment or donor databases) to explore spatial relationships, perform location-based analysis or market research, and integrate systems for better management and resource efficiency.

2.3 Secondary but still important reasons

Though on their own these may not be necessary or sufficient motivations for launching a new program in GIS&T, or sustaining an existing one, these educational and technology rationales warrant consideration as well.

Enhancing learning and teaching through GIS&T

GIS&T is linked to processes of critical thinking, quantitative reasoning, service learning and environmental sustainability, subjects which are *already* part of the mission and concern of many institutions of higher education (Sinton, 2009; Baumann and Gould, 2010). In that sense, GIS&T facilitates both a technological and instructional support role in numerous educational processes.

Critical thinking

Critical thinkers evaluate evidence, distinguish relevancy from the extraneous, recognize patterns, and see issues from multiple perspectives and angles. GIS has the potential to inform this process in several ways, including its ability to break down complex, multi-dimensional phenomena into individual layers of data that can then be evaluated one 'layer' at a time as well as in any other combination. Simple maps may be easy to make and interpret on their face value (and conclude that those two mapped patterns *look* like they are correlated, for example), but GIS further enables quantitative and statistically-based measurements of the relationships among the data sets (and determine that those two mapped patterns *are* related in certain spatial ways). However, embedding critical thinking as a learning outcome within a GIS&T course requires design and forethought that go beyond simple software instruction (Sinton and Bednarz, 2007; Goodchild and Janelle, 2010; SERC, 2010).

Quantitative literacy

A significant proportion of the data within a GIS are numerical, such as measured values, surveyed quantities, or derived calculations. These numbers can be used in simple ways, such as displaying a count, or as more sophisticated contributions to statistical analyses or complex, multi-scale and multi-step models. The availability of those numbers – accessible through a map-based interface that allows visually-informed manipulations of the data – provides an opportunity to build competence in quantitative reasoning and science, technology, engineering and math (STEM) education (National Research Council, 2006; Kolvoord, 2008; Nugent *et al.*, 2010).

Visualization and graphicacy

The digital age has enabled the collection of practically immeasurable amounts of data, through a myriad of sensors, surveys, and other collection methods and devices, and sorting through the data in search of understanding and knowledge is complex in itself (Muigg *et al.*, 2008; Sinha, Winslett and Wu, 2009). Information with a geographic or spatial footprint can be visualized and analyzed through GIS&T, allowing us to 'see' data whose patterns are difficult to appreciate otherwise, such as ground water, or atmospheric temperatures, or how people's measurable attitudes or emotions vary across space. Hence, visualization is a reason to consider GIS&T.

At the same time, our ability to make more maps of more data is a platform to nurture skills of graphicacy. Graphicacy, or the ability to use and understand information in figures, charts, plans, diagrams and maps or other non-text representations of knowledge, is a critical contemporary skill, given the amount of data the world is producing (Aldrich and Shepherd, 2000; Hallisey, 2005). Through graphicacy we add 'critical viewing skills' to the list of other skills cultivated in higher education (critical reading, writing, speaking and thinking). Through GIS, students can rapidly and iteratively generate and modify visual representations of information, appreciating how easily a range of numbers can be cartographically manipulated to display different interpretations of the values and becoming more discerning creators and consumers of information (Perkins, 2003; Hallisey, 2005; Whyatt, Clark and Davies, 2011).

Service learning

Service to communities is a defining characteristic of many institutions, and GIS&T is one form of information technology being used to understand conditions and address needs (Hannon, 2006; Maddux *et al.*, 2006). Engaging in 'real' work of the world, where one can have a direct effect on improving people's lives, is a tremendously powerful motivation to students. Some institutions, including Tufts University (USA) and Clark University

(USA) have aligned their GIS&T curricular programmatic rationales directly with their institutional commitments to international development and humanitarian assistance. Other institutions work much more locally, such as Pace University (USA), which built a GIS 'hub' to support the service work conducted by the university community (Minis and Winkler, 2009). At Wesleyan University (USA), the introductory GIS course is itself a service learning course and each final project addresses a question or need of a local community group. Integrating GIS&T courses tightly with service learning approaches is a model that has been followed at other institutions as well (Gilbert and Krygier, 2007; Elwood, 2009a; Barcus and Muehlenhaus, 2010; Read, 2010).

Environmental sustainability

Spatial analyses and models play a central role in studies of climate change, energy sources and usage, and water availability and distribution. Concerns over these and many other environmental topics motivate students to pursue GIS&T knowledge (Ramamurthy, 2009; Verduyn and Kerski, 2009). Developing knowledge directly through applied learning, in ways that benefit both their own environment and that of others, is an appeal for students in many countries (Baber, 2009; Prüller *et al.*, 2009). Embedding GIS&T instruction within a larger context of policy making to benefit the environment is occurring across several European countries (GI Indeed, 2005).

Spatial analysis also plays a role in the strategic and policy commitments by university administrators to 'green' their campuses and reduce their environmental impacts by carefully evaluating their use of resources. Projects like these can extend far beyond a traditional tree inventory to include data on energy and water use patterns, food waste stream from the cafeteria, recycling, room and building use, etc. Efforts of this kind are ongoing at Simon Fraser University (Canada), the University of California at Santa Barbara (USA) and the Claremont Colleges in California (USA), among other institutions.

Seeking spatial literacy

Throughout history we have evidence of discoveries inspired by a spatial perspective on how the world works and how it is arranged, such as Watson and Cricks's DNA structure, plate tectonic movement, successful journeys by fifteenth and sixteenth century explorers, and even the spatial arrangement of the periodic table. Being able to visualize, analyse and interpret spatial information is fundamental to content understanding and problem solving, especially in STEM disciplines (Carter, LaRussa and Bodner, 1987; Kastens and Ishikawa, 2006; National Research Council, 2006; Sorby, 2009; Wai, Lubinski and Benbow, 2009; Newcombe, 2010). Longitudinal studies by psychologists confirm that 'Spatial thinking is important, probably as important as verbal and mathematical thinking, for success in science, technology, engineering, and mathematics' (Newcombe, 2010). A recent report on STEM innovation recommended that educators 'Expand existing talent assessment tests and identification strategies to the three primary abilities (quantitative/mathematical, verbal, and spatial) so that spatial talent is not neglected' (National Science Board, 2010). Spatial literacy, the confident and competent use of maps, mapping and spatial perspectives to address ideas, situations and problems within daily life, society and the world around us, is clearly an essential educational goal.

'Spatial thinking' itself is a broad idea, defined in multiple ways, yet in surveys of workforce demands it has been the most highly desired geographic 'skill' among employers

(Solem, Cheung and Schlemper, 2008). GIS&T supports spatial thinking processes but the relationship is complex and not automatic, as it represents the nexus of technology, pedagogy and content knowledge questions (Kerski, 2003; National Research Council, 2006; Lee and Bednarz, 2009). While this relationship continues to be studied and measured, its potential strength is a forceful rationale (Tate, Jarvis and Moore, 2005; Goodchild, 2006; Sinton and Bednarz, 2007; Goodchild and Janelle, 2010).

Expectations for use of twenty-first century technologies

Location intelligence has become critical for information sharing and technology development. Applications on smart phones, digital cameras and other devices are 'location-enabled' through GPS or other Wi-fi methods, and the use of GPS for navigation is common-place. Moreover, the public is becoming accustomed to thinking of 'location' as a characteristic of information, and to seeing such information displayed for them on a map of one type or another.

What does this mean for education? A leading group of technology-in-education experts included 'geo-everything' as one of its trends to watch in 2009 (Johnson, Levine and Smith, 2009), and the appeal of geotechnologies is often a recruitment draw for students (Murphy 2007). The ubiquitous nature of digital mapping and related geospatial technologies raises issues about privacy, ethics and how the public understands spatial information, issues that educators can address in GIS&T classrooms (Haklay, Singleton and Parker, 2008; Elwood 2009b; Goodchild and Janelle, 2010).

2.4 Connections between reasons and academic contexts

These reasons represent a snapshot in time, and are certainly not mutually exclusive. As the use and adaptation of GIS&T moves into ever more application areas and penetrates into different industries and disciplines, the types of GIS-relevant jobs will expand and the nature of academic programs necessary to serve those needs will grow as well.

At the same time, each one of these reasons varies by its academic context and there is no one-size-fits-all approach to designing an educational environment for GIS&T. Its curricular programming comes in a great variety of shapes, sizes and flavours around the world, reflecting student expectations, external workforce demands, and cultural stances towards higher education in general. The nature of the institution, the presence or absence of a geography department, and the attitude towards software training are some of the factors that characterize an institution's choices to offer coursework in GISystems, GIScience and/or related geospatial technologies. One of the few consistencies is that the trend for more GIST&T education is likely to be *upward*, across all sectors of learning.

Traditional academic contexts for GIS&T

Departments of geography and geomatics

While GIS&T use is increasingly evident across many academic disciplines, departments of geography have a long and prominent history of working with the technologies (Sui, 1995; Bruun, Cutter and Harrington, 2004; Johnson and Sullivan, 2010; Kawabata *et al.*, 2010).

In the US, only about 20 per cent of institutions offer a geography degree (Bjelland, 2004), but among those that do it is likely that GIS-based coursework is available, and may be required, at the bachelor's level (Murphy, 2007). Among those departments that focus on *applied* geography, GIS&T is a certainty (Boehm and Mohan, 2010). In the US and Canada, GIS&T have become the most common specialty in geography departments at all types of institutions, including community colleges, four-year institutions, and research universities (Association of American Geographers, 2010). This self-reported data does not reflect an exhaustive list, yet it is evidence that the presence of GIS&T within geography in the US is persistent and growing. This trend is consistent with the GIS 'Specialty Group' being the highest subscribed among the Association of American Geographers (Pandit, 2004).

Since the 1990s, geography graduate and undergraduate programs in the US have responded to the growing demand by establishing new GIS&T degree and certificate programs, fueled by the 'exogenous forces of the workplace in both the private and public sectors' and 'consistent with the public's perception of the discipline, the private sector's demand for specific skill sets, and the observations of administrators at many . . . institutions' (Bierly and Gattrell, 2004, 343). However, placing a high value on the demands of the workforce is not consistent across geography departments (Estaville *et al.*, 2006). Boehm and Mohan (2010) suggest that 'Jobs and careers are not the primary focus of university geography departments,' except at institutions with a deliberate focus on applied geography. Moreover, fully embracing GIS&T is not a universal given within geography departments, with notable resistance in the United Kingdom (Tate and Unwin, 2009). Some suggest that failure to appreciate, or disdain for, the public and private sectors' interest in and use of GIS&T is a short-sighted perspective that may ultimately undermine a geography department's success (Richardson and Solís, 2004; Unwin, 2005; Estaville, Brown and Caldwell, 2006; Estaville, 2010).

Doctoral degrees related to GIS&T are more likely to focus on GIScience, and are uncommon outside of geography departments at large universities, worldwide, where research and applied learning rationales drive programmatic decisions. *Directions Media*, a popular online trade magazine for geospatial technologies, posed a question about the value of a 'PhD in GIS' and debated the possible reasons for the existence of an advanced, research-based degree that focused on a *tool* (Francica and Schutzberg, 2008). The programs they profiled, including departments at the State University of New York at Buffalo and the University of Texas, provide degrees in geography or geospatial sciences, using science and interdisciplinary connections as their programmatic rationales. Competition for academic positions in GIST&T and GIScience is fierce, and students completing such programs may find a greater array of career opportunities in the commercial and private sectors (Richardson and Solís, 2004; Solem, Cheung and Schlemper, 2008).

Globally, GIS&T education within geography departments reflects variability of culture, funding and national agendas. In spite of greater European standardization of geography curricula through the Bologna process, individual countries have their own rationales and motivations to influence program design, so different emphases, such as engineering or planning, might prevail (Koutsopoulos, 2008). At the same time, when geography is perceived as 'non-technical' or 'non-scientific,' as it is in some parts of Europe, its departments and laboratories may suffer from chronic under-funding (Donert, 2005). In the United Kingdom, GIS&T is not necessarily an effective recruitment technique for students to geography (Sidaway and Johnston, 2007), while in Spain GIS&T has become a standard and welcome part of the higher education curriculum (Tapiador and Marti-Henneberg, 2007).

In China, a GIS curricular track expanded with such popular success, from being offered at ten different institutions in 1998 to 113 in 2005, that a glut of graduates temporarily overwhelmed the national labour market (Li *et al.*, 2007).

Geography's disciplinary neighbours – planning, natural resources management and the natural sciences

True to its roots, GIS&T education continues to be firmly present in programs of urban planning and landscape architecture, as well as the geosciences, environmental sciences, natural resources management and landscape ecology (Kawabata *et al.*, 2010). As noted earlier, once use of GIS became part of the standard work-flow for practitioners and researchers in these fields, preparation via academic programs became a given demand. 'Understanding core spatial thinking concepts, what the key geospatial technologies are, and how they can be applied to solving urban problems should be a required part of the initial planning education of undergraduate and graduate urban planning students and continuing education for practicing planners' (LeGates, Tate and Kingston, 2009).

While GIS&T has long held value in planning programs, its growth is expanding into neighbouring disciplines of design and architecture, with landscape architecture at the forefront. Competition among students and between programs is strong in these fields, and there is motivation to innovate and advance the design and planning fields with new versions of technologies that facilitate spatial decision making processes, support multi-perspective visualizations and enable 'scientific' planning (LeGates, Tate and Kingston, 2009; Esri, 2010; Zwick, 2010).

Programs in the natural sciences, particularly geology, ecology and environmental studies/science, frequently encourage, or require, that their students take introductory GIS courses so that they become more confident and competent at interpreting their respective discipline's data sets through a spatial lens. The essential role of spatial inquiry and analyses within these fields has now brought GIS&T into their mainstream, and exposure to GIS&T has become a basic necessity to remain competitive in recruitment of students and their successful launching into the workplace or graduate school.

Emerging academic contexts and settings for GIS&T

Disciplines that focus on spatial relationships and interactions within the natural and social world have a natural 'fit' with GIS&T and easily rationalize its instruction. Less predictable is the interesting growth and adoption of GIS&T education into other realms of higher education.

Technical and vocational institutions: community colleges and two-year programs

In the US, certain types of institutions, such as community or vocational colleges with two-year Associates degree programs, are by their nature designed to align their curriculum explicitly with local needs (Sullivan *et al.*, 2008; Johnson and Sullivan, 2010). While the local and regional job markets are therefore paramount, the national and global need for STEM expertise is also a driving institutional factor (Rudibaugh and Ferguson, 2010). Through its Advanced Technology Education (ATE) program, the US National Science Foundation has funded the GeoTech Center, based at Del Mar College in Texas, to support the needs of the 450+ institutions that now offer two-year college geospatial technology programs, almost

40 percent of all such institutions (GeoTech Center, 2010). Workforce demand has clearly been the most compelling rationale for launching new programs, but the demand itself is so highly tied to fluctuating economic conditions that programs are challenged to remain flexible. Because a significant proportion of community college graduates expect to transfer their GIS&T degree credits to other universities for advanced degrees, articulation agreements between institutions continues to be an important and contentious topic. Perceptions and misperceptions about differing course content and learning outcomes are some of the impediments (Johnson and Sullivan, 2010).

Liberal arts and sciences undergraduate institutions

Outside of the US, institutions that focus exclusively on the liberal arts and sciences, explicitly privileging skills of 'critical thinking' over workforce development within their educational mission, are largely unknown. This sector of higher education represents one of the slower markets to achieve GIS&T penetration, though growth has been steadily increasing (Bjelland, 2004; Sinton and Lund, 2007; Cady *et al.*, 2008). Such schools typically do *not* have geography departments; instead, the most common model involves a faculty member, often from the natural sciences, adopting a teaching *with* GIS approach. They may integrate the tools into an existing course, for example, and firmly link to the disciplinary content area with a rationale clearly biased towards GIS&T *as tools* to help *understand* or *solve problems* in *X* (Sinton and Lund, 1997).

Success in teaching *with* GIS often leads to student demand for teaching *about* GIS. In 2006, I conducted a survey of the 90+ US institutions affiliated with the National Institute of Technology and Liberal Education (NITLE) and found that over half of them were offering introductory GIS courses, despite the fact that fewer than six had geography departments. Lead instructors are sometimes sole geographers embedded in other departments, but just as often are other GIS-savvy faculty with the confidence to put themselves in front of students, and curriculum committees (Bjelland, 2004). Stringent and attentive curriculum committees may question such courses in the first place. There are few if any analogous classes in the catalogues of liberal arts colleges that may seem at first glance to focus so exclusively on a suite of software and other tools, and in these settings in particular, this argument has been used as a case *against* GIS&T (Sinton and Baber, 2007).

Instead, within these educational contexts the aforementioned *secondary* rationales of critical thinking, community service and quantitative reasoning have traditionally been as compelling as practical preparation for direct employability (Sinton and Lund, 2007; Sinton, 2009). However, the future of such institutions is threatened by the same factors affecting higher education everywhere: competition for students, constricted funding due to financial downturns, and a perceived disconnect between elite higher education and society at large. Thus, rationales such as career development and graduate school preparation via undergraduate research (Otto, 2007) are likely to become more prominent.

Social sciences

Space and geography are undeniably important to the patterns and processes studied by social scientists, and in disciplines such as archaeology and anthropology the adoption of GIS&T to support research is common (Liverman *et al.*, 1998; Wheatley and Gillings, 2002; Okabe, 2006). Yet other social science disciplines, notably economics and political science, have long overlooked the geographic nature of their data. A 'spatial turn' is now capturing the imagination of a wider range of social scientists and more are asking and answering spatially-based questions within their research agendas (Janelle *et al.*, 2009; Hespanha, Goodchild and Janelle, 2009; Goodchild and Janelle, 2010). Complementing conventional methods of inquiry with spatial statistics and other quantitative and qualitative analytical methods is yielding new powers of inference and insight.

Thus, efforts such as the NSF-funded Center for Spatially-Integrated Social Sciences (CSISS), the Polis Center at Indiana University Purdue University Indianapolis (IUPUI) and the mapping resources provided by the Inter-university Consortium for Political and Social Research (ICPSR) and the National Archive of Criminal Justice Data, have gained ground in expanding awareness. Collectively, they have introduced GIS, as well as associated tools of exploratory spatial data analysis and statistical approaches such as geographically-weighed regression, to hundreds of social science faculty. Confidence in the value of spatial analysis within the social sciences has inspired whole new academic programs, such as the multi-disciplinary PhD program in Spatially Integrated Social Sciences at the University of Toledo (USA).

Humanities

Humanities scholars too are designing innovative applications of GIS&T and exploring spatially-based methodologies in their scholarship (Gregory and Ell, 2007; Knowles, 2008; Bodenhamer, Corrigan and Harris, 2010). Technological advances in the ability to incorporate temporal data, mine databases for spatial relationships, and represent more cartographically nuanced concepts of place and space are some of the reasons that encourage humanities scholars to explore the use of mapping and GIS&T in their teaching and research activities (Staley, 2007; Jessop, 2008). In the US, large-scale efforts are often led by scholars at research-focused institutions in a position to provide significant technical support, such as the Spatial History Project at Stanford University, the Ancient World Mapping Center at the University of North Carolina at Chapel Hill, the Tibetan and Himalayan Historical GIS project at the University of Virginia, and the China Historical GIS project at Harvard University. In the United Kingdom, scholars have contributed to a project focusing on national history (Gregory *et al.*, 2002) and GIS&T in the humanities is a Japanese priority as well (Okabe, 2006).

Formal GIS&T education that has a particular humanities focus is uncommon. An exception is the Master of Arts in Historical Resources Management, launched in 2007 from the Department of History at Idaho State University (USA). To describe the balance of technology and history in their new program, co-designers Jack Owens and Laura Woodworth-Ney state, 'Although the technology permits one to ask new questions about historical processes and develop powerful visual support for one's arguments, any answers come only from the analyst, not the digital tools. Therefore, historians must continue to master the existing techniques for using historical sources because the effective use of GIS requires substantial disciplinary knowledge' (Owens and Woodworth-Ney, 2005).

Schools of education

The use of GIS within K-12 educational settings has been led by a small, loyal set of supporters who maintain a steady optimism about its value and role (Kinniburgh, 2007; Alibrandi and Baker, 2008; Baker, Palmer and Kerski, 2009). These are typically the efforts of individual teachers, or individual school districts, although in the US whole state-wide education systems have brokered arrangements with software companies like Esri (Kerski, *personal communication* 2010). While 'geography' is absent from the state-mandated curriculum in most of the US, some states have standards for 'analysis of geographic information' while others even have standards for GIS and GPS (Milson and Roberts, 2008). Elsewhere in the world, GIS&T use in schools has followed similar patterns, with varying degrees of adoption based on awareness, technology training and support, and effective integration with curricula (Bednarz and van der Schee, 2006; McInerney, 2006; Incekara, 2010).

GIS also links tightly within the 'teaching with technology' programs (Gatrell, 2004). Mapping and GIS are conspicuously present on a list of information, communications and technology literacy skills put together by the leading advocacy group for embedding technology in education, a move that will further drive interest in GIS&T education (Partnership for 21st Century Skills, 2010).

Thus GIS is very slowly catching the attention of Schools of Education, where both pre-service and in-service teachers often turn for their credentials and ongoing professional development. Arguments for its intrinsic *learning* value hold significant weight with this audience, i.e. that it has the capacity to support inquiry, critical thinking and quantitative reasoning, as well as effectively engage young students. Students love the hands-on nature of the tools, the out-of-classroom experiences, and projects that are based in their own communities (Wiegand, 2006). So, faculty within a few Schools of Education are beginning to provide their teacher-students with appropriate methods to be effective and successful with GIS&T (Alibrandi and Palmer-Moloney, 2001; Milson and Alibrandi, 2008). For example, in 2010 the University of Redlands (California, USA) launched a Master's Degree in Education with a Curriculum and Instruction focus in Spatial Literacy, within which GIS will play an important role.

Schools of business

The idea of geographic information is fundamental to the practice of business (Pick, 2008), whether measuring market potential, selecting sites, targeting clients, or designing new distribution networks. Yet very few Schools of Business integrate substantial, if any, GIS education (Ramakrishna *et al.*, 2008; Shepherd, 2009). More often than not, business students are simply unaware of the existence of GIS. 'This provides business schools with an opportunity to fill an important educational gap and a business need. Business schools, therefore, should begin to enhance their efforts in incorporating spatial thinking in their business education curriculum' (Gadish, 2006). Individual early adopter business faculty, or business-minded geographers, have attempted to fill this niche, in marketing (Miller, Holmes and Mangold, 2007) and with the use of geodemographic data (Shepherd, 2009), for example. Broader efforts include the University of Redlands, which has launched a small business 'spatial resource center' and has a four-course GIS emphasis within its MBA program, as well as Kansas State University's new 'GIS in Business' initiative within its College for Business Administration.

Institutional decisions to develop and provide additional educational opportunities in these areas are being driven by the competition to recruit students and the growing demands of adult students, those seeking for themselves the competitive edge that GIS skills will bring them professionally as they seek employment, as well as those seeking to make their own existing businesses or organizations more successful. Given the global emphasis on business, the 'remarkable growth of business geospatial jobs . . . portends a geospatial workforce in the private sector that has the potential of becoming the largest geospatial employment market in the early decades of the 21st century' (Estaville, 2010, 59).

Intelligence – homeland security

GIS software is used in virtually every US federal agency and office, but specialized educational programming has emerged recently for one particular area, intelligence and homeland security. This combination has historically been taught at institutions affiliated with the military, such as the US Military Academy at West Point or the Royal Military Colleges in the United Kingdom, Canada and Australia. Now, at least in the US, these programs are spreading to civilian institutions as well, meeting the public and private sectors' need for students with GIS&T skills as applied to topics of security (Sui, 2008). The US Geospatial Intelligence Foundation (USGIF) has begun to 'certify' curricula within many programs that are aligned with the needs of its geospatial intelligence analysts and this may spur enrollments. These include geospatial intelligence certificates available from programs such as the University of Missouri (USA), Pennsylvania State University (USA) and George Mason University (USA).

2.5 Concluding thoughts

Across higher education, GIS&T education is diversifying and expanding. Reflecting on geography education in the US, Murphy (2007, 133) noted 'Any forward-looking institution of higher education can scarcely avoid serious engagement with GIS/GIScience.' At the same time, institutions are undergoing rigorous scrutiny to demonstrate the return on investment for a university degree. Thus the growth of GIS&T education has implications at many levels of the institution as administrators, departments, faculty and students make educational choices in a very deliberate and strategic manner.

The future of GIS&T in higher education will be strongly affected by current and future trends in academia in general. By definition, programs that are based in digital technologies have steep financial costs that require frequent reinvestments to remain current. Budget reductions will make the competition for public and private grant monies to invest or upgrade hardware, software, and peripheral equipment all the more fierce. As the demand for online education increases, programs that are based on teaching a complex technology will experience additional instructional and technical delivery challenges. Traditional computer labs with local networks for data distribution and software access are being replaced by more complex arrangements involving web-based access to data servers and mapping services. Many institutions are no longer managing GIS&T within one single lab; now there are multiple schools and colleges within a single institution using licenses and data, not to mention satellite and regional campuses with systems networks that must also be coordinated. Having free or low-cost access to current, large-scale, local and national spatial data sets is a luxury not known in many places, and open source GIS software and crowd-sourced data sets will become more prominent.

Thus, making the case for GIS&T in higher education is becoming both easier and more difficult. Evidence continues to mount that GIS&T is an essential component of research processes, provides graduates with a practical suite of skills with which to address social and natural problems of the world, and has become so ingrained within certain disciplines that the need for its ongoing teaching is certain. Yet, as mapping enters the vernacular, traditional mapping tools may lose some of their distinctiveness as technologies, and new paradigms for what maps can do (for whom, by whom, when, how, and why) are emerging. This, coupled with a slowly recovering global economy, creates an interestingly dynamic future for GIS&T in higher education.

References

Aina, Y. A. (2009) Geomatics education in Saudi Arabia: Status, challenges and prospects. *International Research in Geographical and Environmental Education*, 18(2): 111–119.

Aldrich, F., and L. Sheppard (2000) 'Graphicacy': the fourth 'R'? *Primary Science Review*, 64: 8–11.

Alibrandi, M., and J. Palmer-Moloney (2001) Making a place for technology in teacher education with Geographic Information Systems (GIS). *Contemporary Issues in Technology and Teacher Education*, 1(4). http://www.citejournal.org/vol1/iss4/currentissues/socialstudies/article1.htm (accessed November 28, 2010).

Alibrandi, M., and T. Baker (2008) A social history of GIS in Education, 1985–2007, IN: A. J. Milson and M. Alibrandi (eds), *Digital Geography: Geospatial Technologies in the Social Studies Classroom*. Charlotte, NC: Information Age Publishing, Inc., pages 3–37.

Association of American Geographers (2010) *Guide to Geography Programs in the Americas, 2009–2010*. Washington, DC: Association of American Geographers.

Baber, M. (2009) Modeling site suitability for community water access points in the Mayange Sector of Rwanda. *Proceedings of the International Cartographic Conference*, Santiago, Chile, 15–21 November.

Baker, T. T., A. M. Palmer, and J. J. Kerski (2009) A national survey to examine teacher professional development and implementation of desktop GIS. *Journal of Geography*, 108(4): 174–185.

Barcus, H., and B. Muehlenhaus (2010) Bridging the academic-public divide in GIS and cartography: A framework for integrating community partnerships in the classroom. *Journal of Geography in Higher Education*, 34(3): 363–378.

Baumann, J., and M. Gould (2010) The challenges of GIS education today. *GeoConnexion Magazine*, November. http://www.geoconnexion.com/uploads/giseducation_intv9i10.pdf (accessed November 28, 2010).

Bednarz, S. W., and J. van der Schee (2006) Europe and the United States: The implementation of geographical information systems in secondary education in two contexts. *Technology, Pedagogy and Education*, 15(2): 191–206.

Bierly, G. D., and J. D. Gatrell (2004) Structural and compositional change in geography graduate programs in the United States: 1991–2001. *The Professional Geographer*, 56(3): 337–344.

Bishop, B. W., and L. H. Mandel (2010) Utilizing geographic information systems (GIS) in library research. *Library Hi Tech*, 28(4): 536–547.

Bjelland, M. (2004) A place for geography in the liberal arts college? *The Professional Geographer*, 56(3): 326–336.

Bodenhamer, D. J., J. Corrigan, and T. M. Harris (2010) *The Spatial Humanities*. Bloomington, IN: Indiana University Press.

Boehm, R., and A. Mohan (2010) Geospatial technology: Curricular keystone of applied geography. *International Journal of Applied Geographical Research*, 1(1): 26–39.

Brunn, S. D., S. L. Cutter, and J. W. Harrington, Jr. (2004) *Geography and Technology*. New York: Springer-Verlag.

Cady, C., W. Walters, W. Olsen, E. Williams-Bergen, and B. Harloe (2008) Geographic information services in the undergraduate college: Organizational models and alternatives. *Cartographica*, 43(4): 239–255.

Carter, C. S., M. A. LaRussa, and G. M. Bodner (1987) A study of two measures of spatial ability as predictors of success in different levels of general chemistry. *Journal of Research in Science Teaching*, 24: 645–57.

Church, R. L., and A. T. Murray (2008) *Business Site Selection, Location Analysis and GIS*. Hoboken, NJ: John Wiley and Sons.

Davis, P. (2010) Can the Department of Labor's geospatial competency model translate to the needs of the European Union? *Directions Magazine*. http://www.directionsmag.com/articles/

can-the-department-of-labors-geospatial-competency-model-translate-to-/134980 (accessed November 28, 2010).

Devillers, R., A. Stein, Y. Bedard, N. Chrisman, P. Fisher, and W. Shi (2010) Thirty years of research on spatial data quality: Achievements, failures, and opportunities. *Transactions in GIS*, 14(4): 387–400.

DiBiase, D. (2007) Is GIS a Wampeter? *Transactions in GIS*, 11(1): 1–8.

Donert, K. (2005) *Aspects of GIS Education and Geography in European Higher Education: Geographical Information Systems*. Liverpool, UK: Liverpool Hope University Press.

Drummond, W. J., and S. P. French (2008) The future of GIS in planning. *Journal of the American Planning Association*, 74(2): 161–174.

Dunn, C. E., M. J. Blakemore, P. J. Atkins, and J. G. Townsend (1999) Teaching geographical information handling skills for lower-income countries. *Transactions in GIS*, 3(4): 319–332.

Elwood, S. (2009a) Integrating participatory action research and GIS education: Negotiating methodologies, politics and technologies. *Journal of Geography in Higher Education*, 33(1): 51–65.

Elwood, S. (2009b) Geographic information science: New geovisualization technologies – emerging questions and linkages with GIScience. *Progress in Human Geography*, 33(2): 256–263.

Esri (2010) *Changing geography by design*. http://www.esri.com/library/ebooks/geodesign.pdf (accessed November 28, 2010).

Estaville, L. E. (2010) Geospatial workforce trends in the United States. *International Journal of Applied Geospatial Research*, 1(1): 57–66.

Estaville, L. E., B. J. Brown, and S. Caldwell (2006) Geography undergraduate program essentials: Placement. *Journal of Geography*, 105: 93–98.

Florance, P. (2009) Tufts L-SIMS, library spatial information management. Presentation at 2009 Esri Education Users Conference, San Diego, California, 11–14 July. Available on CD Proceedings.

Francica, J., and A. Schutzberg (2008) The GIS PhD dilemma. http://www.directionsmag.com/article.php?article_id=2951 (podcast) (accessed November 28, 2010).

Gadish, D. (2006) Introducing GIS for business in higher education. *International Journal of Information and Communication Technology Education*, 2(2): 62–70.

Gatrell, J. (2004) Making room: Integrating geo-technologies into teacher education. *Journal of Geography*, 103(5): 193–198.

GeoTech Center (2010) Annual national survey of geospatial educators. http://geo.thinkcreative.com/Resources/Publications/2010-Annual-National-Survey-of-Geospatial-Educators-Results (accessed November 28, 2010).

Gewin, V. (2004) Mapping opportunities. *Nature*, 427: 376–377.

GI Indeed (2005) Geo-information in the implementation of net-based distance education for environmental decision-making. http://www.gisig.it/gi-indeed/default.asp (accessed November 28, 2010).

Gibson, C. (2007) Geography in higher education in Australia. *Journal of Geography in Higher Education*, 31(1): 97–119.

Gilbert, M., and J. Krygier (2007) Campus-community collaborations: Integrating partnerships, service learning, mapping, and GIS. IN: D. S. Sinton, and J. J. Lund (eds), *Understanding Place: GIS and Mapping across the Curriculum*. Redlands, CA: Esri Press, pages 63–76.

Goodchild, M., and K. Kemp (1992) NCGIA education activities: the core curriculum and beyond. *International Journal of Geographical Information Systems*, 6(4): 309–320.

Goodchild, M. (2004) GIScience, geography, form, and process. *Annals of the Association of American Geographers*, 94(4): 709–714.

Goodchild, M., and D. Janelle (eds) (2004) *Spatially Integrated Social Sciences*. New York: Oxford University Press.

Goodchild, M. (2006) The fourth R? Rethinking GIS education. *ArcNews*, 28(3): 1, 11.

Goodchild, M., and D. Janelle (2010) Toward critical spatial thinking in the social sciences and humanities. *GeoJournal*, 75: 3–13.

Gregory, I., and P. Ell (2007) *Historical GIS: Technologies, Methodologies and Scholarship*. New York: Cambridge University Press.

Gregory, I., C. Bennett, V. Gilham, and H. Southall (2002) The Great Britain historical GIS project: From maps to changing human geography. *The Cartographic Journal*, 39(1): 37–49.

Haklay, M., A. Singleton, and C. Parker (2008) Web mapping 2.0: The neogeography of the geoweb. *Geography Compass*, 2(6): 2011–2039.

Hall, G. B. (1999) GIS education and infrastructure: Challenges and problems in emerging countries. *Transactions in GIS*, 3(4): 311–317.

Hallisey, E. (2005) Cartographic visualization: An assessment and epistemological review. *The Professional Geographer*, 57(3): 350–364.

Hannon, C. (2006) Service learning in information technology leadership: A natural connection. *Peer Review*, 8(4): 16–19. http://www.aacu.org/peerreview/pr-fa06/pr-fa06_practice2.cfm (accessed November 28, 2010).

Harper, S. J., J. D. Westervelt, and A.-M. Shapiro (2002) Management application of an agent-based model: Control of cowbirds at the landscape scale, IN: H. R. Gimblett (ed), *Integrating Geographic Systems and Agent-based Modeling Technique for Simulating Social and Ecological Processes*. New York: Oxford University Press, pages 105–123.

Hespanha, S. R., F. Goodchild, and D. G. Janelle (2009) Spatial thinking and technologies in the undergraduate social science classroom. *Journal of Geography in Higher Education*, 33(Supplement 1): S17–S27.

Huang, J., Y. Zhan, W. Cui, Y. Yuan, and P. Qi (2010) Development of a campus information navigation system based on GIS. *Proceedings of the International Conference on Computer Design and Applications*, 5: 491–494.

Incekara, S. (2010) The place of geographic information systems in the new geography curriculum of Turkey and relevant textbooks: Is GIS contributing to the geography education in secondary schools? *Scientific Research and Essays*, 5(6): 551–559.

Janelle, D. G., S. R. Hespanha, F. Goodchild, and M. Goodchild (2009) Workshops and national dissemination of geographic analysis in the social sciences: The CSISS experience in the USA. *Journal of Geography in Higher Education*, 33(Supplement 1): S88–S103.

Jessop, M. (2008) The inhibition of geographical information in digital humanities scholarship. *Literary and Linguistic Computing*, 23(1): 39–50.

Johnson, L., A. Levine, and R. Smith (2009) *The 2009 Horizon Report*. Austin, TX: The New Media Consortium.

Johnson, A., and D. Sullivan (2010) Geospatial education at U.S. community colleges: Background, challenges, and opportunities. *Journal of the Urban and Regional Information Systems Association*, 22(2): 5–13.

Johnson, T., M. Bampton, E. Markow, C. Gregory, S. Hood, and C. Colgan (2010) Education for a rural geospatial workforce: The Maine experience. Presentation at the 2010 Esri Education Users Conference, San Diego, California, 10–13 July.

Kastens, K. A., and T. Ishikawa (2006) Spatial thinking in the geosciences and cognitive sciences, IN: C. Manduca, D. Mogk and N. Stillings (eds), *Earth and Mind: How Geoscientists Think and Learn about the Complex Earth*. Boulder, CO: Geological Society of America Special Paper 413, pages 53–76.

Kawabata, M., R. B. Thapa, T. Oguchi, and M.-H. Tsou (2010) Multidisciplinary cooperation in GIS education: A case study of US colleges and universities. *Journal of Geography in Higher Education*, 34(4): 493–509.

Kerski, J. (2003) The implementation and effectiveness of GIS in secondary education. *Journal of Geography*, 102: 128–137.

Kheir, R. B., J. Chorowicz, C. Abdallah, and D. Dhont (2007) Soil and bedrock distribution estimated from gully form and frequency: A GIS-based decision-tree model for Lebanon. *Geomorphology*, 93(3–4): 482–492.

Kinniburgh, J. C. (2007) The future of geographical education in Australia and the role of geographic information systems (GIS), *Geographical Education*, 20: 22–29.

Kistemann, T., F. Dangendorf, and J. Schweikart (2002) New perspectives on the use of geographical information systems (GIS) in environmental health sciences. *International Journal of Hygiene and Environmental Health*, 205(3): 169–181.

Knowles, A. (2008) *Placing History: How Maps, Spatial Data, and GIS are Changing Historical Scholarship*. Redlands, CA: Esri Press.

Kolvoord, R. (2008) Geospatial technologies: Real projects in real classrooms. *Knowledge Quest*, 36(4): 40–45.

Koutsopoulos, K. (2008) What's European about European geography? The case of geoinformatics in Europeanization. *Journal of Geography in Higher Education*, 32(1): 7–14.

Kozak, K. H., C. H. Graham, and J. J. Wiens (2008) Integrating GIS-based environmental data into evolutionary biology. *Trends in Ecology and Evolution*, 23(3): 141–148.

Lee, J., and R. Bednarz (2009) Effect of GIS learning on spatial thinking. *Journal of Geography in Higher Education*, 33(2): 183–198.

LeGates, R., N. J. Tate, and R. Kingston (2009) Spatial thinking and scientific urban planning, *Environment and Planning B: Planning and Design*, 36(5): 763–768.

Leitner, H., R. B. McMaster, S. Elwood, S. McMaster, and E. Sheppard (2002) Models for making GIS available to community organizations: dimensions of difference and appropriateness, IN: W. J. Craig, T. M. Harris, and D. Weiner, D. (eds) *Community Participation and Geographic Information Systems*. London: Taylor and Francis, pages 37–52.

Li, X., Y. Kong, and B. Peng (2007) Development of geography in higher education in China since 1980. *Journal of Geography in Higher Education*, 31(1): 19–37.

Liverman, D., E. F. Moran, R. R. Rindfuss, and P. C. Stern (eds) (1998) *People and Pixels: Linking Remote Sensing and Social Science*. Washington, DC: National Academies Press.

Maddux, H. C., B. Bradley, D. S. Fuller, C. Z. Darnell, and B. D. Wright (2006) Active learning, action research: a case study in community engagement, service-learning, and technology integration. *Journal of Higher Education Outreach and Engagement*, 11(3): 65–78.

Marble, D. (2006) Who are we? Defining the geospatial workforce. *GPS World*. http://www.gpsworld.com/gis/management-and-processes/who-are-we-defining-geospatial-workforce-4686 (accessed November 28, 2010).

McInerny, M. (2006) The implementation of spatial technologies in Australian schools: 1996–2005. *International Research in Geographical and Environmental Education*, 15(3): 259–264.

Miller, F., T. L. Holmes, and W. G. Mangold (2007) Integrating geographic information systems into the marketing curriculum. *Marketing Education Review*, 17(3): 48–63.

Milson, A. J., and M. Alibrandi (2008) *Digital Geography: Geospatial Technologies in the Social Studies Classrooms*. Charlotte, NC: Information Age Publishing, Inc.

Milson, A. J., and J. A. Roberts (2008) The status of geospatial technologies in U.S. high school geography standards, IN: A. J. Milson and M. Alibrandi (eds), *Digital Geography: Geospatial Technologies in the Social Studies Classroom*. Charlotte, NC: Information Age Publishing, Inc., pages 39–59.

Minis, P., and H. Winkler (2009) A GIS hub at Pace University. Cornerstone 2 Reports: Community Outreach and Empowerment through Service Learning and Volunteerism. Paper 11. http://digitalcommons.pace.edu/cornerstone2/11 (accessed November 28, 2010).

Montello, D. (2009) Cognitive research in GIScience: Recent achievements and future prospects. *Geography Compass*, 3(5): 1824–1840.

Muigg, P., J. Kehrer, S. Oeltze, H. Piringer, H. Doleisch, B. Preim, and H. Hauser (2008) A four-level focus and context approach to interactive visual analysis of temporal features in large scientific data. *Computer Graphics Forum*, 27(3): 775–782.

Murphy, A. (2007) Geography's place in higher education in the United States. *Journal of Geography in Higher Education*, 31(1): 121–141.

National Research Council (2006) *Learning to Think Spatially: GIS as a Support System in the K-12 Curriculum.* Washington, DC: National Academies Press.

National Research Council (2010) *Understanding the Changing Planet: Strategic Directions for the Geographical Sciences.* Washington, DC: National Academies Press.

National Science Board (2010) *Preparing the Next Generation of STEM Innovators: Identifying and Developing our Nation's Human Capital.* http://www.nsf.gov/nsb/stem/innovators.jsp (accessed November 28, 2010).

Newcombe, N. S. (2010) Picture this: Increasing math and science learning by improving spatial thinking. *American Educator*, Summer: 29–35, 43.

Nugent, G., B. Barker, N. Grandgenett, and V. I. Adamchuk (2010) Impact of robotics and geospatial technology interventions on youth STEM learning and attitudes. *Journal of Research on Technology in Education*, 42(4): 391–408.

Oberle, A. P., S. A. Joseph, and D. W. May (2010) Geospatial technologies as a vehicle for enhancing graduate education and promoting the value of geography. *Journal of Geography in Higher Education*, 34(4): 475–492.

Okabe, A. (2006) *GIS-Based Studies in the Humanities and Social Sciences.* Boca Raton, FL: CRC Press.

Otto, C. (ed) (2007) GIS across the curriculum. Special issue, *Council on Undergraduate Research Quarterly* 27, no. 3 [March].

Owens, J., and L. Woodworth-Ney (2005) Envisioning a master's degree program in geographically-integrated history. *Journal of the Association for History and Computing*, 8(2). http://hdl.handle.net/2027/spo.3310410.0008.202 (accessed August 4, 2011).

Pandit, K. (2004) Geography's human resources over the past half-century. *The Professional Geographer*, 56(1): 12–21.

Parker, R. N., and E. Asencio (2008) *GIS and Spatial Analysis for the Social Sciences: Coding, Mapping and Modeling.* New York: Routledge.

Partnership for 21st Century Skills (2010). http://www.p21.org/ (accessed November 28, 2010).

Perkins, C. (2003) Cartography and graphicacy, IN: N. J. Clifford and G. Valentine (eds), *Key Methods in Geography.* London: Sage Publications, pages 343–368.

Pick, J. (2008) *Geo-Business.* Hoboken, NJ: John Wiley and Sons.

Prüller, R., J. Scholz, C. Strauss, and T. Achleitner (2009) Teaching GIS in central Asia. *Proceedings from the FIG International Workshop, Professional Education 2009.* Vienna, Austria, 26–28 February. Available at http://www.fig.net/commission2/vienna_2009_proc/ppt/04_prueller_et_al_ppt.pdf (accessed August 4, 2011).

Qin, Q. (2003) GIS curricula in China's higher education. *Annals of GIS*, 9(1&2): 40–47.

Ramakrishna, H., A. Sarkar, and B. Vijayaraman (2008) Infusion of spatial analysis and GIS in business school curriculum: A status report. 39th Annual Meeting of the Decision Sciences Institute, Baltimore, Maryland, 22–25 November.

Ramamurthy, M. K. (2006) A new generation of cyberinfrastructure and data services for earth system science education and research. *Advances in Geosciences*, 8: 69–78.

Read, J. (2010) Teaching introductory geographic information systems through problem-based learning and public scholarship. *Journal of Geography in Higher Education*, 34(3): 379–399.

Reiser, C. (2009) Geospatial information systems market to grow 50% over next five years. ARC Advisory Group. http://www.arcweb.com/AboutARC/Press/Lists/Posts/Post.aspx?ID=111 (accessed November 28, 2010).

Richardson, D., and P. Solís (2004) Confronted by insurmountable opportunities: Geography in society at the AAG's centennial. *The Professional Geographer*, 56(1): 4–11.

Rudibaugh, M., and B. Ferguson (2010) Mapping geospatial education at U.S. community and technical colleges. *Journal of the Urban and Regional Information Systems Association*, 22(2): 40–44.

Scholten, H. J., R. van de Velde, and N. van Manen (eds) (2009) *Geospatial Technology and the Role of Location in Science.* Dordrecht: Springer.

SERC (Science Education Resource Center) (2010) Ideal student outcomes from a GIS or remote sensing course. http://serc.carleton.edu/NAGTWorkshops/gis/outcomes.html (accessed November 28, 2010).

Shepherd, I. D. H. (2009) From geography department to business school: Strategies for transplanting GIS courses between disciplines. *Journal of Geography in Higher Education*, 33(Supplement 1): S28–S45.

Sidaway, J. D., and R. J. Johnston (2007) Geography in higher education in the UK. *Journal of Geography in Higher Education*, 31(1): 57–80.

Sinha, R. R., M. Winslett, and K. Wu (2009) Finding regions of interest in large scientific datasets. *Lecture Notes in Computer Science*, 5566: 130–147.

Sinton, D. S., and M. Baber (2007) Active learning in liberal education with geographic information systems (GIS) technologies. American Association of Colleges and Universities (AAC&U) Annual Conference. New Orleans, Louisiana, 18–20 January. Abstract available at http://www.aacu.org/meetings/annualmeeting/AM07/program.cfm, accessed August 4, 2011.

Sinton, D. S., and S. W. Bednarz (2007) About that G in GIS, IN: D. S. Sinton and J. J. Lund (eds) *Understanding Place: GIS and Mapping across the Curriculum*. Redlands, CA: Esri Press, pages 19–34.

Sinton, D. S., and J. J. Lund (eds) (2007) *Understanding Place: GIS and Mapping across the Curriculum*. Redlands, CA: Esri Press.

Sinton, D. S. (2009) Roles for GIS in Higher Education. *Journal of Geography in Higher Education*, 33(Supplement 1): S7–S16.

Solem, M., I. Cheung, and M. B. Schlemper (2008) Skills in professional geography: An assessment of workforce needs and expectations. *The Professional Geographer*, 60(3): 356–373.

Sorby, S. A. (2009) Educational research in developing 3-D spatial skills for engineering students. *International Journal of Science Education*, 31(3): 459–480.

Staley, D. J. (2007) Finding Narratives of Time and Space, IN: D. S. Sinton and J. J. Lund (eds) *Understanding Place: GIS and Mapping across the Curriculum*. Redlands, CA: Esri Press, pages 35–48.

Sui, D. Z. (1995) A pedagogic fraemework to link GIS to the intellectual core of geography. *Journal of Geography*, 94(6), 578–591.

Sui, D. Z. (ed) (2008) *Geospatial Technologies and Homeland Security: Research Frontiers and Challenges*. Dordrecht: Springer.

Sullivan, D., T. Brase, and A. B. Johnson (2008) A plan for the national coordination of geospatial technology education from a community college perspective. http:www.marinetech.org/workforce/geospatial (accessed November 28, 2010).

Tapiador, F. J., and J. Marti-Henneberg (2007) Best of times, worst of times: A tale of two (Spanish) geographies. *Journal of Geography in Higher Education*, 31(1): 81–96.

Tate, N. J., C. Jarvis, and K. E. Moore (2005) Locating spatial thinking in teaching practice. *Computers, Environment and Urban Systems*, 29: 87–91.

Tate, N. J., and D. J. Unwin (2009) Teaching GIS&T. *Journal of Geography in Higher Education*, 33(Supplement 1): S1–S6.

Teodorescu, D. (2004) *Using Geographic Information Systems in Institutional Research*. San Francisco: Jossey-Bass/Wiley.

Trend, R. (2009) Influences on future UK higher education students' perceptions and educational choices across geography, earth and environmental sciences (GEES), *Journal of Geography in Higher Education*, 33(2): 255–268.

Unwin, D. (2005) Fiddling on a different planet? *Geoforum*, 36(6): 681–684.

US Department of Labor (2010) Geospatial technology competency model. http://www.careeronestop.org/competencymodel/pyramid.aspx?GEO=Y (accessed November 28, 2010).

Valcic, N. (2007) The logistical tracking system (LTS) five years later: What has been accomplished? *New Directions for Institutional Research*, 135: 37–45.

Verduyn, M., and J. Kerski (2009) Shaping our world. *GIM International*, 23(11). http://www.gim-international.com/issues/articles/id1439-Shaping_Our_World.html (accessed November 30, 2010).

Wai, J., D. Lubinski and C. P. Benbow (2009) Spatial ability for STEM domains: Aligning over 50 years of cumulative psychological knowledge solidifies its importance. *Journal of Educational Psychology*, 101(4): 817–835.

Wheatley, D., and M. Gillings (2002) *Spatial Technology and Archaeology: The Archaeological Applications of GIS*. New York: Taylor and Francis.

Whyatt, D., G. Clark, and G. Davies (2011) Teaching geographical information systems in geography degrees: A critical reassessment of vocationalism. *Journal of Geography in Higher Education*, 35(2): 233–244.

Wiegand, P. (2006) *Learning and Teaching with Maps*. New York: Routledge.

Wikle, T. (2010) An examination of job titles used for GIScience professionals. *International Journal of Applied Geospatial Research*, 1(1): 40–56.

Wong, M. S., J. E. Nichol, and K. H. Lee (2009) Modeling of aerosol vertical profiles using GIS and remote sensing. *Sensors*, 9: 4380–4389.

Wright, D. (1999) Getting to the bottom of it: Tools, techniques, and discoveries of deep ocean geography. *The Professional Geographer*, 51(3): 426–439.

Yagoub, M. M., and B. Engel (2009) Remote sensing and geographic information systems in developing countries: Case of the United Arab Emirates (UAE), *The Journal of Terrestrial Observation*, 1(2): 69–88.

Zhou, Y., and J. Wu (2005) Modeling university enrollments with ArcGIS. Esri Educational User Conference, San Diego, California, 23–26 July. http://downloads2.esri.com/campus/uploads/library/pdfs/58255.pdf (accessed November 28, 2010).

Zwick, P. (2010) The world beyond GIS. *Planning*, 76(6): 20–23.

3

The internationalization of Esri higher education support, 1992–2009

Michael Phoenix
Esri, Redlands, California, USA

3.1 Introduction

When I came to Esri in 1992, my background in GIS was not deep. I had just earned a PhD from Clark University at the age of 52, with no courses or training in GIS. In 1991–92, as a new Instructor in the (then called) Department of Geography and Environmental Studies at William Paterson College I taught one GIS course. Despite my lack of background in GIS I agreed to teach the course because students desperately wanted it on their college transcript. The students had begged the department chair to offer a GIS course. Since he had no knowledge or interest in GIS, he came to me as I was the only person in the department who had a personal computer.

And so I met with the students and said, 'Look, I'm no GIS expert, I don't even know this technology. But I'm willing to learn it and work with you on the understanding that we do this together as a group project.' They loved the idea. So we put together a course in GIS. I got the new course approved and added to the catalogue. Twenty-four students enrolled – every graduating senior in geography.

I became a believer in GIS quickly. My students' first assignment was to interview local organizations in the area that were doing GIS. I gave out the assignment in the first week of class, on a Tuesday. One of the more eager students interviewed an aerial photography company on Thursday. He came back to me and said, 'Mike, I want to present this to the class next week.' So he got up in front of the class and said, 'I went in there, and I'm interviewing them, and they're interviewing *me*, I'm asking them questions about what

---

(Editors' note: This chapter is adapted from the transcript of David DiBiase's interview with Mike on 5 January 2010)

Teaching Geographic Information Science and Technology in Higher Education, First Edition.
Edited by David J. Unwin, Kenneth E. Foote, Nicholas J. Tate and David DiBiase.

they are doing with GIS, and they keep asking questions about *me*, about what I'm doing and what I'm studying. Finally, one of the bosses at the company said, 'didn't you come here to interview for a GIS job?' The student then said 'No, I came to find about what you are doing with GIS for a class project.' And the boss said, 'Well, don't you want a job?' And the student said, 'Yes, of course, I'm graduating in May and I need a job.' 'Well, we'd like to hire you,' the man said. And the student replied, 'I don't know GIS yet. I've only been in one class and all we did was get the handouts and talk about what the course is about. I don't have a clue what GIS is.' To which the boss replied, 'That doesn't matter. You're the first person we've talked to who even heard of GIS, who had an interest in it, who had a book on it, and who had a professor. You're the best we can find.' So the student said to his classmates, 'GIS rocks. You know, this is awesome, I love it, I'm going to study this stuff, I'm going to learn it, and I've already got a job for May.' Imagine how excited the class was to hear that. I was excited too, and I was convinced that I wanted a future in GIS myself.

3.2 Higher education support at Esri in the early years

It's hard to start an academic career in your 50s. So I was looking around for alternatives and found that the Esri was looking for something called 'University Programs Manager.' I didn't know what that was but I was very interested in Esri. Everything I had read about Esri fascinated me; it was growing fast and dominating the GIS market. Their annual Map Book showed GIS applications in every conceivable discipline. So, I applied and Esri invited me to interview.

I had never worked for private industry. My background was government, NGOs and academia. During the interview process I kept asking everyone what the job entailed and they said, 'You have to ask Jack Dangermond. This is his idea.' So when I finally got to meet with Jack I confessed that I didn't know what a software company would want from a university programs manager, and asked him what the job entailed. Jack thought about it for a minute and then replied, 'That's backwards. You tell me what you would do with the job if I gave it to you, and I'll tell whether you can have it.'

I had to make something up on the spot. From my academic experience I was aware that most GIS courses at the time were senior or graduate level courses involving very few students. What I said was, 'First of all, we need to find a way to move GIS from being an upper level senior or graduate level course down into undergraduate level, preferably sophomore or freshman. This will be hard, but there is so much content that you simply can't teach enough in a single course or even a single year. Students need multiple courses spread across much of their college careers as possible, and into community colleges and high schools too if possible. Teaching software is one thing, but the hard part is teaching spatial thinking, and that you can't do in a semester. It really takes time, and it's not intuitive to everybody.' Jack liked that idea. I think that he'd already had this same thought, but he liked that I was on the same track.

'The second thing we need to do,' I continued, 'is to move GIS into more academic disciplines.' At that time, according to a survey by Jay Morgan (Morgan and Fleury, 1992), about sixty percent of all GIS courses were taught in the geography departments and a few other disciplines like planning, landscape architecture and surveying. I felt strongly that we needed to promote its use in many other disciplines. I told Jack we needed to go after Colleges of Forestry, Agriculture and Public Health. We needed to show this technology to demographers, business schools and economists. We needed to figure out a way to encourage entire university campuses to embrace GIS, and to embrace spatial thinking. There just weren't enough geography departments to realize the potential impact of spatial thinking and GIS. We needed to reach hundreds of thousands of students to have a real impact.

My third suggestion to Jack was simply that we aim to spread GIS to every university in the US and abroad. The goal I had in mind was to try to get every university in the world to be doing GIS in at least one department. Jack liked my ideas and hired me. He said, 'I don't know how you do all that, but I want you to try.'

Esri had only about 600 employees in 1992. It was still a relatively small company. We had no good numbers at that time about the size of the education market and sub-markets. We did know that at least about 300 universities had our software. We didn't know who at the universities had the software or what they are doing with it. Finding that out became part of my job.

I wasn't the first Esri liaison to higher education, by the way. In the early years (the 1970s to 1990), Jack handled university relations himself. He knew all the GIS professors personally. But by the 1980s he was getting too busy and really needed someone else to play this role.

The first person Jack hired for higher education support was Bob Maher, who worked in that role from 1989 until 1991. Bob came from the College of Geographic Sciences (COGS) in Nova Scotia. COGS offered an intensive one-year postgraduate diploma program that was one of the first GIS programs of real significance. They did the best ArcInfo training in the world at the time. There were more people from COGS working at Esri than from any other college or university. We were hiring half the graduating class of this small college in rural Nova Scotia. Bob's job was different from what I ended up doing. One of his tasks was building good relationships with the leading GIS programs and schools like those participating in the National Center for Geographic Information Analysis (NCGIA), and with the Michael Goodchilds and Duane Marbles of the world. Bob was perfect for that role because he knew all those people. He had been in GIS education forever in Canada and has a tremendous knowledge of the field and connections with everyone. Bob did a great job of getting GIS into the big university programs; my job was to get it to everyone else. Bob's second challenge was to create a product that helped students learn GIS using Esri software. He spent a lot of his time creating a product called the 'GIS Concepts Kit.' He was well suited for that because he knew the software really well, he was an experienced GIS classroom instructor, and he had helped to set up the GIS program at COGS.

By the time I joined Esri in 1992 it seemed like there was a lot going on in higher education, but it was really very little compared to today. There were quite a few software courses being taught, even at community colleges like Lansing Community College. And software companies were doing training. In general there was a lot of training happening, but relatively little education. I saw universities as a marketing opportunity, not a sales opportunity. Our competition didn't get that. We wanted everyone teaching everything with our software. They thought our approach was naïve and uncompetitive. But I think that helping educators was strategic and very productive in the long run. We eventually had about 90 per cent of students learning GIS with Esri products. That had to have an impact on the sales of the software to government and industry.

3.3 Domestic focus in the 1990s

Once people heard that there was someone at Esri whose job was to help universities, the calls never stopped. It was hard to keep up with the demand, particularly for free software. I was putting in 60 hours a week and more. I spent about four hours every day on the phone. I had so much trouble keeping up that on weekends I'd come in and just do e-mail all day long. I had a database of about 3,000 people I'd helped in some way. I tried to build good relations with GIS professors, and to get the word out that Esri wanted to help. I did a lot of conferences where I could meet dozens of GIS professors at a single meeting. During that time I did give away a lot of software. I probably gave away something like 15,000 to 20,000 licenses of ArcGIS in the mid 1990s. I particularly liked helping newly hired Assistant Professors fresh out of their PhD programs. A big donation and a letter telling them that the donation had a commercial value of $200,000 made a big impression on their Chairpersons and Deans. Esri and I made many, many friends this way.

Helping universities get free software became a personal mission for me. When I started at Esri, I thought of it as kind of a subversive program, helping universities covertly, perhaps being more helpful than the company might like. But I was wrong; it wasn't subversive because I was doing exactly what Jack wanted. That made my job really fun. But there were a lot of people in the company who didn't like what I was doing at all. Some salesmen and the legal department had a problem with it. For instance, in 1993 I had just signed a statewide site license with the University of Georgia and they ordered 1,200 ArcInfo licenses. The legal department saw ArcInfo licenses at that time being worth an average about $10,000. So they saw this as 12 million dollars worth of software that I sold for a few thousand dollars. The head of the legal department sent a memo to Jack saying that Mike Phoenix was out of control. Jack called me in and said, 'What's all this about?' After I told him he said, 'Wow, that's great, let's put it in *ArcNews*.'

Jack wanted everyone to use the software. He was as concerned about spatial thinking as I was. He understood that decision makers too often ignore the spatial context of the problems they need to solve. At the same time, because he was aware that things change over time as well as space, and he felt strongly that spatio-temporal analysis needed to be added to decision-making processes too. So spatial thinking was key to Jack's thinking right from the beginning. But he was also keenly aware that however good your tools are, you don't change behavior with tools or technology. You change behavior with education. That's why education has always been important to him and to Esri.

We had so much work in the early years that our focus was mainly domestic. Universities were calling Esri all the time asking for help in setting up GIS courses and programs. I was getting about two or three new contacts every day. I never had to do a 'cold call' where I didn't know the person who I was calling and they weren't expecting me to call. Then, in 1998 Karen Kemp got me involved in a conference in Europe – the European Union GIS Education Seminar (EUGISES). I loved that conference. The EUGISES meetings became one of my favorite venues. Everyone at EUGISES was interested in GIS education and there was so much going on in GIS education in Europe that we in the US needed to know about.

It's a very different educational environment and very dynamic. Still, I really didn't have the time to work on international GIS education.

3.4 The 2000s and emerging international emphasis

In 2000 I turned 60, and I had always wanted to retire at 60. So, I told Jack that I was going to retire. I found a position in Bhutan as a United Nations Volunteer. It was a political job, not a technical one – bringing ministries together to create a GIS agency and set GIS policy. That sounded perfect to me. Bhutan is one of the most special, most pristine places in the world. I had wanted to go there all my life. When I went there one thing that really struck me was the lack of educational opportunities for professionals in GIS.

There were twelve GIS labs in Bhutan. The people working in these labs only had access to a few weeks of software training. They didn't really know how to think spatially. Most of the few people in developing countries like Bhutan who had received in-depth GIS education did so at ITC in the Netherlands. But ITC was teaching hundreds of students while many thousands more needed education in GIS and spatial thinking. I did some lectures on spatial analysis in Bhutan 2001 and it was new to them. Some of these people had been working with GIS for ten or twelve years and had no idea what spatial analysis was. These were bright, hard working people who simply lacked access to appropriate educational opportunities. As part of a GIS needs assessment I did for the government, I talked to every single person who worked with GIS in the Bhutanese government – about forty people altogether. None of them really understood what the potential of GIS was. I helped organize Bhutan's first GIS conference and Jack came to give the keynote speech. The purpose of the conference was to let the Bhutanese GIS practitioners present to each other, and that was very effective. Most of the people who presented had never heard what the other labs were doing, and had never seen a presentation of other ministries' GIS programs.

For example, the Ministry of Agriculture (MoA) was amazed at the rural cadastre of farms that the Survey of Bhutan had done. The Survey had created a land use map of the whole country that included every single land parcel and farm. The GIS staff at MoA had never seen it, and had no access to it until that conference. MoA itself had done an agricultural census and interviewed every farmer in the country, but it was not geographically referenced nor connected in any way to the rural cadastral map. And the main user of the rural cadastre was the judiciary for adjudicating tax claims. The judiciary had no previous relationship with MoA. As these organizations began to talk and work together I realized that there must also be unfulfilled potential in many other poorer countries like Bhutan. Thanks in part to the international donor community, GIS was being used even in the world's poorest places. Organizations like the United Nations, the World Bank and the Asian Development Bank were using GIS, as were donor agencies from almost every developed country and non-governmental organization. But these organizations introduced GIS as part of limited-term projects that usually included minimal GIS software training. Rarely did the local participants in these projects have any access to the kinds of educational opportunities that would prepare them to realize the potential of GIS. This realization is what made me want to go back to work at Esri, but with a new emphasis on the developing world.

3.5 Working with distributors

I returned to Esri in 2002 with a mission. I wanted to take GIS education global, particularly to the poorer parts of the world. Part of the challenge in going global was to convince Esri's international distributor network that GIS that education is important. Years before, when Esri began pursuing sales overseas, it did so by finding industry partners and offering them exclusive distribution rights in return for marketing and supporting Esri products. The international distributors are diverse – they include perhaps 100 different kinds of companies. They tend to be very sales-oriented, and they operate with a fair bit of autonomy. So, I couldn't sell software licenses (or give them away free) to colleges and universities in other countries without the permission of our partner in that particular country. If I wanted to give away software to a Chinese university, for example, I'd have to work through Esri China. Some distributors shared Jack's and my goals for higher education, but some didn't. One of the distributors that 'gets it' is Esri China.

3.6 China

Jack had long been interested in China. He'd visited many times and had given lectures at universities, receiving very enthusiastic responses. He sent some others on lecture tours too, including David Maguire and Michael Goodchild. But clearly we needed to do more, since China has over 2000 universities. So then he asked me to visit China in May, 2009. I talked at about ten universities and one high school. I went to eight cities in fifteen days, which was exhausting. I followed this up with a trip in 2010 were I visited five cities in ten days. I had expected to be impressed as I knew a lot about what was happening China, but I was totally amazed at what's going on there. The growth of GIS education is incredible. For example, about 1,200 students are pursuing GIS-related degrees at Wuhan University – 400 graduate students and 800 undergraduates led by 50 faculty members. I was impressed by how intense these students were. I presented a talk to about 400 people at Wuhan, and at the end of the talk there must have been 100 or 200 hands up. And the questions were amazing too. One student asked, 'How do I put these different things together in my personal life so that I become a spatial thinker?' Another asked, 'How can I implement these ideas in an environment where access to data is so restricted?' And another asked, 'Can this technology be used by foreign countries militarily against China?' There were more questions than I could hope to answer.

Every talk I gave in China was in English without translation. You know, you can tell whether an audience is listening and understanding. Their nods and smiles and head-shakes, and of course their questions, convinced me that they not only understood spoken English, but they also grasped the concepts I was trying to convey. I noticed that the high school students' English was even better than the University students'.

The young people I met in China were alive with curiosity and interest and hungry for information. The audience for my talk at Wuhan included students from four different universities. 'We heard about your talk on the internet,' they told me. 'There is a lot of chatter about your lectures.' Later I received an e-mail message from China saying that 'I was so disappointed that you didn't come to my city, but we read about your talk on the internet.' I don't find the same level of excitement and interest among US students. Then, someone in Taiwan who viewed one of my lectures online invited me to come there and present. So I did another lecture tour there. Before one of my first talks, some professors warned me that I shouldn't expect questions from their students, since they never asked questions in class. With this in mind I challenged the Taiwanese students. I said, 'I hope you guys are as good as the students in China. They asked a lot of good questions.' There were three students from mainland China who were in Taiwan for a semester abroad. At the end of the lecture each of those three students asked a question and that broke the ice. Hands began to go up all over the lecture hall. I answered questions for an entire hour. The intensity of interest in spatial thinking in both Taiwan and China really impressed me. Unfortunately, despite even the large programs like Wuhan, only students in disciplines like geography, landscape architecture, planning, or survey engineering have access to GIS education. In most of the world, including China, GIS education still hasn't spread to many academic disciplines. Worldwide, access to GIS education remains grossly inadequate.

3.7 The numbers

In 2000 I estimated that about 500,000 individuals in the US used Esri software products as part of their jobs, and that some 50,000 worked as full-time GIS specialists (Phoenix, 2000). At the same time I estimated that 'the shortfall in producing individuals with an advanced level of GIS education is around 3,000 to 4,000 [annually] in the US alone,' (Phoenix, 2000, p. 13) and concluded that 'the few graduate programs now in place cannot meet the needs of the marketplace.' Later (Phoenix, 2004) I estimated that about 150,000 higher education students worldwide were taking at least one course in GIS each year, that close to 10,000 students worldwide were pursuing some sort of academic certificate of achievement that involves GIS, but that only about 500 students were pursuing master's degrees in GIS or GIScience. My visit to China convinced me that we have come a long way, but that there's still so much to do.

How did I come up with those estimates? Our customer database included information about how many universities were teaching GIS, or at least had GIS software. Around 96 percent of larger US universities had Esri software on campus. I estimated student enrolments by talking to many of the faculty members I knew were teaching GIS. Then there was a publication called *Schwendemann's Directory of College Geography in the United States* that listed enrolments in geography courses at every geography department in the country. Finally I had access to a database produced by a company called Market Data Retrieval (MDR). MDR's database lists college and university professors and the courses they teach. The company sells the data to textbook companies, among others. The first year I saw their list, in 1993 or 1994, I recall that there were six people listed as teaching a course called 'GIS.' But over the years it grew to 2,000 or 3,000 professors. I knew from my own list that there were many more people who were teaching GIS-related courses but weren't included in MDR's list. Some of these courses were hard to identify because of course titles that didn't specify GIS explicitly – titles like 'Field Methods in Soil Science' or 'Spatial Methodologies' for example. MDR couldn't easily place these in the GIS category. In addition, Jay Morgan of Towson State University in Maryland did several excellent surveys of the academic GIS community that were very useful (Morgan and Fleury 1992, Morgan, Fleury and Becker, 1996). In the end I had a pretty good idea of GIS enrollment counts for about half of the universities that I knew taught GIS. For the other half, where I didn't know enrollments, I estimated twenty students per campus. This was conservative since I knew campus enrollments in GIS classes elsewhere averaged about forty. I was really confident in these estimates. I can only guess what the numbers are like today, but based on what I've seen in China, and in India, the number of students who enroll in GIS courses annually must be greater than 300,000 worldwide.

3.8 Looking ahead

After I retired for a second time in 2009, Jack hired Mike Gould as my successor. Mike and I talked a lot about the future. Mike's got a lot more management responsibility than I had. There's now a nine-person team to support projects in higher education, K-12 education, and museums and libraries. Each of these has its own education projects going on. All that needs to be managed now, so Mike doesn't have the freedom that I had when Jack hired me in 1992. Mike and his team do an annual strategic planning exercise to coordinate their activities and make the most of the resources available. If I was coming in from the outside now, seeing all Esri is doing and all that's happening in the education community, I think I would find it a bit overwhelming. It was pretty simple back in 1992. But I think some of the original challenges remain.

One of the things that I have always enjoyed, and I think that made the biggest differences, is helping people make connections with one another. Some of things that I am the most proud of happened by simply connecting two people and seeing them do something creative, even if I didn't participate further. One of the most important things that Esri is doing is bringing together thought leaders in various fields. An example is the new GeoDesign summit.

If I were starting over today, I'd want to promote GIS within lots of disciplines in which GIS and spatial thinking haven't yet taken root. For example, we still need to find a strategy to interest business schools in this technology. I'd also love to see a meeting with professors of agriculture who are interested in GIS. I read recently that enrollments in schools of agriculture are declining nationwide. Maybe part of the problem is that students think agriculture is old fashioned or low-tech. I think we should go to the schools of agriculture and show them our technology and help them understand the potential of spatial thinking.

I would also like to find a way to interest Schools of Education and Educational Psychologists in doing more research on spatial thinking. One of the things that the National Research Council's *Learning to Think Spatially* (National Research Council, 2006) report called for was more research into how we learn and teach spatial thinking. More needs to be written on the importance of spatial literacy. Couldn't we convince education professors to encourage more of their graduate students to study spatial thinking? Might these students respond with the kind of excitement I saw in China?

References

Morgan, J. M., III, and B. B. Fleury (1992) Academic GIS directory: GIS in higher education. *GeoInfoSystems*, 2(5): 51–73.

Morgan, J. M., III, B. B. Fleury, and R. A. Becker (1996) *1996 Directory of Academic GIS Education.* Dubuque, Iowa: Kendall/Hunt Publishing Co.

National Research Council (2006) *Learning to Think Spatially: GIS as Support System for K12 Education.* Washington, DC: National Academies Press.

Phoenix, M. (2000) Geography and the demand for GIS education. *Association of American Geographers Newsletter*, June: 13.

Phoenix, M. (2004) Global GIS education. *GIM International*, February: 35–37.

4

Reflections on curriculum development in the US and abroad: from core curriculum to body of knowledge

Karen K. Kemp

Spatial Sciences Institute, University of Southern California, Los Angeles, California, USA

4.1 Introduction

It is a daunting task to be asked to write a reflection on GIS curriculum development in the US when there are many other people who are equally qualified to prepare a suitable reflections paper, not the least of which are the editors of this volume. Indeed, three of the editors have produced individually more than one such article (for example DiBiase, 2006; Foote, 2005; Unwin, 1991). All of us were involved in contributing to significant projects over the past 20 years that have had enormous impacts on how GIS is taught today in the US and worldwide. Providing a comprehensive review of all of the projects we and our colleagues have collectively undertaken would be overwhelming and perhaps unnecessary. The following, then, is a personal reflection on the path I have strolled along with many friends in the GIS academic community over two decades, moving through an evolving series of projects, technologies and approaches to GIS curriculum development and teaching.

4.2 Early days of GIS

While some might now identify the beginning of the 'modern' age of GIS education to date from the work of the *National Center for Geographic Information and Analysis* (NCGIA) in the 1980s, before it was formed there was a lively, albeit small, international community concerned about how GIS might be taught better. In the 1950s, influential research in

Teaching Geographic Information Science and Technology in Higher Education, First Edition.
Edited by David J. Unwin, Kenneth E. Foote, Nicholas J. Tate and David DiBiase.

computer mapping and quantitative geography at the University of Washington engaged a number of talented graduate students including Duane Marble, Waldo Tobler, John Nystuen and Brian Berry who are now widely regarded as pioneers in our field (Chrisman, 1998). In the 1960s the focus moved to the *Laboratory for Computer Graphics and Spatial Analysis* at the Graduate School of Design, Harvard University where Howard Fisher's team of faculty and graduate students created SYMAP, SYMVU and CALFORM, early software packages that ran on mainframe computers fed with punch cards. Other notable products such as MAP, GRID and POLYVRT emerged shortly afterwards at Harvard and elsewhere. These early academic cauldrons provided the first opportunities for GIS education, but it was limited to a very few advanced students, and very much in the style we now call project-based learning.

Funding from US government agencies that supported these Harvard projects included commitments to undertake 'the widest possible distribution of the mapping programs' (Shepard, 1970) and the development of instructional manuals including, in one case at least 'an introductory correspondence course about SYMAP' (Shepard, 1970). Thanks to this deliberate dissemination of the source code for these FORTRAN IV programs and typewritten user guides, by the early 1970s hundreds of universities had acquired the Harvard programs. At that time, these instructional manuals and course materials provided the only available curriculum resources for faculty developing courses in computer mapping. For those of us who became the 'second generation' of GIS students as undergraduates in the decade after the graduate students who participated in the creation of the fundamental tools, these dense programming manuals were remarkable learning resources to have available. Without them, it would have been impossible for the large number of junior faculty at universities that had acquired the Harvard programs to become sufficiently knowledgeable in a new field to be able to teach this rapidly spreading technology to non-computer science students.

The technology that we now think of as GIS grew steadily in popularity and ease of use through the 1980s. Recognition of the need to expand education in the field also grew. In 1980, the US *National Research Council Committee on Geodesy* produced the report *Need for a Multipurpose Cadastre* (National Research Council, 1980). The report recommended that the federal government establish a center of excellence that would provide 'a program that develops scholars and professionals. The curriculum should include direct experience with land-data-systems problems' (National Research Council, 1980, p. 96). In 1984, Duane Marble presented a paper at a conference in Sioux Falls SD (reprinted in Marble, 1990) that reported the findings of a small working group gathered by NASA in the spring of 1983 to discuss critical research needs in the area of 'spatial data handling'. The list is quite long and, surprisingly, in many ways still current. Although it was deeply buried in that paper, the list includes recognition that 'model curricula should be devised and publicized' (Marble, 1990, 16). The idea of a model or core curriculum reverberated for a few years before becoming realized in the US in the form of the *NCGIA Core Curriculum in GIS* (Goodchild and Kemp, 1990) which began wide distribution in 1990 and in the UK in the form of an RISC/Auto-Carto syllabus (Unwin *et al.*, 1990).

By the mid-1980s a number of important journal papers had been published that provide insightful discussions about how GIS might be structured and integrated into university programs (see for example Goodchild, 1985; Maher and Wightman, 1985; Marble, 1981; Poiker, 1985). Several workshops and conferences in the late 1980s and early 1990s continued to build the GIS education community. David Unwin was a key player in this community in the UK along with Duane Marble in the US. Both organized and led important workshops on both sides of the Atlantic that helped articulate what was needed to encourage faculty colleagues and universities to get started teaching GIS.

At the start of the 1990s, Unwin nicely summarized the state of the GIS education context at the time, noting that 'the rapidity with which the demand for GIS education and training has arisen has meant that there are very few academics competent to teach it' (Unwin, 1991, 82). At that time, we recognized many problems facing those who wanted to see GIS education improve and increase quickly. They included the facts that few of the faculty who were teaching or planned to teach GIS had any formal education in the subject, there were very few teaching materials and the technology was expensive and difficult to operate. The first GIS textbook was published in 1986 (Burrough, 1986) and the second one in common use did not appear until 1991 (Aronoff, 1991).

4.3 Defining content: the NCGIA Core Curriculum

In 1987 the US *National Science Foundation* issued a solicitation for proposals to create a *National Center for Geographic Information and Analysis* (NCGIA). This stressed the need to provide funding to create a national center that would be able to, among other goals, 'augment the United States' supply of experts in GIS and geographical analysis' (Abler, 1987, 303). The proposal to the NSF by the consortium that eventually won the award addressed this goal by outlining the usual set of educational components such as enhanced teaching of undergraduate and graduate students at the three member universities, the involvement of graduate students in the proposed research, and the offer of some workshops and short courses. Importantly, however, the consortium also proposed to undertake the 'development and dissemination of a standardized curriculum for teaching the basics of GIA (geographic information analysis) to undergraduate and graduate students elsewhere' (University of California Santa Barbara, State University of New York at Buffalo and University of Maine, 1988). Thus began the NCGIA *Core Curriculum.*

Funding for the NCGIA began in August 1988, just as I arrived in Santa Barbara to begin my PhD studies. I was immediately engaged to work as the Education Coordinator and to take the lead in the effort involved in pulling together the Core Curriculum. That fall we began working internally in the Consortium to develop a list of seventy-five lecture topics organized into three courses to be called *Introduction to GIS, Technical Issues in GIS* and *Applications of GIS.* Next, we invited thirty-five academics in the US and elsewhere to contribute one or more sets of lecture notes for individually assigned topics. Topics for which we were unable to find authors were assigned to NCGIA members. Deadlines were short with submissions due in the spring of 1989. Each set of notes was heavily edited and rigorously structured to fit a standard format to produce the draft version of the Curriculum (Kemp and Goodchild, 1991). One hundred carefully selected universities around the world were engaged to test the draft version in the 1989–90 school year. The evaluation program, reported in Kemp and Goodchild (1992), was rigorous and produced considerable input for the preparation of the final version which began distribution in the summer of 1990.

The Curriculum was produced as a three volume spiral bound set of photocopied lecture outlines, each accompanied by a number of black and white graphics for preparation as overhead transparencies and a brief set of exam and discussion questions. Each hard copy version was accompanied with floppy disks containing all materials in digital format. The final set of lecture topics is shown in Table 4.1. Although a preliminary collection of

Table 4.1 Topics in the NCGIA Core Curriculum in GIS (1990)

Introduction to GIS

A. Introduction
  1. What is GIS?
  2. Maps and map analysis
  3. Introduction to computers

B. A first view of GIS
  4. Raster GIS
  5. Raster GIS capabilities

C. Data acquisition
  6. Sampling the world
  7. Data input
  8. Socio-economic data
  9. Environmental data

D. Spatial databases
  10. Models of reality
  11. Spatial objects and database models
  12. Relationships among spatial objects

E. Vector view of GIS
  13. Vector GIS
  14. Vector GIS capabilities

F. Using the GIS
  15. Spatial analysis
  16. Output
  17. Graphic output design issues
  18. Modes of user/GIS interaction
  19. Generating complex products
  20. GIS for archives

G. Past, present and future
  21. Raster/vector debate
  22. Object/layer debate
  23. History of GIS
  24. GIS marketplace
  25. Trends in GIS

Technical Issues in GIS

H. Coordinate systems & geocoding
  26. Common coordinate systems
  27. Map projections
  28. Affine & curvilinear transformations
  29. Discrete georeferencing

I. Vector data structures & algorithms
  30. Storage of complex spatial objects
  31. Storage of lines: chain code
  32. Simple algorithms I – line intersection
  33. Simple algorithms II – polygons
  34. Polygon overlay

J. Raster data structures & algorithms
  35. Raster storage
  36. Hierarchical data structures
  37. Quadtree algorithms, spatial indexes

K. Data structures & algorithms for surfaces, volumes & time
  38. Digital elevation models
  39. TIN data model
  40. Spatial interpolation I
  41. Spatial interpolation II
  42. Temporal and 3D databases

L. Databases for GIS
  43. Database concepts I
  44. Database concepts II

M. Error modeling & data uncertainty
  45. Accuracy of spatial databases
  46. Managing error
  47. Fractals
  48. Line generalization

N. Visualization
  49. Visualization of spatial data
  50. Color theory

Application Issues in GIS

O. GIS application areas
  51. GIS application areas
  52. Resource management applications
  53. Urban planning and management
  54. Cadastral records and LIS
  55. Facilities management
  56. Demographic and network applications

P. Decision-making in a GIS context
  57. Multiple criteria methods
  58. Location-allocation on networks
  59. Spatial decision support systems

Q. System planning
  60. System planning overview
  61. Functional requirements analysis
  62. System evaluation
  63. Benchmarking
  64. Pilot project
  65. Costs and benefits

R. System implementation
  66. Database creation
  67. Implementation issues
  68. Implementation strategies for large organizations

S. Other issues
  69. GIS standards
  70. Legal issues
  71. Development of a national GIS policy
  72. GIS and global science
  73. GIS and spatial cognition
  74. Knowledge based techniques
  75. The future of GIS

laboratory materials was distributed with the draft version, the final introductory materials to support the Curriculum were produced in 1991 (Dodson *et al.*, 1991).

A copy of the original Curriculum is still maintained on-line at the University of British Columbia (www.geog.ubc.ca/courses/klink/gis.notes/ncgia/toc.html). From my biased perspective as co-editor with Michael Goodchild, the *NCGIA Core Curriculum in GIS*, presented as a framework from which educators could construct their own courses, was a major watershed in the quality and content of GIS education in the US. Thanks in large part to the pent-up demand for teaching materials and the building stature of the NCGIA, the Core Curriculum diffused very rapidly through the academic community. The NCGIA itself printed and mailed out over 1500 copies to more than 70 countries, with approximately 60% sent to addresses in North America. More numerous are the uncounted numbers of copies that were informally distributed digitally as well as through the fourteen officially designated international distributors (Kemp, 1997). For example, by early 1995 Béla Márkus and his colleagues in Hungary had distributed at least 800 copies of the Hungarian language version (Márkus, 1993). Translations into Russian, Chinese, Korean, Hungarian, Czech, Portuguese, Japanese, Polish, French and Arabic were authorized, though not all materialized. So, while the Core Curriculum was certainly US centric, its effectiveness in international contexts is clear.

It is important to acknowledge that there were a few other GIS curriculum development projects underway around this same time (see for example Maher and Wightman, 1985; Nyerges and Chrisman, 1989; Poiker, 1985; Unwin, *et al.*, 1990). Apart from the Royal Institute of Chartered Surveyors/Auto-Carto proposal created in the UK (Unwin *et al.*, 1990), none of the others were as detailed in regards to content, focusing instead on articulating the larger context of GIS education within a university.

Assessments of the Core Curriculum

Soon after the Curriculum was distributed, Heywood called it 'arguably the most comprehensive curriculum project undertaken in higher education to date' (Heywood, 1990, p. 851). As described above, the Curriculum was initially distributed as a draft version in 1989 only to a select group of 100 universities who committed to evaluating the materials and contributing suggestions for revisions and enhancements. This was an extremely successful approach as it not only provided important revisions to the materials, it also ensured open participation and led to a shared sense of ownership. It gave permission for public review and criticism of the materials. Although it was not at the time a conscious choice, there were no intellectual property rights nor copyright restrictions imposed on these documents, a condition which, of course, aided in the Curriculum's widespread diffusion.

As one of the evaluators of the draft version in the UK, Unwin became a strong and vocal supporter of the Curriculum. At the same time he provided thoughtful and thorough critiques. Most important was his assertion that the materials were inappropriately named as they are not a fully articulated curriculum *per se*, but simply a set of lecture materials. Morgan echoed this assessment, suggesting that the materials should not be organized into three courses, and thus appear as a prescription for curricula, but rather they should be seen as a 'smorgasbord of information' (Morgan, 1990, 22). This is a valid criticism and I frequently advocated in public presentations and in print that the materials were not intended to be adopted directly 'as is,' but rather that they were intended to guide those who were creating new GIS courses, providing some direction with regards to the full range of knowledge that ought to be considered for inclusion in GIS course syllabi. In fact, in a review of the development of the Curriculum, Goodchild and I noted that 'we have chosen the term 'core curriculum' rather than 'model curriculum', as the latter suggests an ideal, rather than a core around which one can build a specialised programme' (Kemp and Goodchild, 1991, 127).

Other valid criticisms note that while comprehensive, the Curriculum does not completely cover the field. There was most definitely a problem with depth and balance of coverage given that some topics, such as project management, received unusually more focus than would have been expected for academic instruction contexts at the time (Coulson and Waters, 1991). There was also some discussion about how best to balance the need for training versus education. Can theory be taught purely in practical hands-on exercises, as the students might have preferred, or does it need to be handled separately in the style of the Curriculum? Despite these unresolved questions, by articulating a preliminary 'straw-man' version of, what we would now call, a body of knowledge for the field, concerted attention to determining what we should teach in GIS began in earnest.

What difference did the NCGIA Core Curriculum make?

While the Curriculum might correctly be criticized for leading to a tendency amongst less experienced GIS instructors to frame GIS only within the constrained collection of notes rather than creatively from the large and growing domain of grey and eventually published literature on GIS, it provided substantial fuel for initiating courses and programs worldwide in the early 1990s. In an article reviewing their experience at the University of Calgary during the evaluation year, Coulson and Waters noted 'involvement in the NCGIA core curriculum has been a truly exceptional experience for our department. It has allowed us, despite limited teaching resources and even less exemplary hardware and software, to become a legitimate player in teaching one of the few quality GIS programs in Western Canada' (Coulson and Waters, 1991, 101). They concluded 'the core curriculum has begun the process of identifying and structuring standards of course content in GIS. It has caused us to reconsider not only what we are teaching, but what should have preceded the GIS course and when' (Coulson and Waters, 1991, 101).

It was also put to use in contexts beyond lecturing and course design. An early analysis of the demand for the Curriculum showed that only 58 percent of the 736 copies distributed by the NCGIA by the end of 1991 were sent to educational institutions. The remainder went to commercial organizations, government agencies, libraries, research institutes and bookstores (Goodchild and Kemp, 1992). Raper and Green used the Curriculum as a reference for the development of their tutorial software *GISTutor* (Raper and Green, 1992). In their annual GIS education surveys during the 1990s, Morgan and colleagues used it as a benchmark against which to compare course content (see for example Morgan, 1992). The effort to translate the Curriculum into Hungarian involved the creation of new technical vocabulary in that language and became the catalyst for the formation of a large and active community of GIS educators in Hungary (Márkus, 1993).

Even ten years after the release of the Curriculum, in an article that reviewed GIS curricula developments, Montagu states 'The Core Curriculum now represents the *de facto* standard of the model [sic] curricula approach, and although others have produced similar models, most exist as alternatives or complements to the NCGIA Core' (Montagu, 2001, 186). As recently as 2009, the incorporation of materials from the original curriculum is proudly noted in course syllabi and lecture PowerPoint slides readily found on the internet (examples found at http://www.telecarto.com/content/maincontent/DownLoad/GIS/chapter%201.pdf and http://www.nmsu.edu/~geoweb/BrownSyllabi/Geog_521_Syllabus_Sp_2009.pdf)

Despite its many shortcomings, the NCGIA Core Curriculum has had a lasting, and I believe, positive, impact on GIS education in the US and in many countries worldwide.

The international post-graduate course on GIS

It is worthwhile mentioning briefly another early curriculum development project I participated in during my brief European sojourn in 1992–93. This one was funded by the European Union (EU) and undertaken at the Technical University of Vienna. The objective was to build upon our experience with the NCGIA Core Curriculum to create a European-based GIS curriculum. Distinct from the NCGIA approach, the EU funding provided an opportunity to undertake a Delphi survey though which we could elicit opinions from the academic community about what should be taught in GIS (Kemp and Frank, 1996). A Delphi survey uses an iterative approach, first asking each person for their personal opinions and then returning to participants two further revisions of their collective inputs to culminate in a consensus. Over sixty academics and GIS professionals in Europe were invited to participate in the three rounds of the survey with returns at 79%, 66% and 44% for the first, second and third rounds respectively.

The survey resulted in yet another list of suggested lecture topics, this time collected into eighteen units. These are listed in Table 4.2. Rather than producing lecture outlines, the EU funding supported the authorship of extended essays that expand in detail on each of the unit themes. The end result was a set of three volumes published as a collection of course notes by the Department of Geoinformation at the Technical University of Vienna (Frank, 1995).

Although the product of this project did not have the same wide uptake as the NCGIA curriculum and is little known outside of Europe, the effort was important in its initial focus on defining the domain of knowledge. At the NCGIA, this stage was a quick initial step undertaken internally in the consortium with little contribution from the community at large. It was comforting to note the similarities between the lists of topics in Table 4.1 and Table 4.2 as this gave confirmation that our first attempt, though originating from a smaller team, was valid. However, as awareness of the field expanded, it became increasingly clear that determining in a credible manner what should be taught began to be more important than simply offering prepared lecture materials.

4.4 Collections and portals: curriculum materials on the web

While the NCGIA Core Curriculum was very successful as an initiator for GIS courses in the early 1990s, very quickly it became outdated in design and comprehensiveness. Even more important in hastening the evolution of our efforts, however, was the emergence

Table 4.2 Units in the international post-graduate course (1993)

Spatial Information for GIS

1. Spatial concepts and the representation of spatial knowledge
2. Determining and representing location
3. Modelling reality in an information system
4. Spatial concepts as implemented in GIS
5. Data sources for GIS
6. Traditions and use of GIS
7. Needs analysis and feasibility studies for GIS

Information systems for GIS

8. Technical aspects of information systems
9. Special information system requirements of GIS
10. Database issues
11. Technical aspects of digital spatial data
12. Spatial analysis
13. Methodologies for system design and selection

Using GIS in the organization

14. Communicating spatial information
15. Economics of geographical information
16. Project management
17. Implementing GIS in an organization
18. GIS in society

of the World Wide Web and its immediately recognized potential for organizing and distributing information for educators. At that time we saw its greatest value was in being able to link teaching materials in a multi-connected web framework that broke us free of the constraints of the linear progression of knowledge construction required in printed books with sequentially numbered topics. We were no longer required to enforce a single sequence on the learning process, nor to struggle with establishing which topics were the prerequisites of other topics. Below I review a few of the key early web-based GIS curriculum projects.

The NCGIA on-line curricula

As the web became widely accessible and relatively easy to use, the NCGIA developed several new curriculum projects in rapid succession during the 1990s: the *GIS Core Curriculum in Technical Programs* aimed at two-year technical colleges; the *Remote Sensing Core Curriculum* and the *Core Curriculum in Geographic Information Science* which was intended as an expansion of original curriculum under the new GISc umbrella (Kemp, 1997). Structured simply as html versions of the by then familiar lecture note format, they are rich and useful products, still alive on the web at www.ncgia.ucsb.edu. However, none of these later curricula were as successful as the first, nor were any of them completely finished as interest in contributing to such efforts lagged. The timing for the first curriculum was a perfect balance of opportunity and need. Later curricula had to compete with a rapidly growing volume of published literature and increasing expertise amongst GIS educators.

The Geographer's Craft

A project similar to the set of NCGIA on-line curricula, but one that was fleshed out more completely, is Ken Foote's *Geographer's Craft* (Foote, 1997). These materials are still available on-line at www.colorado.edu/geography/gcraft/contents.html, though the courses are no longer taught. Structured, like the Core Curriculum, as a set of lecture materials for a pair of introductory courses in 'modern geographical techniques', it made better use of the first generation web structure and provides excellent content and links.

Interoperable GIS education

There were many important and useful conferences, workshops and conference sessions about GIS education held throughout the 1990s. While all of them gave us the opportunity to explore new ideas and approaches to teaching, one workshop in particular attempted to lay the foundation for yet another approach to organizing all the information we might want to teach about and with. *The International Workshop on Interoperability for Distributed GIScience Education* (IGE '98) was held in Soesterberg, The Netherlands in 1998.

At this meeting, participants envisioned the creation of a knowledge base of 'interoperable education objects' that could be snapped together in any number of ways to create personalized education opportunities. These might be used by faculty for traditional on-campus courses, or, more likely, in distance education contexts where students and instructors are able to organize learning in flexible and rapidly evolving organic sequences. It was a grand vision, but like so many that emerged in the 1990s post-curriculum, the vision was there, but the momentum failed us. The conclusions and list of action items from the meeting are recorded in Heywood, Kemp and Reeve (1999) and materials from the meeting are still posted on the NCGIA website. Revisiting this wish list is still informative and may yet motivate some innovations.

What is important about this project was the recognition that, in the context of the web and increasing demand for innovation in learning contexts, a structured set of lecture notes is not the best way to support the rapidly expanding world of GIS education. This idea of an interoperable knowledge base might be seen as one of the steps towards the emergence of the GI Science and Technology Body of Knowledge described below.

Digital library for Earth system education

Although it is not strictly a GIS&T-oriented project and is directed towards K-12 education, the NSF funded *Digital Library for Earth System Education* (DLESE, see www.dlese.org) is an excellent example of the kind of education materials portal that would be required to host a collection of interoperable education objects as envisioned at IGE '98. Continued support for DLESE under the NSF's *National Spatial Digital Library* program since 2000 has permitted this portal to maintain a rigorous framework and excellent quality control. It is well supported by the community and now acts as a clearinghouse of links, contacts and activities for teachers in relevant disciplines. Sustained funding is clearly a key determinant in the longevity of such portal projects.

4.5 The AAG/UCGIS body of knowledge

In the late 1990s, as evidenced in part by the deliberations at the IGE '98 meeting, interest grew again in developing a coordinated framework for defining the scope of the field of geographic information science. Echoing the demand for teaching materials that set in motion the development of the NCGIA Core Curriculum, discussions at the second summer assembly of the University Consortium for Geographic Information Science (UCGIS) in 1997 to define the organization's education priorities, included a call for the organization to 'identify the various constituencies who will benefit from education in GIScience' and 'identify specific sets of key concepts and skills required by each constituency' (Kemp and Wright, 1997, p. 17). It is important to note that this did not suggest the UCGIS create a curriculum, but rather that the organization focus on the identification of education needs.

UCGIS moved quickly to undertake this effort. Under the guidance of Duane Marble, and returning to language from his 1984 proposal mentioned above, in 1998 the UCGIS formed the 'Model Curriculum Task Force'. Over a period of eight years, this group evolved through several different permutations of membership, venue and title. Once it was recognized that a single curriculum in the field is impossible and inappropriate, pluralizing the noun it was renamed the 'Model Curricula Task Force' and, once we finally figured out how to get to an achievable end product, the group eventually concluded its work as the 'Body of Knowledge Editorial Committee'.

The result of this massive community effort is the *UCGIS Geographic Information Science and Technology Body of Knowledge* (GIS&T *BoK*) which was published in 2006 by the Association of American Geographers (DiBiase *et al.*, 2006). It is structured in ten knowledge areas, containing more than 330 topics organized into 73 units. Critically, in an important and welcome innovation for the GIS domain, topics are defined in terms of educational objectives.

This project and its outcomes are covered at length elsewhere in this volume so I restrict my comments to those relevant in the context of the evolution of GIS curriculum development. The GIS&T *BoK* represents a remarkable and satisfying culmination of two decades of work aimed at trying to articulate what it is we believe should be taught in our field. Like the Vienna project, it involved a large sector of the GIS academic community in defining the domain of the field and, among many intended outcomes, it provides a comprehensive ontological framework for future web-based knowledge repositories. It is hoped that funding and motivation will be found to use this ontology to support the creation of a GIScience semantic network, the next generation of portals which expand on earlier projects such as DLESE. In addition, the *BoK* is the first comprehensive attempt in GIS&T to establish intended learning outcomes as a means of stating content widely across the domain.

In addition to plans for the development of a second edition and a number of related projects on the agenda at UCGIS, the *BoK* is finding its place in many other activities. Like the NCGIA Core Curriculum, it appeared at the perfect time to fill a number of niches and has recently been used in such diverse applications as:

- a checklist of knowledge topics used to assess the academic qualifications of applicants for professional certification at the *GIS Certification Institute* (www.gisci.org);
- a list of topics that must appear collectively in the contents of courses offered by educational institutions seeking accreditation for offerings of *Geospatial Intelligence Certificate* programs through the *US Geospatial Intelligence Foundation* (www.usgif.org);
- a framework for organizing the contents of the *Sage Encyclopedia of Geographic Information Science* (Kemp, 2008).

Bookending the two decades of large-scale community efforts in GIS curriculum development, the NCGIA Core Curriculum (first conceived in 1987) and the UCGIS Body of Knowledge (published in 2006) have been essential, perhaps even inevitable, stages in the growth and continuing vitality of GIS&T. While GIS&T researchers drive our field forward, creating the innovations that sustain our field's continuing contributions to helping support our increasingly complex, critically populated world, we educators might be seen as the field's dogged chroniclers, plodding along, not far behind, trying to make sense of it all so that we can inform the masses of our progress.

4.6 Where are we now and what's next?

The technology we teach now is several generations removed from what we worked with twenty years ago. It is much easier to use, inexpensive, even free, and incorporates a much richer tool box that can be used to illustrate many fundamental aspects of GIS&T without too much training overhead. For those who care to try, it is now much easier to teach students about our science without it becoming obscured by the technology. Students are now two or three generations ahead of current older GIS faculty in terms of their facility with computers and digital information. Much of what we struggled to teach in the early years has evaporated. For example it is no longer necessary to teach basic computer operation and file access procedures. We can get down to the real details about geographic information and analysis very quickly. While this is a great leap forward in our ability to cover the vast richness of our domain within a small curriculum, the field continues to expand rapidly so that for whatever we no longer need to teach, there always seems to be twice as much new material to add.

Educators are now much more aware of the role of educational objectives in designing learning experiences. In the US this is becoming increasingly critical as the university accreditation organizations move towards requirements for fully documented objectives and assessments of achievement in their re-accreditation processes. Education is now an extremely competitive marketplace with informed learners demanding accountability and quality learning experiences. Efforts such as these in our field have been critical in providing compelling and credible foundations for many of our current education offerings whether they are traditional classroom programs, distance learning opportunities, or web-based podcasts for those with internet-induced short attention spans.

There is now an immense amount of information available within the realm of GIS&T. Innumerable textbooks are now available, including many written for domains as diverse as marketing, ecology, surveying, archaeology and marine science. There is so much spatial data available (particularly in the US) that sending students out to find and assess their own data is now how we often begin laboratory exercises. Our path over the past twenty years reflects this expansion of the information universe that we currently experience as information overload. At the beginning, we were delighted to find that we could transmit documents digitally by sending $5^1/_4$ inch floppy disks through the mail, the digital format making them easy to edit for publication and to customize for use in our own classes. This was quickly followed by the ability to take advantage of web-based information nets, and has culminated in the development of a shared ontological framework for our field with which we can organize, tag and seek out knowledge wherever it exists in the cloud.

It has been a fascinating and rewarding journey so far, and as we continue to work together into the future as effectively as we have in the past as a cohesive and collegial GIS&T education community of practice, or what might now be called our professional social network, sustained growth is likely as we address the new challenges of professional education, professional certification and e-learning. The field continues to expand into the new realms including spatial thinking, spatially integrated social science, the full integration of remote sensing with GIS, place-based decision making, the spatial humanities, climate change impact forecasting, and many other geographically informed enterprises. Our experience in these mutual efforts to date will surely enhance our future work.

References

Abler, R. F. (1987) The National Science Foundation National Center for Geographic Information and Analysis. *International Journal of Geographical Information Science*, 1(4): 303–326.

Aronoff, S. (1991) *Geographic Information Systems: A Management Perspective*. Nepean, Ontario: WDL Publications.

Burrough, P. A. (1986) *Principles of Geographical Information Systems for Land Resources Assessment*. Oxford: Oxford University Press.

Chrisman, N. R. (1998) Academic Origins of GIS. IN: T. W. Foresman (ed), *The History of Geographic Information Systems: Perspectives from the Pioneers*. Upper Saddle River NJ: Prentice Hall, pages 33–46.

Coulson, M. R. C., and N. M. Waters (1991) Teaching the NCGIA Curriculum in practice: Assessment and evaluation. *Cartographica*, 28(3): 94–102.

DiBiase, D. (2006) Why is a GIS&T Body of Knowledge needed? IN: D. DiBiase, M. DeMers, A. Johnson, K. Kemp, A. T. Luck, B. Plewe and E. Wentz (eds), *Geographic Information Science and Technology Body of Knowledge*. Washington, DC: Association of American Geographers, pages 9–25.

DiBiase, D., M. DeMers, A. Johnson, K. Kemp, A. Luck, B. Plewe, and E. Wentz (eds) (2006) *Geographic Information Science and Technology Body of Knowledge*. Washington, DC: Association of American Geographers and University Consortium for Geographic Information Science.

Dodson, R. F., K. K. Kemp, and S. D. Palladino (1991) *The NCGIA Guide to Laboratory Materials – 1991*. National Center for Geographic Information and Analysis, University of California, Santa Barbara, CA.

Foote, K. E. (1997) The geographer's craft: Teaching GIS in the web. *Transactions in GIS*, 2(2): 137–150.

Foote, K. E. (2005) Education and training in GIScience: New challenges, innovative solutions. *Directions Magazine*, 7 March.

Frank, A. U. (ed) (1995) *Geographic Information Systems – Materials for a Post-Graduate Course*. Vienna, Austria: Department of Geoinformation, Technical University of Vienna.

Goodchild, M. F. (1985) Geographic Information Systems in undergraduate Geography: A contemporary dilemma. *The Operational Geographer*, 8: 34–38.

Goodchild, M. F., and K. K. Kemp (1990) *NCGIA Core Curriculum in GIS* (University of California at Santa Barbara: National Center for Geographic Information and Analysis.

Goodchild, M. F., and K. K. Kemp (1992) NCGIA education activities: The Core Curriculum and beyond. *International Journal of Geographical Information Systems*, 6(4), 309–320.

Heywood, D. I., K. K. Kemp, and D. E. Reeve (1999) Interoperable education for interoperable GIS. IN: M. F. Goodchild, M. J. Egenhofer, R. Fegeas and C. A. Kottmann (eds), *Interoperating Geographic Information Systems*. New York: Kluwer, pages 443–458.

Heywood, I. (1990) Commentaries. *Environment and Planning A*, 22: 849–854.

Kemp, K. (ed) (2008) *Encyclopedia of Geographic Information Science*. Thousand Oaks, CA: Sage Publications.

Kemp, K. K. (1997) The NCGIA Core Curricula in GIS and remote sensing. *Transactions in GIS*, 2(2): 181–190.

Kemp, K. K., and A. U. Frank (1996) Toward consensus on a European GIS curriculum: The international post-graduate course on GIS. *International Journal of Geographical Information Systems*, 10(4): 477–497.

Kemp, K. K., and F. M. Goodchild (1992) Evaluating a major innovation in higher education: the NCGIA Core Curriculum in GIS. *Journal of Geography in Higher Education*, 16(1): 21–35.

Kemp, K. K., and M. F. Goodchild (1991) Developing a curriculum in Geographic Information Systems: The National Center for Geographic Information and Analysis Core Curriculum project. *Journal of Geography in Higher Education*, 15(2): 121–132.

Kemp, K. K., and R. Wright (1997) UCGIS identifies GIScience education priorities. *Geo Info Systems*, 7: 16–20.

Maher, R. V., and J. F. Wightman (1985) A design for geographic information systems training. *Operational Geographer*, 8: 43–46.

Marble, D. F. (1981) Toward a conceptual model for education in digital cartography. Paper presented at the 41st Annual Meeting of the American Congress on Surveying and Mapping.

Marble, D. F. (1990) Geographic Information Systems: An overview, IN: D. J. Peuquet and D. F. Marble (eds), *Introductory Readings in Geographic Information Systems*. London and Bristol, PA: Taylor & Francis, pages 8–17.

Márkus, B. (1993) NCGIA Core Curriculum in the Hungarian GIS/LIS Education. *Computers, Environment and Urban Systems*, 17(3): 277–283.

Montagu, A. S. (2001) Repackaging the Revolution: Making GIS instruction relevant to planners. *Journal of Planning Education and Research*, 21(2): 184–195.

Morgan, J. M., III (1990) Education and training: Meeting the rising demand for GIS personnel. *Geo Info Systems*, Nov/Dec Issue: 22.

Morgan, J. M., III (1992) Academic GIS directory: GIS in higher education. *Geo Info Systems*, 2: 51–73.

National Research Council. (1980) *Need for a Multipurpose Cadastre*. Washington DC: National Academies Press.

Nyerges, T. L., and N. R. Chrisman (1989) A framework for model curricula development in cartography and GIS. *The Professional Geographer*, 41(3): 283–293.

Poiker, T. K. (1985) Geographic information systems in the geographic curriculum. *Operational Geographer*, 8: 38–41.

Raper, J. F., and N. Green (1992) Teaching the principles of GIS: Lessons from the GISTutor project. *International Journal of Geographical Information Systems*, 6(4): 279–290.

Shepard, D. S. (1970) SYMAP interpolation characteristics. IN: J. C. Goodrich (ed), *Computer Mapping as an aid in air pollution studies*, Vol. 2. Cambridge MA: Graduate School of Design, Harvard University, pages v and vi.

University of California Santa Barbara, State University of New York at Buffalo and University of Maine (1988) *National Center for Geographic Information and Analysis: A Proposal*. Santa Barbara CA: University of California.

Unwin, D. J. (1991) The academic setting of GIS. IN: D. J. Maguire, M. F. Goodchild and D. W. Rhind (eds), *Geographical Information Systems: Principles and Applications*, Harlow, Essex, UK: Longman Scientific & Technical, pages 81–90.

Unwin, D. J., M. J. Blakemore, P. Dale, R. G. Healey, M. Jackson, D. J. Maguire, D. Martin, H. Mounsey, and J. Willis (1990) A syllabus for teaching geographical information systems. *International Journal of Geographical Information Systems*, 4(4): 457–465.

Section II

Issues in curriculum and course design

5

Using the GIS&T Body of Knowledge for curriculum design: different design for different contexts

Steven D. Prager

Department of Geography, University of Wyoming, Laramie, Wyoming, USA

5.1 Introduction

The rapid proliferation of geographic information science and technology (GIS&T) in government, industry and academic settings is raising the bar with regard to expectations associated with students exiting college, university and professional training programs. Not only have the expectations changed, so has the manner in which students become acquainted with geographic information and spatial thinking. Engagement often begins in the K-12 learning process (Alibrandi and Palmer-Moloney, 2001; Baker, 2005) and progresses in earnest into the community college and university undergraduate and graduate landscapes (Wikle and Finchum, 2003). The educational process extends beyond the 'traditional' settings and, in fact, is often initiated in a post-collegiate setting with either professional training courses, non-traditional students reentering the community college and university environments, and even individual and self-taught learning.

Early in the advent of GIS&T education, Wikle (1999) observed that university curricula tended toward a 'one size fits all' model that was, perhaps, offset by the establishment of specialized certificate programs. Unfortunately, as Marble (1998) noted, the increasing ubiquity of GIS&T as well as the increasingly 'user-friendly' nature of the software seemed to be undermining aspects of the GIS&T educational process and, consequently, eroding the expertise in the discipline. Use of GIS&T in a supporting role in the educational arena is also expanding; geographic information systems and science are, with increasing frequency, being used as a complement or tool supporting the teaching and learning of another discipline. For example, Bednarz (2004) notes how geographic information systems (GIS) can support geography and environmental education, but identifies a number of barriers

Teaching Geographic Information Science and Technology in Higher Education, First Edition.
Edited by David J. Unwin, Kenneth E. Foote, Nicholas J. Tate and David DiBiase.

that limit the effectiveness of GIS-based pedagogy. Finally, just as educational needs vary in context of the teaching and learning process, they also vary in context of the application of learned material. Dunne *et al.* (1999) illustrate how learners from lower-income countries have the potential to benefit from a teaching and learning process geared to their specific context and the needs of the developing world.

The varying institutional contexts of GIS&T teaching and learning in combination with the varying expected outcomes of the teaching and learning process conspire to present GIS&T educators with a bedeviling question. The question, albeit approached from many different directions, generally reduces to, 'How can I develop a GIS&T curriculum that works?' This question is more easily asked than answered; hopefully this chapter will help guide readers toward a methodological approach for designing successful GIS&T teaching and learning experiences.

5.2 The GIS&T Body of Knowledge

One of the most significant advancements in GIS&T education in recent years is the *Geographic Information Science and Technology Body of Knowledge* (DiBiase *et al.*, 2006). Produced by the University Consortium for Geographic Information Science and published by the Association of American Geographers, the *Body of Knowledge* (colloquially referred to as the *BoK*), is a milestone for the discipline in that it is the first time an attempt has been made to systematically and thematically catalog the depth and breadth of the discipline. In a departure from the earlier developed National Center for Geographic Information Analysis (NCGIA) core curriculum (Kemp and Goodchild, 1992), the *BoK* was conceived as a means to support development of curricula rather than as curriculum itself (DiBiase *et al.*, 2007).

The *Body of Knowledge* is structured hierarchically into 10 knowledge areas, 73 units, 329 topics, and over 1,600 formal educational objectives. The educational objectives were designed around a cross-section of knowledge dimensions and cognitive levels after the work of Anderson *et al.* (2001). The purpose of this approach was to provide users of the *BoK* with exemplar objectives that are, 'applicable across the undergraduate, graduate, and postbaccalaureate/professional sectors of the GIS&T education infrastructure' (DiBiase *et al.*, 2006). Prager and Plewe (2009) demonstrate that, indeed, the educational objectives provide a useful framework for assessment and evaluation of GIS&T curricula but conclude that curriculum design necessarily exceeds the scope of the *BoK*. DeMers (2009) also notes that, though there is relative consistency in the distribution of the expressed cognitive levels across the learning objectives in the *BoK*, the level of the stated objectives is generally biased towards lower cognitive skills. This chapter addresses these issues more directly by suggesting ways in which the *BoK* can play a role in the design process in a manner that considers the curriculum as a whole and in relation to appropriate educational contexts.

5.3 Different contexts require different design

Higher education serves many purposes in society. Importantly, stakeholders in higher education include educational institutions, the students pursuing an education and the employers of those students following their educational experience. The educational experience is also, however, not a one-time occurrence; rather it is a process that develops and changes for each individual as well as for society as a whole. Individuals often enter a particular GIS&T curriculum with their expected outcomes in mind as suggested by Diana Sinton in Chapter 2 (e.g., to get a job, to learn a specific skill, etc.). Institutions developing curricula have a variety of managerial concerns that also affect the curriculum development process. Biggs and Tang (2007, 3) point out that the new agenda for universities is often to, 'sell education and to provide for market needs,' and, consequently, that the educational endeavor requires built-in accountability and quality assurance.

With this in mind, it is important to consider both the individual and institutional contexts prior to considering curriculum design. Whereas the institutional context of curriculum evolves from the intersection of teacher-as-agent and managerial structures and their influences (e.g., state and national standards) (Sowell, 2000), the individual context is value driven and motivated by a desired outcome in relation to personal or professional goals (Table 5.1).

The incongruity between institution and individual contexts is further compounded because the higher education homes for GIS&T degree programs tend to vary geographically. In the United States, GIS&T programs tend to be more widely distributed across academic disciplines than in Europe (cadastral sciences), Australia (computer sciences), and the United Kingdom (geography) (Wikle and Finchum, 2003). The challenge, then, is to design and develop curricula in a way that clearly articulates the expected outcomes of those curricula in a manner that can be easily digested by both institutional and individual consumers.

Another important context for consideration during a curriculum development process is that institutions of higher education are one of the primary conduits of future employees entering the job market. The proliferation of GIS&T in both industry and government is such that employers are increasingly vested in the educational process. At the same time, however, employers are not simply looking for technically savvy individuals, they are looking for individuals with competencies that span GIS&T, specific disciplinary expertise, creative thinking, problem-solving skills, leadership, and much more (Gaudet *et al.*, 2003). In a similar vein, Fink and Ganus (2009) illustrate that significant learning also requires developing both the learners' sense of their relationship to their area of expertise as well as to the world around them. It thus benefits the curriculum development process to consider technical competencies as well as more general competencies, 'soft skills,' and authentic experiences as the intended learning outcomes.

5.4 Beginning with the end

Given the variety of both individuals and institutions with vested interest in GIS&T education, it should be no surprise that a 'one size fits all' approach to developing curricula is not feasible. Early in the advent of GIS&T as an academic discipline, it was clear that a number of educational challenges lay ahead for GIS&T education. These included, for Unwin (1991, 82), that:

1. GIS&T is a set of enabling technologies so it is possible to teach both with it and about it.
2. GIS&T is interdisciplinary and, consequently, its corresponding teaching and learning is not in the province of any single discipline.

Table 5.1 Institutional and individual contexts affecting curriculum content in the US

Institution type	Institutional context	Individual context
Community College and Vocational Training	Tends to provide technology oriented education and job specific training (e.g., photogrammetry, survey GPS, etc.)	Individual enters program often with little direct GIS&T experience. Expectation is often employability following training. Individual will often not have any direct GIS&T prior knowledge or experience
Undergraduate Education	May be either technology oriented education or general theory and application, often with discipline specific GIS&T (e.g., anthropology, civil engineering)	Individual engages GIS&T from interest based on prior knowledge or disciplinary relevance (e.g., geomatics for surveying). Often associated with 'traditional' university student, but may also be 'non-traditional' student seeking either retraining or a first time degree. May also encounter GIS&T as a pedagogical tool in non-GIS&T learning experiences
Graduate Education	Theory and application, research and development, discipline specific	Similar to undergraduate education but with increased expectations with regard to knowledge dimension and cognitive levels. May also include deeper understanding of GIS&T theory or the application of GIS&T to more complex problems
Non-degree Post Baccalaureate Education and Professional Training	Discipline and job oriented training, new technology familiarization, continuing education	Some overlap with community college and vocational training. Individuals may often possess deep domain expertise but little GIS&T experience. Expectations may include increased marketability, increased job performance and/or increased potential for promotion
Labor Force Evaluation	Employee level assessment of knowledge, skills and abilities	Knowledge, skills and abilities inventories are increasingly common for individuals entering the workforce. Individuals may be expected to perform at a certain level on a standardized assessment
Certification and Accreditation	Program-level internal and external evaluation of course and program content and curricular alignment	An accredited program or a program with a certificate may be perceived as more valuable

3. GIS&T can be viewed as technology requiring training or as a group of concepts for which education in their underpinnings is required.
4. GIS&T is neutral to the scale and problems to which it is applied and thus both what and how it is taught is subject to a great deal of variation.
5. GIS&T is rapidly expanding, changing, and expensive to deliver.

This cross-section of challenges, in combination with the variety of individual and institutional contexts, immediately underscores the need for curricula that are *designed* around a specific set of desired outcomes, rather than a single curriculum that is simply adopted wherever GIS&T education is needed. For example, a teacher needs to determine if they are treating GIS&T as a pedagogical tool (i.e., as a set of enabling technologies) or as the subject of the learning experience (e.g., as might be the case in courses that are exploring the underlying principles of spatial analysis). Simultaneously, consideration of the relevant soft-skills and learning that goes beyond foundational knowledge is required. As our view of a given curriculum – and curricula in general – becomes more holistic, it becomes evident that we must think about overall desired outcomes. In turn, we must then consider the outcomes associated with individual courses or modules and build up a cross-section of learning experiences wherein the cumulative learning outcomes address the overarching outcomes at the curricular level (Biggs and Tang, 2007, 5).

Unfortunately, when presented with a resource such as the *Body of Knowledge*, another significant challenge arises. It is easy to move away from thinking in terms of what one wishes to accomplish with a learning experience to one where the consideration is simply which topics should be covered. Fink and Ganus (2009) suggest that 'integrated course design' for significant learning experiences should build learning skills, foundational knowledge, understanding of application, integration (i.e., connections and relationships), an appreciation of the human dimension of a subject, and caring (i.e., development of interests or values associated with the subject). In essence, a document such as the *BoK* has such obvious relevance to the curriculum design process that it can easily be mistaken as curriculum. This leads to a possibility wherein a curriculum created based on the identification of relevant topics and objectives in the *BoK* develops a focus which emphasizes content mastery rather than who (the learner), what (the intended outcome) and how (the designed teaching and learning process).

An alternative strategy that effectively employs the *Body of Knowledge* is to focus on the design of clearly articulated intended learning outcomes prior to selecting content from the *BoK*. It is important that this process consider outcomes at the institutional level, the program level and the course level (Biggs and Tang, 2007, 65). An outcomes statement at the institutional level is a characterization of what graduates of educational institution are expected to be able to do. Similarly, outcome statements at the program and course levels are intended to articulate what a student should get out of a degree program or an individual course.

5.5 Integrated course design

Integrated course design (Fink and Ganus, 2009) and the intended learning outcomes approach (Biggs and Tang, 2007, 65) philosophically align in that the first step in the process is to understand the context and purpose of the teaching and learning activity. Understanding the purpose of a curriculum or of a course within a curriculum also entails the understanding and clarification of the larger purpose of the course or curriculum. Wiggins and McTighe (2005, 256) put this in terms of clarifying, 'what does the content enable you to do in the real world if you master it,' and in emphasizing how 'specific, complex, real-world tasks' facilitate the achievement of such understanding. The development of a purpose statement also allows the teacher to begin reflecting on both the portions of the *Body of Knowledge* and the broader dimensions of learning that are relevant to the teaching and learning process.

The creation of an integrated course design or intended learning outcomes (ILOs) that appropriately consider the institutional and individual contexts is a challenging task. This philosophical adjustment in the curriculum design process requires a shift away from the idea of simply 'covering' the curriculum to an approach that emphasizes understanding (Wiggins and McTighe, 2005) and learning that goes beyond content mastery (Fink and Ganus, 2009). Wiggins and McTighe emphasize 'backwards design' that first identifies desired results, determines acceptable evidence for performance and then plans the learning experience that places the greatest emphasis on understanding. This is very similar to Fink and Ganus who suggest beginning with an identification of situational factors *then* desired student-centered learning goals (to include the behavioral goals). They then recommend the selection of feedback and assessment activities and other supportive activities. Fink and Ganus incorporate a mid-stream 'integration check' to make sure that identified activities support the learning goals (this is also a useful stage to evaluate overall alignment of a course within a broader curriculum). Finally, they suggest designing the instructional strategy, consideration of self-evaluation and peer review, and the development of a specific lesson plan and assignment rubrics.

The choice of the integrated course design or backwards design emphasizes specific consideration of learning associated with elements other than disciplinary content: how to learn, caring and other non-content oriented learning experiences. The ideas of human dimension, caring and learning how to learn are the behavioral aspects of the learning experience. To many, the idea of integrating 'caring' and other apparent soft skills into the instructional design process may seem somewhat alien. Nevertheless, this is something good teachers do, in many cases, without even realizing it. In a recent report on the integration of teaching and research as a means to enhance the teaching and learning process, Jensen *et al.* (2007) highlight four strategies for improving the integration of teaching and research: (1) develop students' understanding of the role of research in their discipline, (2) develop students' abilities to carry out research, (3) progressively develop students' understanding, and (4) manage students' experience of research. In many regards, these strategies can ONLY be accomplished if the instructional design goes beyond content and emphasizes integration, the human dimension, caring and learning how to learn as much as it does content mastery. The establishment of curricular priorities (i.e., determining what elements of the curriculum are most important to learners in a particular institutional context) is an important step in calibrating the design process. This step allows for direct acknowledgement of both the individual and institutional contexts and fosters alignment of the outcomes associated with individual curricular components with the desired outcomes of the overall curriculum.

Initiating the design process

In order to start the curriculum design process, consideration of desired curricular outcomes must be integrated with consideration of the more discrete course and activity outcomes. In essence, the broader intended learning outcomes help orient *what* is important and *why* a particular learning activity is to be incorporated. It then becomes necessary to consider the relative importance of each activity both in terms of the curricular and course priorities. More important topics should be taught in a way that facilitates higher levels of understanding for those topics than for less important topics; this is necessarily different than simply spending more or less time on a topic (DeMers, 2009).

When developing curricula, Biggs and Tang (2007) reason that we need to ask why we are teaching the topic. They pose several model questions that ask whether a topic is intended to:

- Delineate boundaries, giving a broad picture of what is there.
- Inform on a current state of play, to bring students up to date on the state of the topic or discipline.
- Stockpile knowledge, of no perceived use for the present, but likely to be needed later.
- Inform the decision making process, such as with problem-based learning.

While the above questions are important, they do tend to emphasize understanding of content or the application of that understanding to a problem. This is where the value of thinking in terms of intended learning outcomes and the basis for choosing an integrated course design or ILO-based backwards design process becomes readily apparent; the outcomes of the design process need not be exclusively focused on content mastery. As mentioned, significant learning experiences span six aspects of learning: foundational knowledge, application, integration, human dimension, caring and learning how to learn (Fink and Ganus, 2009). Though the *GIS&T Body of Knowledge* is an excellent resource supporting learning of foundational knowledge and, to a degree, application and integration of concepts, it leaves something of a void in terms of integration, human dimension, caring and learning how to learn.

Characterizing learning activities in terms of levels of understanding may seem like a daunting task, there are several learning taxonomies that serve as useful guides. The 'structure of the observed learning outcome' or SOLO taxonomy is a very good starting point (Biggs and Tang, 2007). In the SOLO taxonomy, the verbs used to describe the level of engagement associated with a learning activity escalate with each level and range from verbs such as memorize and identify at the lowest level to theorize, generalize and hypothesize at the most advanced level. The SOLO taxonomy treats learning as occurring in four phases (unistructural, multistructural, relational and extended abstract), the first two of which are quantitative and more about 'doing,' the latter two are qualitative, or more about understanding.

In a similar vein, the taxonomies offered by Wiggins and McTighe (2005) in *Understanding by Design,* and Anderson *et al.* (2001) in, *A Taxonomy for Learning, Teaching and Assessing*, serve as useful guides for structuring learning outcomes for different levels of understanding. In their taxonomy, Wiggins and McTighe relay six 'facets' of understanding while Anderson *et al.* characterize six cognitive levels. Regardless of the taxonomy used, the resultant intended learning outcomes benefit from a simple structure requiring a verb at the appropriate level of understanding, the content the verb is to address, and the context of content discipline in which the verb is used (Biggs and Tang, 2007). Properly selected, the verbs become the basis for both the design of an ILO and the assessment of whether or not the ILO has been met (Table 5.2).

In highlighting both the different levels of understanding and the corresponding verbs, Table 5.2 clearly illustrates how different frameworks for curriculum design overlap in

Table 5.2 Three methods of articulating levels of student understanding, ordered downward by increasing engagement

(Biggs and Tang, 2007)	(Wiggins and McTighe, 2005)	Anderson *et al.* (2001)
Unistructural – memorize, identify, recognize, count, define	*Can explain* – generalize, justify, connect	*Remember* – define, describe, draw, find, label
Multistructural – classify, describe, list, report, discuss	*Can interpret* – translate, reveal, model	*Understand* – classify, compare, exemplify, demonstrate, explain
Relational – Apply, integrate, analyze, explain, predict	*Can apply* – use, adapt, do	*Apply* – Apply, change, choose, compute, implement
Extended abstract – theorize, hypothesize, generalize, reflect, create	*Has perspective* – critique, contextualize, contrast	*Analyze* – Analyze, characterize, classify compare, contrast
	Can empathize – value, perceive, experience	*Evaluate* – Appraise, argue, assess, conclude, prove
	Has self-knowledge – reflect, realize, discover	*Create* – Construct, design, develop, generate, hypothesize

terms of intent even though the underlying basis for design tends to vary. It is thus critical to select an approach to curriculum design that best supports the overall design process given the institutional context and accountability needs. For example, whereas the SOLO and *Understanding by Design* taxonomies are developed around creating aligned curricula and learner-centered experiences, Bloom's original taxonomy was designed to reduce the labor associated with the development of comprehensive examinations (Krathwohl, 2002). Regardless of the taxonomy used, a structured and deliberate approach to curriculum design will yield more consistent results than simply orienting curriculum around some previously specified content.

Integrated course design, the backwards design process, and the development of intended learning outcomes require careful consideration of both the purpose of a given curriculum and the courses or activities within a curriculum wherein significant learning is expected to occur. The backwards design process begins with an evaluation of content standards for key nouns, key understandings and key verbs that are useful for describing performance levels. In the following sections, I illustrate the process of evaluating the content standard relation to different institutional contexts and different desired outcomes.

5.6 Example 1: map algebra for introduction to GIS

Though specifically not prescriptive, the *Body of Knowledge* characterizes approximately 25 'core units' that the editors consider central to general understanding of GIS&T. These units, in combination with the remaining units, serve as a useful content standard. Once the content standard is evaluated and important content areas are identified, the overarching purpose and associated auxiliary learning associated with the chosen content must be clarified. For this example, consider the 'Introduction to GIS' course at the University of Wyoming (UW). This course is the first GIS course for many students and therefore students are not required to have any prior knowledge regarding GIS&T, though some prior geography knowledge is required through prerequisite geography courses. Because this course is the primary course for UW students to familiarize themselves with GIS, there tends to be a wide range of students with backgrounds ranging from anthropology and geography to wildlife management and zoology. These situational factors thus dictate that the scope and sequence of the course must be constructed in a way that exposes students to a 'broad picture' of GIS&T understanding while, at the same time, allowing more advanced students to delve deeper into the content. This course directly deals with many of the challenges presented earlier (Unwin, 1991), due to the range of students and the need to lay the foundation for more advanced learning that some portion of students will encounter later in the curriculum. The purpose of the course is to thus introduce students to some of the fundamental concepts of GIS&T while simultaneously opening the door for later discovery and reflection on how GIS can be used to solve a variety of real-world problems.

With respect to map algebra, the *BoK* lists several relevant learning objectives under the heading of analytical methods (AM), basic analytical operations (AM4), map algebra (AM4-4) (DiBiase *et al.*, 2006, 37):

- Describe how map algebra performs mathematical functions on raster grids.
- Describe a real modeling situation in which map algebra would be used (e.g., site selection, climate classification, least-cost path).
- Explain the categories of map algebra operations (i.e., local, focal, zonal and global functions).
- Explain why georegistration is a precondition to map algebra.
- Differentiate between map algebra and matrix algebra using real examples.
- Perform a map algebra calculation using command line, form-based and flow charting user interfaces.

These items articulate a series of increasingly challenging learning objectives spanning both declarative and functional knowledge, yet they offer only limited insight as to why the objectives are important or in what context understanding will be developed. Similarly, there is only passing reference to the 'real-world' aspect that supports authentic learning and deeper understanding.

Another important factor to consider are the linkages and sequencing between prior declarative and functioning knowledge. For example, the first objective regarding how map algebra works on raster datasets is predicated on declarative knowledge regarding the raster data model. The final objective of performing a map algebra calculation, which also happens to be the one that most innately supports deeper understanding, requires prior functioning knowledge on the use of a geographic information system. Thinking in terms of curriculum as a 'system of inter-related parts' (Unwin, 1997) is thus an important mindset when building up ILOs in this context.

Building the ILOs

Using the SOLO taxonomy and the previously discussed approaches for thinking about course design along with the AM4-4 learning objectives as a guide, the first step in developing

Table 5.3 Identifying aims of the ILOs for map algebra

Kind of knowledge	Topic	Level
Declarative	Relationship between map algebra and the raster data model	Describe
Declarative	Applications of map algebra	Discuss
Declarative	Categories of operations	Classify
Functioning	Perform map algebra operations in a GIS	Apply

intended learning outcomes is to identify the kinds of knowledge, the topics and the desired levels of understanding or performance. An important consideration at this stage of the curriculum design process is to realize that while the *BoK* spans much of the depth and breadth of GIS&T, it does so in a non-prescriptive manner. In designing ILOs, it is up to the teacher to select the appropriate kinds of knowledge given the context of the overarching scope and sequence of the curriculum. In the case of the introductory course discussed here, the requirement is to provide students with introductory level familiarity with a broad range of topics while also providing them the opportunity to exercise some of that knowledge in an applied context. This 'mission' leads to the identification of four relevant objectives that are reorganized in terms aimed to support the intended learning outcomes (Table 5.3).

Once the aims of the ILOs are identified based on the *BoK* and matched with appropriate performance verbs, the intended learning outcomes must be written. In moving from identified aims to the actual ILOs, the action verb is simply combined with the topic in a way that relates the topic to the level of understanding. In translating the aims into ILOs, students should be able to:

- Describe the relationship between map algebra and the raster data model.
- Discuss applications of map algebra.
- Classify categories of map algebra operations.
- Apply map algebra operations to solve a real-world problem in a GIS.

These ILOs thus serve as an adaptation of the *BoK* topic and objectives given the scope and sequence of the course in which they occur. The combination of the kind of knowledge (with the emphasis here on the declarative) and the level of performance (with an emphasis on multistructural and relational understanding) serve to tie the intended learning outcomes to the overarching purpose of the course.

5.7 Example 2: data quality for K-12 teachers

In this example, I address the issue of data quality in the context of a teacher-training workshop conducted in Greybull, Wyoming for the Bighorn School District. The school district received a large state grant supporting science, technology, engineering and math (STEM) education via the integration of global positioning systems and geographic information systems across the curricula at each grade level. The teachers neither desired nor needed to become highly technically competent, but did need to have a sufficient general level of understanding with the principles and processes of both GIS and GPS so as to create interesting exercises for K 12 students. The emphasis in this teaching and learning process was to help the teachers gain basic technical skills (enough to understand and interact with the requisite software systems) and to provide them with highly local and authentic experiences on which they could build for purposes of curriculum development and planning.

The *Body of Knowledge* has an entire unit in the 'geospatial data' (GD) knowledge area dedicated to data quality (GD6). Given the situational context, the balance of instruction was more on authentic and experiential learning than on deep understanding of the theoretical issues associated with data quality. As such, the relevant learning objectives selected from the content standard emphasize objectives that are conducive to active learning and the STEM education context.

First, from the geometric accuracy (GD6-1) topic, a basic objective regarding the accuracy of GPS is useful:

- Explain the factors that influence the geometric accuracy of data produced with Global Positioning System (GPS) receivers.

Second, it is also useful to differentiate primary and secondary data sources. From the primary and secondary sources (GD6-5) topic, the two objectives are quite ideal given the situational context:

- Explain the distinction between primary and secondary data sources in terms of census data, cartographic data and remotely sensed data.
- Describe a scenario in which data from a secondary source may pose obstacles to effective and efficient use.

These three objectives facilitate a substantial set of interesting learning opportunities. First, the GPS objective is accompanied with actual GPS-based data collection. In turn, the teachers were able to load their GPS data into either a GIS or virtual earth application such as Google Earth and compare their collected data with both primary and secondary data sources. This direct experience enabled the teachers to understand the limitations of the data collection process and the tradeoffs regarding the expense of collecting primary data versus using secondary data that might be more readily available.

Building the ILOs

The teacher-training example is interesting in that the development of ILOs must emphasize the active learning aspects and STEM contexts of the training activity. Context becomes an especially important consideration in the teacher training setting – the intended learning outcomes should balance basic theory and practical work in order to ensure teachers understand enough to correctly incorporate the material into their own classroom (Table 5.4).

When converted to intended learning outcomes, the emphasis on declarative and functioning knowledge foster the above-mentioned balance on practical theory and corresponding practice. In that teachers must then transpose the ILOs to their own teaching and

Table 5.4 Identifying the aims of the ILOs for data quality

Kind of knowledge	Topic	Level
Declarative	Factors influencing geometric accuracy of GPS data	Describe
Declarative	Distinctions between primary and secondary data	Discuss
Declarative	Costs and benefits of collecting primary data versus using secondary data	Discuss
Functioning	Collect GPS data (points, lines, or polygons) for a problem of interest	Apply
Functioning	Compare GPS data collected to secondary data source of same phenomena	Evaluate

learning setting, they can further refine the depth of engagement depending on the grade level of the students. In the context of this unit, teachers should thus be able to:

- Describe how different factors may influence the geometric accuracy of GPS data.
- Discuss and differentiate primary and secondary data.
- Discuss the costs and benefits of collecting primary data versus using secondary data.
- Use a GPS to collect data for local geographic phenomena of interest.
- Evaluate the quality of collected GPS data relative to a secondary dataset.

As illustrated, the ILOs directly engage the student (teachers in this case) in a learning process that involves some basic theory and then underscores that theory with a 'boots on the ground' field experience. The ILOs gently encourage the teacher as student to consider the ideas of data quality at a relational level, but accomplish this objective by allowing them to explore something that is specifically of interest to them. This latter point underscores how the ILO can be structured in such a way that enables students to learn more about something relevant to their own interests and, consequently, enhances the 'caring about' aspect of significant learning.

5.8 Example 3: ethics and the certification process

An important aspect of the *GIS&T Body of Knowledge* is that it includes a knowledge area characterizing GIS&T and society (GS). The GS knowledge area addresses a wide range of issues ranging from legal and economic aspects of GIS&T (GS1 and GS2) to ethics issues and critical GIS (GS6 and GS7).

For numerous reasons, ethics is an area that is particularly interesting with respect to geographic information science and technology. Geographic information systems and databases are nearly ubiquitous in both government and business sectors, yet data workflows, quality assurance and quality control procedures are often unspecified or unmonitored. Second, information regarding location of individuals is increasingly available through mobile devices and, consequently, there are an increasing number of opportunities to expose this information, both inadvertently and willfully. Finally, geographic analyses may have unanticipated results that are in contradiction to developed plans or desired operational outcomes (see the caribou migration routes case study (Wright, DiBiase and Harvey, 2008) for one excellent example).

One instance where ethics issues take center stage is in the certifications associated with various professional organizations. For example, the GIS Certification Institute (www.gisci.org) requires applicants for its GIS Professional (GISP) certificate to sign and agree to a code of ethics. In addition, the GISP application process asks the applicant to use the *BoK* to help justify the relevance of coursework to the certification. In examining the ethical aspects of geospatial information and technology (GS6) unit, however, it becomes immediately evident that, while the *BoK* addresses ethics, the range of ethical areas it touches on is relatively narrow. The two topics are:

Ethics and geospatial information (GS6-1), including:

- Describe a variety of philosophical frameworks upon which codes of professional ethics may be based.
- Discuss the ethical implications of a local government's decision to charge fees for its data.
- Describe a scenario in which you would find it necessary to report misconduct by a colleague or friend.
- Describe the individuals or groups to which GIS&T professionals have ethical obligations.

Codes of ethics for geospatial professionals (GS6-2), including:

- Compare and contrast the ethical guidelines promoted by the GIS Certification Institute (GISCI) and the American Society for Photogrammetry and Remote Sensing (ASPRS).
- Describe the sanctions imposed by ASPRS and GISCI on individuals whose professional actions violate the codes of ethics.
- Explain how one or more obligations in the GIS code of ethics may conflict with organizations' proprietary interests.
- Propose a resolution to a conflict between an obligation in the GIS code of ethics and organizations' proprietary interests.

An interesting circular reference arises if an applicant for certification were attempting to use the *BoK* for determining whether or not a course met an ethics requirement. The *Body of Knowledge* coverage for ethics is very general in one sense (GS6-1), and very specific in another in that it refers the reader back to the GISCI code of ethics (GS6-2).

Building the ILOs

From the perspective of either preparing students for the certification process or from the certification process itself, it is useful to consider the types of activities that could be used to help accomplish the *BoK* objectives in a meaningful way. The Caribou case study (Wright, DiBiase and Harvey, 2008) and related case studies are one obvious way to accomplish this goal. For example, the Caribou study offers a clear case wherein an organization's proprietary interest is seemingly in conflict with a code of ethics. Depending on the level of engagement (e.g., an introductory course versus an advanced course on GIS in the enterprise), ILOs

Table 5.5 Identifying the aims of the ILOs for ethics

Kind of knowledge	Topic	Level
Declarative	Purpose of a disciplinary code of ethics	Describe
Declarative	Steps in ethical decision making	Describe
Functioning	Relevant factors in a case study	Identify
Functioning	Options for ethical decision making in a case study	Create
Functioning	Defensibility of options	Analyze
Functioning	Proactive reduction of future ethical issues in similar cases	Reflect

can be written to address unistructural engagement (i.e., identify the ethical issue in a case study) through the extended abstract (i.e., generalize types of ethical issues associated with a case study and develop a new case study that addresses similar issues in relation to a problem of interest) (Table 5.5). In addition, given the well-developed literature on the subject of ethical decision making, it is important to supplement the *BoK* with appropriate practices such as the steps to ethical decision making proposed by Davis (1995).

Reasoning through ethical questions is an active process. Commensurately, developed ILOs regarding ethics tend to lean toward functioning knowledge over declarative knowledge. While appropriate declarative knowledge is required to understand the full parameters of a GIS&T ethics question, this knowledge will vary depending on the problem. In addition, the declarative aspects of many ethics questions may actually require the experience and perspectives of individuals coming from a variety of different backgrounds. In this way, the structure of the developed ILOs has the potential to foster thinking that necessarily entails deeper levels of engagement with the material. Given the developed aims based on the *BoK* content, the supplementary case study (Wright, DiBiase and Harvey, 2008) and ethics literature (Davis, 1995), students who complete a module based on these outcomes should be able to:

- Describe the purpose of a GIS&T disciplinary code of ethics.
- Describe the steps required for ethical decision making in GIS&T.
- Identify the relevant factors in an ethics case study involving geographic information.
- Create a series of options for responding to the ethical questions presented in a case study.
- Analyse the different options created for defensibility given a set of different tests for each option.
- Reflect on how a chosen option can be used to revise a business process or workflow to present future ethical issues in similar situations.

Clearly, the above ILOs require students to engage one another and think about problems with which they may not understand or be aware of every aspect of the problem. Ethics cases thus provide instructors with important tools for moving the teaching and learning process out of the area of content mastery and into the areas of integration, human dimensions, caring, and learning how to learn. As with other ILOs, the ILOs associated with ethics must be considered in context, and must account for the level of prior learning and prior experience with the different technical and theoretical areas of GIS&T.

5.9 Integrated design and alignment

The above ILOs reflect the range of possibilities associated with using the *BoK* and additional design effort to incorporate more behaviorally oriented outcomes. One of the challenges in thinking in terms of more behaviorally oriented learning goals is that they are interactive rather than hierarchical as with the SOLO and Bloom's taxonomies (Fink and Ganus, 2009). In addition, behavioral outcomes tend to span both course and curriculum and, consequently, are more meta-oriented in nature. As such, it is useful to think about how the structure of activities has the potential to model the desired behaviors. In the case of the map algebra content described above, I incorporate active learning activities that allow the students to group together and identify questions for which map algebra approaches are useful in solving. This approach leads to students finding problems that interest them and provides the opportunity for me, as the instructor, to model the process of reasoning through *how* map algebra might be used to solve the particular problem. The same process can be used to decide on a GPS data collection project on the decision-making process in an ethics case study. By modeling the process of reasoning and then asking students to model the process on another example, students are reinforcing patterns of learning how to learn. Students who learn to learn are better able to ask questions of new material, break complex ideas and tasks into more manageable units, and tend to devise informal feedback mechanisms and peer interaction for self-assessment (Heiman, 1987).

The above process is, of course, iterative (Figure 5.1). As ILOs are developed, purpose statements are refined, new ILOs are developed and the process continues. Additionally, whether thinking in terms of integrated design or the SOLO taxonomy, a driving concern is that the curriculum is working as a whole in achieving overarching institutional objectives as well as meeting the needs of the learner. While the process is somewhat cyclical, periodic feedback between the phases improves the overall design. Often, an integrated course design is a progressive process and lessons learned from any phase of the design implementation process may indicate the necessity of revisiting a prior phase. This iterative process is

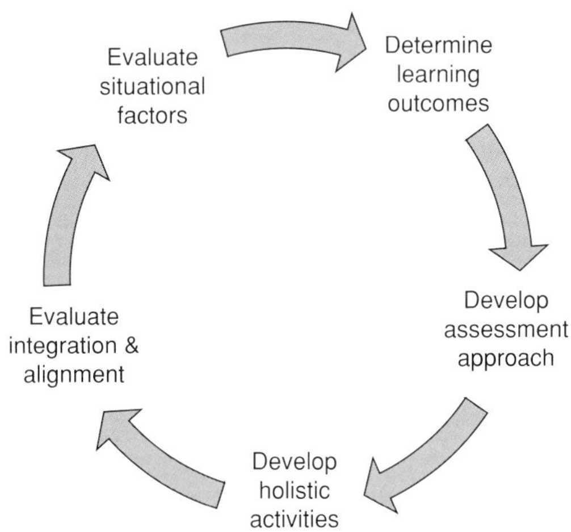

Figure 5.1 The iterative process for developing an integrated course design (Fink and Ganus, 2009)

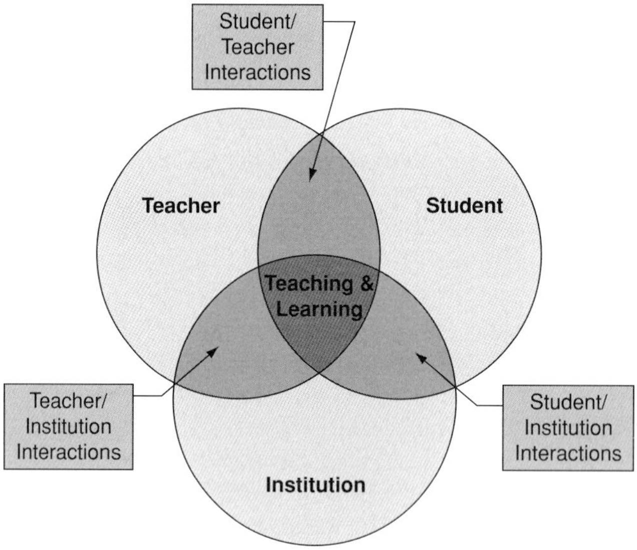

Figure 5.2 The relationships formed between the agents involved in the curriculum development process. Adapted from (Biggs and Tang, 2007, 248)

also elemental to ensuring that a course stays current, and continues to be aligned with overarching curricular goals as the curriculum develops and changes over time.

The need to define curricular priorities positions the curriculum design process at the crux of the relationship among students, teachers and educational institutions. The net teaching and learning process happens at the intersection of student, teacher and institutional priorities but, at the same time, is shaped by the relationships between the student and teacher, the teacher and institution and the student and institution (Figure 5.2).

Even though actual teaching and learning activities are fundamentally in the domain of the student-teacher interactions, the curriculum design process does not occur in isolation. The teacher-institution context evolves because teachers are increasingly responsible for 'assessing' the effectiveness of their efforts. In illustrating how the *BoK* can be used to help facilitate the assessment process, Prager and Plewe (2009) underscore the point that for the assessment to be effective, it must occur in the context of the curriculum as a whole, or else there is no assurance that individual courses are contributing to larger programmatic goals.

Student-institution interactions are also an important consideration during curriculum design. As students leave the educational setting and progress into the workforce, they are well positioned to reflect on the relevance of their education to their chosen career path. In the case of GIS&T education, students who leave their educational institution having had 'significant learning experiences' and who have developed a sense of caring about the subject matter (Fink, 2003) are in much better positions to provide feedback to the institution about the strengths and weakness of their educational experience in relation to their chosen career path. Similarly, students who are equipped during their educational experience to be reflective and adaptive – critical traits given the rapidly changing technological side of GIS&T – are more likely to respond positively to the dynamic nature of the GIS&T workplace.

The development of an aligned curriculum thus:

- Proactively articulates the goals of individual teaching and learning activities.
- Provides a mechanism for the assessment of the success of the teaching and learning process in meeting both course and program goals.
- Promotes the development of deep understanding, reflection and caring about the subject matter.

By accelerating and systematizing the specification of intended learning outcomes, the *GIS&T Body of Knowledge* is a very important element supporting the creation of an aligned, cohesive curriculum.

5.10 Conclusion: reconciling needs and context

The increased emphasis on learner-centered educational experiences is changing the educational landscape. Interestingly, GIS&T education can benefit significantly from the incorporation of learner-centered teaching and learning because of the diversity of needs that accompany the many individual contexts of learners entering GIS&T programs. The diverse and evolving nature of GIS&T also underscores the need for thinking about curricula in a holistic manner; those who adapt to the rapid changes occurring in GIS&T (including teachers, students, employers and educational institutions) will be successful over the long run.

The *Body of Knowledge* can be viewed as the first phase of an evolving characterization of the depth and breadth of geographic information science and technology. If we build our curricular systems using the *BoK*, the process of thinking about and designing a curriculum is not only easier, it has the potential to help facilitate summative evaluation of individuals and programs (e.g., for hiring, for program evaluation). The utility of the *BoK* in facilitating the development of intended learning outcomes is also an important component of creating aligned curriculum and thus has the potential to improve other aspects of the teaching and learning process. In order to enhance the teaching and learning experience, however, we must look beyond simple, content-oriented approaches to course and program design. The illustrated approach offers practitioners a guide for integrated the *Body of Knowledge* into the curriculum planning process.

References

Alibrandi, M., and J. Palmer-Moloney (2001) Making a place for technology in teacher education with geographic information systems (GIS). *Contemporary Issues in Technology and Teacher Education*, 1(4): 483–500.

Anderson, L. W., D. R. Krathwohl, and B. S. Bloom (2001) *A Taxonomy for Learning, Teaching, and Assessing: A Revision of Bloom's Taxonomy of Educational Objectives.* New York: Longman.

Baker, T. (2005) Internet-based GIS mapping in support of K-12 education. *The Professional Geographer*, 57(1): 44–50.

Bednarz, S. (2004) Geographic information systems: A tool to support geography and environmental education? *GeoJournal*, 60(2): 191–199.

Biggs, J., and C. Tang (2007) *Teaching for Quality Learning at University*, 3rd ed. Maidenhead, UK: SRHE and Open University Press.

Davis, M. (1999) *Ethics and the University*. London: Routledge.

DeMers, M. (2009) Using intended learning objectives to assess curriculum materials: The UCGIS Body of Knowledge. *Journal of Geography in Higher Education*, 33(Supplement 1): S70–S77.

DiBiase, D., M. DeMers, A. Johnson, K. Kemp, A. Luck, B. Plewe, and E. Wentz (eds) (2006) *The Geographic Information Science and Technology Body of Knowledge*. Washington, DC: Association of American Geographers and University Consortium for Geographic Information Science.

DiBiase, D., M. DeMers, A. Johnson, K. Kemp, A. Luck, B. Plewe, and E. Wentz (2007) Introducing the first edition of Geographic Information Science and Technology Body of Knowledge. *Cartography and Geographic Information Science*, 34(2): 113–120.

Dunne, C. E., P. J. Atkins, M. J. Blakemore, and J. G. Townsend (1999) Teaching geographical information handling skills for lower-income countries. *Transactions in GIS*, 3(4): 319–332.

Fink, L. D. (2003) *Creating Significant Learning Experiences: An Integrated Approach to Designing College Courses*. San Francisco: Jossey-Bass.

Fink, D., and M. Ganus (2009) Designing significant learning experiences, IN: M. Solem, K. Foote and J. Monk (eds) *Aspiring Academics: A Resource Book for Graduate Students and Early Career Faculty*. Upper Saddle River, NJ: Prentice Hall, pages 68–83.

Gaudet, C. H., H. M. Annulis, and J. C. Carr (2003) Building the geospatial workforce. *URISA Journal*, 15(1): 21–30.

Jenkins, A., M. Healey and R. Zetter (2007) *Linking Teaching and Research in Disciplines and Departments*. York: The Higher Education Academy. http://www.heacademy.ac.uk/assets/York/documents/ourwork/EvidenceNet/Summaries/jenkins_healey_zetter_linking_tandr_in_disciplines_and_departments_summary.pdf (accessed August 4, 2011).

Kemp, K. K., and F. M. Goodchild (1992) Evaluating a major innovation in higher education: The NCGIA. *Journal of Geography in Higher Education*, 16(1): 21–35.

Krathwohl, D. R. (2002) A revision of Bloom's taxonomy: An overview. *Theory into Practice*, 41(4): 212–218.

Marble, D. F. (1998) Rebuilding the top of the pyramid. *ArcNews*, 20(1): 28–29.

Heiman, M. (1987) Learning to learn: A behavioral approach to improving thinking, IN: D. W. Perkins, J. Lochhead, and J. C. Bishop (eds) *Thinking: The Second International Conference*, London: Routledge, pages 431–452.

Prager, S. D., and B. Plewe (2009) Assessment and evaluation of GIScience curriculum using the Geographic Information Science and Technology Body of Knowledge. *Journal of Geography in Higher Education*, 33(Supplement 1): S1–S24.

Sowell, E. (2000) *Curriculum: An Integrative Introduction*. Upper Saddle River, NJ: Prentice Hall.

Unwin, D. J. (1991) The academic setting of GIS, IN: D. W. Rhind, M. F. Goodchild, and D. J. Maguire (eds) *Geographic Information Systems: Principles and Applications*. Harlow: Longman, pages 81–90.

Unwin, D. J. (1997) Unit 159: Curriculum design for GIS. *NCGIA Core Curriculum in GIScience* http://www.ncgia.ucsb.edu/giscc/units/u159/u159.html (accessed January 8, 1998).

Wiggins, G., and J. McTighe (2005) *Understanding by Design*. Alexandria, VA: Association for Supervision & Curriculum Development.

Wikle, T. (1999) GIS education through certificate programs. *URISA Journal*, 11(2): 53.

Wikle, T. A., and G. A Finchum (2003) The emerging GIS degree landscape. *Computers, Environment and Urban Systems*, 27(2): 107–122.

Wright, D., D. DiBiase, and F. Harvey (2008) *Case study: Caribou migration routes*. https://www.e-education.psu.edu/files/sites/file/caribou_routes_case.pdf (accessed August 4, 2011).

6

Scope and sequence in GIS&T education: learning theory, learning cycles and spiral curricula

Kenneth E. Foote

Department of Geography, University of Colorado at Boulder, Boulder, Colorado, USA

6.1 Overview

This chapter focuses on the value of using the concepts of scope and sequence to design and improve GIS&T curricula. Scope refers to depth of knowledge about a given concept or skill introduced at each stage of the learning process. Sequence refers to the order in which concepts are introduced. Research in educational psychology indicates that both the scope and the sequencing of learning tasks can have a great influence on a student's mastery of concepts and skills (Ritter *et al.*, 2007) and, in particular, that 'order effects are more pervasive and important than they have previously been treated, and . . . learning order affects the final outcome of learning' (Ritter and Nerb, 2007, 4). Topics introduced out of sequence often end up having to be repeated. Too broad a scope at too early a stage may overwhelm students if they are not given time to practice and assimilate new material. In the worlds of primary and secondary education, scope and sequence are widely recognized as important parts of instructional design (Gagné, Briggs and Wager, 1992, 165–184; Hume and Boehm, 2001; Morrison, Ross and Kemp, 2001, 110–121; Smith and Ragan, 2005, 290–293) and they are charted grade by grade and subject by subject in considerable detail as has been done by Vender (2003) for the US national geography standards.

Such detailed consideration of scope and sequence is less common in higher education. Introductory classes lead to intermediate and advanced offerings and the scope of each is intended to build on the last and support the next, but often this alignment of content is difficult to achieve. Here, however, I argue that effort invested in articulating scope and sequence can improve GIS&T curricula for at least three reasons. First, GIS&T is a

Teaching Geographic Information Science and Technology in Higher Education, First Edition.
Edited by David J. Unwin, Kenneth E. Foote, Nicholas J. Tate and David DiBiase.

demanding area of study involving a number of complex and interlocking concepts, theories and skills. Careful attention to how these are introduced, at what level, and in how much detail can influence how quickly and how well students master these concepts, theories and skills. Or, as Reigeluth (1999, 430–431) points out, sequencing makes the most difference when there 'is a strong relationship among the topics of a course' or curriculum (as is the case with GIS&T) and as 'the size of the course increases.' Or as he notes in the same passage: 'When the content requires more than a couple of hours to learn, sequencing is likely to begin to make a significant difference in the learners' ability to master it, because most learners will have a difficult time organizing so much content logically and meaningfully if it is poorly sequenced.' Second, constructivist theory indicates that effective sequencing of learning activities strengthens and improves student achievement. Third, careful attention to scope and sequence can create opportunities for learners to revisit and practice basic ideas until these concepts are mastered.

Most instructors are aware of the importance of scope and sequence without, perhaps, using those particular terms. They realize that foundations need to be laid for students to understand some complex concepts and that mastery of some skills often involves regular, systematic practice over extended periods of time. My point is not to discount these intuitive notions of how students learn, but rather to consider systematically how the concepts of scope and sequence can be used to strengthen GIS&T curricula. Certainly other issues are critical to curriculum design. As Prager notes (Chapter 5, page 63), setting clear learning outcomes and aligning these outcomes with classroom and laboratory activities and assessments are fundamental to good pedagogy. My point is that, along with the alignment of courses around learning outcomes, issues of scope and sequence can make an important difference to curriculum design.

6.2 The importance of scope and sequence in curriculum design

Focusing on scope and sequence suggests carefully considering how ideas and concepts interlock and support one another across an individual class or an entire curriculum. It is not that textbook authors and instructors overlook this issue but, rather, there is little consensus. As can be seen by comparing the tables of contents of several popular textbooks (Table 6.1), the sequencing varies book by book and sometimes these differences are considerable. Database concepts and design tend to come early; other topics vary considerably in their placement including cartography and visualization; datums and map projections; and modeling. The scope of these chapters also varies considerably from one textbook to the next with some including advanced topics early, others not. Only one or two of the books include chapters on the conceptual foundations of GIS&T like DeMers's chapter entitled 'Spatial learner's permit.' Also absent for the most part are linkages between GIS&T and issues of spatial learning.

The UCGIS *Body of Knowledge* (*BoK*) (DiBiase *et al.*, 2006), as another example, does an excellent job of inventorying core and advanced units, but does not order them in a particular sequence. Eighteen of the twenty-six core units are contained in four of the ten knowledge areas – geospatial data; cartography & visualization; analytical methods; and data modeling – suggesting the foundational importance of these four areas, but no sequence is specified (Table 6.2). The authors do recognize the value of considering 'multiple pathways

Table 6.1 Sequence of topics employed by a number of popular introductory GIScience textbooks. The sequencing of topics varies from book to book, sometimes greatly as is apparent from the chapter titles listed below

	Bolstad (2008)	Chang (2009)	DeMers (2008)	Heywood *et al.* (2006)	Lo and Yeung (2007)	Longley *et al.* (2005)
0.			Spatial learner's permit			
1.	Introduction	Introduction	Introduction to digital geography	What is GIS?	Introduction	Systems, science & study
2.	Data models	Coordinate systems	Basic geographic concepts	Spatial data	Maps and geospatial data	Gallery of applications
3.	Geodesy, datums & map projections	Vector data model	Map basics	Spatial data modeling	Digital representation & organization of geospatial data	Representation geography
4.	Maps, data sources & data entry	Raster data model	GIS computer structure basics	Database management	Geospatial data quality & standards	Nature of geographic data
5.	GPS/GNSS	GIS data acquisition	GIS data models	Data input & editing	Raster geoprocessing	Georeferencing
6.	Aerial & satellite images	Geometric transformation	GIS input	Data analysis	Vector geoprocessing	Uncertainty
7.	Digital data	Spatial data editing	Data storage & editing	Analytical modeling in GIS	Geovisualization & geospatial information products	GIS software
8.	Attribute data & tables	Attribute data management	Query & description	Output: from new maps to enhance decisions	Remote sensing & GIS integration	Geographic data modeling
9.	Basic spatial analysis	Data display & cartography	Measurement	Development of computer methods for handling spatial data	Digital terrain modeling	GIS data collection

(continued)

Table 6.1 (*Continued*)

	Bolstad (2008)	Chang (2009)	DeMers (2008)	Heywood *et al.* (2006)	Lo and Yeung (2007)	Longley *et al.* (2005)
10.	Raster analysis	Data exploration	Classification	Data quality issues	Spatial data analysis, modeling & mining	Creating & maintaining geographic databases
11.	Terrain analysis	Vector data analysis	Statistical surfaces	Human & organizational issues	GIS implementation & project management	Distributed GIS
12.	Spatial estimation	Raster data analysis	Terrain analysis	GIS project design & management	GIS issues & prospects	Cartography & map production
13.	Spatial models & modeling	Terrain mapping & analysis	Spatial arrangement	Future of GIS		Geovisualization
14.	Data quality & standards	Viewsheds & watershed	Map overlay			Query, measurement, & transformation
15.	New developments	Spatial interpolation	Cartographic modeling			Descriptive summary, design, & inference
16.		Geocoding & dynamic segmentation	Cartography and visualization			Spatial modeling with GIS
17.		Path analysis and network applications	GIS design			Managing GIS
18.		GIS models and modeling				GIS & management, the knowledge economy & information
19.						Exploiting GIS assets & navigating constraints
20.						GIS partnerships
21.						Epilog

Table 6.2 Core units (italicized in bold) within the 'knowledge areas' of the UCGIS *Body of Knowledge* (2006)

Geospatial Data
1. ***Earth geometry***
2. Land partitioning systems
3. ***Georeferencing systems***
4. ***Datums***
5. ***Map projections***
6. ***Data quality***
7. ***Land surveying & GPS***
8. Digitizing
9. Field data collection
10. ***Aerial imaging & photogrammetry***
11. ***Satellite & shipboard remote sensing***
12. ***Metadata, standards, & infrastructures***

Cartography & Visualization
1. History & trends
2. ***Data considerations***
3. ***Principles of map design***
4. Graphic representation techniques
5. Map production
6. ***Map use & evaluation***

Analytical Methods
1. Academic & analytical origins
2. Query operations & query languages
3. ***Geometric measures***
4. ***Basic analytical operations***
5. ***Basic analytical methods***
6. Analysis of surfaces
7. Spatial statistics
8. Geostatistics
9. Spatial regression & econometrics
10. Data mining
11. Network analysis
12. Optimization & location-allocation modeling

Data Modeling
1. Basic storage & retrieval structures
2. ***Database management systems***
3. ***Tessellation data models***
4. ***Vector & object data models***
5. Modeling 3D, temporal, & uncertain phenomena

Conceptual Foundations
1. Philosophical foundations
2. Cognitive & social foundations
3. ***Domains of geographic information***
4. ***Elements of geographic information***
5. Relationships
6. Imperfections in geographic information

Data Manipulation
1. ***Representation transformation***
2. ***Generalization & aggregation***
3. Transaction management of geospatial data

Organizational & Institutional Aspects
1. Origins of GIS&T
2. Managing GIS operations & infrastructure
3. Organizational structures & procedures
4. GIS&T workforce themes
5. ***Institutional & inter-institutional aspects***
6. ***Coordinating organizations (national & international)***

Design Aspects
1. The scope of GIS&T system design
2. Project definition
3. Resource planning
4. ***Database design***
5. Analysis design
6. Application design
7. System implementation

GIS&T & Society
1. Legal aspects
2. Economic aspects
3. Use of geospatial information in the public sector
4. Geospatial information as property
5. Dissemination of geospatial information
6. ***Ethical aspects of geospatial information & technology***
7. Critical GIS

Geocomputation
1. Emergence of geocomputation
2. Computational aspects & neurocomputing
3. Cellular automata models
4. Heuristics
5. Genetic algorithms
6. Agent-based models
7. Simulation modeling
8. Uncertainty
9. Fuzzy sets

to diverse outcomes' (DiBiase *et al.*, 2006, 34), implying differing sequences through the *BoK*, but they set this task aside for the future.

Consideration of the issues of scope and sequence also implies planning classes and curriculum in somewhat finer detail than is usually done. Rather than a week-by-week topical approach – 'This week we focus on map projections, next week database design' – it suggests that effective learning may involve weaving some core topics into classes and the curriculum in several places. This may involve a step-wise, linear progression (easy-to-difficult) for some topics such as datums and map projections, but other approaches are possible. For example, some topics like database design might be mastered by using a variety of contrasting case studies each highlighting different coding strategies. By using contrasting cases, students could also gain experience of the transfer and application of database concepts to different domains of knowledge.

6.3 Theoretical rationales

Perhaps the most important justifications for thinking carefully about scope and sequence come from research into how people learn, particularly the broad domain of constructivist learning theories. Rather than viewing learning as the transmission of knowledge from teacher to learner, or as a purely behaviorist process, constructivism considers how learners 'construct' or build their understanding of new experiences, phenomena and processes based upon their existing knowledge, motivations and preconceptions. Constructivism positions the teacher as a facilitator of learning, as a guide who develops activities or realistic problems which promote student exploration and discovery. Constructivism is not a new theory but rather one based on decades of research in cognitive development, psychology, education and philosophy (Bransford *et al.*, 1999, 10–11; Driscoll, 2005, 384–406; Jarvis, 2005, 18–40). Its pedigree can be traced back to the work of Dewey, Piaget and Vygotsky and, through time, the principles of constructivism have provided the foundations for a number of pedagogical innovations including experiential learning, active pedagogy and discovery learning. At the same time, learning theories and constructivism, in particular, have grown into very broad fields with many varied and contrasting schools of theoretical and applied research. Rather than review all of this research, I want to call attention here to just a few points from these varied sources – those with the greatest bearing on scope and sequence – recognizing that my discussion only scratches the surface of the ways learning theory can help improve GIS&T learning and teaching.

The work of Kolb (1984; Healey and Jenkins, 2000) is a good starting place for considering the importance of scope and sequence in GIS&T education. Kolb is best known for his work on experiential learning which, drawing upon its constructivist roots, stresses the importance of personal and environmental experiences in learning – 'Learning is the process whereby knowledge is created through the transformation of experience' (Kolb, 1984, 38). His key innovation was, however, to draw attention to the ways in which learners must experience new phenomena in order to gain understanding. He posits a four-step cycle involving: concrete experience, reflective observation, abstract conceptualization, and active experimentation (Figure 6.1). The order in which these steps are undertaken is less important in some ways than having learners step through all four. It seems natural in many cases to start with concrete experience, but some topics in GIS&T can, after all, be introduced from a highly theoretically perspective (abstract conceptualization) such as many

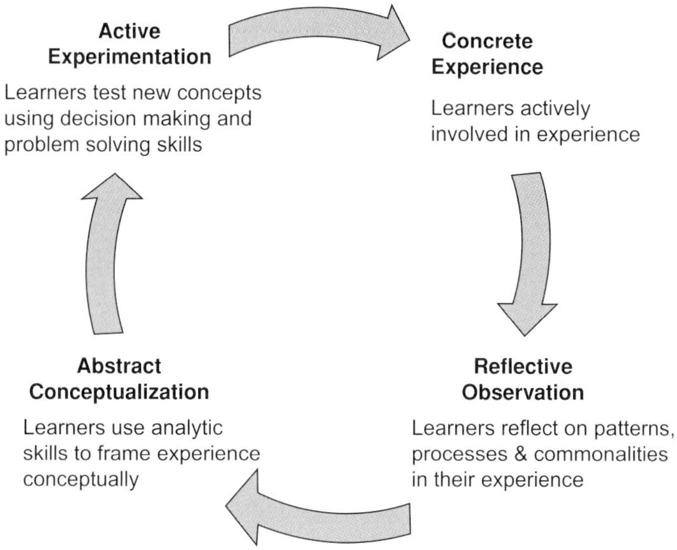

Figure 6.1 Kolb's (1984) experiential learning cycle. Mastery of a concept involves all four activities

geostatistical methods, others through active experimentation such as some principles of map design.

This four-step cycle is also related to Kolb's 'learning style inventory' based on the idea that individuals have preferred points of entry into the learning cycle. Some prefer learning through concrete experience, others through abstract conceptualization. Kolb distinguishes four general types of learner based upon their preferences along two continua (active experimentation-reflective observation and abstract conceptualization-concrete experience). The terms he applies to these learning styles are less important than the observation that learners, left to their preferences, may not attempt to move through the entire learning cycle and, as a consequence, may never completely master a topic or concept. From Kolb's perspective the teacher plays an important role as a facilitator of student learning, both in recognizing a learner's preferred style and in challenging the learner to move around the cycle even when the steps are unfamiliar. Kolb's learning style inventory is also a reminder that instructors often teach in ways that favor their own learning style and need to be encouraged to vary their approaches to suit the full range of learners in their classes.

The key issue here is that simply moving through a GIS&T course from topic to topic using lectures, demonstrations and labs, does not necessarily move all the way around the learning cycle, unless the activities are carefully interlocked and together offer exposure to concrete experience, reflective observation, abstract conceptualization and active experimentation.

There are three additional reasons for focusing on scope and sequence in GIS&T education. The first is its value in addressing explicitly the issue of 'transfer,' that is a student's ability to apply concepts from one learning domain to another (Bransford *et al.*, 1999, 39–66). This is an important issue in GIS&T education since students who master a GIS&T curriculum are often assumed to be able to apply their conceptual knowledge effectively across a broad range of disciplines and research problems. But research seems to indicate that transfer does not come automatically to students. Their understanding of theory or abstract concepts does not mean they will recognize the applicability of these theories and concepts outside the situations in which they are learned. Instead, it is better to see transfer as a learnable skill, one which needs to be addressed explicitly in the curriculum if learners are going to gain facility in applying GIS&T concepts and theories across application domains.

A second reason comes from research focusing on differences between novice learners and experts (Bransford *et al.*, 1999, 19–38). This research indicates substantial difference not just between the amount of knowledge experts bring to bear on a problem, but substantial differences in how they 'see' or recognize problems in the first place. Experts often are able to recognize patterns and problems more holistically than beginners. Experts seem able to view new situations in terms of analogies with comparable problems they have solved in the past. That is, they recognize the broad outline of a situation first, consider how they addressed similar problems previously, and then begin by applying this previously successful approach to the new problem. Without the same depth of experience, beginning learners tend to employ rule-based approaches, working up from the details of a given problem to a solution. Instructors often forget that their understanding of a subject is based on years of experience. Their expert knowledge of GIS&T topics may mean that they skip over steps in their reasoning assuming that learners in their classes see the material in the same holistic way. By addressing scope and sequence explicitly in curriculum design, instructors are perhaps likelier to view what they teach from the learner's perspective, rather than from their own expert perspective.

A final reason for considering scope and sequence is that it allows instructors to focus more carefully on how to address difficult concepts and issues. There is a growing body of research into what are termed 'troublesome knowledge' and 'threshold concepts' and how they influence the learning process, particularly at points when students report 'being stuck' (Meyer and Land, 2005; Land *et al.*, 2009). As Matthew Bampton points out (Chapter 8, page 117), troublesome knowledge is something that, to the learner, seems at first to be counter-intuitive, inconsistent with previous knowledge, or alien to a student's understanding, say for instance, the concept of randomness or of 'imaginary' numbers (Perkins, 1999). A threshold concept is 'akin to a portal, opening up a new and previously inaccessible way of thinking about something. It represents a transformed way of understanding, or interpreting, or viewing something without which the learner cannot progress' (Meyer and Land, 2006, 3). Of course, in practice, both 'troublesome knowledge' and 'threshold concepts' are related to preconceptions and misconceptions students hold which may block learning (Francek *et al.*, 1993) as well as the cognitive schemata they have developed. Although Bampton's work is a first step in understanding how 'troublesome knowledge' and 'threshold concepts' should be addressed in GIS&T curricula, it does already suggest that they must be designed to address more carefully concepts which are difficult to master.

6.4 From learning cycles to spiral curricula

Research into learning cycles, transfer, expert vs. novice learning and threshold concepts, though coming from varied sources, all suggest the value of considering scope and sequence seriously in GIS&T curriculum planning. The question remains, however, about how to proceed. Kolb's learning cycles suggests steps toward mastery of particular concepts and topics, but how can such cycles be linked together within a class or curriculum? One

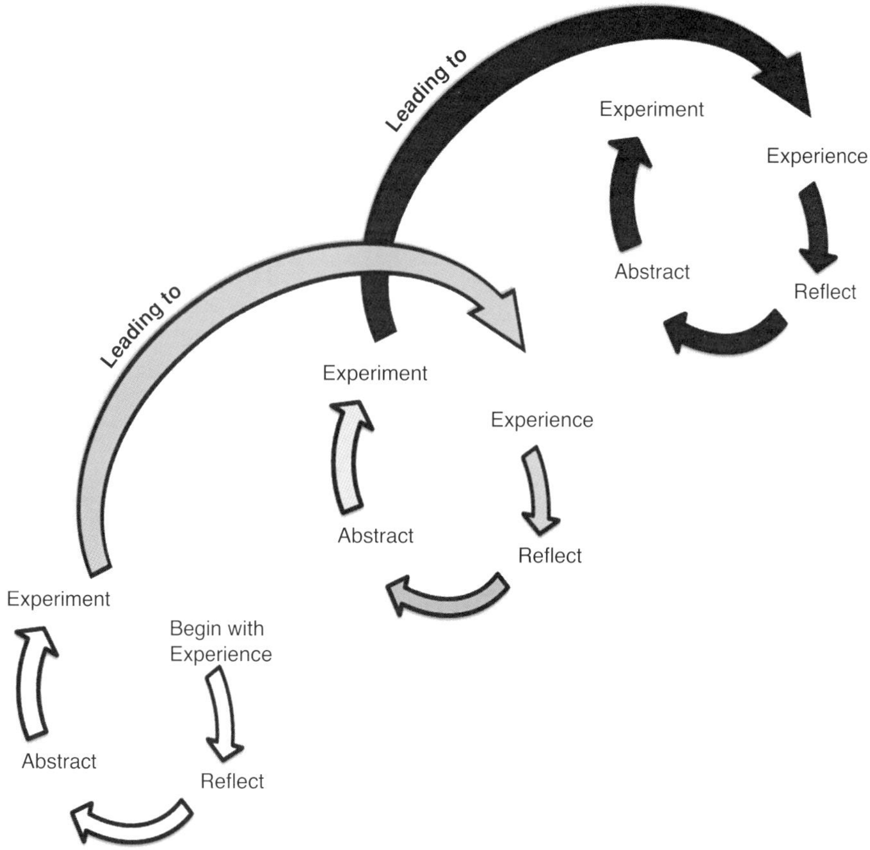

Figure 6.2 A spiral curriculum promotes mastery through the repeated use of key concepts in increasingly challenging learning experiences

particularly useful idea in this context is Bruner's (1960) construct of a spiral curriculum, suggesting that key ideas and concepts need to be revisited throughout a class or curriculum so that learners can build on these until they reach mastery (Figure 6.2). This means, implicitly that learners move around something like Kolb's learning cycle but, with each rotation, the level of the curriculum moves to a progressively higher level of challenge. This means that important concepts and skills are practiced regularly, providing reinforcement as well as opportunities for learners to extend their knowledge and deepen their confidence through time.

The spiral curriculum is different from more conventional approaches to teaching GIS&T. If we use a textbook to order a class (Table 6.1), then the week arrives for 'raster data models.' Learners have that week to master these concepts irrespective of background, comprehension or motivation. Next week will be time to learn 'vector data models,' and so forth through the term. The obvious weakness of this approach is that, even among a small group of learners, background, comprehension or motivation can vary dramatically. So the learning in that particular week will vary accordingly – with some learners moving quickly toward mastery, others perhaps slowed by troublesome knowledge.

Instead, by focusing on one topic for an extended period of time, a spiral curriculum involves introducing a wider variety of topics at once, but also revisiting them regularly through time. This means if learners do not understand a concept the first time around – or are not prepared or motivated to learn it – they will have a chance the next time around. For those learners who understood the first time around, they can move further along the spiral. The spiral approach addresses two of Bruner's key points about learning. First, that learners of all ages and abilities can be taught a subject if it is built up in appropriate steps. Second, that motivation – engagement with a subject or what is called 'readiness to learn' – is critical to learning, but is unpredictable. By spiraling the curriculum there is a greater chance of catching the interest of students when they are ready to learn.

A spiral suggests a unidirectional movement through the curriculum, but this may not always be the case. Reigeluth's 'elaboration theory' of sequencing highlights the importance of matching scope and sequence to the conceptual structure of the subject being learned. Often, as in GIS&T, there are causal or explanatory relationships among concepts (datums, map projections, national coordinate systems) or there are natural- or social-process relationships which need to be mastered together (for instance in understanding the hydrological cycle and its relationship to groundwater modeling). In such cases, hierarchical, nested, multiple strand, or other organizing principles may be more appropriate than a spiral. The point is not to follow Bruner's spiral slavishly, but rather to recognize the value of sequencing concepts through a course or curriculum.

6.5 An example

My argument implies that no single scope and sequence pathway will work in all curricula settings. Nor would it be particularly profitable or feasible to develop a single scope and sequence for the entire *BoK*. The curricular path chosen in a particular setting will hinge on the needs of a particular department or university, and multiple pathways are possible (DiBiase *et al.*, 2006, 34). Some universities offer courses covering the majority of knowledge areas including specialized courses in advanced topics, others offer only the core areas. But once one or two pathways are chosen, then their scope and sequence can be considered in the curriculum planning process. This means that scope and sequence can be used at the level of the individual class or department curricula, as I do in my department for one of the introductory classes I teach regularly.

I have used the ideas of scope and sequence in my own work since the early 1990s when I developed the *Geographer's Craft Project* at the University of Texas (Foote, 1997). This was a year-long undergraduate class in GIScience and computer research techniques. I sought to offer an integrated approach to geographical techniques by having students solve real-life research problems. Rather than focusing on each technique separately, the students were offered problems that involved applying several techniques together to address a single research question. Through the year the problems became more complex and allowed students more latitude to pursue their own strategies culminating in a final, independent research project. I found the notions of scope and sequence of tremendous value in course planning so that I could link research techniques from problem to problem and increase the challenge of the problems through the year.

The curriculum is organized differently where I now teach, but I still teach an introductory course in GIS&T which I model, in part, on the first semester of the *Geographer's Craft*.

Mapping a Changing World (GEOG 2053) is a sophomore (second year) class designed to provide geography and environmental studies students with a hands-on overview of modern cartography and the important roles maps play in today's world. It fulfills our cartography requirement for undergraduate majors but, for those continuing their studies, it serves as a prerequisite for our intermediate and advanced courses in GIScience and cartography. I list the learning outcomes in three groups to indicate that, by the end of the semester, students should be able:

1. To create maps for research, term papers, professional reports, and personal use
   a. To know the basic principles of cartographic and visual communication and apply them to the composition of classroom projects (CV2, CV3, CV4).
   b. To understand the strengths and weaknesses of various methods of displaying data and statistical information on maps (CV2, CV3, CV4).
   c. To gather, organize and symbolize geospatial data so as to depict spatial relationships of location, direction, distance, and movement (CV3, GD7, DA4, DM4).
   d. To understand the relationships between principles of cartographic and graphic visualization (CV4).
   e. To develop the confidence and ability to create maps for personal and professional use (DA4).
2. To gather and analyse information from maps for study, research, and personal use and enjoyment
   a. To be familiar with the coordinate systems used most widely where you live and within the US as a whole (GD1, GD3, GD4).
   b. To understand why and how map projections effect the display of distance, direction, area, and shape (GD5).
   c. To be familiar with methods for using maps in exploratory data analysis (CV6, AM3, AM4, AM5).
   d. To develop a personal interest in mapping and GIS&T.
   e. To become careful, critical readers of maps in popular and scholarly media (GS7).
   f. To identify the factors influencing the 'social construction' of particular maps (GS7).
3. To master the digital skills needed to find, compile, edit and create useful maps
   a. To know the most important starting places for cartographic information on the web (GS4, GS6).
   b. To create useful maps using simple graphics and word processing software.
   c. To develop a working knowledge of at least one software package for cartography and GIS.
   d. To build confidence in digital and GIS&T skills.

The parentheses indicate the *BoK* units to which the learning outcomes relate. Notably, not all of my learning goals 'map' onto the *Body of Knowledge*. This is because, although the *BoK* does an excellent job of covering the basic concepts of the course, it does not encompass all of my goals for student learning. In the terminology of curriculum designer Fink (2003), my intent is also to help students 'learn how to learn,' 'caring' (that is, developing an interest in GIScience), and the 'human dimension' of building self confidence in GIS&T. Also, although I include at least a brief introduction to most of the *BoK* core units in this course, I also touch on a few more advanced topics. These topics, although not absolutely essential to the course, fit well with other course topics.

Table 6.3 Scope and sequence chart for *Mapping a Changing World*, an introductory cartography and GIS&T class (* indicates *BoK* core unit)

Body of Knowledge units	Favorite place	Election campaign	Tectonics	Flood risk	Cost of eating	Maps to persuade
CV2 Data considerations*	Introduce	Practice	Practice	Master	Master	
CV3 Principles of map design*	Introduce	Practice	Practice	Master	Master	
CV4 Graphic representation	Introduce	Practice	Practice	Master	Master	
CV6 Map use and evaluation*		Introduce	Practice	Practice	Practice	Master
DA4 Database design*		Introduce	Introduce	Master	Master	
GD1 Earth geometry*			Introduce	Practice	Practice	
GD3 Georeferencing systems*			Introduce	Practice	Practice	
GD4 Datums*			Introduce	Practice	Practice	
GD5 Map projections*			Introduce	Practice	Practice	
GD6 Data quality*		Introduce	Introduce	Practice	Practice	Practice
GD7 Land surveying and GPS*					Introduce	
GD9 Field data collection					Introduce	
GD10 Aerial imaging*				Introduce		
AM3 Geometric measures*				Introduce	Extend	
AM4 Basic analytical operations*		Introduce	Extend			
AM5 Basic analytical methods*		Introduce	Extend			
GS4 Geoinformation as property	Introduce		Practice	Practice		
GS6 Ethical aspects*						Introduce
GS7 Critical GIS						Introduce
DM4 Vector and object models*		Introduce	Practice	Practice		
Other Learning Goals						
Build confidence in digital and GIS&T skills	Introduce	Extend	Practice	Practice	Master	Master
Develop working knowledge of software	Introduce	Extend	Practice	Practice	Master	
Develop interest in maps & GIS&T						
Transfer GIS&T concepts across domains		Addressed in topical range of problems presented thoughout semester				

But *Mapping a Changing World*, like the *Geographer's Craft*, is not organized sequentially by topic, but instead around active-learning assignments emphasizing how GIScientists address real-world research issues. The problems have again been selected so that they raise a variety of technical and methodological problems, allow students lots of practice with difficult concepts, and relate to issues which, I hope, will perk student interest in geography. In this way, students learn by experience how GIScientists gather and weigh evidence about natural and human processes, employ maps and databases to represent and model real-life situations, analyse spatial, temporal and functional relationships, and communicate findings cartographically and graphically and in written and spoken presentations.

Table 6.3 illustrates how the activities are matched to *BoK* units and learning outcomes in terms of both scope and sequence. Although the *BoK* details learning outcomes for each unit in more detail, I have used only the unit headings to limit my chart to a single page. The chart lists the activities I used in the most recent offering of the course, although I rotate these activities from year to year:

1. Favorite place: Create a digital sketch map from available sources showing the route from the classroom to the student's favorite place anywhere in the world.
2. Presidential campaign trail: Use data from past presidential elections to create a geographic campaign strategy for either party.
3. Tectonic hotspots: How are current earthquakes and volcanism related to plate boundaries at the global scale and, at a larger scale, how does activity along the San Andreas fault threaten the San Francisco Bay area.
4. Boulder flooding: How many people would be threatened by a 100-year flood in Boulder, Colorado and what would be the best place for siting rescue and evacuation stations.
5. Cost of eating: How much do retail food prices vary across the county and what factors account for the differences? Students compile data from local stores using GPS.
6. Maps to persuade and mislead: Search newspapers, magazines, or the web to find one map used either to: persuade consumers to purchase something; or to sway readers to a particular point of view, mislead them, or inflame debate.

Table 6.3 indicates how I have developed the projects so that students revisit some of the more difficult core topics in several problems to gain practice. Other topics are introduced in only one or two projects, just enough for students to understand basic principles which they practice in subsequent courses.

6.6 Conclusion

My suggestions about using the concepts of scope and sequence in GIS&T curricula can be only tentative. Future research may help to provide guidelines for organizing elements of GIS&T curricula. This would involve testing particular implementations of scope and sequence to see if they affect learning and mastery in a variety of settings among students with varying learning styles and levels of background. Such research would be particularly helpful with respect to the core units of the *BoK* including, among others, earth geometry, georeferencing systems, datums, and map projections, which are fundamental to overall student mastery. But research on other related topics would also be valuable. Investigating situations in which 'troublesome knowledge' and 'threshold concepts' block learning is also important. Similarly, focusing on effective ways to promote transfer of GIScience knowledge across a broad range of disciplines and research problems would also be useful. Finally, research into the differences between novice and expert GIScience users could help improve advanced courses and graduate curricula.

In the meantime, the concepts of scope and sequence may prove useful in two ways. First, it may be valuable in the short term to develop a scope and sequence for the core units of the *BoK* perhaps over a two or three-term timeline. Although there is wide variation in how GIS&T classes are organized university by university and in different countries, the core units need to be addressed early in any curriculum. Some guidance on sequencing would therefore help instructors considerably, perhaps in the same way the NCGIA Core Curriculum helped many instructors organize GIS courses for the first time at their universities. Second, the concepts of scope and sequence can be employed productively in any situation where GIS&T classes and curricula are being developed or refined. They focus attention on the learning cycle from the student's perspective and point the way to how the insights of learning theory and instructional design can improve GIS&T teaching and learning.

References

Bolstad, P. (2008) *GIS Fundamentals. A First Text on Geographic Information Systems*, 3rd ed. White Bear Lake, MN: Eider Press.

Bransford, J. D., A. L. Brown, and R. R. Cocking (eds) (1999) *How People Learn: Brain, Mind, Experience, and School*. Washington, DC: National Academy Press.

Bruner, J. S. (1960) *The Process of Education*. Cambridge, MA: Harvard University Press.

Chang, K. (2009) *Introduction to Geographic Information Systems*, 5th ed. Boston: McGraw Hill.

DeMers, M. N. (2008) *Fundamentals of Geographic Information Systems*, 4th ed. New York: Wiley.

DiBiase, D., M. DeMers, A. Johnson, K. Kemp, A. Luck, B. Plewe, and E. Wentz (eds) (2006) *The Geographic Information Science and Technology Body of Knowledge*. Washington, DC: Association of American Geographers and University Consortium for Geographic Information Science.

Driscoll, M. P. (2005) *Psychology of Learning for Instruction*, 3rd ed. Boston: Pearson Allyn and Bacon.

Fink, L. D. (2003) *Creating Significant Learning Experiences: An Integrated Approach to Designing College Courses*. San Francisco: Jossey-Bass.

Foote, K. E. (1997) The geographer's craft: Teaching GIS in the web. *Transactions in Geographic Information Systems*, 2: 137–150.

Francek, M. A., B. D. Nelson, R. H. Aron, and W. J. Bisard (1993) The persistence of selected geographic misconceptions: A survey of junior high through undergraduate college students. *Journal of Geography*, 93(6): 247–253.

Gagné, R. M., L. J. Briggs, and W. W. Wager (1992) *Principles of Instructional Design*, 4th ed. Fort Worth, TX: Harcourt Brace Jovanovich.

Healey, M., and A. Jenkins (2000) Kolb's experiential learning theory and its application in geography in higher education. *Journal of Geography*, 99: 185–195.

Heywood, I., S. Cornelius, and S. Carver (2006) *An Introduction to Geographical Information Systems*, 3rd ed. Upper Saddle River, NJ: Prentice Hall.

Hume, S. E., and R. G. Boehm (2001) A rationale and model for a scope and sequence in geographic education, grades K-12. *Social Studies*, 92(1): 16–21.

Jarvis, M. (2005) *The Psychology of Effecting Learning and Teaching*. Cheltenham, UK: Nelson Thomas.

Kolb, D. A. (1984) *Experiential Learning: Experience as the Source of Learning and Development*. Englewood Cliffs, NJ: Prentice-Hall.

Land, R., J. H. F. Meyer, and J. Smith (eds) (2009) *Threshold Concepts within the Disciplines.* Rotterdam: Sense Publishers.

Lo, C. P., and A. K. W. Yeung (2002) *Concepts and Techniques of Geographic Information Systems.* Upper Saddle River, NJ: Prentice Hall.

Longley, P. A., M. F. Goodchild, D. J. Maguire, and D. W. Rhind (2005) *Geographic Information Systems and Science*, 2nd ed. Hoboken, NJ: Wiley.

Meyer, J. H. F., and R. Land (2005) Threshold concepts and troublesome knowledge (2): Epistemological considerations and a conceptual framework for teaching and learning. *Higher Education*, 49(3): 373–388.

Meyer, J. H. F., and R. Land (eds) (2006) *Overcoming Barriers to Student Understanding: Threshold Concepts and Troublesome Knowledge.* London: Routledge.

Morrison, G. R., S. M. Ross, and J. E. Kemp (2001) *Designing Effective Instruction*, 3rd ed. New York: John Wiley.

Perkins, D. (1999) The many faces of constructivism. *Educational Leadership*, 57(3): 6–11.

Reigeluth, C. M. (1999) The elaboration theory: Guidance for scope and sequence decisions, IN: C. M. Reigeluth (ed), *Instructional-design Theories and Models. Volume 2: A New Paradigm of Instructional Theory.* Mahwah, NJ: Erlbaum, pages 425–453.

Ritter, F. E., and J. Nerb (2007) Call to order: How sequence effects in humans and artificial systems illuminate each other, IN: F. E. Ritter, J. Nerb, E. Lehtinen, and T. M. O'Shea (eds), *In Order to Learn: How the Sequence of Topics Influences Learning.* New York: Oxford University Press, pages 3–15.

Ritter, F. E., J. Nerb, E. Lehtinen, and T. M. O'Shea (eds) (2007) *In Order to Learn: How the Sequence of Topics Influences Learning.* New York: Oxford University Press.

Smith, P. L., and T. J. Ragan (2005) *Instructional Design*, 3rd ed. New York: Wiley.

Vender, J. C. (2003) *Teaching the Standards: A K-12 Scope and Sequence in Geography.* Washington, DC: Geographic Education National Implementation Project.

7

Building dynamic, ontology-based alternative paths for GIS&T curricula

Marco Painho and Paula Curvelo

Institute for Statistics and Information Management, New University of Lisbon, Lisbon, PORTUGAL

7.1 Knowledge production and the demand for new curricula

The impetuses for curriculum innovation and revision in higher education include advancements in knowledge domains, the combination of different subject fields forming new branches of knowledge, and uncertainties associated with social, economical and environmental problems that tend to promote innovative research methods and expand traditional investigative approaches. The impacts of these factors on knowledge production, associated with the use of information and communication technologies and a tendency towards a personalized learning, suggest that the demand for new curricula will persist.

In geographic information science and technology (GIS&T) those impetuses are particularly forceful. A wide variety of disciplines and industries are using these technologies to share, promote and value their knowledge, activities and products, spawning a growing and increasingly diverse set of GIS&T tool users. Specialists with different backgrounds and expectations, non-experts and occasional users pose, together with GIS professionals, new challenges for GIS&T curricula construction and development.

To answer these new challenges, in 2005 the Institute for Statistics and Information Management of the New University of Lisbon (ISEGIUNL) began a process of curricular revision in Geographic Information Systems and Science (Painho *et al.*, 2006, 2007). This initiative led to the establishment of a curriculum development model prototype based on a notion of curriculum as a process or praxis. With the purpose of framing the results stemming from this initiative, next we will discuss some of the concepts and underlying principles guiding its execution.

Teaching Geographic Information Science and Technology in Higher Education, First Edition.
Edited by David J. Unwin, Kenneth E. Foote, Nicholas J. Tate and David DiBiase.

Conceptions of curriculum

The notion of 'curriculum' is not new. However, definitions vary and a consensus about its meaning remains elusive (Posner and Rudnitsky, 1982; Smith, 1996; Glatthorn *et al.*, 2001; Pacheco, 2001). For many education specialists the question of curriculum conception is of central importance (Tanner and Tanner, 1995). Generally speaking, the different definitions of curriculum may be arrayed along a continuum framed by prescriptive and descriptive conceptions (Glatthorn *et al.*, 2006). Prescriptive definitions translate that which is necessary to attain; thus, the curriculum usually takes the shape of a study plan, an intended syllabus, or, sometimes, it reflects the vision of specialists on how a study program should take place. The curricular development model that embodies this tradition emphasizes curricular organization and systematization, and is strongly dependent on the definition of behavioral objectives. This tradition leads in turn to straightforward learning outcomes, unambiguous contents and clear teaching methods, therefore ensuring the attainment of prescribed learning outcomes.

But what is valued as clarity by some is problematic for others. One problem follows from the realization that outcomes exist both prior to and after learning experiences. A second issue is linked to the nature of outcomes, taking for granted that students' performance is amenable to objective evaluation. In many cases, this presupposition has created the need to subdivide learning experiences into increasingly smaller units, containing long descriptive lists of competencies and skills to be developed, and so creating a compartmentalization of the curriculum, conferring more importance to the parts rather than the whole. A third criticism pertains to the difficulties felt when attempts are made to put in practice a curriculum model heavily inspired in technological and industrial environments, out of touch with the school universe and the environment of a classroom. A fourth problem follows from the difficulty of anticipating results. Emphasis on the pre-specification of learning outcomes may drive teachers and students alike to neglect important aspects of the teaching-learning process, which are not summed in the objectives of the curriculum and that are dependent on students' learning routes and the knowledge and competencies previously acquired (Smith, 1996).

In contrast to prescriptive definitions of 'curriculum,' those that are descriptive try to incorporate the experiences that take place while the learning activities occur. This view translates an informal and flexible idea of curriculum, conceived as a 'set of educational experiences and a dynamic, probabilistic and complex system without a predetermined structure' (Pacheco, 2001).

This curriculum conception embodied in descriptive definitions seemed more appropriate to frame ISEGIUNL's GIScience curricula revision process, since it recognizes the encircling nature of knowledge, the complexity associated with the teaching-learning experiences and the components considered essential for the activities of curricular work (Tanner and Tanner, 1995; Hewitt, 2006).

Curriculum development – theory and praxis

In the broadest sense, curriculum development may be defined as a dynamic and continuous process integrating different stages, from the curriculum justification to its evaluation, including the conception/creation and implementation of its activities. In a less encompassing definition, which seems the more commonly used, curriculum development denotes only the construction phase of the study plan, considering the context and justification that supports it as well as its specific institutional implementation requirements (Ribeiro, 1999).

The conception of curriculum as a product or study plan to be transmitted emphasizes the planning process and activities of a prescriptive nature which exist before the realization and experimentation of the curriculum itself. The distance between intent and action stresses the dichotomies *curriculum/instruction*, *design/implementation*, *curriculum planning/curriculum praxis* (Pacheco, 2001). These dichotomies tend to be sustained by those who find it useful to maintain a dualism between curriculum and teaching, specifying 'what to teach?' separately from considerations about 'how to do it?' (Petrina, 2004).

In contrast, the notion of curriculum as a process or as a praxis highlights the integration of the curriculum-teaching-learning *continuum*, emphasizing the reciprocity of the mechanisms that interact with the curriculum. This conception of curriculum in permanent construction, in an endless search for the links between means and ends, is apparent in recent curriculum development proposals. A good example in GIScience is the Geographic Information Science and Technology Model Curricula initiative (DiBiase *et al.* 2006). In reality, the proposed applications of the *GIS&T Body of Knowledge* (curriculum planning, program accreditation, program evaluation and assessment, program articulation, professional certification and employee screening) denote the necessity of fostering synergies among the components of the education infrastructure.

Curriculum construction

Curriculum development, in a strict sense, corresponds to the process of curriculum design, structuring and organization, and integrates all activities connected to the creation and production of its determining elements. Those elements are: objective; subject matter; method and organization, and evaluation (Tanner and Tanner, 1995). As a result of educational policies, of changes in cognitive structures and of shifts in educational paradigms (Delanty, 2001) it is possible to find a set of tendencies that influence current curriculum construction approaches (González and Wagenaar, 2003; Meyer *et al.*, 2005; Shepherd, 2005; Fisher, 2006; Tomkins and Case, 2006):

- Institutionalization of technology.
- Increased value of the role played by language and communication in educational processes.
- Growing interest in constructivist curricula, which appeal to 'dynamic knowledge' towards critical reasoning and metacognition activities (Glatthorn *et al.*, 2006).
- Development of educational strategies centered on the student and exploration of interactive behaviors.
- Decrease in specialization of academic teaching.
- Higher flexibility in study programs, to meet different basic training and students' expectations.

- Curriculum integration in an academic teaching perspective as well as in a professional/vocational one (classroom-based learning and work-based learning).
- Higher emphasis on 'application fields' which contribute to render the academic curriculum more relevant in professional terms.
- Integration of measures promoting lifelong learning and fostering self-learning.
- Development of quality evaluation and accreditation systems in higher education institutions.
- Adoption of measures promoting mobility, based on academic degrees recognition systems and on credit systems or similar processes.

In a more or less systematic way, all these tendencies were considered in the GIScience curriculum process revision conducted by ISEGIUNL. By addressing the issues regarding curriculum content organization, knowledge acquisition, and meaningful learning, many of these tendencies were taken into account while exploring the potentialities offered by information and communication technologies in curriculum organization, presentation and exploitation.

Models of curriculum organization

Some elements of curriculum development tend to be considered more important than others. Objectives, teaching methods and educational strategies typically receive the most attention, while content organization aspects and the global structure of the curriculum tend to receive less (Tanner and Tanner, 1995; Harden and Stamper, 1999; Daniel, 2004; Tomkins and Case 2006). Although there are several ways to classify curriculum organization models, most authors recognize two essential dimensions along which approaches may be arrayed.

The first of these dimensions corresponds to the processes of *vertical curriculum articulation*, i.e., the way contents tend to be organized according to the sequence and continuity of learning tasks within a given field of study over time. Successful vertical articulation requires a thematic coherence based on epistemic and cognitive structures associated with a certain knowledge area. The second dimension corresponds with *horizontal articulation* and refers to the scope and integration of curricular contents and knowledge domains, from different disciplinary areas, within a particular grade level. When trying to improve coherence of studies in a given discipline, what is at stake is the vertical articulation or the sequence of the curriculum. When trying to develop the relation between disciplines or knowledge areas we are facing questions of horizontal articulation or the balance of the curriculum (Tanner and Tanner, 1995).

Figure 7.1 represents these two fundamental dimensions of curriculum organization process which were considered in ISEGIUNL's GIScience curriculum development model (GISc-CDM). The scope of the curriculum is a product of both the extent of the horizontal articulation and the depth of the vertical articulation. In the illustration, 'sequence' refers to the continuity of the curriculum, and 'balance' points to a full and even integration of the different disciplines or subject fields within the curriculum. In accordance with these three properties, the process of curricular organization represents an effort to improve the balance, scope, and sequence of curriculum contents, so as to give a harmonious proportion, a complete scope and a systematic continuity to the study program.

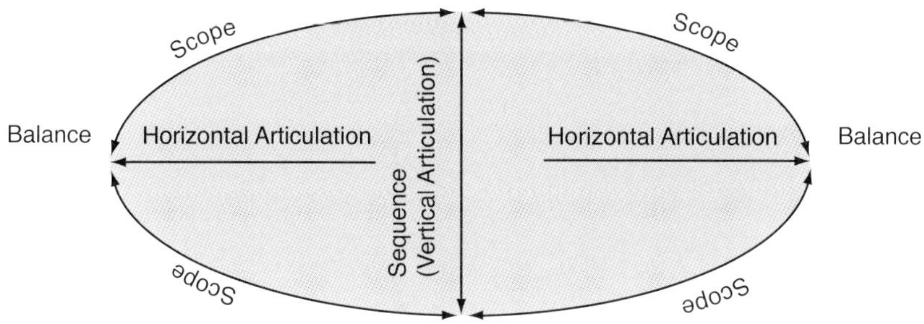

Figure 7.1 Curriculum organization: scope, sequence and balance of the curriculum (adapted from Tanner and Tanner, 1995, 374)

The spiral curriculum

The vertical and horizontal articulation of the curriculum are present in the concept of 'spiral curriculum' introduced by Jerome Bruner (Bruner, 1977). Founded on constructivist theories and on the idea that behaviors, learning processes and acquired competencies are the result of a progressive construction of the subject in its interaction with the environment (*interactionism*), the spiral curriculum, thus, presents a sequential structure that amplifies revisiting and deepening of concepts throughout the teaching and learning process. In that manner it confers an iterative nature to the modes in which knowledge is constructed/reconstructed as shown in Figure 7.2.

As stated by Harden and Stamper (1999) the four main characteristics of the spiral curriculum are:

- Topics and concepts revisiting – reiteration, as opposed to redundancy, leads to the deepening of subjects founded on knowledge and competencies previously acquired.
- Growing levels of complexity – as topics and concepts are revisited, the complexity of the subjects taught tends to increase. The recursive character of the learning process implies that each revisiting/iteration is associated with new learning objectives and the development of higher level competencies.
- Learning sequencing – the continuity of the learning process is ensured by a logical sequence, leveraging the cognitive and epistemological structures that sustain the process of producing new knowledge and competencies. In most cases this knowledge and competencies are directly linked to the ones preceding them (earlier in the spiral) setting the stage for mastery of subsequent competencies acquisition (further along the spiral).
- Augment students' competencies – the level of competencies that must be acquired increases as the topics and subjects are revisited, until the global learning objectives of the curriculum are attained. This gradual addition of competencies may easily be seen by students using self-evaluation mechanisms.

Though the advantages and characteristics of the spiral curriculum are apparent, it is not straightforward how to achieve the systematization and organization of its contents, so that the curriculum components are aligned with the curriculum work activities, including knowledge making, policy making, development, planning, management, assessment

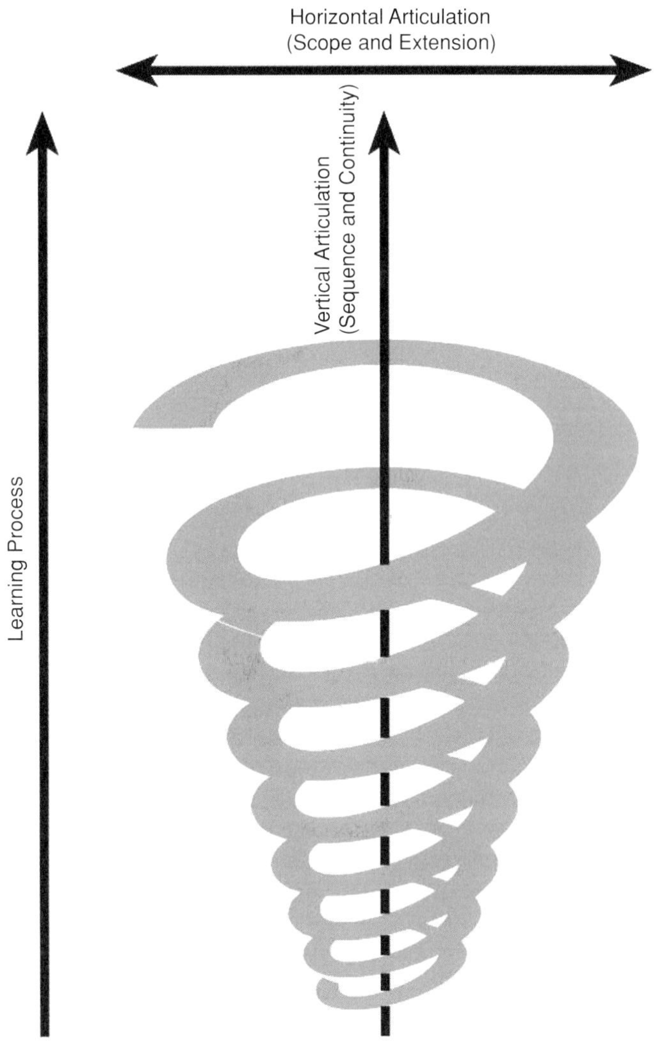

Figure 7.2 Curriculum articulation: horizontal and vertical articulation under a spiral concept of curriculum

and evaluation/research (Hewitt, 2006). This process of systematizing and organizing curriculum contents varies from one setting to another, with implications on scope, depth, integration and sequence of the curriculum. In the process, discrete disciplinary boundaries may or may not be preserved. In our view, curriculum synthesis is attained when divisions amongst different subject areas are no longer perceptible.

Developing curriculum for personalized and meaningful web-based learning

One of the greatest potential affordances of web-based education is personalized learning environments (Heller *et al.*, 2005; Heller *et al.*, 2006; Steiner and Albert, 2008).

Personalized e-learning systems aim at adjusting teaching and learning to the needs, interests and aptitudes of each individual learner, resorting to a diagnosis of its relevant characteristics and employing that information during the process of curriculum implementation. From this perspective, the process of curriculum development should accommodate the need to assess learners' prior knowledge and abilities and guide him or her towards the course of desired learning outcomes.

Some authors suggest that the mathematical psychological framework of Knowledge Space Theory (Doignon and Falmagne, 1985; Falmagne *et al.*, 1990; Doignon and Falmagne, 1999) and its competence-based extensions provide a sound basis to promote learning personalization (Heller *et al.*, 2006; O'Conlan *et al.*, 2006; Steiner and Albert, 2008). In this approach, defined relationships between learning objects, problems and skills of a specific knowledge domain are used to structure learning and knowledge assessment processes. Different approaches to identify prerequisite relationships have been put forth in order to create a prerequisite structure through which personalized web-based learning environments may be founded. Some of these approaches propose the use of concept maps and semantic networks to present, in an intuitive visual way, the dependencies amongst learning objects and the problems and competencies to be acquired in a given knowledge domain (Heller *et al.*, 2005; Heller, *et al.*, 2006; Cañas and Novak, 2008; Suthers, 2008).

The concept of meaningful learning, which played an important role in Ausubel's assimilation learning theory (Ausubel, 1962, 1963, 1968), provides a theoretical framework adequate to the challenges nowadays facing personalized curricula (Novak, 1998; Cañas and Novak, 2008). In actuality, the three principles sustaining this concept seem to encapsulate in themselves a set of presuppositions favoring the process of curriculum development fitting learners' different needs and expectations:

- The importance given to the choices made by learners so as to relate new concepts and propositions with an already established knowledge structure.
- The relevance that preexisting concepts assume in consolidating and expanding a cognitive structure.
- The way in which curriculum design favors the exploration of that cognitive structure, consequently promoting the continuity and sequence of learning processes.

In this context, the process of curricular development should present the flexibility needed to help learners bridge the gap between knowledge they already possess and knowledge yet to be acquired, taking into account not only their expectations, but mostly, their cognitive structures, inevitably related to their prior instruction (Novak, 1998). Whilst their expectations may directly link to the explicit identification of curriculum learning outcomes, their cognitive structures will tend to position them within the curricula contents, ideally represented through an ontological semantic network (Pirrone *et al.*, 2005; Painho *et al.* 2007; Sherborne, 2008).

In a constructivist view of knowledge, the possibility of making students part of the process of meaningful learning depends on the extent to which concepts and propositions already attained can be illuminated, and on the learning paths which, from then on, may be generated. The use of technologies that promote interaction between learner and curriculum are, therefore, particularly important. Besides, the possibility of visually exploring a curriculum thus organized may bring numerous benefits to learners, of which we underline the following:

- Favor the understanding of that which is already known, but mostly, of that which is still to be learned.
- Identify gaps in the learning process that may hinder or prevent their progression.
- To know and select their own learning paths, in accordance to expected outcomes and in-depth exploring of various degrees of preexisting cognitive structures.

According to this view, the construction of an ontological-based curriculum development model, based on a semantic network and supported by a technological solution that facilitates exploration, is a potentially effective approach to the challenges of creating meaningful and personalized learning.

7.2 Challenges to GIS&T curriculum development

Numerous factors have contributed for the expansion of GIS&T, today seen as 'critical components of the global cyberinfrastructure, both in university and in society' (Richardson, 2006, vii). The fast growth of computational technology, a favorable quality/price relation of software and hardware, the development of friendly interfaces and emergent availability of spatial data, have all had great impact in extending geographic information technologies to many scientific domains. These factors also contribute to the recent growth in graduate and postgraduate GIS&T courses. However, the growing recognition of the crucial role geographic information plays within the knowledge society entails a number of issues that are being discussed by many of those closely related to GIS&T education (Kemp and Goodchild, 1991; Sui, 1995; Kemp and Wright, 1997; Unwin, 1997; Solem, 1998; Wikle, 1998; Forer and Unwin, 1999; Painho, 1999; Wikle, 1999; DiBiase, 2003; Brown *et al.*, 2003; Harris, 2003; Kemp, 2003; Kemp and Wiggins, 2003). Among those issues, we highlight those that are especially relevant to GIS&T curriculum development:

- GI technology versus GI science – Apparent conflict between the need for teaching to boost the exploration of concepts and theories underlying Geographic Information Science, while covering a vast domain of GIS applications subject to continuous technological innovations.
- 'GIS teaching' versus 'GIS training' – The tension between industry needs for skills-based workforce training and academic predilection toward research-based scholarship. Because no pedagogical approach can reconcile such diverse educational objectives, GIS&T curricula in higher education are often less cohesive than they could be.
- Accreditation and certification – Need to ensure quality in GIS&T education, which presupposes the identification of conditions and specific competencies by which educational effectiveness may be evaluated and improved.
- Flexibility and scope of GIS&T curricula – If GIS&T is conceived broadly, it follows that GIS&T curricula will vary from institution to institution, responding to the varied needs of a diverse field.
- Access and equity – As GIS becomes a pervasive technology in the knowledge society, it is essential that access to knowledge and skills acquisition is widespread and equitable.

- Web-based learning and GIS&T teaching – The internet is rapidly changing the way educators, investigators and GIS professionals gather materials, perform background research, collaborate and communicate with colleagues and share the results of their researches. At the same time, the possibilities of multimedia, hypertext and asynchronous learning are currently being integrated into GIS&T teaching. The development of GIS&T curricula should involve the analysis of factors that permit the exploration of these resources and technologies through adequate teaching methods.

All these topics involve a set of challenges and opportunities that must be considered in GIS&T curricula development proposals. These topics reflect crucial aspects connected to curriculum contents, sequence and articulation which, when analysed from the perspective of diversifying and personalizing teaching experiences, offer numerous possibilities for innovation.

7.3 The GIScience curricula development model – GISc-CDM

In 2005 ISEGIUNL began work on a prototype model for a new GIScience curricula (Painho *et al.*, 2006, 2007). The curricula development model we adopted embodies a broad conception of GIScience, here defined as an experimental research field that has geographic space as its object of study, geographic information as its means of representing knowledge and geographic information systems as its disciplinary tool. GIScience is a field that develops a systematic scientific investigation both on the representation and description of geographic objects, and the understanding and explanation of spatial phenomena. This definition points to the manner in which GIScience relies on the way it reconciles the technological, cognitive and conceptual aspects of geographic information, generates new contexts for the (re)formulation of geographical problems and (re)creates environments that stimulate and challenge spatial reasoning and spatial knowledge. These contexts and environments depend on the way in which theory, concepts, data, technology, methods, individuals and society interact and go through the *continuum* GIScience – GISystems (Painho *et al.*, 2007).

The GISc-CDM project leveraged teaching strategies honed over several years of teaching GIS (Painho, 1995, 1999). The overall purpose of the initiative was to define a documented and well-structured curriculum development process approach, able to support the construction of flexible study programs, adaptable to the different needs and students expectations, and allowing for an effective and on-going process of curriculum improvement and innovation. Ultimately, we hoped that the project would produce a set of web-based tools to promote effective knowledge production, representation and dissemination using basic terminology of the field and relating it with curricula contents.

The ontological approach

The first prototype of the GISc-CDM appeared in 2006 (Painho *et al.*, 2006, 2007). This model, based on the conception of curriculum as a process or praxis, centered upon the design of an ontology which would establish knowledge structures relevant to a variety of curricula proposals in GIS&T studies. Key to this approach was the development of an extensive bibliography and identification of categories and relationships among themes, concepts, technologies, methods, initiatives and institutions linked to GIScience domain. Based on this activity, an ontology model and a terminological knowledge base were conceived, serving three main purposes:

1. Construct a linguistic data repository for access to specialized terms and the creation of domain specific-dictionaries.
2. Create learning units, i.e., the pedagogical structures that form the basis of curricula construction and that, oriented to specific objectives, foster interaction and significant learning experiences.
3. Make available, a GIScience cognitive map based upon an ontology and supported by a semantic network linking multiple linguistic expressions to conceptual GIScience curricula structure.

To ensure that the terminological resources encompassed the full range of topics germane to geographic information technologies, the first bibliographic survey started with reference and general literature. This approach made it possible to establish a first frontier of the GIScience body of knowledge, and thus determine the scope and extension of the curricula. After that, another bibliographical survey was made, this time taking into consideration works of a more specific nature, in order to allow for a more in-depth analysis of themes and subjects identified in the previous stage.

A key challenge associated with development of learning units was the aggregation of related topics. This process allowed the association of a set of objectives, topics and concepts to each learning unit, thus providing the conditions for the modular construction of curricula. To ensure the adequacy of the curricula to the Bologna Process (Wächter, 2004), for each of the learning unit the following descriptors were identified: denomination, topics, objectives (learning outcomes) and student workload in hours that can easily be converted into the European Credit Transfer and Accumulation System (ECTS).

The attempt to describe knowledge of a specialized nature led to many adjustments to our initial boundary of the GIScience domain, highlighting its interdisciplinary nature. On the other hand, the identification of topics and subjects connected with different, but closely related domains (such as remote sensing, geostatistics, geocomputation, cartography and topography, among others), challenged the process of curricula organization, since these related elements are not easily subsumed under one disciplinary or compartmentalized vision. Additionally, it suggested the necessity to anticipate further developments, mainly concerned with the cognitive and epistemological issues that must be considered in order to define and limit the scope of the analysis, as well as in the identification of terminological units and conceptual relations necessary to model the corpus and to delimit the domain.

These activities reflected the need to frame the GISc-CDM in the context of the recent terminology research, increasingly orientated towards facilitating information retrieval and knowledge engineering (Casati *et al.*, 1998; Guarino, 1998). Actually, the possibility of establishing multidimensional networks of concepts, by constructing a repository of terms related to the problem in a given learning unit, advances a non-compartmentalized and non-hierarchical idea of knowledge, via the establishment of semantic networks, which tend to bring the association process and the mechanisms of knowledge exploration closer to the ways of human memory and thinking.

Semantic organization tools help students analyse and organize what they know or are in the process of learning, enveloping them in a process of critical reasoning regarding the topics taught (Vilela *et al.*, 2004; Novak and Cañas, 2006; Jonassen *et al.*, 2007; Cañas and Novak, 2008; Steiner and Albert, 2008). Jonassen *et al.* (1998) argue that the two most common ways for semantic organization of knowledge are databases and semantic network tools. Of these, semantic network tools involve construction of concept maps which facilitate not only exploration of contents and the gathering and sharing of information, but also help the process of acquiring structural knowledge (Jonassen *et al.*, 1993).

There is a certain consensus that the associative non-linear and non-hierarchic hypermedia structure of the web (with unlimited capabilities for hyper linking) resembles the associative memory mechanism (that tends to form an organization of the human memory as a net). However, there are different ideas about how to leverage this similarity (Miller and Miller, 2000). The GISc-CDM project took into account the way the brain stores information, both to facilitate personalized web-based instruction in GIS&T, and to ensure that components of the GIScience would be configured and represented appropriately.

Conceived as an instrument for the support of teaching and learning activities, the GISc-CDM sought to support a knowledge organization structured on networks of concepts and on the construction of concept maps. The integration of information visualization tools was one of the innovative aspects of the project, encouraging an intuitive representation, understanding and exploration of the GISc-CDM components.

Exploring the links within the GISc-CDM components

Centered on learning units, the GISc-CDM takes advantage of a modular organization that promotes the construction of differentiated study programs, accommodating the diverse learning needs that characterize the GIS&T field. To support, organize and manage a vast amount of information related with the different GISc-CDM's components, we developed the relational database shown in Figure 7.3.

The database design embodies the conception of curriculum as a process or praxis, and, by facilitating the sequence and organization of its contents, ensures a focus on intellectual processes (Marzano *et al.*, 1988). The database structure addresses the need for information manipulation associated with the learning units, considered a critical component of the GIScience curriculum development model, since it assumes a leading position between curriculum design and instructional design. Thus, and by focusing our attention on the learning units, it is possible to identify the following meaningful relations between the different GISc-CDM components:

- Learning units – Cognitive Skills: This association translates the meaning given to the learning outcomes in order to promote specific cognitive skills. It is based on the assumption that cognitive skills increase as students are led to revisit, explore in depth, and encounter topics and subjects of varied domains of knowledge and in different pedagogical contexts.
- Learning units – Evaluation: This relation reinforces the importance of adequate evaluation methods (summative and formative evaluation) according to the cognitive skills being promoted.

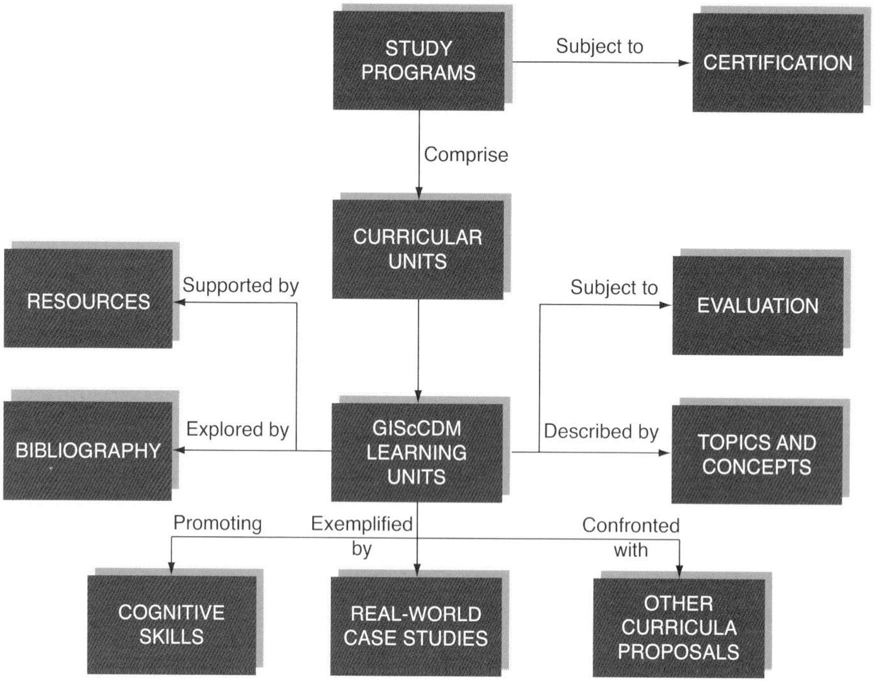

Figure 7.3 The GISc-CDM relational database: major links among curriculum components

- Learning units – Bibliography: Refers to the set of works from which the content of a given learning unit derived. Thus, the bibliography is mentioned as the prime resource of a learning unit.
- Learning units – Topics and Concepts: Focused on the contents to be taught, the topics and concepts are the key terms that summarize/describe the main subjects/topics to be explored on a specific learning unit and that sustain the meaningful learning approach. This relation also constitutes the core of the ontological approach supporting the idea behind the GIScience knowledge cartography (Painho *et al.*, 2006).
- Learning units – Other curricula proposals: Correspondence between GISc-CDM learning units and reference curricula units (Kemp and Goodchild, 1990; NCGIA, 2000; Strobl and Shahnawaz, 2004; DiBiase *et al.*, 2006); offering different analysis opportunities for curricula evaluation, innovation and improvement.
- Learning units – Resources: Instructional material (eBooks, presentations, data, films, exercises, etc) to support the teaching/learning activities.
- Learning units – Case studies: Projects, events and institutions connected with the subjects of a learning unit, considered useful to establish relations between the unit contents and the real world contexts of its applications.
- Learning units – Curricular units: This relation concerns the aggregation of learning units under a specific course or discipline. It reflects the main foci or problems on which subjects and contents tend to be focused.

The relational database at the heart of GISc-CDM that binds the semantic organization of the GIScience curriculum development model is accessible on the web through two distinct

Figure 7.4 The web interface to access the GISc-CDM database, which allows the management of diverse curriculum components. The introduction and definition (in Portuguese and English) of a new concept: volunteered geographic information – VGI

interfaces. The first, directly related to curriculum management activities, offers educators a set of tools to ensure the continuity of the curriculum innovation process. The ability to update contents, redefine concepts, or find matches to other curricula proposals are but a few examples of functionalities available there (Figure 7.4).

As shown in Figure 7.5, the second interface uses THINKMAP visualization software to enable exploration of the semantic network built on the connections established between learning units and concepts.

Using these two key components of the GISc-CDM it is possible to go through the GIScience knowledge domain and explore the associated remaining components of the curriculum, particularly by:

- Enabling the representation and visualization of the GIScience ontology.
- Supporting the design of instructional sequences and the compilation of a set of exemplar pathways.
- Responding to searches by providing users (students and teachers) with hierarchies of data and information.
- Creating concept maps or, in a broader sense, cognitive maps, that facilitate the acquisition of structural knowledge related to GIS&T, as well as the understanding of curriculum development processes and the notion of curriculum as praxis.

The ability of GISc-CDM users to perform different kinds of searches, to visualize contents associated with their searches in a dynamic and interactive way (Figure 7.6), to design flexible curricula, to construct learning pathways that accommodate student expectations and prior knowledge (Figure 7.7), to take control of the process of conceiving their study programs (Figure 7.8), to visualize or to see represented, in an intuitive way, a given domain or problem, or simply to explore knowledge linked to an area close to GIScience, are some of our objectives for the GISc-CDM.

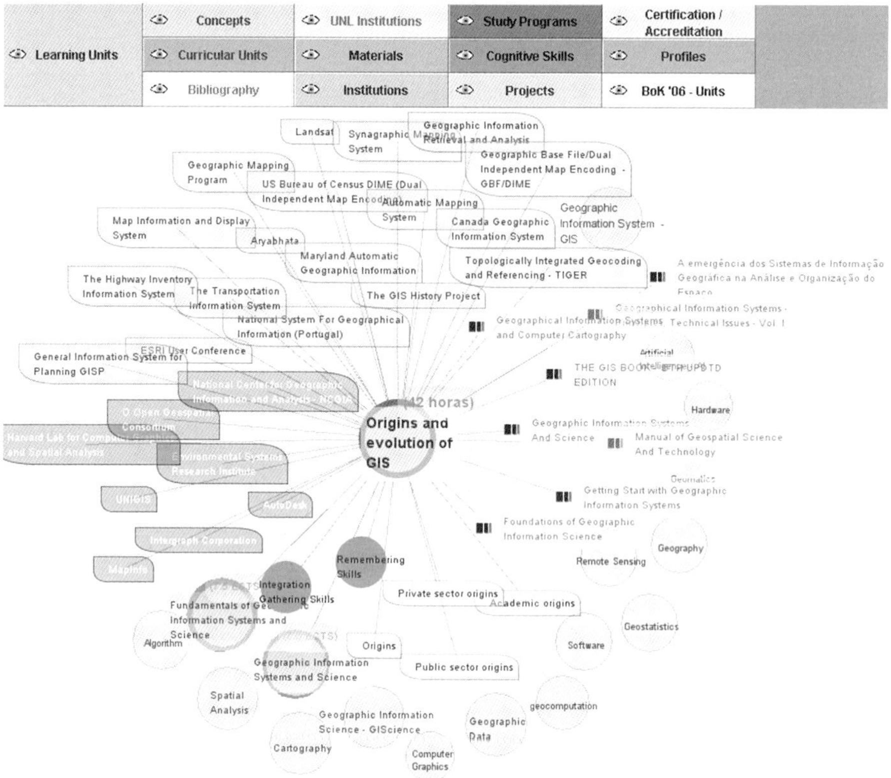

Figure 7.5 Components of the GISc-CDM. Exploring the 'Origins and evolution of GIS': concepts, bibliography, institutions, projects/milestones and the four GIS&T *BoK*'s topics related with this learning unit

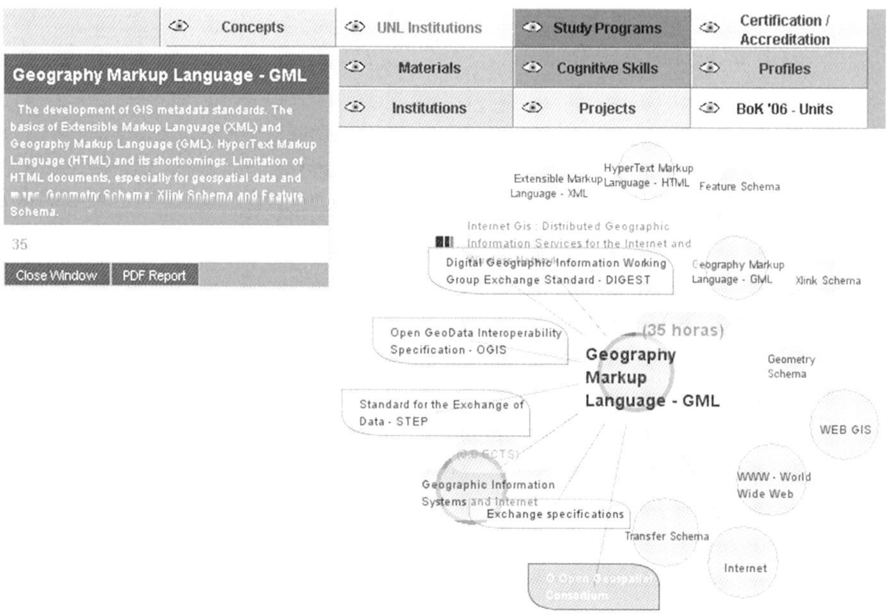

Figure 7.6 The GISc-CDM's interface: searching information about GML. What is GML? Brief description in the text box. What are the curricular units that deal with this subject? What are the related concepts?

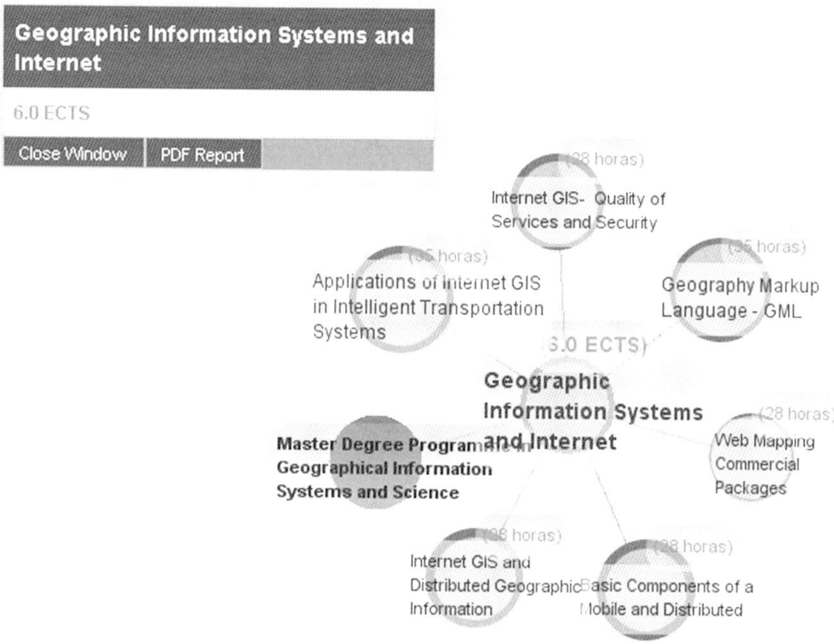

Figure 7.7 Creating a curricular unit related to GIS and internet

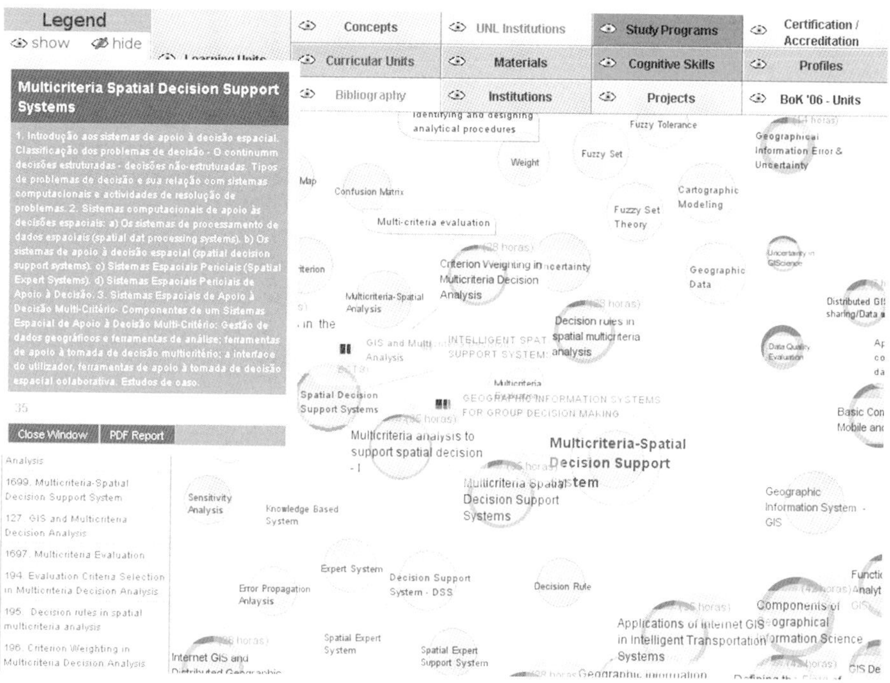

Figure 7.8 The student's ability to take informed decisions about her/his learning path. The expected student workload required to complete the learning unit: 'Multicriteria spatial decision support systems'

7.4 Conclusions

The ongoing development of geographical information technologies, the growing need for professionals in these areas and the use of new information and communication technologies in teaching, are all factors which call for a broad dissemination of the GISc-CDM within the framework of web-based learning. We believe that the GISc-CDM enables flexible integration and articulation of the components considered essential for curriculum design and construction. As a meta-representation of the contents in GIScience, the GISc-CDM may facilitate acquisition of structural knowledge, essential for the development of meta-cognitive competencies and for the promotion of meaningful learning, driven by learners' cognitive structures and oriented by prior knowledge structures. Furthermore, we expect that the ontological approach will help cultivate a GIS learning community by providing access to shared information, contents and knowledge, improving learning opportunities and promoting personalized learning pathways. With this approach we expect to facilitate access to GIS&T learning opportunities for diverse groups in varied settings. We envisage a future in which the framework described here may influence the ways in which higher education institutions worldwide respond to the new and emerging challenges posed by GIS&T education. The development of a GISc-CDM prototype adapted to the GIS&T Body of Knowledge reinforces these possibilities (Painho and Curvelo, 2008). However, the expected repercussions of this project are not limited to the GIScience domain. Extending this approach to other knowledge areas has the potential to nurture within those fields greater access to and support for effective lifelong learning, the 'guiding principle for provision and participation across the full continuum of learning contexts' (EC, 2000).

Acknowledgment

The authors wish to thank David DiBiase for his thoughtful and constructive review.

References

Ausubel, D. P. (1962) A subsumption theory of meaningful verbal learning and retention. *The Journal of General Psychology*, 66: 213–244.

Ausubel, D. P. (1963) Cognitive structure and the facilitation of meaningful verbal learning. *Journal of Teacher Education*, 14: 217 222.

Ausubel, D. P. (1968) *Educational Psychology: A Cognitive View*. New York: Holt, Rinehart & Winston.

Brown, D. G., G. Elmes, K. Kemp, S. Macey, and D. Mark (2003) Geographic information systems, IN: G. L. Gaile and C. J. Willmott (eds), *Geography in America at the Dawn of the 21st Century*. Oxford: Oxford University Press, pages 353–375.

Bruner, J. S. (1977) *The Process of Education*. Cambridge: Harvard University Press.

Cañas, A. J., and J. D. Novak (2008) Concept mapping using CmapTools to enhance meaningful learning, IN: A. Okada, S. B. Shum, and T. Sherborne (eds), *Knowledge Cartography: Software Tools and Mapping Techniques*. London: Springer-Verlag, pages 25–46.

Casati, R., B. Smith, and A. C. Varzi (1998) Ontological tools for geographic representation, IN: N. Gaurino (ed), *Formal Ontology in Information Systems*. Amsterdam: IOLS Press, pages 77–85.

Daniel, E. (2004) Curriculum notes. University of North Carolina at Chapel Hill, School of Information and Library Science. http://www.ils.unc.edu/daniel/242/CurrNotes.html (accessed August 16, 2011).

Delanty, G. (2001) *Challenging Knowledge: The University in the Knowledge Society*. Buckingham, England: Open University Press.

DiBiase, D. (2003) On accreditation and the peer review of education in geographic information systems and science. *URISA Journal*, 15(1): 7–14. http://www.urisa.org/files/Dibiasevol15no1.pdf (accessed August 16, 2011).

DiBiase, D., M. DeMers, A. Johnson, K. Kemp, A. Luck, B. Plewe, and E. Wentz (eds) (2006) *The Geographic Information Science and Technology Body of Knowledge*. Washington, DC: Association of American Geographers and University Consortium for Geographic Information Science.

Doignon, J., and J. Falmagne (1985) Spaces for the assessment of knowledge. *International Journal of Man-Machine Studies*, 23: 175–196.

Doignon, J., and J. Falmagne (1999) *Knowledge Spaces*. Berlin: Springer-Verlag.

European Commission (2000) *Commission Staff Working Paper: A Memorandum on Lifelong Learning*. Commission of the European Communities. http://www.bologna-berlin2003.de/pdf/MemorandumEng.pdf (accessed August 16, 2011).

Falmagne, J. C., M. Koppen, M. Villano, J. Doignon, and L. Johannesen (1990) Introduction to knowledge spaces: How to build, test and search them. *Psychological Review*, 97: 201–224.

Fisher, T. (2006) *Trends and Challenges in Higher Education*. http://www.cala.umn.edu/letters/TrendsInHigherEducation.pdf.

Forer, P., and D. Unwin (1999) Enabling progress in GIS and education, IN: P. A. Longley, M. F. Goodchild, D. J. Maguire, and D.W. Rhind (eds), *Geographic Information Systems: Techniques, Applications and Management*, Chichester, UK: John Wiley & Sons, pages 747–756.

Glatthorn, A. A., F. Boschee, and B. M. Whitehead (2006) *Curriculum Leadership – Development and Implementation*. Thousand Oaks, California: SAGE Publications.

Glatthorn, A. A., J. F. Carr, and D. E. Harris (2001) *Planning and Organizing For Curriculum Renewal – Thinking About Curriculum*. Association for Supervision and Curriculum Development (ASCD). http://webserver3.ascd.org/handbook/demo/planning2.html (accessed August 16, 2011).

González, J., and R. Wagenaar (2003) *Tuning Educational Structures in Europe – Final Report: Phase One*. University of Deusto and University of Groningen http://tuning.unideusto.org/tuningeu/.

Guarino, N. (1998) Formal ontology and information systems. *Paper read at Formal Ontology in Information Systems – FOIS'98, 6–8 June, Trento, Italy*.

Harden, R. M., and N. Stamper (1999) What is a spiral curriculum? *Medical Teacher*, 21(2): 141–143.

Harris, R. (2003) Building a GIScience community in cyberspace: reflections on GIScOnline. *Journal of Geography in Higher Education*, 27(3): 279–295.

Heller, J. C., B. Mayer, C. Hockemeyer, and D. Albert (2005) Competence-based knowledge structures for personalised learning: Distributed resources and virtual experiments. Paper read at 1st ELeGI Conference, 14–16 March, Vico Equense.

Heller, J., C. Steiner, C. Hockemeyer, and D. Albert (2006) Competence-based knowledge structures for personalised learning. *International Journal on E-Learning*, 5(1): 75–88.

Hewitt, T. W. (2006) *Understanding and Shaping Curriculum – What We Teach and Why*: London: Sage Publications, Inc.

Jonassen, D. H., K. Beissner, and M. Yacci (1993) *Structural Knowledge: Techniques for Representing, Conveying, and Acquiring Structural Knowledge*. Hillsdale, NJ: Lawrence Erlbaum Associates.

Jonassen, D. H., C. Carr, and Y. Hsiu-Ping (1998) Computers as mindtools for engaging learners in critical thinking. *TechTrends*, 43(2): 24–32.

Jonassen, D. H., J. Howland, R. M. Marra, and D. P. Crismond (2007) *Meaningful Learning with Technology*. Upper Saddle River, New Jersey: Merrill Prentice Hall.

Kemp, K. K. (2003) Guest Editorial: Why GIS professional certification matters to all of us. *Transactions in GIS*, 7(2): 159–163.

Kemp, K. K., and M. F. Goodchild (1991) Developing a curriculum in geographic information systems: The National Center for Geographic Information and Analysis Core Curriculum Project. *Journal of Geography in Higher Education*, 15(2): 121–132.

Kemp, K. K., and M. F. Goodchild (1990) *NCGIA Core Curriculum in GIS*. University of California. http://www.geog.ubc.ca/courses/klink/gis.notes/ncgia/toc.html (accessed August 16, 2011).

Kemp, K. K., and L. Wiggins (2003) Introduction to the special issue on GIS education. *URISA Journal*, 15(1): 4–6.

Kemp, K. K., and R. Wright (1997) UCGIS identifies GIScience education priorities. *Geo Info Systems*, 7(9): 16–20. http://www.ncgia.ucsb.edu/other/ucgis/ed_priorities/geoinfosystems.html (accessed August 16, 2011).

Marzano, R. J., R. S. Brandt, C. S. Hughes, B. F. Jones, B. Z. Presseisen, S. C. Rankin, and C. Suthor (1988) *Dimension of Thinking: A Framework for Curriculum Instruction*. Alexandria, VA: Association for Supervision and Curriculum Development.

Meyer, J. W., F. O. Ramirez, D. J. Frank, and E. Schofer (2005) Higher education as an institution. http://www.sess.smu.edu.sg/events/Paper/ramirez_0505.pdf.

Miller, S. M., and K. L. Miller (2000) Theoretical and practical considerations in the design of web-based instruction, IN: B. Abbey (ed), *Instructional and Cognitive Impacts of Web-Based Education*. Hershey, PA: Idea Group Publishing, pages 156–177.

NCGIA (2000) *The NCGIA Core Curriculum in GIScience*. National Center for Geographic Information and Analysis. http://www.ncgia.ucsb.edu/education/curricula/giscc/ (accessed August 16, 2011).

Novak, J. D. (1998) *Learning, Creating, and Using Knowledge: Concept Maps as Facilitative Tools for Schools and Corporations*. Mahwah, NJ: Lawrence Erlbaum Associates.

Novak, J. D., and A. J. Cañas (2006) The theory underlying concept maps and how to construct them. IHMC – Institute for Human and Machine Cognition. http://cmap.coginst.uwf.edu/info/.

Conlan, O., I. O'Keeffe, C. Hampson, and J. Heller (2006) Using knowledge space theory to support learner modeling and personalization. Paper read at World Conference on E-Learning in Corporate, Government, Healthcare, and Higher Education 2006 at Chesapeake, VA.

Pacheco, J. A. (2001) *Currículo: Teoria e Práxis*. 2nd ed. *Colecção Ciências da Educação*. Porto: Porto Editora.

Painho, M. (1995) Teaching GIS to multiple background students: ISEGI's experience. Joint European Conference and Exhibition on Geographical Information – from Research to Application through Cooperation, *Proceedings Vols 1 and 2*, B281–B285.

Painho, M. (1999) *Teaching GIS in information management – A discussion of teaching methods, options and curricula for GIS teaching in higher education*, Lisboa: Instituto Superior de Estatística e Gestão de Informação da Universidade Nova de Lisboa.

Painho, M., and P. Curvelo (2008) *BoK* e-Tool prototype – An ontological-based approach to the exploration of Geographic Information Science & Technology Body of Knowledge. *Proceedings of the 2008 Sixth EUGISES 2008 – European GIS Education Seminar*. Royal Agricultural College, Cirencester.

Painho, M., P. Curvelo, and I. Jovani (2006) Designing GIS curriculum for higher education. Paper read at Seventh Annual Esri Education User Conference, August 7–11, 2006, San Diego, California, USA.

Painho, M., P. Curvelo, and I. Jovani (2007) An ontological-based approach to geographic information science curricula design, IN: S. Fabrikant and M. Wachowicz (eds), *The European Information Society: Leading the Way with Geo-information*. Berlin: Springer-Verlag, pages 15–34.

Painho, M., P. Curvelo, I. Jovani, M. Peixoto, and M. Neves (2006) The cartography of GIScience ontology. Esri User Conference, August 7–11, 2006. http://www.isegi.unl.pt/unigis/mpeixoto/pages/poster06.html (accessed August 16, 2011).

Petrina, S. (2004) The politics of curriculum and instructional design/theory/form: Critical problems, projects, units, and modules. *Interchange*, 35(1): 81–126.

Pirrone, R., G. Pilato, R. Rizzo, and G. Russo (2005) Learning path generation by domain ontology transformation, IN: S. Bandini and S. Manzoni (eds), *AI IA 2005: Advances in Artificial Intelligence*. Berlin: Heidelberg: Springer, pages 359–369.

Posner, G., and A. Rudnitsky (1982) *Course Design: A Guide to Curriculum Development for Teachers*. New York: Longman, Inc.

Ribeiro, A. C. (1999) *Desenvolvimento Curricular*. 8th ed, *Colecção Educação Hoje*. Lisboa: Texto Editora.

Richardson, D. (2006) Transforming science and society, IN: DiBiase, D., M. DeMers, A. Johnson, K. Kemp, A. Luck, B. Plewe, and E. Wentz (eds) (2006) *The Geographic Information Science and Technology Body of Knowledge*. Washington, DC: Association of American Geographers and University Consortium for Geographic Information Science, vii–x.

Shepherd, Q. D. (2005) Integration of technology: A look at curricular trends, current research, and best practices in education. http://www.nelliemuller.com/CurriculumTrends.doc (accessed August 16, 2011).

Sherborne, T. (2008) Mapping the curriculum: How concept maps can improve the effectiveness of course development, IN: A. Okada, S. B. Shum, and T. Sherborne (eds), *Knowledge Cartography: Software Tools and Mapping Techniques*. London: Springer-Verlag, pages 183–198.

Smith, M. K. (1996) Curriculum theory and practice. http://www.infed.org/biblio/b-curric .htm (accessed August 16, 2011).

Solem, M. N. (1998) Position statement on interoperability for GIScience education. Paper read at the International Workshop on Interoperability for GIScience Education-IGE '98, May 18–20, 1998, Soesterberg, Netherlands.

Steiner, C. M., and D. Albert (2008) Personalising learning through prerequisite structures derived from concept maps. *Advances in Web Based Learning – ICWL, Lecture Notes in Computer Science*, 4823: 43–54.

Strobl, J., and S. Shahnawaz (2004) *InterGIS standard curriculum for GIScience education (first draft)*. http://www.giscampus.org/intergis/OutComes/InterGIS%20Curriculum%20Summary.pdf.

Sui, D. Z. (1995) A pedagogic framework to link GIS to the intellectual core of geography. *Journal of Geography*, 94(6): 578–591.

Suthers, D. D. (2008) Empirical studies of the value of conceptually explicit notations in collaborative learning, IN: A. Okada, S. B. Shum, and T. Sherborne (eds), *Knowledge Cartography: Software Tools and Mapping Techniques*. London: Springer-Verlag, pages 1–23.

Tanner, D., and L. Tanner (1995) *Curriculum Development: Theory into Practice*. Columbus, Ohio: Merrill – Prentice Hall.

Tomkins, G. S., and R. Case (2006) Curriculum development. The Canadian Encyclopedia. http://www.thecanadianencyclopedia.com/index.cfm?PgNm=TCE&Params=A1ARTA0002083 (accessed August 16, 2011).

Unwin, D. J. (1997) Unit 159: Curriculum design for GIS. NCGIA Core Curriculum in Geographic Information Science. http://www.ncgia.ucsb.edu/giscc/units/u159/u159.html (accessed August 16, 2011).

Vilela, R. B., L. Austrilino, and A. C. Costa (2004) Using concept maps for collaborative curriculum development. First International Conference on Concept Mapping, Pamplona, Spain.

Wächter, B. (2004) The Bologna process: developments and prospects. *European Journal of Education*, 39(3): 265–273.

Wikle, T. A. (1998) Continuing education and competency programmes in GIS. *International Journal of Geographical Information Science*, 12(5): 491–507.

Wikle, T. A. (1999) GIS Education through certificate programs. *URISA Journal*, 11(2): 53–60.

8

Addressing misconceptions, threshold concepts, and troublesome knowledge in GIScience education

Matthew Bampton

Department of Geography-Anthropology, University of Southern Maine, Gorham, Maine, USA

8.1 Overview

This chapter focuses on the role of misconceptions, threshold concepts and troublesome knowledge in determining the efficacy of teaching and learning GIScience. Many instructors are familiar with fundamental problems of understanding that bedevil students. For example correctly understanding scale, projection and orientation is often difficult even for advanced students. Failure to master these ideas early in the course of study undermines these students' ability fully to grasp more advanced topics. Physics students grapple with the concept of force and biology students are challenged by the mechanics of inheritance with similar consequences. Misconceptions are those persistent incorrect ideas that students bring with them into the classroom, that prove difficult or impossible to dislodge (Hestenes, Wells and Swackhammer, 1992); threshold concepts are the essential ideas which must be grasped for students to advance their understanding of particular topics (Meyer and Land, 2003), and; troublesome knowledge is that knowledge which students find counter-intuitive and which stands as a barrier to their ability to grasp threshold concepts (Perkins, 1999).

A large body of science, technology, engineering and mathematics (STEM) education research, conducted over several decades, indicates many students bring misleading ideas or misconceptions into the classroom with them, and that these misconceptions hamper their ability to master materials essential to learning the subject they are being taught. Misconceptions identified by STEM researchers are diverse, occurring in a variety of fields and at a variety of instructional levels. There are good indications that a failure to recognize

Teaching Geographic Information Science and Technology in Higher Education, First Edition.
Edited by David J. Unwin, Kenneth E. Foote, Nicholas J. Tate and David DiBiase.

and address misconceptions in introductory-level classes accounts for a higher failure rate among introductory students and diminished comprehension among advanced students. Misconceptions not appropriately addressed early in a student's career frequently manifest themselves as conspicuous gaps in knowledge later on. More recently, a number of social science researchers have demonstrated that the related notions of troublesome knowledge and threshold concepts inhibit a student's ability to develop a complete understanding of materials. There is significantly more empirical research into misconceptions than there is into troublesome knowledge and threshold concepts. However, the two sets of ideas are closely linked: troublesome knowledge, as described by Perkins (1999), is essentially troublesome because of the misconceptions students bring with them into the classroom. Threshold concepts are the ideas students must grasp to circumvent misconceptions. Regardless of the terminology employed, in both STEM disciplines and the social sciences, evidence indicates that timely intervention and curricula designed to target misconceptions help students master threshold concepts and resolve the difficulties posed by troublesome knowledge, and can significantly improve overall student success rates and comprehension. Accurately identifying and appropriately addressing these difficulties increases student success within and beyond the classroom. A review of the existing literature indicates that a significant number of the misconceptions STEM researchers have identified match core concepts routinely identified as essential to GIScience in textbooks (e.g. Lo and Yeung, 2007), the model GIScience Curricula (e.g. UCGIS, 2003), and the GIS&T *Body of Knowledge* (DiBiase *et al.*, 2006).

As misconceptions have been identified in a wide variety of fields matching many essential theoretical underpinnings of GIScience, it is likely that the difficulties encountered by students in these fields also occur among GIScience students, and account for some of the problems they encounter as they strive to understand GIScience. GIScience curricula informed by an understanding of misconceptions are more likely to succeed than those that ignore the issue. The importance of exploring and overcoming GIScience students' misconceptions is apparent in the context of Unwin's (1997, sec. 159.1) observation that:

- Defining and delivering an effective curriculum is the most important professional responsibility for GIS instructors.
- GIS instructors in higher education have shown an almost exemplary concern for teaching.
- Concern for education in GIS goes back a long way.
- Concern and care for education in GIS has been a major factor in allowing the technology to diffuse so rapidly into geography and related sciences as well as into industry and commerce.

There are significant challenges to overcome: to date GIScience educators have not undertaken any empirical work to explore this question; GIScience students have a variety of different educational objectives; GIScience is taught both in traditional college environments and in a variety of professional and vocational settings, and; GIScience students' prior knowledge, aptitudes, age and cultural profiles are highly varied.

Educators and the GIScience industry have developed extensive and varied model curricula, sophisticated multimedia teaching materials, and many textbooks. These materials are designed around specific educational outcomes, but are largely uninformed by empirical study of the process of GIScience learning. Thus, although the existing materials succeed in supporting some students, GIScience courses are typically seen as intellectually and technically challenging (Goodchild and Kemp, 1992). Laboratory work is time-consuming, theories are arcane, and it is difficult to link the theoretical abstractions to the practical business of 'learning the clicks.' The resulting problems are outlined by Marble (1997, 1), who notes that those not grasping this connection are:

> at best, able to apply perhaps 10 percent of the power of the technology, and this often incorrectly. The present attitude, if it is allowed to continue, will turn out to be unfortunate for the discipline of geography, for the GIS industry and for those hundreds of thousands of individuals and organizations who are finally finding out that 'geography really matters' through their exposure to the highly useful products of GIS technology.

Numerous educational resources are available, a cadre of dedicated and capable instructors exists, and yet GIScience students frequently fail to meet the learning outcomes instructors seek. To address this problem effectively we must acknowledge that there are areas of the discipline where, much of the time, despite our best efforts our students do not learn what we are trying to teach them. To solve this problem we must design a strategy that will identify and remedy misconceptions. Here I suggest that we identify the topics where students are most likely to hold misconceptions by matching the existing empirical studies from other STEM fields to the core concepts of GIScience. Once a preliminary map of likely misconceptions is devised we can then start to design research instruments that will identify the problems facing GIScience students. With an improved understanding of why students have difficulties with some topics it will be possible to design more effective curricula, and teach GIScience more successfully.

8.2 Misconceptions, troublesome knowledge, threshold concepts and STEM education

The documentary *A Private Universe* (Schneps and Sadler, 1988) explores one common misconception – that summers are warmer than winters because the Earth's orbit brings the planet closer to the Sun in the summer, rather than because the Earth's rotational axis is tilted with respect to its orbital plane. In the film, twenty-one out of twenty-three faculty and graduating students from Harvard University – perhaps one of the top universities in the US – gave the incorrect explanation. The film goes further, demonstrating how difficult such misconceptions are to dispel. Subsequent interviews with students and teachers at a nearby high school show that, even when students are offered extra tutoring to correct this misconception, it is so deep-seated that the students revert to the incorrect explanation almost immediately. Equally disturbing were signs that this misconception may actually be caused or worsened by textbooks which typically show Earth-Sun relationships in oblique views which exaggerate the elliptical appearance of the Earth's orbit and makes it look as though it passes substantially closer to the Sun during part of its orbit.

Misconceptions, sometimes referred to as 'mind bugs' (VanLehn, 1990), are the misleading ideas and theories about how the world works that students bring with them into the classroom, or that are introduced into students' thinking by poorly designed teaching materials. These preconceived notions frequently remain in place, despite teachers' best efforts to dislodge them and replace them with proper understanding. There is a large body of empirical material that has identified a strong relationship between misconceptions and subsequent student under-performance (Bransford *et al.*, 2000, Richardson, 2004). Hestenes, Wells, and Swackhammer (1992, 141), in their study of physics education summarize thus: 'Every student begins physics with a well-established system of commonsense beliefs about how the physical world works derived from years of personal experience. Over the last decade physics education research has established that these beliefs play a dominant role in introductory physics. Instruction that does not take these beliefs into account is almost totally ineffective.' Threshold concepts, sometimes called 'conceptual gateways' are those essential ideas that lead students to new ways of understanding what they are studying. Meyer (2005, 373–74) describes them as:

> *transformative* (occasioning a significant shift in the perception of a subject), *irreversible* (unlikely to be forgotten, or unlearned only through considerable effort), and *integrative* (exposing the previously hidden interrelatedness of something). In addition they may also be *troublesome* and/or they may lead to troublesome knowledge for a variety of reasons.

The idea that misconceptions pose a barrier to learning is now well established and used to inform pedagogical strategies in a range of disciplines. Physics educators have developed the Force Concept Inventory (Hestenes, Wells, and Swackhamer 1992) as a tool for evaluating student misconceptions. In areas as different as astronomy (CAER, 1999; Hufnagle, 2002; Zeilik, 2002), chemistry (Zoller, 1990; Gaddis, 2001), physiology (Michael *et al.*, 1999), and geology (LiBarkin *et al.*, 2005) misconceptions have been shown to be common, and relatively consistently held. Further, this appears to be true across multiple levels of education from K-12 classrooms, through community colleges, private universities, state universities and liberal arts colleges. Alarmingly, some misconceptions can be present even when the evidence needed to overcome them is within arm's reach: physiology instructors have reported that certain misconceptions persist although students could easily falsify them by an examination of their own bodies' function. Michael *et al.* (1999, 2002) note that of 400 students surveyed, almost half incorrectly believed that as respiration rate increased during exercise, the volume of air inhaled with each breath either remained the same, or decreased.

Given their importance as a barrier to learning, identifying the origins of misconceptions is of some interest. Four possibilities are suggested in the literature. First, experiences: students' incorrect interpretations of everyday phenomena and processes can cause difficulties. Halloun and Hestenes (1985b) note that many novice physics students maintain 'commonsense' beliefs concerning force, momentum and motion that are closer to those of Aristotle than of Newton. Second, language: instructors' use of, and students' understanding of, both technical and everyday vocabulary can be at variance. Marais and Jordaan (2000) found that students encountered significant problems in interpreting both symbolic and narrative language that hampered their ability to understand written explanations and test questions. Third, visual representations: diagrams are necessary simplifications of reality, and sometimes create more confusion than they eliminate. Schonborn, Anderson and Grayson (2002) show that after studying a text-book diagram of immunoglobulin G, students demonstrated three distinct classes of misconception directly attributable to the image: process type difficulties, structure-type difficulties and DNA-type difficulties. Fourth, faulty analogies: teachers of science habitually use models of scientific phenomena, either verbal or physical, to help explain their material. Evidence indicates that oversimplified, imprecise, or poorly chosen models may contribute to creating misconceptions rather than clarifying material. Hammer (2000) notes the difficulties associated with the commonly used analogy of flowing water to explain the behavior of a beam of sunlight. As the interactions between water and a static object and sunlight and a static object have far more dissimilarities than they have commonalities, the evidence indicates students are frequently misdirected by the analogy.

Misconceptions research is summarized in Bransford, Brown and Cocking (2000, 14–18) as follows:

1. Students come to the classroom with preconceptions about how the world works. If their initial understanding is not engaged they may fail to grasp the new concepts and information that are taught, or they may learn them for the purpose of a test but revert to their preconceptions outside the classroom.
2. To develop competence in an area of inquiry, students must (a) have a deep factual knowledge, (b) understand facts and ideas in the context of a conceptual framework, and (c) organize knowledge in ways that facilitate retrieval and application.
3. A 'metacognitive' approach to instruction can help students to take control of their own learning by defining learning goals and monitoring their progress in achieving them.

8.3 Do GIScience students have misconceptions?

There are misconceptions identified in other STEM fields that have a bearing on GIScience. Francek *et al.* (1996) identified a consistent pattern of geographical misconceptions across grade levels from junior high school through to undergraduates. Gautier, Deutsch and Rebich (2006) demonstrated that students have problems understanding the mechanics of climate change and labor under significant misconceptions concerning the mechanics of radiation and the greenhouse effect. Anderson and Leinhardt (2002) demonstrated novice students had encountered significant problems with the interpretation of projected maps. Libarkin *et al.* (2005) demonstrated that students encountered significant difficulties with many fundamental earth science concepts, including several requiring spatial visualization skills. It is highly probable that if these types of misconceptions have been identified in student populations in other areas of STEM education comparable misconceptions will be present in GIScience students.

There is as yet no published empirical research that identifies the misconceptions held by GIScience students. However, GIScience instructors frequently describe persistent difficulties encountered by their students when presented with some types of materials. Gersmehl and Gersmehl (2007) have argued that at least some of the challenges students face in mastering spatial concepts may be 'hard wired,' requiring special attention and novel approaches to teach. In my own teaching I find that students typically encounter difficulties with such ideas as projections, coordinate systems and data classification. The problems experienced by many GIScience teachers striving to help their students reach desired learning outcomes, coupled with empirical literature from other STEM fields, suggests that it is highly probable that misconceptions exist in GIScience students.

8.4 Survey of existing research

Capturing an accurate impression of student misconceptions is a complex undertaking. Typically, researchers identify a topic area of concern and then conduct extensive interviews, 'talk-throughs,' and focus group sessions with both students and experienced instructors. They then use these discussions to construct an objective instrument that can be deployed in a classroom environment. In recent years this technique has been evolving to incorporate more sophisticated tools and techniques. The Bioliteracy Project (http://www.bioliteracy.net/) uses a latent semantic analysis (LSA) tool to identify and classify misconceptions (Deerwester *et al.*, 1990; Landauer and Dumais, 1997; Garvin-Doxas *et al.*, 2005) by coding written content from students' qualitative answers to open-ended questions. The results of the LSA provides the raw material for developing the objective instrument. Libarkin and Anderson (2006) extend the scope of the typical technique, using student input to develop the initial topic areas for study rather than by defining them in advance. These topic areas are then refined with reiterative testing in a wide array of school and college settings before the instrument is developed. The product of the development process is an objective instrument that can be employed as a pre- and post-test, with analysis of the results giving insight into students' thought processes.

Here I draw on two bodies of material to identify probable areas of misconceptions in GIScience students. First, the work of researchers broadly interested in geography education who have gathered data indicating that key spatial concepts essential to the understanding of GIScience are lacking in students at many levels. Studies of this kind have already demonstrated that a variety of student populations from K-12 upwards have flawed cognitive models of spatial concepts that impair their ability to master essential spatial ideas (Downs and Liben, 1991; Golledge, 2002; Battersby, Golledge and Marsh, 2006). Other researchers have started to observe characteristics of learning behaviors that indicate the possibility of misconceptions (e.g. Kankaanrinta, 2006). Downs and Liben (1991, 304) summarize this saying 'What is logical, reasonable, and self-evident from the perspective of an expert geographer may not match the psychological realities of a student.' Second, I have reviewed the literature documenting misconceptions in other undergraduate STEM fields showing a close match to key GIScience concepts.

I have referred to the *Body of Knowledge* to identify key GIScience concepts, as this work provides a useful digest of the current GIScience canon. It is divided into 10 main categories with twenty-six 'core' units, and forty-seven 'elective' units. The main units are sub-divided into 118 'core' sub-units and 214 'elective' sub-units listing over 1600 educational objectives. To test the notion that misconceptions may present problems in GIScience education I reviewed each of the units in the *Body of Knowledge*, and identified the key words in each item. I cross-referenced each item with the education research literature of a variety of STEM fields using journal indexes, databases and specialized search engines, noting those cases in which a key concept outlined in one of the *Body of Knowledge* sub-units had already been identified as problematic in another discipline.

The fields surveyed include geography, physics, chemistry, biology, astronomy, geosciences, economics, computer and information sciences, engineering, math, and statistics. Three types of study are included in the data: those in which some objective instrument was used to gather empirical data and the results systematically analysed; those in which systematic classroom observation was used to gather empirical data, and; meta-studies, in which a body of literature containing significant empirical findings was reviewed to identify problem areas.

Within these data, studies were selected using the following criteria:

1. Those in which empirical evidence documents student misconceptions or problems with ideas that are close matches for specific elements listed in the *Body of Knowledge* sub-units. For example Hazzan and Hadar's (2005) study of computer science students' difficulties constructing Hamilton and Euler paths indicates potential problems with AM11-2 graph-theoretic (descriptive) measures.
2. Those in which empirical evidence documents student misconceptions or problems with ideas that are good analogues for specific elements listed in the *Body of Knowledge* sub-units. For example, Engelhardt and Beichner's (2004) study of engineering students' difficulties with circuit design indicates potential problems with AM11-1 networks defined.
3. Those in which empirical evidence documents student misconceptions closely matching or analogous to the more broadly defined knowledge areas and units listed in the *Body of Knowledge*. For example Trumper's (2000) study of basic astronomy concepts, such as the shape, position and movements of the Earth indicates potential problems with GD1 Earth geometry.

I include two additional bodies of work in my study that are exceptions to this rubric. First, the spatial cognition work led by Golledge (2002) and his co-authors (Battersby, Golledge and Marsh, 2009), in which they explore the ways in which people understand geography. Some of this work concerns K-12 education, but much is driven by a broader concern with spatial cognition in non-university populations. This work adds to overall understanding of how people perceive and process spatial information. Most recently some of these studies have explored the challenges that spatial cognition problems pose for the teaching and learning of conceptual foundations in GIScience. Secondly, Mark Monmonier's (1991) work, while not explicitly driven by an educational research agenda, provides a meta-study of cartographic design and map use. Several of his observations on cartographic communication closely match the educational research definition of misconceptions. The examples Momonier uses to support his account of the myriad problems facing anyone designing, publishing, or using a map, closely match the concepts outlined in CV3 principles of map design and CV6 map use and evaluation.

Most of the studies I have reviewed draw their subjects from entry-level undergraduates. There are two exceptions to this rule. First, a few studies include advanced undergraduates, and graduate students. I have assumed that if problems exist among advanced students, then problems can reasonably be predicted among the undergraduate populations from which they are drawn. Second, in three of the studies some or all of the subjects studied were non-university students. Golledge (2002) and Dehaene *et al.* (2006) both included the general public in their research (in Dehaene's case, some of the subjects were indigenous people of the Amazon basin, though I have not based any conclusions on this segment of his work). In these cases I assumed that if a proportion of the adult population displayed a misconception it is reasonable to assume that a similar proportion of entry-level undergraduates will hold a similar misconception. Finally, in a few cases either some or all of the subjects were drawn from high-school populations. In these cases, as the subjects were from advanced placement programs, or specialist technical high schools (e.g. Fessakis, Dimitracopoulou and Komis, 2005). I assumed that problems manifesting themselves among these populations would persist into the entry-level undergraduate classroom, given that typically no further educational intervention exists between advanced placement high-school courses and university entry. This last assumption is supported by the findings of Hestenes, Wells and Swackhammer (1992) who note that physics misconceptions identified in high-school students consistently manifest among students at higher educational levels.

I should note here that most of the research on which I have been able to draw comes from the US and UK, with some additional work from Israel and Australia. This, of course, serves to qualify the results of my survey, at least to some extent.

8.5 Findings

There is documented evidence of misconceptions or related cognitive difficulties for twenty-seven of the seventy-three units of the *Body of Knowledge*. Of these, thirteen problems match units identified in the *Body of Knowledge* as 'core', or containing 'core' sub-units (Table 8.1). Table 8.2 shows the disciplines in which misconceptions studies originated and the *Body of Knowledge* units to which they are relevant.

The disciplines that produced the greatest number of matches between observed misconceptions and components of the *Body of Knowledge* were math and statistics (nine), computer science (eight), geography (seven), engineering (five) and physics (three). Other disciplines producing relevant results were astronomy, chemistry, education, business, economics, general education and medical informatics.

The five units in which the greatest number of problems have been identified are AM – Analytical methods (nine), DA – Design aspects (eight), DM – Data modeling (four), GD – Geospatial data (four) and GC – Geocomputation (three) (Table 8.2). Other units show one or two problems each, with the exception of OI – Organizational & institutional Aspects, which has been spared misconceptions.

If the data are grouped by STEM discipline in which the study was conducted, or by section in the *Body of Knowledge*, (1) math, statistics, and geometry; and (2), and then information science concepts emerge as the two areas of greatest difficulty for students. Sixteen misconceptions studies focusing on mathematical and geometric concepts match concepts from Analytical Methods, Geospatial Data and Geocomputation. Twelve misconceptions studies focusing on computer science concepts show up in the knowledge areas of DA – Design aspects and DM – Data modeling. As noted previously, the majority of these studies were drawn from math and statistics, computer science, geography and engineering. There are some intriguing outliers in the pattern of misconceptions documented in the STEM literature – for example those reported around ethical behavior by Chou, Chan and Wu (2007), which match some of the 'ethics' concepts. However, at present it is most productive to focus on the fundamental building blocks of the core skill-set of GIScience.

Table 8.1 The units of *The Body of Knowledge* for which a keyword in either the unit or the sub-unit matches a misconception or related cognitive difficulty which has been identified by STEM education researchers. Superscript numbers match authors from column 'Citations' to disciplines from column 'Field' where multiple studies exist

Unit	Source	Core	Field
AM2 Query operations and query languages	Chin and Tall 2001[1], Dorier 2001[2], Kearns 1997[3], Mitrovic 1998[4]	Y	Math[1 & 2], Computer Science[3 & 4]
AM3 Geometric measures	Engelhardt 2004		Physics and Engineering
AM7 Spatial statistics	Cohen *et al.*, 1996[1], Callert 2004[2], Moore 1997[3]		Math and Statistics[1, 2 & 3]
AM11 Network analysis	Engelhardt 2004[1], Hazzan and Hadar 2005[2], Sanger 1996[3]		Physics and Engineering[1], Computer Science[2], Chemistry[3]
CF2-1 Perception and cognition of geographic phenomena	Montello, Lovelace, Golledge and Self 1999		Geography
CF6 Imperfections in geographic information	Garfield 1995, 1988		Math and Statistics
CV2 Data considerations	Downs 1991	Y	Geography
CV3 Principles of map design	Monmonier 1991	Y	Geography
CV6 Map use and evaluation	Monmonier 1991	Y	Geography
DA1 The scope of GIS&T system design	Weber 2001		Math and Statistics
DA4 Database design	Batra 1990[1], Shanks 1997[2], Greca and Moreira 2001[3]	Y	Business[1], Computer Science[2], Engineering and Math[3]
DA5 Analysis design	Ben-Ari and Yeshno 2006		Science Education
DA6 Application design	Jadud 2006		Computer Science
DA7 System implementation	Kaplan 2001[1], Clarke, Thomas, Adams 2005[2]		Medical Informatics[1], Computer Science[2]
DM1 Basic storage and retrieval structures	Fessakis, Dimitracopoulou, Komis 2005		Computer Science
DM2 Database management systems	Fessakis, Dimitracopoulou, Komis 2005	Y	Computer Science
DM4 Vector and object data models	Dehaene 2006[1], Engelhardt[2] (2004)	Y	Math[1], Physics and Engineering[2]
DN1 Representation transformation	Downs 1991[1], Scherr 2001[2]	Y	Geography[1], Physics[2]
GC2 Computational aspects and neurocomputing	Kozma 1997		Chemistry
GC4 Heuristics	Garfield 1995		Math and Statistics
GC8 Uncertainty	Deardorff 2001		Engineering
GD1 Earth geometry	Trumper 2000[1], Galili and Lehavi 2003[2]	Y	Astronomy[1], Physics[2]
GD3 Georeferencing systems	Downs 1991	Y	Geography
GD5 Map projections	Downs 1991[1], Anderson 2001[2]	Y	Geography[1], Education[2]
GD11 Satellite and shipboard remote sensing	Bardar *et al.*, 2005	Y	Astronomy
GS6 Ethical aspects	Chou, Chan, and Wu 2007	Y	General Education
GS7 Critical GIS	Brickhouse 1999		Astronomy

Table 8.2 The units with the greatest number of misconceptions already identified (left), and the disciplines in which the studies identifying those difficulties were conducted (right)

Unit	Number of studies pertaining to each unit	Discipline	Number of studies generated from selected disciplines
AM (Analytical Methods)	9	Math and Statistics	9
DA (Design Aspects)	8	Computer Science	8
DM (Data Modeling)	4	Geography	7
GD (Geospatial Data)	4	Engineering	5
GC (Geocomputation)	3	Physics	3
GS (GIS&T and Society)	3	Astronomy	3

8.6 Examples of math, statistics and geometry misconceptions

The most frequently reported misconceptions centered on basic mathematical, statistical and geometric reasoning critical to understanding fundamental GIScience operations. Thus, Cohen, reporting on a study of the effectiveness of instructional software for teaching students statistics, noted that students' misconceptions included interpreting the y-axis on a histogram as if it were a y-axis on a scatter plot. He says 'One pattern of errors appears to stem from students' attaching inappropriate importance to certain features of graphical displays. This may result from a confusion of the display of a function and the function itself, or from a tendency to confuse displays of probability distributions with displays of data.' (Cohen *et al.*, 1996, 48). Similarly Deardorf (2001, 105–106), studying students' understanding of basic statistical principles, observed: 'Students often make arbitrary judgments about the agreement between results and fail to consider the uncertainty estimates when making these comparisons... tend to avoid specific statements that quantify the uncertainty of a measurement, even when they are explicitly instructed to include an uncertainty estimate... have difficulty identifying the primary source of error in an experiment... tend to overstate precision (too many significant figures) of calculated values, and understate the precision of directly measured values.'

Downs and Liben (1991, 315, 322) noted that undergraduate students significantly underperformed on both projection and coordinate transformation tasks, remarking: 'The finding that subjects performed far from ceiling even on the selection task [as opposed to a drawing task] is compelling.' With regard to coordinates 'Although the present data do not allow us to assert confidently that some adults lack the competence to understand coordinate axes... they do allow us to state that at the very least, a significant proportion of the adult population has difficulty calling upon these competencies when they are relevant.'

8.7 Examples of information sciences misconceptions

The next-ranked area of misconceptions is in information sciences. Mitrovic (1998, 1), when studying the difficulties encountered by students studying SQL, notes 'Students experience many problems.... Some errors come from the burden of having to memorize database schemas; incorrect solutions may contain incorrect table or attribute names. Other errors come from misconceptions in the student's understanding of the elements of SQL and the relational data model in general. Some of the concepts students find particularly difficult to grasp are grouping and restricting grouping. Join conditions and the difference between aggregate and scalar functions are another two common sources of confusion.' Fessakis *et al.* (2005) note that the most fundamental principles of relational database design, identified as core concept DM2-2, elude many students: 'Most students designing digital relational databases using ER [entity relationship] in the conceptual level face difficulties with the relationship concept understanding and representation ... we found that most students either ignore relationships representation or treat them "syntactically"', (Fessakis, Dimitracopoulou and Komis, 2005, 192).

8.8 Discussion and conclusions

The number of matches between documented STEM education misconceptions and concepts identified as essential to GIScience education indicates there are some significant problems confronting GIScience educators. The existing research from other fields indicates that there are some consistent patterns in the way misconceptions divert teachers' efforts to teach their students, summarized by Halloun and Hestenes (1985a, 6) thus:

1. Pretest scores are *consistent* across different student populations
2. Mechanics and mathematics pretests assess *independent* components of a student's initial knowledge state
3. The two pretests have higher *predictive validity* for student course performance than all other documented variables combined.

If this pattern holds in GIScience, as it does across other areas of STEM education we can make the following predictions:

1. Problems consistent with those seen in students from other STEM disciplines will occur among GIScience students
2. GIScience students will experience difficulties with misconceptions in independent, unrelated and diverse topic areas
3. GIScience students' performance on misconceptions tests will be a stronger predictor of overall success than other variables.

Students with very different backgrounds frequently perform to a similar standard when tested for misconceptions. Consequently, although GIScience is a multi-disciplinary field drawing students from a wide variety of backgrounds, it is probable that similar misconceptions will emerge as problems regardless of these students' prior training.

Accurately identifying misconceptions is the first step toward developing successful strategies to correct them (Modell *et al.*, 2004; Modell, Michael and Wenderoth, 2005). Yet the problem is persistent, and complex. Even those fields such as physics, in which considerable progress has been made, still confront challenges. Thus the Force Concept Inventory (FCI), based on an analysis of a single major set of problems centered on the fundamental concept of force, is now widely used in physics and engineering education, and is credited with significantly improving undergraduate performance. But despite the success of the FCI, problems related to misconceptions persist in physics education, leading scholars such as Dancy and Beichner (2002) to argue for continuing research into both the problem and effective solutions. Active learning seems to be one key to correcting the problem. Traditional experiments in a laboratory setting or virtual experiments may give students the opportunity to build and test models of adequate complexity so that they gain a correct understanding of scientific phenomena (Yang, Greenbow and Andre 2004; Michael *et al.*, 2002; Michael, 2007). Even paper and pencil exercises that help students sort and order concepts may overcome misconceptions (Machin, Varleys and Loxley, 2004).

GIScience educators are at the beginning of a process that physics educators have been engaged in for over two decades. The list of GIScience misconceptions identified here is certainly not exhaustive, and misconceptions are by no means the only challenge facing GIScience educators. However, all available evidence points to misconceptions as a significant, enduring, and as yet unstudied problem facing GIScience students and teachers. Whether there are also threshold concepts and troublesome knowledge remains to be determined. If we take the lessons from other STEM disciplines that have successfully addressed misconceptions and apply them to GIScience education, our next logical step is to develop a concept inventory of topic areas at high risk for 'mind bugs.' The *Body of Knowledge* clearly outlines a set of learning outcomes and, like many of the other curricular materials available to GIScience educators, it provides a rubric for curriculum design. However none of these resources codifies the concepts necessary for achieving the outcomes described. To teach successfully to the largest number of undergraduates we must identify the essential underlying concepts, test students studying them for misconceptions, and develop curricula explicitly designed to overcome the difficulties that stand in the way of comprehension. Three successful efforts to approach this problem that are already underway in other disciplines, the Force Concept Inventory, the Astronomy Concept Inventory and the Bioliteracy Project, indicate that this is a challenging task, requiring a collective effort by a wide array of educators over a significant period of time. In these cases, where the effort is already advanced, researchers have used a reiterative process of instructor workshops, instrument design, objective testing of students, student talk-throughs, structured observation, and re-testing. The work is time-consuming, and can require significant investment of resources in developing and administering survey instruments and interviews, and then analysing and interpreting results. GIScience education researchers must incorporate the approaches successfully used in other STEM fields, and develop an understanding of student misconceptions in our own field. Only in the light of structured research into undergraduate misconceptions in GIScience, will it be possible to develop diagnostic tools to identify misconceptions, and to build improved curricula and teaching and learning materials that effectively address the problem.

Acknowledgment

This work was supported by NSF CCLI 083729. *Creating and Implementing a Concept Inventory-Based Diagnostic Tool to Improve Undergraduate GIS Education.*

My thanks to Joe Szakas of the University of Maine, Augusta, Brad Dearden, Cathleen McAnneny and Matt McCourt of the University of Maine, Farmington, Dave Hobbins of the University of Maine Fort Kent, Tora Johnson, of the University of Maine, Machais, and Chunzeng Wang, of the University of Maine, Presque Isle, all of whom have contributed significantly to this project.

References

AAG (Association of American Geographers) (2005) *Spatial Skills Test* Washington, DC: Association of American Geographers. http://genip.tamu.edu/spatial_skills_test.pdf (accessed August 8, 2011).

Anderson, K. C., and G. Leinhardt (2002) Maps as representations: Expert novice comparison of projection understanding. *Cognition and Instruction*, 20(3): 283–321.

Bardar, E. M., E. E. Prather, K. Brecher, and T. F. Slater (2006) The need for a light and spectroscopy concept inventory for assessing innovations in introductory astronomy survey courses. *Astronomy Education Review*, 2(4): 20–27.

Battersby, S. E., R. G. Golledge, and M. J. Marsh (2006) Incidental learning of geospatial concepts across grade levels: map overlay. *Journal of Geography*, 105(4): 139–146.

Batra, D., J. A. Hoffer, and R. P. Bostrom (1990) Comparing representations with relational and EER models. *Management of Computing*, 33(2): 126–139.

Ben-Ari, M., and T. Yeshno (2006) Conceptual models of software artifacts. *Interacting With Computers*, 18(6): 1335–1350.

Bransford, J. D., A. L. Brown, and R. R. Cocking (eds) (2000) *How People Learn: Brain, Mind, Experience and School*. Washington, DC: National Academies Press.

CAER (Collaboration for Astronomy Education Research) (1999) Astronomy Diagnostic Test V 2.0. http://solar.physics.montana.edu/aae/adt/ (accessed November 30, 2010).

Chou, C., P.-S. Chan, and H.-C. Wu (2007) Using a two-tier test to assess students' understanding and alternative conceptions of cyber copyright laws. *British Journal of Educational Technology*, 38(6): 1072–1084.

Clark, S., R. Thomas, and M. Adams (2005) Developing case studies to enhance student learning. *Proceedings of the 7th Australasian Conference on Computing Education*, 42: 101–108.

Cohen, S., G. Smith, R. A. Chechile, G. Burns, and F. Tsai (1996) Identifying impediments to learning probability and statistics from an assessment of instructional software. *Journal of Educational and Behavioral Statistics*, 21(1): 35–54.

Dancy, M. H., and R. J. Beichner (2002) But are they learning? Getting started in classroom evaluation. *Cell Biology Education*, 1(3): 87–94.

Deardorf, D. L. (2001) *Introductory Physics Students' Treatment of Measurement of Uncertainty*. PhD diss. North Carolina State University, Raleigh.

Deerwester, S., S. T. Dumais, G. W. Furnas, T. K. Landauer, and R. Harshman (1990) Indexing by latent semantic analysis. *Journal of the American Society for Information Science*, 41(6): 391–407.

Dehaene, S., V. Izard, P. Pica, and E. Spelke (2006) Core knowledge of geometry in an Amazonian indigene group. *Science*, 311(5759): 381–384.

DiBiase, D., M. DeMers, A. Johnson, K. K. Kemp, A. T. Luck, B. Plewe, and E. Wentz (eds) (2006) *The Geographic Information Science and Technology Body of Knowledge*. Washington, DC: Association of American geographers and University Consortium for geographic Information Science.

Dorier, J.-L., and A. Sierpinska (2001) Research into the teaching and learning of algebra. IN: D. Holton (ed), *The Teaching and Learning of Mathematics at University Level: An ICMI Study*. Dordrecht: Kluwer Academic, pages 255–273.

Downs, R. M., and L. S. Liben (1991) The development of expertise in geography: A cognitive-developmental approach to geographic education. *Annals of the Association of American Geographers*, 81(2): 304–327.

Engelhardt, P. V., and R. J. Beichner (2004) Students' understanding of direct current resistive electrical circuits. *American Journal of Physics*, 72(1): 98–115.

Fessakis, G., A. Dimitracopoulou, and V. Komis (2005) Improving database design teaching in secondary education: action research implementation for documentation of didactic requirements and strategies. *Computers in Human Behavior*, 21(2): 151–194.

Francek, M. A., D. M. Burton, R. H. Aron, and W. J. Bisard (1996) The persistence of selected geographic misperceptions: A survey of junior high through undergraduate college students. *Journal of Geography*, 92(6): 247–253.

Galili, I., and Y. Lehavi (2003) The importance of weightlessness and tides in teaching gravitation. *The American Journal of Physics*, 71(11): 1127–1135.

Garfield, J., and A. Algren (1988) Difficulties in learning basic concepts in probability and statistics: Implications for research. *Journal for Research in Mathematics Education* 19(1): 44–63.

Garfield, J. (1995) How students learn statistics. *International Statistical Review/Revue Internationale de Statistique*, 63(1): 25–34.

Garvin-Doxas, K., and M. Klymkowsky (2005) Understanding randomness and its impact on student learning: Lessons learned from building the biology concept inventory (BCI). *Cell Biology Education – Life Sciences Education*, 7: 233–277.

Gaddis, B. (2001) *Conceptual Change in an Organic Chemistry Laboratory: A Comparison of Computer Simulations and Traditional Laboratory Experiments*. PhD diss., University of Colorado at Denver.

Gautier, C., K. Deutsch, and S. Rebich (2006) Misconceptions about the greenhouse effect. *Journal of Geoscience Education*, 54(3): 386–395.

Gersmehl, P. J., and C. A. Gersmehl (2007) Spatial thinking by young children: Neurologic evidence for early development and 'educability'. *The Journal of Geography*, 106(5): 181–191.

Golledge, R. (2002) The nature of geographic knowledge. *Annals of the Association of American Geographers*, 92(1): 1–14.

Goodchild, M. F., and K. K. Kemp (eds) (1992) *NCGIA Core Curriculum in GIS*. Santa Barbara, CA: National Center for Geographic Information and Analysis, University of California, Santa Barbara.

Greca, I. M., and M. A. Moriera (1997) The kinds of mental representations – models, propositions and images – used by college physics students regarding the concept of field. *International Journal of Science Education*, 19(6): 711–724.

Greca, I. M., and M. A. Moreira (2001) Mental, physical, and mathematical models in the teaching and learning of physics. *Science Education*, 86(1): 106–121.

Halloun, I., and D. Hestenes (1985a) The initial knowledge state of college physics students, *American Journal of Physics*, 53(11): 1043–1048.

Halloun, I., and D. Hestenes (1985b) Common sense concepts about motion. *American Journal of Physics*, 53(11): 1056–1065.

Hammer, D. (2000) Student resources for learning introductory physics. *Physics Education Research, American Journal of Physics Supplement*, 68(7): 52–59.

Hazzan, O., and I. Hadar (2005) Reducing abstraction when learning graph theory. *Journal of Computers in Mathematics and Science Teaching*, 24(3): 255–272.

Hestenes, D., M. Wells, and G. Swackhammer (1992) Force concept inventory. *The Physics Teacher*, 30(3): 141–158.

Hufnagel, B. (2002) Development of the astronomy diagnostic test. *Astronomy Education Review*, 1(1): 47–51.

Kankaanrinta, I. K. (2006) Teaching and learning geographical information systems effectively – reflections in teachers' pedagogical diaries, IN: T. Johansson (ed), *Geographical Information Systems Applications for Schools – GISAS*. Helsinki: Yliopiston Maantieteen Laitoksen Julkaisuja Publicationes Instituti Geographici Universitatis Helsingiensis a 141, pages 31–38.

Kaplan, B. (2001) Evaluating informatics applications – clinical decision support systems literature review. *International Journal of Medical Informatics*, 64(1): 15–37.

Kearns, R., S. Shead, and A. Fekete (1997) A teaching system for SQL. *Association for Computing Machinery International Conference Proceeding Series; Vol. 2. Proceedings of the 2nd Australasian Conference on Computer Science Education*, Melbourne, Australia, 2–4 July, pages 224–231.

Kemp, K. K., and F. M. Goodchild (1992) Evaluating a major innovation in higher education: the NCGIA Core Curriculum in GIS. *Journal of Geography in Higher Education*, 16(1): 21–35.

Kozma, R. B., and J. Russell (1997) Multimedia and understanding: Expert and novice responses to representations of chemical phenomena. *Journal of Research in Science Teaching*, 34(9): 949–968.

Landauer, T. K., and S. T. Dumais (1997) A solution to Plato's problem: The Latent Semantic Analysis theory of acquisition, induction and representation of Knowledge. *Psychological Review*, 104(2): 211–240.

Libarkin, J. C., and S. W. Anderson (2007) Development of the geoscience concept inventory. IN: D. Deeds and B. Callen. *Proceedings of the National STEM Assessment Conference*, Washington, DC., National Science Foundation, October 19–21, 2006, pages 148–158.

Libarkin, J. C., S. W. Anderson, J. Dahl, M. Beilfuss, W. Boon, and J. Kurdziel (2005) College students' ideas about geological time, Earth's interior, and Earth's crust. *Journal of Geoscience Education*, 53(1): 17–26.

Lo, C. P., and A. K. W. Yeung (2007) *Concepts and Techniques in Geographic Information Systems*, 2nd ed. Saddle River, NJ: Prentice-Hall.

Machin J., J. Varleys, and P. Loxley (2004) Exploring the use of concept chains to structure teacher trainees' understanding of science. *International Journal of Science Education*, 26(12): 1445–1475.

Marble, D. (1997) Rebuilding the top of the pyramid: Structuring GIS education to effectively support GIS development and geographic research. GIS in Higher Education Conference 1997, Chantilly: Virginia, October 30 to November 2. http://www.ncgia.ucsb.edu/conf/gishe97/program_files/papers/marble/marble.html (accessed November 30, 2010).

Marais, P., and F. Jordaan (2000) Are we taking symbolic language for granted? *Journal of Chemical Education*, 77(10): 1355–1357.

Meyer, J. H. F., and R. Land (2005) Threshold concepts and troublesome knowledge (2): Epistemological considerations, and a conceptual framework for teaching and learning. *Higher Education*, 49: 373–388.

Michael, J. A., D. Richardson, A. Rovick, H. I. Modell, W. Cliff, B. Horowitz, M. Hudson, D. Silverton, S. Whitscarver, and S. Williams (1999) Undergraduate students' misconceptions about respiratory physiology. *Advances in Physiology Education*, 22(1): S127–S135.

Michael J. A., M. P. Wenderoth, H. I. Modell, W. Cliff, B. Horowitz, M. Hudson, D. Silverton, S. Whitscarver, and S. Williams (2002) Undergraduates' understanding of cardiovascular phenomena. *Advances in Physiology Education*, 26(2): 72–84.

Mitrovic, A. (1998) A knowledge-based teaching system for SQL. *Proceedings of EDMEDIA/ED-TELECOM '98*, Freiburg 1998, pages 1027–1032.

Modell, H. I., J. A. Michael, T. Adamson, and B. Horowitz (2004) Enhancing active learning in the student laboratory. *Advances in Physiology Education*, 28(3): 107–111.

Modell, H. I., J. A. Michael, and M. P. Wenderoth (2005) Helping the learner to learn: The role of uncovering misconceptions. *American Biology Teacher*, 67(1): 20–26.

Montello, D. R., K. L. Lovelace, R. G. Golledge, and C. M. Self (1999) Sex-related differences and similarities in geographic and environmental spatial abilities. *Annals of the Association of American Geographers*, 89(3): 515–534.

Monmonier, M. (1991) *How to Lie With Maps*. Chicago: University of Chicago Press.

Moore, D. S. (1997) Bayes for beginners? Some pedagogical questions, IN: S. Panchapakesan and N. Balakrishnan (eds), *Advances in Statistical Decision Theory*. Boston: Birkhauser, pages 3–17.

Perkins, D. (1999) The many faces of constructivism. *Educational Leadership*, 57(3): 6–11.

Richardson, J. (2004) Concept inventories: Tools for uncovering STEM students' misconceptions, IN: S. Cunningham and Y. S. George (eds), *Invention and Impact: Building Excellence in Undergraduate*

Science, Technology, Engineering and Mathematics (STEM) Education. Washington, DC: American Academy of Sciences and the National Science Foundation, pages 19–25.

Sanger, M. J., and T. J. Greenbowe (1997) Common student misconceptions in electrochemistry: galvanic, electrolytic and concentration cells. *Journal of Research in Science Teaching*, 34(4): 377–398.

Schneps, M. H., and P. M. Sadler (1988) *A Private Universe.* Santa Monica, CA: Pyramid Films.

Scherr, R. E. (2001) *An Investigation of Student Understandings of Special Relativity.* PhD diss., University of Washington, Seattle, WA.

Schonborn, K. J., T. R. Anderson, and D. J. Grayson (2002) Students' difficulties with the interpretation of a textbook diagram of immunoglobulin G (IgG) *Biochemistry and Molecular Biology Education*, 30(2): 93–97.

Toulmin, S. (1972) *Human Understanding.* Princeton: Princeton University Press.

Trumper, R. (2000) University students' conceptions of basic astronomy concepts. *Physics Education*, 35(1): 9–15.

Tyson, T. M., G. J. Venville, A. G. Harrison, and D. F. Treagust (1996) A multidimensional framework for interpreting conceptual change events in the classroom. *Science Education*, 81(4): 387–404.

UCGIS (2003) *Development of Model Undergraduate Curricula for Geographic Information Science & Technology: The Strawman Report Task Force on the Development of Model Undergraduate Curricula.* Washington, DC: UCGIS.

Unwin, D. J. (1997) Unit 159: Curriculum design for GIS. *NCGIS Core Curriculum in Geographic Information Science.* http://www.ncgia.ucsb.edu/giscc/units/u159/u159.html (accessed November 30, 2010).

VanLehn, K. (1990) *Mind Bugs: the Origins of Procedural Misconceptions.* Cambridge, MA: MIT Press.

Weber, K. (2001) Student difficulty in constructing proofs: the need for strategic knowledge. *Educational Studies in Mathematics*, 48(1): 101–119.

Yang, E. M., T. J. Greenbowe, and T. Andre (2004) The effective use of an interactive software program to reduce students' misconceptions about batteries. *Journal of Chemistry Education*, 81(4): 587–595.

Zeilik, M. (2002) The birth of the astronomy diagnostic test: Prototest evolution. *Astronomy Education Review*, 1(2): 46–52.

Zoller, U. (1990) Student's misunderstandings and misconceptions in college freshman chemistry (general and organic), *Research in Science Education*, 27(10): 1053–1065.

9

Active pedagogy leading to deeper learning: fostering metacognition and infusing active learning into the GIS&T classroom

Richard B. Schultz

Department of Geography and Geosciences, Elmhurst College, Elmhurst, Illinois, USA

9.1 Introduction

It is no secret that traditional teaching methods, in which professors lecture and students listen and take copious notes, tends to dominate college and university classrooms. This passive learning approach, although effective in the past, could, depending upon the delivery and communication skill sets of the instructor, tend to lack stimulation and engagement on the part of the student, and does not necessarily include kinesthetic activities that can ultimately lead to deeper thinking. It is, therefore, important to discuss the nature of *active* learning, why students should be thinking about *how* they are learning while processing spatial information, and what empirical research exists on its implementation in the GIS&T classroom. Some of the common obstacles and barriers that give rise to faculty resistance in the implementation of active instructional techniques will also be presented. Finally, how faculty, facilitators and course development specialists can make use of an active and engaging framework in the GIS&T classroom or via the online format will be offered. Although some faculty may feel that simple button pushing is considered to be under the realm of active pedagogy, this chapter will illustrate that, in addition to mouse clicks, active engagement with data and hypothesis testing promotes a richer learning experience for GIS&T students. GIS&T instructors should be encouraged to experiment with active pedagogy for the betterment of their students' learning and advocate that their students think about how it is that they best learn about spatial concepts.

Teaching Geographic Information Science and Technology in Higher Education, First Edition.
Edited by David J. Unwin, Kenneth E. Foote, Nicholas J. Tate and David DiBiase.

9.2 The importance of active learning and metacognition in the higher education classroom

Although a variety of definitions exist for active learning (Meyers and Jones, 1993; Fern *et al.*, 1993; Allen and Tanner, 2005; McKinney, 1995), 'active learning,' in the context of GIS&T, can be thought of as an approach to create a student-centered experience for the purpose of engaging students at a higher level of engagement and to promote active processing of spatial information via efficient cognition.

Analysis of educational literature (Chickering and Gamson, 1987) suggests that students must do more than simply just listen to the words of wisdom from the professor. They must also read closely, communicate effectively, discuss their observations and conclusions intelligently, and be engaged in problem solving activities for a practical purpose. Most importantly, to be actively involved, students must engage in such higher-order thinking tasks as peer instruction, critical analysis, content synthesis and peer evaluation, stemming from Bloom's taxonomy (Bloom and Krathwohl, 1956). Within this context, it is herein proposed that pedagogical techniques promoting active learning in GIS&T be under the realm of any active learning processes involving students in kinesthetic activities and consciously analysing spatial data while thinking about their learning (e.g., metacognition).

The term 'metacognition' refers to an individual's awareness of his or her cognitive processes and strategies. Metacognitive knowledge is considered to be a higher order of thinking because it concerns a more 'executive function of coordinating learning behaviors' (Myers and Paris, 1978, 680). Having a metacognitive awareness means an individual not only has knowledge about themselves and how their learning modality is particularly well suited to them, but it also includes knowledge about the strategies used to solve problems (Fisher, 1998). Any knowledge that is actively used as part of a strategy to meet a cognitive learning goal can be considered metacognitive in nature (Livingston, 1997) and active on the part of the learner. Vos and de Graaff (2009) claim that students learn more and acquire different abilities when they think about their learning than in traditional formats of teaching where that may not be entirely true. GIS&T is particularly well suited to active learning and metacognitive activities since it is, by nature, a visual learning experience that benefits from tactile/kinesthetic practice and reflection.

In addition to fostering metacognitive activities, the use of active learning techniques in the higher education classroom is essential because of the potentially powerful impact upon students' learning. A number of studies, including Ruhl, Hughes and Schloss (1987) have shown that students actually prefer classroom activities promoting active learning in comparison to traditional lectures. Bonwell and Eison (1991, 1) contend that strategies promoting active learning are 'comparable to lectures in promoting the mastery of content, but, in effect, are actually superior to lectures in promoting the development of students' skills in thinking and writing.' Further, cognitive research (Penner, 1984) has shown that a significant number of individuals have learning modalities best accommodated by pedagogical techniques other than lecturing. Therefore, a thoughtful and scholarly approach to employing skillful pedagogical techniques in the GIS&T classroom requires that faculty become knowledgeable about the many ways in which teaching strategies promoting active learning have been successfully used across campus, including in the GIS&T classroom.

Strategies for carrying out active learning vary depending on learning outcomes, but have traditionally tended to focus on critical thinking and problem-solving activities (Dewey,

1916). In order to facilitate active learning, course development experts such as Fink (2003) argue that not only does active pedagogy on the part of the facilitator ultimately lead to deeper learning and build the framework for active learning to occur, but it also encourages students to actually construct their knowledge base (a *constructive* versus a *receptive* process) (Gijselelaers, 1996), and focus their attention on the pragmatic aspects of learning as opposed to learning simply for the sake of learning. One could argue that there are many theoretical and practical reasons for using active learning and metacognition as a construct for learning subject matter, but the best arguments relate to increased learning and engagement on the part of the students ultimately leading to better understanding of the subject matter with the ability to apply what they have learned to solving problems. It seems that spatial awareness leading to spatial perception would be an outcome of employing active learning and metacognition in the GIS&T educational arena.

9.3 How active pedagogy leads to deeper learning

Pedagogically speaking, the teacher in an active learning classroom should utilize pre-structured exercises which challenge the student to increase their own learning skills and encourage the student to think about their learning in their approach to problem solving. This occurs by way of a deliberate process. A simple, straightforward exercise can be followed by a more difficult exercise until the instructor has fulfilled the design criterion for an active learning technique and the student has a working knowledge of the exercise, often referred to as 'scaffolding.' The student walks away with a better understanding of the overall concept and the instructor has fostered the students into fulfilling the predetermined learning outcomes. In essence, a process has been structured to fulfill the course learning outcomes. However, one could argue that pre-structured exercises may coincide with inquiry-based learning and deviate slightly from this structured pathway. When this occurs, the facilitator can emphasize the 'real world' aspects of investigating a particular problem or issue and underscore the practical nature of problem solving. This doesn't decouple the pedagogy, but rather supplements the learning experience for the student.

It has been shown that if students can discuss and reflect on their lessons with other students in the classroom, then the student engagement is of such a level to be helpful in solving problems related to the student's assignment (Bonwell and Eison, 1991). Furthermore, this interactive approach enables the students to find common understanding about the problems which exist in the assigned lesson. Van de Bogart (2009) suggests that active pedagogy on the part of the instructor leads to deeper learning by way of such techniques as scaffolding, 'think-pair-share' activities, and collaborative learning. Active learning takes on a more robust engagement for the teacher as well because even though the teacher may not be fully involved in the student learning activities *per se*, the lesson that is designed using one or more of the active learning techniques is considered first before giving an assignment to the student (Bonwell and Eison, 1991). Another point the teacher needs to consider is how to anticipate a cognitive routine that the student will utilize when attempting to learn a task which the teacher has designed in the lesson. Taking into account that a learner may not always find the exercise easy to perform necessitates the teacher to vary the active learning techniques to optimize the cognitive development of the student. Not all students have the same learning style even though each student must learn the same exercise. Some students are reflective and prefer to work alone and others are active in their learning by asking questions and working in pairs or groups and this is a factor the teacher must consider when helping their students learn a lesson. By its spatial nature, GIS&T lends itself to visual learners and problem solvers, but this does not necessarily indicate that without planning and forethought, students possessing various learning modalities cannot benefit from active learning.

Chet Meyers from the Community College of Aurora, Colorado has developed a program titled 'Overcoming Impediments to Active Learning.' His approach is to develop the basic assumptions of the active learning pedagogy and then outline the corollary principles to guide the practice of active learning. Meyers feels that the purpose of active learning is to get students to interact in ways that cause reflection, thus promoting cognition and deeper learning. Students can more readily synthesize new learning with old when they have worked with the new information and somehow made it their own. Faculty can capitalize on this by openly discussing metacognitive observations by students and leading class discussions on how students are learning content from a spatial perspective.

It has been my experience that if students are allowed to work together in groups, it increases their opportunity to communicate with one another leading to a deeper understanding of how best to complete the lesson or assignment. When a similar lesson is then presented to a student on an individual basis, the ability of the student to complete the lesson is usually enhanced. Thus, collaborative learning has promoted cognition and deeper thinking by way of reflection and encouraging autonomy.

9.4 How active pedagogy can be more widely used in the higher education GIS&T classroom

Active learning methodology has become a preferred way to change the traditional teacher-centered classroom into the more contemporary student-centered approach to learning. Relatively little development, however, on the pedagogy of active learning or the assessment of student retention using active learning methodology in the GIS&T classroom, is available in the scholarly literature. Currently, most published articles on active learning have been descriptive accounts rather than empirical investigations. Many are outdated, either chronologically or methodologically, and a large number of important conceptual issues have never been explored in any sort of detail from an empirical investigation. New qualitative and quantitative research should examine strategies that enhance students' learning from presentations; explore the impact of previously overlooked, yet educationally significant, characteristics of students, such as gender, different learning styles, and stage of intellectual development; and be disseminated in journals widely read by faculty (Bonwell and Eison, 1991). Lacking empirical evidence, how then can active-learning methodology be implemented in the GIS&T classroom to ensure student learning within a well-grounded pedagogical framework?

In a previous section, various definitions from a number of educational researchers were provided for active learning. In all of these definitions there is a sense that an approach to teaching can offer the student an environment where their talents and/or learning modality can be utilized to foster the learning process. Yet, there is one hurdle to overcome. How does one teach the teacher the process of implementing active learning (the pedagogy) such that learning can proceed in such a way so that learning outcomes can be met?

Each GIS&T faculty member should engage in self-reflection, exploring his or her personal willingness to experiment with alternative approaches to instruction. What works for one faculty member may not work for another. In much the same way that students have different learning modalities, faculty can also have different pedagogical tendencies. This may be where barriers to invoking active learning into the GIS&T classroom may rear their ugly head. Rather than discuss those here, however, the barriers will be examined in a subsequent section.

The unquestioned fact that the 'geospatial technology revolution' is accelerating will require accommodation across multiple disciplines in the educational realm. Academic administrators must come to understand that institutions of higher education must not only adapt continually to such change, but promote ground breaking research and educational reform to promote the sustainability and development of such integral educational skill sets revolving around spatial thinking. Educators, too, must shy away from the comfortable familiarity of lectures and passive instruction in order to adapt to the changing technological advances and use active pedagogy and dynamic teaching methodologies to promote spatial conceptual thinking processes and critical problem-solving prowess.

There are many active learning techniques which have been designed to encourage independent learning for the students in the classroom. The notion that the classroom is no longer a teacher-centered classroom but rather a *student-centered* classroom is a significant change in the way knowledge is transferred to the student. Most teachers throw up their hands when they are told to let the students figure out what they are supposed to learn, leaving the teacher wondering what they are supposed to teach. In other words, if a teacher initiates an active learning technique such as scaffolding, or think-pair-share, what is the teacher's role in the classroom while the students are busily involved in their active learning exercise? This question gets even more complex when evaluation and testing of what the student has learned is initiated.

Modifying traditional lectures (Penner, 1984) is one way to incorporate active learning in the classroom. With technological tools that can produce screencasts and online lectures, faculty members can post voice-over lectures online and ask students to listen and take notes prior to attending classroom sessions. Students can be assessed on their learning outside the classroom via a conceptual quiz during the subsequent class meeting. For online or hybrid format courses, this can be accomplished entirely online. Alternatively, when a faculty member allows students to consolidate and reflect on their notes by pausing for reflection and discussion during a lecture, or asking students to pause screencasts, students will learn significantly more information (Ruhl, Hughes and Schloss, 1987).

Two other simple yet effective ways to involve students during a classroom session are to insert brief demonstrations, video clips, or discussion pieces, and ask students to complete short, ungraded writing exercises followed by class discussion. Certain alternatives to the traditional lecture format can also increase student level of engagement, as noted by Bonwell and Eison (1991): 1) the feedback lecture, which consists of two mini lectures separated by a small group study session built around a study guide, and 2) the guided lecture, in which students listen to a twenty- to thirty-minute presentation without taking notes, followed by their writing for five minutes about what they remember and spending the remainder of the class period in small groups clarifying and elaborating the material. It all culminates in a one-minute paper, as made famous by Angelo and Cross (1993) as a means of assessment, feedback for the facilitator, and reflective learning for students. In the GIS&T classroom, learning may be assessed by way of a practical spatial problem using software in which students are asked to solve a simple problem pertaining to a given set of spatial skill sets. With an increasing awareness of spatial skill sets relating to proficiency in using GIS software, this approach meshes well with active learning techniques.

Lastly, class discussions, both inside the classroom as led by the faculty member, and outside the classroom, as facilitated on a discussion board is one of the most common strategies promoting active learning with good reason. If the objectives of a course include promoting long-term retention of information or course content, motivating students toward deeper learning and application thinking, allowing students to apply course content to new real-world settings, and developing students' critical-thinking skills, then discussion is preferable to lecture (Sviniki and McKeachie, 2010). Research has suggested, however, that to achieve these goals faculty must be knowledgeable of the aforementioned alternative techniques and strategies for questioning and discussion (Hyman, 1980), whether that be in the classroom or in an online discussion board, and create a supportive intellectual and emotional environment which encourages students, as well as faculty members, to take calculated risks (Lowman, 1984).

Several additional pedagogical strategies promoting active learning have been shown to favorably influence students' attitudes about learning and their overall achievement. Visual-based instruction, for example, can provide a helpful focal point for other interactive techniques. GIS&T seems to fit well with this type of instruction. Maps, charts, graphs and spatial concepts can all be explained and demonstrated most effectively via a visual approach, perhaps through a whiteboard or face-to-face discussion. In-class writing about GIS&T concepts is another productive way to involve students in doing things and thinking about the skills they are learning. Two popular instructional strategies based on problem-solving models include the case-study method of instruction and guided design (Bonwell and Eison, 1991). Other active-learning pedagogies worthy of mention include collaborative learning or team-based activities, formalized debates, role playing, simulations or board games, and peer teaching. In short, the published literature on alternatives to traditional classroom presentations provides a rich menu of different approaches faculty can readily add to their repertoire of instructional skills, especially within the realm of GIS&T where visuals and issues of global concern tend to be in the forefront.

9.5 What are the barriers to implementing active learning in the GIS&T classroom?

Employing active learning and fostering a GIS&T classroom relying entirely upon student-centered activities is obviously not without risks to both faculty member and student. In order to address these risks and identify barriers to the inception of this type of GIS&T classroom, one must examine both faculty and student perspectives.

Educational tradition has a stronghold on senior faculty members who do not always embrace change and progress with open arms. Simply jumping onto the active learning bandwagon is not a simple transition and often creates anxiety, self-doubt and lack of self-worth to faculty. After all, those lectures have been structured with great care over the past few decades and throwing them out seems counterproductive. However, adapting to educational reform and learning to change along with contemporary learning and technology is of foremost importance in higher education. The leap to fully trusting active learning is risky business for the faculty member, but can provide its rewards if employed properly.

Students have come to expect an active classroom. As noted previously, students don't want to be lectured *at* in the classroom, they want to experience deep learning and walk away with a sense of accomplishment, renewed vitality towards learning, and a sense that they can use what they have learned to succeed in the workplace. They want to sharpen their spatial skill sets for the purpose of being able to market themselves as not only GIS&T users, but geospatial community members who are actively involved and engaged with the community. To do this, they must feel as if they are part of the GIS&T community and network with professionals as well as practice their spatial skills in the setting of a nurturing learning environment. This happens by way of careful planning on the part of the faculty member and knowledge of the best practices in which to create this type of learning community.

Barriers to employing active learning in a student centered GIS&T classroom include, but are not limited to: limited class time in which to promote and establish active learning; a steep increase in class preparation time; the potential obstacles of using active learning in large lecture halls; and a lack of needed materials, equipment, or resources to carry out the activities necessary to attain the learning outcomes. Perhaps the single greatest barrier of all, however, is the notion that faculty members' efforts to employ active learning involve risk. This includes the potential risks that students will not actively participate in class discussions, they will not use 'higher-order thinking' or attempt to learn sufficient course content. Additionally, faculty members will feel a loss of control in the classroom, the thought that students will lack the necessary spatial skills, or faculty members will be criticized for teaching spatial concepts in unorthodox ways. Each obstacle or barrier and type of risk, however, can be successfully overcome through careful, thoughtful planning and clear communication with students.

9.6 Case studies: how various institutions use active learning in the GIS&T classroom

Meitner *et al.* (2005), examine active learning techniques in their presentation to the GIS&T users community by exploring four issues in the GIS&T classroom: (1) step-by-step instructions from a structured exercise versus that of exploration and inquiry-based learning, (2) individual responsibility versus that of collaborative efforts, (3) instructor-led classrooms (and often instructor-centered) versus student-led classrooms (tending towards student centered), and (4) invoking more spatial content versus less content with a tendency to emphasizing skill sets as opposed to underscoring theory. Meitner *et al.* (2005) go on to examine the details of the four issues and conclude by noting that they strongly encourage colleagues to think deeply about pedagogical issues. They claim 'answers lie not in a one size-fits-all solution but rather in careful consideration of paths to take. This is a complex task and one should never discount the power of creativity and insight to aid in the navigation of this rocky terrain' (Meitner *et al.*, 2005, 7). Their preferences based on their experiences in the GIS&T classroom at the University of British Columbia gravitate towards implementing an active learning environment based primarily on inquiry based approaches grounded in collaborative student efforts emphasizing the practicing of spatial skill sets for the purpose of deeper learning and solving practical spatial problems.

Oyana (*personal communication* 2009) utilizes a 'case study teaching model' in his approach to teaching an advanced course in GIS studies. He contends that the teaching model is extremely powerful and is used to teach advanced GIS principles and concepts of geospatial technologies. The model engages and teaches students the 'nuts and bolts' of GIS, thus promoting students' desire to learn and discover new concepts (Boehrer and Linsky, 1990; Williams, 1992; Barnes *et al.*, 1994; Herreid, 1994), and, most importantly, to actively engage students in providing real-world GIS&T solutions. Following this teaching model, the instructor primarily focuses on two key topics: (1) 3-D imaging and visualization, and (2) GIS design, while students focus on the remaining four mini-case studies whereby direct ownership is passed onto them. Oyana groups the class into four learning groups who undertake to research and present on one of the four cases. The learning groups reflect each individual's prior GIS knowledge/skills or potential interests in one of these areas:

1. Enterprise GIS – ERP Learning Group.
2. Geospatial Data Structure, Algorithms, and Data Mining – GDF Learning Group.
3. GIS Programming – G-Code Learning Group.
4. Cognitive Science, Geospatial Ontologies – Concept-GIS Learning Group.

Each learning group is then provided with a summary of learning goals, list of topics to be investigated and expectations. The instructor provides each group with key instructional materials, research articles and references to study. Prior to each presentation, the instructor reviews the material to be presented and offers written feedback and suggestions. Upon receiving a revised presentation, the instructor posts a PDF version of the presentation on the course website two days in advance for the rest of the students to review in preparation for the classroom discussions. After the presentation, students are encouraged to peer review the material and offer suggestions. At the end of the semester, a log report and a twenty-page term paper are required for all the groups. The log report provides students with an opportunity for self-evaluation (i.e., metacognition), a peer evaluation, and also documents the participation, attendance and contribution of each group member. Oyana (*personal communication* 2009) encourages each learning group to approach the above issue from different vantage points, thus constructing a 'jigsaw-type' exercise.

9.7 What conclusions should be drawn and recommendations made?

With the 'geospatial technology revolution' accelerating, as I've mentioned, academic administrators must come to understand that institutions of higher education must not only adapt continually to such change, but promote ground breaking research and educational reform to promote the sustainability and development of such integral educational skill sets revolving around spatial thinking. Progress has its rewards, but not without its sacrifices. Educators, too, must stray away from the comfortable familiarity of lectures and passive instruction in order to adapt to the changing technological advances and use active pedagogy and dynamic teaching methodologies to promote spatial conceptual thinking processes and critical problem-solving prowess.

Because our societal issues are so critical, we must formulate an acceptable and progressive foundation for advancing the enormous applications of geospatial technologies. Recent developments in geospatial interoperability enable users to model seemingly unrelated aspects of problems and visualize new connections and associations. In short, this synergy amplifies the power of spatial thinking. To fully take advantage of this recent geospatial revolution, it is germane to critically examine how educators can use active teaching methodologies to enrich the development of spatial thinking towards educating the next generation of the geospatial workforce. The sustainability of the GIS community rests on the academic community's effort to promote deeper spatial learning in future GIScientists or geospatial scientists.

It comes then as no surprise that active learning and GIS&T go hand in hand. Fluency using the complex tools of GIS&T can only be learned through an active desire to fully experiment and learn by doing. Active learning, problem-based learning (PBL) and inquiry-based learning (IBL) mesh together to create the mechanism by which GIS&T can be practiced to a point of fluency. Johnson and Hobbs (2000) asked, 'how do we determine the best methods and technologies to use when little study has been done on what does, and does not, work?' This section strives to shed some light on what works and also why active learning may, at first glance, appear to be easier, i.e., less work, for the instructor – fewer formal lectures and more 'participation' time by students. Active learning in practice is not easier, but considerably more difficult, especially for the new practitioner or for one who is also learning GIS technology along with active learning methodology. Fundamental to active and project-based learning methodologies, instructors have to be comfortable with the notion that they cannot possibly know all things about GIS and must be willing to learn alongside their students. Again, progress has its rewards and its sacrifices.

Jackson *et al.* (2009) contends that 'geospatial educators should track and contribute to this progress in geospatial research as it unfolds, because it is essential to their future success.' Students' exposure to new geospatial technologies will both prepare them for a wide variety of professional pathways that involve geospatial information and inspire educators to evolve new branches of geospatial research and study.

It is inevitable that certain inertia does indeed exist among academics who use and teach GIS concepts in moving beyond easy-to-use familiar tools to embrace web-based geoprocessing, yet the traditional curricular emphasis will have to make this paradigm shift as young students, indoctrinated into the Web 2.0 environments early in life, become impatient with less powerful and diverse tools and knowledge techniques (Jackson *et al.*, 2009).

Few can have a complete appreciation for the development of their disciplines without fully understanding why the discipline developed as it did. The implication for geospatial academia and the commercial GIS&T community is that we, as a society, have currently attained a point where the objectives of the two pursuits have become divergent, thus defining a necessary academic distinction between the traditional study of geography as a human and cultural discipline and the scientific study of geospatial information technology as a true science. Many progressive thinkers and educators in the geospatial realm believe that it is of crucial importance to not only appreciate, but cultivate, this distinction in order to more precisely and productively define these fields in the academic environment. Jackson *et al.* (2009) pose the question: 'can we find the institutional will – in academia . . . to make changes that enable societies around the world to make the most of these new tools?'

It seems obvious that faculty members should be able to adapt to changes in technology and curricular developments. However, this is not always true for a variety of reasons. To address this issue, an appropriate first step is to select strategies promoting active learning that a faculty member can feel comfortable with. Such low-risk (and often high impact) strategies are typically designed for short-duration exercises and activities which center on subject matter that is neither too abstract nor too controversial, and somewhat familiar to both the faculty member and the students. The use of active learning should become both the subject matter of professional development workshops and the instructional method used to facilitate such programs. The emphasis on pedagogy over technology is a key here for both the success of the developmental process and the means by which content is presented. Technology is to be viewed as a tool to assist with pedagogy and not vice versa.

In retrospect, it appears that previous classroom initiatives and written materials about active learning have all too often been isolated and fragmented. The resulting pedagogical efforts have therefore lacked coherence, and the goal of interactive classrooms has remained unfulfilled. Through the coordinated efforts of individual faculty, faculty developers, academic administrators and educational researchers, however, higher education in the coming decade can make real the promise of active learning a reality in the GIS&T classroom. Because of the visual and spatial nature of GIS&T, it seems an obvious fit that active learning in the classroom can promote better spatial cognition as well as deeper learning.

References

Allen, D., and K. Tanner (2005) Infusing active learning into the large enrollment biology class: Seven strategies, from the simple to complex. *Cell Biology Education*, 4: 262–268.

Angelo, T. A., and K. P. Cross (1993) *Classroom Assessment Techniques: A Handbook for College Teachers*. San Francisco: Jossey-Bass.

Barnes, L. B., C. R. Christensen, and A. J. Hansen (1994) *Teaching and the Case Method: Text, Cases, and Readings*, 3rd ed. Boston, MA: Harvard Business School Press.

Bloom, B. S., and D. R. Krathwohl (1956) *Taxonomy of Educational Objectives, Handbook 1: Cognitive domain*. New York: Longmans.

Boehrer, J., and M. Linsky (1990) Teaching with cases: Learning to question. *New Directions for Teaching and Learning*, 42: 41–57.

Bonwell, C. C., and J. A. Eison (1991) Active learning: Creating excitement in the classroom. ERIC Digests, ERIC Identifier ED340272. http://www.oid.ucla.edu/about/units/tatp/old/lounge/pedagogy/downloads/active-learning-eric.pdf (accessed May 1, 2010).

Chickering, A. W., and Z. F. Gamson (1987) Seven principles for good practice in undergraduate education. *American Association for Higher Education Bulletin*, 39(7): 3–7.

Dewey, J. (1916) *Democracy and Education*: New York: Macmillan. http://www.ilt.columbia.edu/publications/dewey.html (accessed May 1, 2010).

Fern, V., K. Anstrom, and B. Silox (1993) Active learning and the limited English proficient student. *Directions in Language and Education*, 1(2). http://www.sube.com/uploads/95/T5/95T5NfBOBiBZf0cgMBAhTQ/Active_Learning_and_the_LEP_student.pdf (accessed May 7, 2010).

Fink, L. D. (2003) *Creating Significant Learning Experiences: An Integrated Approach to Designing College Courses*. San Francisco, CA: Jossey-Bass.

Fisher, R. (1998) Thinking about thinking: developing metacognition in children. *Early Child Development and Care*, 141(1): 1–15.

Gijselaers, W. H. (1996) Connecting problem-based practices with educational theory, IN: L. Wilkerson, and W. H. Gijselaers (eds), *Bringing Problem-based Learning to Higher Education: Theory and Practice.* San Francisco: Jossey-Bass, pages 13–21.

Herreid, C. F. (1994) Case studies in science: A novel method of science education. *Journal of College Science Teaching,* 23(4): 221–229. http://sciencecases.lib.buffalo.edu/cs/pdfs/Novel_Method.pdf (accessed June 9, 2010).

Hyman, R. T. (1980) *Improving Discussion Leadership.* New York: Columbia University, Teachers College Press.

Jackson, M., D. Schell, and D. R. Fraser Taylor (2009) The evolution of geospatial technology calls for changes in geospatial research, education and government management. *Directions Magazine* (April 6, 2009). http://www.directionsmag.com/articles/the-evolution-of-geospatial-technology-calls-for-changes-in-geospatial-rese/122571 (accessed August 8, 2011).

Johnson, A., and G. Hobbs (2000). GIS access – Active learning pedagogy. Presented at the Second European GIS Education Seminar, Budapest, Hungary. http://www.geo.info.hu/eugises/papers_pdf/JohnsonHobbs.pdf (accessed August 8, 2011).

Livingston, J. A. (1997) Metacognition: An overview. http://www.gse.buffalo.edu/fas/shuell/cep564/Metacog.htm#Top (accessed February 15, 2010).

Lowman, J. (1984) *Mastering the Techniques of Teaching.* San Francisco: Jossey-Bass.

McKinny, K. (1995) Active learning. http://www.cat.ilstu.edu/additional/active.php (accessed May 7, 2010).

Meitner, M. J., J. Gonzalez, R. Gandy, and J. Maedel (2005) Critical thinking, knowledge retention, and strife: Reflections on active-learning techniques. Paper presented at the Esri Education Users Conference, San Diego, California, 23–26 July. http://proceedings.esri.com/library/userconf/educ05/papers/pap1081.pdf (accessed June 10, 2010).

Meyers, C., and T. B. Jones (1993) *Promoting Active Learning: Strategies for the College Classroom.* San Francisco: Jossey-Bass.

Myers, M., and S. G. Paris (1978) Children's metacognitive knowledge about reading. *Journal of Educational Psychology,* 70(5): 680–690.

Oyanna, T. (2009). GEOG 420 Advanced GIS studies syllabus. http://www.geography.siu.edu/people/oyana/Teaching/Geog_404/Advanced_GIS.html (accessed August 8, 2011).

Penner, J. G. (1984) *Why Many College Teachers Cannot Lecture.* Springfield, IL: Charles C. Thomas.

Ruhl, K. L., C. A. Hughes, and P. J. Schloss (1987) Using the pause procedure to enhance lecture recall. *Teacher Education and Special Education,* 10: 14–18.

Sviniki, M., and W. J. McKeachie (2010) *McKeachie's Teaching Tips: Strategies, Research, and Theory for College and University Teachers,* 13th ed. Stamford, CT: Wadsworth Publishing.

Van de Bogart, W. G. (2009) Developing a pedagogy for active learning: Part I, including a brief history of active learning in Thailand. http://www.earthportals.com/Portal_Messenger/ActiveLearning.html (accessed May 2, 2010).

Vos, H., and E. de Graaff (2004) Developing metacognition: A basis for active learning. *European Journal of Engineering Education,* 29(4): 543–548.

Williams, S. M. (1992) Putting case-based instruction into context: Examples from legal and medical education. *Journal of Learning Sciences,* 2(4): 367–427.

10

Where to begin? Getting started teaching GIS&T

Eric West

Department of Geography, Southern Connecticut State University, New Haven, Connecticut, USA

10.1 Overview – the challenges of getting started

This chapter is designed to help faculty teaching GIS&T for the first time. It makes concrete suggestions about how to begin teaching GIS&T, indicates additional resources in the literature to which the reader may want to turn, and provides guidance for some future concerns that first-time teachers of GIS&T are likely to have irrespective of their career stage and reasons for teaching GIS&T. Experience gained in other subjects is useful but doesn't always translate directly into teaching GIS&T. For this reason, I will focus on some of the important considerations in fostering good learning experiences in the GIS&T classroom.

Ten or twenty years ago I might have written a very different chapter about getting started. In those days, the lack of computer resources in many departments and the lack of instructional materials were major hurdles for first-time instructors. Now the issues have changed. Few first-time instructors have to create a computer facility from scratch, and teaching resources are more plentiful, but obstacles remain, including:

- the expense of updating and improving hardware and software;
- obtaining and maintaining adequate IT support for the installation of and upgrades to software, networking and other vital functions;
- the difficulty of learning to use GIS as a tool and how to make GIS work with other software;
- the difficulty of learning other geospatial or supporting technologies such as global positioning systems (GPS);
- the difficulty of finding and managing laboratory assistants.

The consequence is that, although some barriers have been reduced, there is a steep hill for first-time teachers of GIS&T to climb.

Teaching Geographic Information Science and Technology in Higher Education, First Edition.
Edited by David J. Unwin, Kenneth E. Foote, Nicholas J. Tate and David DiBiase.

There is now an impressive variety of resources for first-time instructors – excellent textbooks and laboratory manuals, sourcebooks of data and exercises, and rich data sources available from a wide array of commercial, educational and governmental organizations. There is also a substantial and growing body of literature in books and academic journals, such as the *Journal of Geography in Higher Education, International Journal of GIS* and *Journal of Geography* to provide guidance as well as many list servers that disseminate useful information. This plethora of resources means that new teachers do not have to start from scratch, but knowing how to get the most out of these resources and how to use them to create an effective learning environment can still present challenges. The point of this chapter is to provide some tips about where to start in this process, how to: put into practice some of the ideas presented in this section of the book; how to quickly gain the experience and confidence needed to teach GIS&T effectively; and to innovate and improve your teaching in the future. I see effective teaching as a process of continuous and iterative improvement, and I highlight these issues in the following sections:

- Establishing learning objectives and achieving alignment between objectives, teaching methods and assessments.
- Designing laboratory instruction to enhance learning (and suggesting expository methods for first-time teachers of GIS&T).
- Using other methods for instruction once the teacher has become accustomed to the GIS&T classroom.
- Applying some practical suggestions.

10.2 Learning objectives, assessment and alignment

Setting clear goals for courses and curricula is the most important first step. Although learning objectives are usually implicit in the organization of most popular GIS&T textbooks, care must still be taken to adapt them to particular classes. As Prager emphasizes in his chapter in this book, learning objectives are important, because they point toward how student learning will be assessed as well as the types of activities which are most useful in reaching those particular objectives. Prager and Plewe (2009, S47), relying on Biggs (1999) and the *Commission on the Future of Higher Education* (2006, 23), explain the importance of achieving and reflecting upon results in higher education this way:

> outcomes-based assessment of student learning in conjunction with evaluation of curriculum alignment [which is] the idea that curriculum should be specified as a series of clear objectives with an associated level of understanding . . . is rapidly becoming the preferred form of educational quality control in the United States and has been successfully demonstrated through programmes of normative and summative assessment abroad.

Determining how and to what extent individual instructors are fostering student learning hinges on clearly defining learning outcomes and developing the language and conceptual framework to relate teaching, learning and assessment at the classroom level. Learning outcomes should indicate mandatory requirements for student learning and factor in how students think, understand and mature in their sensitivities and attitudes in addition to what they will be able to do with their educations (Doherty, Riordan and Roth, 2002).

An important tool for first-time teachers of GIS&T should be the recently published book *The Geographic Information Science and Technology Body of Knowledge* (*BoK*, DiBiase *et al.*, 2006). The *BoK* establishes ten knowledge areas within GIS&T, which are further divided into multiple units, each composed of discrete topics. These topics can translate directly into learning objectives (DeMers, 2009). For example, the list below includes examples of learning objectives that I have adopted in GIS&T courses, paired with knowledge area, unit and topic designations from the *BoK*:

- Define a city, an urbanized area, and a metropolitan area; identify the commonalities and differences between these three regions and formal, functional, cultural and physical regions; and, explain what data would be necessary in a GIS to represent them (conceptual foundations (CF) 5-7).
- Diagnose and repair an incorrectly defined spatial reference (datum or map projection) that results in a misalignment between two map layers (geospatial data (GD) 4-1 and 5-3).
- Join two attribute tables with inconsistent formatting (data modeling (DM) 2-2 and data manipulation (DN) 3-1).
- Write out an adjacency matrix, an incidence matrix, a left/right list, and a polygon/arc list for a given graph (DM 4-3).
- Calculate and interpret the nearest neighbor statistic for a small number of points using pencil, graph paper and a calculator and then apply the statistic and interpret it for a much larger data set using GIS (analytical methods (AM) 5-1).
- Identify the visual variables and measurement scale for each type of feature on a map and then evaluate the appropriateness of the visual variables used (cartography and visualization (CV) 3-2).
- Apply the Goodness of Variance Fit Assessment Index using a spreadsheet program to evaluate the strength of various data classifications (CV 2-2).

The reader can easily locate each learning objective within the *BoK*, construct similar or reinforcing learning objectives, and identify knowledge areas and units that remain unaddressed. Prager's chapter in this volume includes several references that suggest ways that learning outcomes can be defined and makes valuable suggestions for aligning learning activities with learning outcomes.

After setting learning goals, it is easier to develop assessments. Prager and Plewe (2009) provide a competency scale, informed by Marble's (1998) competency pyramid, which was a part of the *BoK* (DiBiase *et al.*, 2006) and appeared earlier in the ESRI's *ARC News*. They have adapted their scale for measuring the learning of topics rather than for indicating levels of mastery for individual learners. It includes five ratings on an ordinal scale, including unfamiliar, familiar, competent, expert, and research and development. If it is desired that students assess their own learning outcomes, Smith and Zhou (2005) provide an example of student self assessments, which was published the year before the *BoK*.

In turning to assessment, two important concepts are worth stressing, the value of *authentic assessment* and the difference between *auditive* and *educative* assessment. Authentic assessment is the notion that students be assessed in ways that reflect how a concept, idea, or method is actually used. For example, inauthentic assessment would emphasize whether a student could simply list the properties of a map projection, while authentic assessment would reinforce the student actually knowing and deciding what map projection to use in a particular situation. The terms auditive and educative assessment were coined by Wiggins (1998) to differentiate between assessments which are largely backward-looking and aimed at auditing what students have mastered to a particular point in the course and those which are forward-looking and attempt to assess and provide feedback which will be of the greatest value later in the course – hence, its 'educative' value. Many of us are more familiar with auditive assessments, such as a few midterms and a final, for the purpose of auditing learning and generating final grades. This model for assessment is backward-looking, because it retrospectively evaluates whether students have internalized material that was introduced in the course prior to the assessment. Educative assessment, on the other hand, seeks to assist students in their learning.

To educative assessment, Fink (2003, 83) adds his concept of 'FIDeLity feedback' in that feedback should be: Frequent, Immediate, Discriminating (based on clear criteria and standards), and delivered Lovingly (in positive, respectful ways). The result is a model that recognizes at least two axioms:

1. Appropriately constructed feedback is essential to support the maturation of students into self-directed learners. The inevitable shift from teacher-driven assessment to self-assessment depends on feedback so that students are cognisant of when they are learning correctly.
2. Back-and-forth communication between teacher and student is necessary at the time of feedback.

Feedback plays a crucial role in Fink's (2003) twelve-step model for integrated course design which he argues can accomplish the following six goals of significant learning:

- Foundational Knowledge.
- Application.
- Integration (linking ideas within the course to each other and with fields outside of the direct concern of the course).
- Human Dimensions (vital knowledge about the self and about working with others).
- Caring.
- Learning How to Learn.

Significant learning is noteworthy, because the *BoK* only emphasizes the knowledge base of GIScience. However, we often include, as part of our courses, goals that relate to other important aspects of professional or academic practice, such as making sure that students develop the ability to give effective presentations as well as skills in research, writing and using computers and other forms of technology. Thus, significant learning is valuable, because it helps us to build six types of learning objectives into the learning experiences we create in addition to those contained in the *BoK*. Recently, spatial thinking has become an important learning objective. Readers who wish to consider extending their learning outcomes to include spatial thinking may refer for additional information on the topic to Hespanha, Goodchild and Janelle (2009), National Research Council (2006), Lund and Sinton (2007), Sinton and Bednarz (2007) and Rhind (1992).

Chickering and Gamson (1987) are cited by scholars in GIS&T and geographers in general in texts about pedagogy and assessment for their *Seven Principles for Good Practice in Undergraduate Education* (Unwin, 1997b; Estaville, Brown and Caldwell, 2006). Their good practices especially emphasize interaction between the instructor and students and also include:

- Collaborative learning between students.
- Active learning.
- Timely feedback from teachers.
- Prioritizing the effective use of time.
- Emulating high standards.
- Consideration for various types of ability and styles of learning.

While also distinguishing between feedback and assessment and indicating even earlier than Wiggins (1998) and Fink (2003) that assessment without prompt feedback is of little assistance to the learning process, they also argue for self-assessment, particularly at the outset of a class. Self-assessment at this time promotes students to evaluate the knowledge and abilities they already bring to the table. Chickering and Gamson (1987) reinforce the tenet that awareness of knowledge and gaps in knowledge, derived from classroom performance, is essential to improving performance and learning. Assessments are important instruments in this process. Additionally, my experience is that properly framed feedback can help to redirect students who seem to have performance-based goals (learning for the sake of a high grade) to instead adopt mastery-goals (learning for the sake of self-improvement). The implication here is that there is a two-way relationship not just between assessment and learning but between assessment and students' attitudes.

Boice (2000) points out that the teaching of early-career faculty is all too often guided by a dearth of understanding about how audiences learn. Several initiatives have emerged that promote the linking of teaching, learning and assessment in geography and GIS&T, including a workshop series by the Geography Faculty Development Alliance at the Department of Geography at the University of Colorado, Boulder (http://www.colorado.edu/geography/gfda/gfda.html), the *Spatial Perspectives on Analysis for Curriculum Enhancement* at the Center for Spatially Integrated Social Science at the University of California, Santa Barbara (http://www.csiss.org/SPACE/workshops/), and most recently, the T3G Institute of ESRI (http://edcommunity.esri.com/community/institute2010/index.cfm). These projects incorporate an emphasis on learning in their contributions to quality in pedagogy. One core notion is that the initial establishment of learning goals enables the design of activities and assessments that support them. The *BoK* contains a large number of learning objectives, organized by the ten knowledge areas, and strongly supports the development of course materials and assessment instruments around those goals. Regardless of the learning objectives adopted, first-time teachers of GIS&T should exercise care to keep good records of assessments and be prepared to provide departmental assessment coordinators with at least some evidence that will function as data to speak to the above considerations.

10.3 The relationship between teaching, laboratory exercises and learning outcomes

In no other area of instruction is it as important to match what is done with the intended learning outcomes as it is for laboratory classes (Unwin, 1997a, 1). One of the greatest challenges in getting started in GIS&T teaching is coordinating lectures, readings, discussions and laboratory learning. More than in many subjects, working with GIS software is a key way for students to master both theoretical concepts and practical skills. The chief objectives of laboratory exercises are explained by Brown and Atkins (1988) based on Brock and Meadows (1977) and Phillips (1981):

1. Teaching manual and observational skills relevant to the subject.
2. Improving understanding of methods of scientific enquiry.
3. Developing problem-solving skills.
4. Nurturing professional attitudes (which should include social skills, collegiality and the ability to communicate and work in groups).

For laboratory exercises in particular, they list eighteen discrete learning objectives, which support the alignment of time in class with the establishment of learning objectives. I have reorganized them in Table 10.1 below.

Fink's (2003) conceptualization of significant learning has remarkable connections to Brown and Atkins' (1988) learning objectives. Foundational knowledge and integration are partially represented, while the other four categories of significant learning are emphasized more. Coppola (2002, 235) comments that students who participate in laboratory sections should theoretically exhibit better rates of retention, better skills at observing or working with data, and heightened abilities to apply their learning and design experiments.

In addition to determining learning objectives and aligning assignments with them, the complexity of teaching GIS&T for the first time is also an outgrowth of using GIS software, which requires students to learn various functions for managing spatial databases and analysing spatial data, for generating and interpreting feedback regarding the steps undertaken and tasks related to the system itself, and for display and output (National Research Council, 2006). The need to teach the additional skills of using computer operating systems, navigating site-specific computer networks, personal habits for organizing and backing up data, and other skills that contribute to success in GIS&T further encumbers first-time teachers of GIS&T. Combined with the expense and logistical considerations for providing the GIS hardware and software to begin with, the cost and effort to create teaching materials may seem overwhelming. First-time teachers of GIS&T are fortunate because of the quantity of teaching materials (textbooks, laboratory books, exercises and spatial data sets) that are already available to them. I highly recommend that they take advantage of these resources in order to lighten the burden of teaching those first few classes. Doing so implies that first-time teachers of GIS&T will employ the controlled exercise laboratory format.

The controlled exercise style, which involves vigorously controlled exercises that reproduce known results, usually involves using GIS to teach what Unwin (1997a) calls GIS theory. Unwin (1997a) indicates that this is perhaps the most prevalent approach to GIS laboratories and that its strength lies in connecting theory to the use of a GIS itself using actual data. This style of laboratory instruction has the advantage of creating a ready-made course, which only needs a syllabus and a classroom of students to get off the ground, thus sparing the instructor from having to assemble data and write laboratory exercises. Coppola's (2002) model of expository instruction matches the use of extant laboratory materials, which encourage the student to follow directions and produce predetermined results. This model helps to develop skills in manipulation and is broadly usable by any student, regardless of variations in the particular qualities of students. Additionally, the controlled exercise style can be implemented easily on a wide scale.

Table 10.1 Components of the overarching objectives of laboratory exercises adapted from Brown and Atkins (1988), Phillips (1981), and Meadows (1977)

CHIEF GOAL:	Teaching Manual and Observational Skills Relevant to the Subject		Improving Understanding of Methods of Scientific Enquiry		Nurturing Professional Attitudes		Developing Problem-Solving Skills
SUB-GOALS:	Working with Equipment	Working with Phenomena Under Study	Communication of Information	Membership in a community	Care and Maturity	Intellectual Autonomy	Critical Thinking
Learning Objective	To familiarize students with instruments and apparatus	To develop observational skills	To illustrate and amplify lecture material	To develop the ability to work effectively as a member of a team	To develop ability to follow instructions	To develop the capacity for self-directed learning	To stimulate thought through experimental interpretation
Learning Objective	To familiarize students with the design and construction of experimental equipment	To develop skills in gathering and interpreting data	To teach theoretical material not presented elsewhere	To develop skill in communicating experimental results	To develop personal responsibility and reliability for experimentation	To encourage independent thinking	To develop the student's skill in problem-solving with a wide number of variables and many possible solutions
Learning Objective		To develop manipulative skills		To develop the ability to write coherent and well-argued reports	To encourage enterprise, initiative, resourcefulness		
Learning Objective					To develop a concern for accuracy		

Controlled exercises, through the opportunities for interaction that they afford, present many opportunities for the instructor to provide feedback to students and have the potential to reinforce many of the principles of good practice. Ultimately, Unwin (1997a) reminds us, 'the essential idea behind a laboratory class is that, rather than being taught, students teach themselves and each other.' In addition to other types of feedback, feedback for the instructor is also vital, and it is incumbent on the instructor to create opportunities for students to generate feedback during the GIS&T course. McKeachie (2002) offers suggestions for improving feedback from students and for obtaining feedback from other sources as part of career growth.

Furthermore, the demonstration of procedures and inculcation of particular habits in students concomitant with this exercise style are important, because the correct application of these techniques is understood as crucial to arrive at correct answers. Moreover, students' interest in laboratory work is directly tied to their perceptions of it as relevant and meaningful (Brown and Atkins, 1988). Stating learning objectives at the beginning of an assignment reinforces relevance and helps the student to take ownership over the learning process. The student's internal establishment of learning objectives related to the mastery of the subject rather than to performance relative to other students is also an underlying factor in motivation (see Hespanha, Goodchild, and Janelle, 2009; Hofer, 2002).

Once the first-time teacher of GIS&T has gained some experience teaching the subject, then consideration of exercise formats other than the controlled exercise style is appropriate. Unwin (1997a) does indicate, presumably with this laboratory format in mind, that the student's approach to and thought process in solving laboratory problems may be difficult to ascertain. There is the potential for students to perform steps uncritically or to focus disproportionately on the tools of GIS&T rather than the understandings that it creates. My experience has been that writing assignments that promote critique of the results of laboratory work and the connection of laboratory work with reading can help to direct students' attention toward more important concepts. Finally, as the teacher gains experience, it may be desirable to involve students in data collection, which is often weakly emphasized by the controlled exercise format.

10.4 Keeping learning alive

As Schultz discusses in his chapter in this book, active learning processes in GIS&T incorporate three components: kinesthetic activities, the conscious analysis of spatial data, and reflection on learning. Schultz recommends that faculty acquaint themselves with a variety of means by which active learning is promoted across disciplines as they learn to make their own classes more engaging and student-centered. He makes several suggestions for increasing active pedagogy in GIS&T classes, starting simply with adjusting traditional lectures. Pre-structured exercises that require students to develop their learning skills are pedagogically appropriate. Bearing this in mind, I draw on Coppola (2002) and Domin (1999) to present two additional models for promoting active learning in the GIS&T classroom: inquiry instruction and discovery instruction. However, fieldwork deserves some special consideration here. It clearly includes all three components of the active learning process and has great potential, in combination with geospatial technology, to promote significant learning and achieve important learning outcomes.

Fieldwork and GPS can be used either in combination with lectures and controlled exercises or in replacement of them. For example, I implemented a class project related to campus-safety in which students measured the intensity of light at night for particular areas on campus around dormitories and commuter parking lots using instruments borrowed from the physics department. The class then used spatial interpolation to develop trend surfaces and draw conclusions about where lighting conditions might have been unsafe. I used several lectures and controlled exercises to develop skills that the students needed in order to complete the project, but they were not the focus of the class. Refer to Holmes (2007) for more ideas for projects that combine GIS and GPS. Dobson (2001) provides rich anecdotal evidence and practical advice for taking GIS and GPS into rugged and remote environments.

Carlson (2007) treats the combination of GIS and GPS as a substitute for the lecture and lab model. He elaborates on the architecture of his students' fieldwork to support accuracy-assessment of zoning, streets and other features as well as the stages before and after fieldwork. He presents evidence that the combination of GIS with GPS promoted student engagement and learning. Favier and van der Schee (2009) test four different student research projects that leverage the field-based collection of quantitative data using GPS with GIS skills. They draw conclusions about the essential features of successful projects for student learning and project viability. They also make observations about what students enjoy and dislike, safety, and which types of geographic thinking occur in which phases of the projects.

Returning to classroom methods, inquiry instruction is a model that incorporates students being given alternative ways to solve a problem. Students are allowed to perform those various methods and to weigh them against one another. The instructor can end the exercise by asking questions or by having the students do the same. This model for instruction gives students a feeling of choice and accountability for their results and requires them to confront the question of whether others can independently verify their results. A disadvantage of this model for instruction is that students can transmit solutions to problems from earlier classes to later ones.

GIS does indeed offer alternative ways to answer the same problem. Vector and raster data structures can represent the same phenomena differently and allow for different manipulations to arrive at similar products. Selections and spatial analysis operations can be used in differing combinations to answer the same question, especially when working with large datasets, yet processing requirements and results may vary. Additionally, GIS oftentimes provides several algorithms to accomplish similar tasks, such as line smoothing, interpolation, or resampling raster data. Questions that require the joining or relating of attribute tables may often be solved in more than one way. Sometimes solutions may be created either analytically or cartographically. This list of possibilities is certainly not exhaustive. Inquiry instruction definitely can be used to teach GIS&T and is an alternative to cookbook procedures.

A second model for laboratory instruction is called discovery instruction, to which Coppola (2002) also refers as guided inquiry. This model stages a pre-planned experiment as the spontaneous outcome of a structured discussion between an instructor and students that builds on understandings and observations that arose earlier in the class. To start the discussion before the laboratory exercise, the instructor poses a question. The students 'make predictions, formulate hypotheses, and then design experiments' (Coppola, 2002, 238). The instructor regulates the discussion and makes connections with information introduced in prior classes. He assesses the approaches that the students come up with and helps to develop their interest while steering them toward the planned activity. Students may work alone or in groups, but their approaches and results will vary enough in order for them to be compared and contrasted with one another at the finish. A discussion at the end of the experiment directs the students toward the intended conclusions. This model of laboratory instruction sets students up to be more dedicated toward their learning and to generate richer understandings.

Discovery instruction is applicable when teaching GIS&T insofar as learning objectives are constructed around spatial thinking and conclusion drawing from the data. An instructor might ask the class any number of questions that require spatial thinking, such as to determine the most effective way to visualize a complex collection of data and then to produce the result; to predict the conditions during which certain features are clustered and then to test their guesses; or, to imagine which of several methods of spatial analysis will most favor a certain conclusion and then to conduct the analysis and whatever other steps for data display are necessary in order to make a case for that finding. Golledge, Marsh and Battersby (2008) emphasize the role of geospatial technologies in teaching students geospatial concepts and how to think spatially. They point to two National Research Council committees that advanced the rationale for and importance of spatial thinking in social science research and for the benefit of society as a whole (National Research Council, 1997, 2006). Many other academics, most recently Tate and Unwin (2009), Hespanha, Goodchild, and Janelle (2009), and Sinton (2009) have made important connections between teaching GIS&T and spatial and critical thinking. Discovery instruction clearly requires learning objectives that transcend teaching practical skills in the operation of GIS software, and instructors may find this model for instruction useful when teaching GIS as a support system for spatial thinking (National Research Council, 2006).

In addition to using data included with textbooks, instructors can share data from their own research with students or acquire secondary data in order to develop pre-structured exercises. Large quantities of secondary data can be found readily on the web at no cost and also through various arrangements requiring payment. Additional resources and instructional materials are available online, including:

- A searchable database of GIS lessons using ESRI's ArcGIS software spanning a large number of topics, in addition to case studies for GIS in education, a community blog, and forums at http://edcommunity.esri.com/.
- Copies of course syllabi at http://www.csiss.org/learning_resources/content/syllabi/.
- Links to open source GIS at http://opensourcegis.org/.
- A form to request one-year site licenses for ArcGIS and extensions for your students at no cost at http://www.esri.com/industries/apps/education/offers/promo/index.cfm.

10.5 Theory into practice – first steps, next steps

When beginning the process of preparing to teach GIS&T for the first time, the instructor should keep a few of the overarching, particular challenges of teaching and learning GIS&T in mind:

1. Learning GIS&T requires a delicate balance between theory and practice. Students can't usually understand some of the theoretical concepts without practicing them with the software, but practicing the software doesn't alone reveal underlying theories.
2. Mastering areas of GIS&T involves learning together and all at once elements of graphic, visualization and database software, also often programming, statistical and mathematical software. This integration of software is the strength of GIS, but also a particular challenge to students and teachers.

3. Like point 2, the multidisciplinary nature of the field means that students must learn many concepts from many fields. This means that students may also be heir to misconceptions and preconceptions which may block learning. Matthew Bampton's chapter in this volume explores the role of misconceptions in GIScience education.

With these challenges in mind, a few guiding principles make sense. First, first-time instructors of GIS&T should use the complexity and diversity of GIS&T as an opportunity to design and refine courses that build on the history and strength of their own professional experiences and interests. No single instructor can have strengths in every unit in the *BoK*. Keeping teaching interesting for yourself is essential in order for your teaching to be interesting for your students. Second, the discoveries that instructors make by teaching courses in GIS&T are authentic. Those discoveries are even more valuable in light of the work of Boyer (1990) and Glassick, Huber and Maeroff (1997), who respectively extend the conceptualization of scholarly work to include teaching and identify as an international phenomenon the scholarship of teaching and learning (SoTL). Participating in SoTL is increasingly important. Solem, Foote and Monk (2009, ix) conceive of SoTL as a crucial mechanism to connect teaching with research and as a cornerstone of effective professional development, which should be 'anchored in a solid knowledge of relevant research and literature on learning theory, curriculum design, student assessment, curriculum alignment, project evaluation, instructional technology, and other topics.' Third, teaching improves in slow, incremental steps. Solem, Foote and Monk (2009) maintain that as a scholarly pursuit, teaching develops over time as do other aspects of the academic profession. Relying on Bass (1999), Solem, Foote and Komoto (2009) invite teachers to imagine the stepwise resolution of barriers in teaching as an 'ongoing investigation.' Instructors informed by this notion of teaching should not attempt to implement more pedagogical innovations than they are comfortable with at any one time. Hespanha, Goodchild and Janelle (2009) caution that new instructors of GIS&T should test their ideas before trying to implement major changes. The following shortlist summarizes the priorities discussed in this chapter:

1. Teaching and learning go together, but think of the learner first and the teaching second.
2. Don't reinvent the wheel. Use existing texts and materials to get started. Get help from colleagues and network with other GIS&T educators.
3. Effective learning requires a balance of lectures, labs, and other activities, all with feedback.
4. The *BoK* is very helpful, but think broadly about learning outcomes beyond it.
5. Keep learning alive. Active pedagogy supports effective learning.
6. Develop a strategy for introducing students to collecting secondary and primary data.
7. Help build collegiality and cooperation among learners.
8. Use and contribute to SoTL to improve your teaching.
9. Experiment and improve.

References

Bass, R. (1999) The scholarship of teaching: What's the problem? http://doit.gmu.edu/Archives/feb98/randybass.htm (last accessed March 4, 2011).

Biggs, J. (1999) What the student does: Teaching for enhanced learning. *Higher Education Research & Development*, 18(1): 57–75.

Boice, R. (2000) *Advice for New Faculty Members.* Needham Heights, MA: Allyn & Bacon.

Boyer, E. (1990) *Scholarship Reconsidered: Priorities of the Professoriate.* Princeton, NJ: Carnegie Foundation for the Advancement of Teaching.

Brock, W., and A. Meadows (1977) Physics, chemistry and higher education in the UK. *Studies in Higher Education*, 2(2): 109–124.

Brown, G., and M. Atkins (1988) *Effective Teaching in Higher Education.* New York: Methuen.

Carlson, T. (2007) A field-based learning experience for introductory level GIS students. *Journal of Geography*, 106: 193–198.

Chickering, A. and Z. Gamson (1987) Seven principles of good practice in undergraduate education. *American Association of Higher Education Bulletin*, 39(7): 3–7.

Commission on the Future of Higher Education (2006) *A Test of Leadership: Charting the Future of US Higher Education.* http://www.ed.gov/about/bdscomm/list/hiedfuture/reports/pre-pub-report.pdf (last accessed 4 March 2011).

Coppola, B. (2002) Laboratory instruction: Ensuring an active learning experience, IN: W. J. McKeachie (ed), *McKeachie's Teaching Tips: Strategies, Research, and Theory for College and University Teachers.* Boston, MA: Houghton Mifflin Company, pages 235–244.

DeMers, M. (2009) Using intended learning objectives to assess curriculum materials: The UCGIS *Body of Knowledge. Journal of Geography in Higher Education*, 33(supp. 1): S70–S77.

DiBiase, D., M. DeMers, A. Johnson, K. Kemp, A. Luck, B. Plewe, and E. Wentz (eds) (2006) *The Geographic Information Science and Technology Body of Knowledge.* Washington, DC: Association of American Geographers and University Consortium for Geographic Information Science.

Dobson, J. (2001) Fieldwork in a digital world. *Geographical Review*, 91(1–2): 430–440.

Doherty, A., T. Riordan, and J. Roth (eds) (2002) *Student Learning: A Central Focus for Institutions of Higher Education. A Report and Collection of Institutional Practices of the Student Learning Initiative.* Milwaukee, WI: Alverno College Institute.

Domin, D. (1999) A review of laboratory instruction styles. *Journal of Chemical Education*, 76: 543–547.

Estaville, L., B. Brown, and S. Caldwell (2006) Geography undergraduate program assessment. *Journal of Geography*, 105(6): 239–248.

Favier, T. and J. van der Schee (2009) Learning geography by combining fieldwork with GIS. *International Research in Geographical and Environmental Education*, 18(4): 261–274.

Fink, L. D. (2003) *Creating Significant Learning Experiences.* San Francisco: Jossey-Bass.

Glassick, C., M. Huber, and G. Maeroff (1997) *Scholarship Assessed: Evaluation of the Professoriate.* San Francisco: Jossey-Bass.

Golledge, R., M. Marsh, and S. Battersby (2008) A conceptual framework for facilitating geospatial thinking. *Annals of the Association of American Geographers*, 98(2): 285–308.

Hespanha, S., F. Goodchild, and D. Janelle (2009) Spatial thinking and technologies in the undergraduate social science curriculum. *Journal of Geography in Higher Education*, 33(Supplement 1): S17–S27.

Hofer, B. (2002) Motivation in the college classroom, IN: W. J. McKeachie (ed), *McKeachie's Teaching Tips: Strategies, Research, and Theory for College and University Teachers.* Boston, MA: Houghton Mifflin Company, pages 118–127.

Holmes, D. (2007) Using GIS in geographical fieldwork. *Geography Review*, 20(4): 6–7.

Lund, J. J., and D. S. Sinton (2007) Critical and creative visual thinking, IN: D. S. Sinton and J. J. Lund (eds), *Understanding Place: GIS and Mapping across the Curriculum.* Redlands, CA: ESRI Press, pages 1–18.

Marble, D. F. (1998) Rebuilding the top of the pyramid. *ArcNews*, 20(1): 28–29.

McKeachie, W. J. (ed) (2002) *McKeachie's Teaching Tips: Strategies, Research, and Theory for College and University Teachers.* Boston, MA: Houghton Mifflin Company.

National Research Council (1997) *Rediscovering Geography: New Relevance for Science and Society.* Washington, DC: National Academies Press.

National Research Council (2006) *Learning to Think Spatially*. Washington, DC: National Academies Press.

Phillips, M. (1981) Early history of physics laboratories for students at the college level, *American Journal of Physics*, 49: 522–527.

Prager, S. and B. Plewe (2009) Assessment and evaluation of GIScience curriculum using the *Geographic Information Science and Technology Body of Knowledge*. *Journal of Geography in Higher Education*, 33(Supplement 1): S46–S69.

Rhind, D. (1992) Why GIS? *ARC News*, 11(3): 1–4.

Sinton, D. S. (2009) Roles for GIS within higher education. *Journal of Geography in Higher Education*, 33(Supplement 1): S7–S16.

Sinton, D. S., and S. Bednarz (2007) About that G in GIS. IN: D. S. Sinton and J. J. Lund (eds), *Understanding Place: GIS and Mapping across the Curriculum*. Redlands, CA: Esri Press, pages 19–34.

Smith, B., and Y. Zhou (2005) Assessment of learning outcomes: the example of spatial analysis at Bowling Green State University. *International Research in Geographical and Environmental Education*, 14(3): 211–216.

Solem, M., K. Foote, and C. Komoto (2010) scholarship of teaching and learning in geography, IN: M. Solem, and K. Foote (eds), *Teaching College Geography*. Upper Saddle River, NJ: Prentice Hall, pages 65–68.

Solem, M., K. Foote, and J. Monk (2009) Creating a climate of support of graduate students and early career faculty: A call for change and action, IN: M. Solem, K. Foote, and J. Monk (eds), *Aspiring Academics: A Resource Book for Graduate Students and Early Career Faculty*. Upper Saddle River, NJ: Prentice Hall, pages vii–xvii.

Tate, N., and D. J. Unwin (2009) Teaching GIS&T. *Journal of Geography in Higher Education*, 33(Supplement 1): S1–S6.

Unwin, D. J. (1997a) Unit 160: Teaching and learning GIS in laboratories. *NCGIA Core Curriculum in Geographic Information Science*. Available at http://www.ncgia.ucsb.edu/giscc/units/u160/u160.html, accessed March 4, 2011.

Unwin, D. J. (1997b) Unit 159: Curriculum design for GIS. *NCGIA Core Curriculum in Geographic Information Science*. http://www.ncgia.ucsb.edu/giscc/units/u159/u159.html (accessed March 4, 2011).

Wiggins, G. (1998) *Educative Assessment: Designing Assessments to Inform and Improve Student Performance*. San Francisco: Jossey-Bass.

11

Issues in curriculum and course design: discussion and prospect

Kenneth E. Foote

Department of Geography, University of Colorado at Boulder, Boulder, Colorado, USA

11.1 A record of progress and innovation

The chapters in this section draw attention to the dramatic contrast between GIS&T education twenty or so years ago and its situation today. As Karen Kemp notes in her chapter in the previous section, when the Core Curriculum in GIS was proposed in the late 1980s, the key challenge was that 'few of the faculty who were teaching or planned to teach GIS had any formal education in the subject, there were very few teaching materials and the technology was expensive and difficult to operate' (page 49). As she points out, the situation today is almost completely reversed:

> The technology we teach now is several generations removed from what we worked with twenty years ago. It is much easier to use, inexpensive, even free, and incorporates a much richer tool box that can be used to illustrate many fundamental aspects of GIS&T without too much training overhead. For those who care to try, it is now much easier to teach students about our science without it becoming obscured by the technology . . . Much of what we struggled to teach in the early years has evaporated (page 57).

To reach this point, GIS&T education moved through a remarkable period of innovation. As Kemp details, GIS&T educators were ahead of their time in developing the Core Curriculum as well as in efforts to create an international post-graduate course on GIScience; online curriculum resources and educational web portals; interoperable materials for GIS&T education; digital libraries; and, most recently, the GIS&T *Body of Knowledge (BoK)*. The result is that:

> it is no longer necessary to teach basic computer operation and file access procedures. We can get down to the real details about geographic information and analysis very quickly. While this is a great leap forward in our ability to cover the vast richness of our domain within a small curriculum, the field continues to expand rapidly so that for whatever we no longer need to teach, there always seems to be twice as much new material to add (page 57).

Teaching Geographic Information Science and Technology in Higher Education, First Edition.
Edited by David J. Unwin, Kenneth E. Foote, Nicholas J. Tate and David DiBiase.

This supports the assertion in Chapter 1 that 'Repeatedly, GIS&T educators have been among the first to implement new educational strategies and take advantage of new developments' (page 7) and has helped to push a number of pedagogical innovations into the mainstream – active pedagogy, problem-based learning, open educational resources, e-, v-, m-learning, and certification. We would argue further that, as a result of these innovations, GIS&T educators have been at the forefront of engaging a broader array of important theoretical and practical issues in curriculum design, educational psychology, cognitive theory, and other fields as well.

Steven Prager's chapter on using the *BoK* for curriculum design is a good example of this engagement. The *BoK* stands out as one of the most significant accomplishments of the GIS&T education community in the past decade – an attempt, as we noted in the first chapter, to formalize and publicize the knowledge of its practitioners in terms of learning outcomes. Prager pushes this frontier forward in two ways. First, he demonstrates theoretically and with examples how the *BoK* can actually be used to develop courses and curricula. Despite the effort its authors put into developing detailed learning outcomes across the entire GIS&T curriculum, one difficulty with the *BoK* is that its great breadth makes it difficult to know what to include in any particular course, module or curriculum. Given the scope of the *BoK*, most instructors feel pressured to force concept after concept into their courses hoping, perhaps, to cover as much of it as possible in a given term. Prager addresses this issue by showing how it can be used as a framework for identifying key concepts and weaving them together into coherent courses and curricula. This means being sensitive to the individual and institutional contexts within which GIS&T is situated, rather than feeling compelled to cover everything. Furthermore, rather than serving as a 'one size fits all' model, Prager argues that the *BoK* helps – as it was intended – to articulate multiple paths to diverse learning outcomes.

Secondly, Prager shows how the *BoK* can be used with integrated course design to create these multiple paths, and his overview of integrated course design is one of the strengths of the chapter. Relatively few GIS&T educators have either extensive training or background in the theory and practice of course and curriculum design. Using the insights of major figures like Wiggins, Biggs and Fink, Prager demonstrates what can be gained from such systematic approaches. He draws especially on Wiggins's notion of 'backward design,' i.e. setting intended learning outcomes first and then planning backward to the start of the course to align activities and assessments with these goals. Integrated course design also allows Prager to caution against focusing too much on the knowledge base of GIS&T – the facts, concepts and principles outlined in the *BoK* – and to acknowledge that most teaching (and learning) involves a broader array of goals, such as encouraging students to: 1) learn how to learn, i.e. to develop the ability to learn on their own throughout their lives; 2) integrate the growing expertise in GIS&T with their knowledge, skills and abilities in other fields; 3) apply or transfer what they have learnt about GIS&T to other contexts and disciplines; and 4) develop a wide variety of teamwork, communication, critical reasoning and spatial thinking skills which will help them in their careers.

In my chapter, I take Prager's chapter a step further. I call attention to the vital issues of scope and sequence in GIS&T education. Scope is the detail or depth of knowledge about a given concept or skill introduced at each stage of the learning process and sequence is the order in which concepts are introduced. As I note, based on a review of recent research, that 'both the scope and the sequencing of learning tasks can have a great influence on a student's mastery of concepts and skills... Topics introduced out of sequence often end up having to be repeated. Too broad a scope at too early a stage may overwhelm students if they are not given time to practice and assimilate new material' (page 81). The point here is that, even given Prager's overview of backward design, more attention needs to focus on how students develop their mastery of GIS&T at the 'micro-scale,' so to speak. My arguments do not necessarily imply major changes in GIS&T education, but they do suggest how the concepts of learning cycles and spiral curricula may help improve courses and curricula.

The chapter by Marco Painho and Paula Curvelo extends my discussion of learning cycles and spiral curricula in a very important direction. As Painho and Curvelo write:

> One of the greatest potential affordances of web-based education is personalized learning environments. Personalized e-learning systems aim at adjusting teaching and learning to the needs, interests and aptitudes of each individual learner, resorting to a diagnosis of its relevant characteristics and employing that information during the process of curriculum implementation. From this perspective, the process of curriculum development should accommodate the need to assess learners' prior knowledge and abilities and guide him or her towards the course of desired learning outcomes (page 102–3).

Such personalized learning environments – the trend in e-learning – coupled with the recursive nature of spiral curricula offer a powerful model for learning. Such a model offers students progressively more challenging learning sequences tailored to their knowledge and competencies at any given point in the learning process.

Matthew Bampton's chapter on misconceptions, troublesome knowledge and threshold concepts is another good example of how GIS&T educators can gain from engagement with the broader educational literature and theories. His focus is on the misconceptions and preconceptions that students bring into the classroom and which inhibit and block learning. Though studied in science, mathematics and engineering education, such misconceptions and troublesome knowledge are only beginning to gain attention in GIS&T and geography education. His research indicates that, of the numerous troublesome concepts already identified in nearby disciplines, many play a role in GIS&T and that 'all available evidence points to misconceptions as a significant, enduring, and as yet unstudied problem facing GIScience students and teachers' (page 128). His conclusion suggests not just the need for more research, but also the implicit irony that, for educators, preconceptions and misconceptions about how students learn may block their efforts to improve pedagogy. This, again, is a forceful argument for more engagement by GIS&T educators with research in educational psychology and cognitive theory.

Similarly Richard Schultz's chapter offers clear theoretical reasons for employing active learning techniques in GIS&T curricula. His point is that techniques such as problem-based learning (PBL) and inquiry-based learning (IBL) are far more than simple add-ons to traditional lecture/laboratory formats. Instead, Schultz argues that 'that active learning and GIS&T go hand in hand. Fluency using the complex tools of GIS&T can only be learned through an active desire to fully experiment and learn by doing. Active learning, problem-based learning (PBL) and inquiry-based learning (IBL) mesh together to create the mechanism by which GIS&T can be practiced to a point of fluency' (page 141). His point is that considerable research points to the fact that active pedagogy leads 'to increased learning and engagement on the part of the students ultimately leading to better understanding of the subject matter with the ability to apply what they have learned to solving problems. It seems that spatial awareness leading to spatial perception would be an outcome of employing active learning and metacognition in the GIS&T educational arena' (page 135).

Finally, Eric West's chapter answers a very practical question: given the plethora of resources available now – textbooks, lab manuals, web and e-learning resources, journal articles and books on teaching and learning GIS&T – where should first-time GIS&T educators begin? His chapter provides a useful 'carry away' message for this section of the book. His points bear emphasis here (page 162):

1. Teaching and learning go together, but think of the learner and learning outcomes first and the teaching second.
2. Don't reinvent the wheel. Use existing texts and materials to get started, but advance by getting help from colleagues and networking with other GIS&T educators.
3. Effective learning requires a balance of lectures, labs, and other activities, all aligned with learning outcomes and all providing prompt feedback.
4. The *BoK* is very helpful, but think broadly about learning outcomes beyond the *BoK*.
5. Keep learning alive. Active pedagogy supports effective learning.
6. Develop a strategy for introducing students to collecting secondary and primary data.
7. Help build collegiality and cooperation among learners.
8. Use and contribute to the Scholarship of Teaching and Learning (SoTL) to improve your teaching.
9. Experiment and improve.

11.2 Issues and questions ahead

The chapters in this section present a picture of continuing and deeper engagement than ever before with issues of curriculum design as well as with educational theory and practice, but what remains ahead? At least three areas deserve more attention.

First, as the chapters on misconceptions, scope and sequence, and ontology-based curricular paths imply, more research is needed into some of the questions and issues explored in this book. Over the past decade research into GIS&T has come to assume a dominate position within key educational journals like the *Journal of Geography in Higher Education* and *Journal of Geography*, yet this is only a start. Much of what we know is based largely on case studies of individual projects for improving practice within very particular institutional contexts and educational settings. These are informative, but there is a wide range of issues relating to curriculum and course design and learning theory in GIS&T that would benefit from sustained, broad-based research attention – perhaps a project like the *BoK*, but directed toward the underlying research questions in GIS&T education. For a number of years the University Consortium for GIScience (UCGIS) developed and discussed educational priorities. Perhaps the time is ripe to revive such discussion in light of some of the issues presented in this section. Furthermore, as West argues, supporting the scholarship of teaching and learning (SoTL) movement more explicitly in GIS&T education can help to encourage this engagement between theory and practice.

Second, the issues discussed in this section have an increasing saliency for those topics discussed in Section Three – what this all means for professional training in GIS&T, for certification and professional master's programs and in situations where GIS&T education is developed or implemented in contexts with no connection to academic geography. GIS&T has long been taught in diverse settings but, as Diana Sinton, Karen Kemp and Mike Phoenix make clear in their chapters in Section One of this volume, this diversity continues to grow and presents certain tensions for GIS&T education. There is still a push from many for increased rigor within existing curricula so that GIS&T students have the programming and computational abilities to develop cutting-edge GIS&T applications and tools for the next generation. On the other hand, a key trait of contemporary GIS&T education is the many bridges it is building into and across other curricula in the sciences and humanities. Given these contrasting paths into the future, the question is how to both strengthen the 'core' of the GIS&T curriculum while extending its interdisciplinary range. This tension between balancing depth and breadth in GIS&T education merits continued discussion.

Finally, collaboration has been a hallmark of GIS&T education for years, but that doesn't mean that more isn't needed. One of the important themes of this book is the value of cultivating communities of practice in GIS&T education. Given the changes underway in the politics, economics and demographics of higher education around the world, building stronger collaborations will be vital to sustaining this community. This applies to not only the research needs discussed above, but also to the sharing of know-how and common practice about educators. West's chapter stresses how much can be gained by such collaborations, particularly for first-time GIS&T educators. It is still the case that many first-time educators find themselves 'reinventing the wheel' or, perhaps rather, rediscovering by accident decades worth of pedagogical research in GIS&T and other fields when, in fact, such knowledge could – and should – be communicated through communities of practice and more effective channels of collaboration.

11.3 Into the cloud

In closing, it might be worth asking – even given the chapters in this section – whether what we have to offer in the domain of curriculum and course design is too little and too late? As some of the developments in the fourth and fifth sections of this book suggest, GIS&T courses and curricula of the future may look far different from even the most advanced programs of today. Is education, like computing moving into the 'cloud,' where resources will be provided 24/7 sometimes for free, sometimes at cost, but available increasingly for mashups suited to the needs of individual users?

There is certainly much to suggest that change is imminent. The open educational resources movement continues to gain traction, offering more and more instructional materials independently of any sort of accredited educational program. And it may be the case that neogeography is not just a movement involving the informal use of GIS&T by non-professionals, but also an issue of how GIS&T will in the future be learned in a world where the idea of teaching and 'being taught' yields to andragogy and the notion of the self-directed learner. So, are we looking over the horizon to consider how this new world of education might be effectively and creatively served in the next generation? Much attention seems to be focusing right now on improving GIS&T learning in relatively traditional, formal educational settings. This is not to discount the work needed to create effective certification programs, online and distance degrees, or the efforts now directed to extend GIS&T into new disciplinary territories, but perhaps we also need to take a more radical view and re-envision the relationship between learner and teacher in a world of cloud-computing. This is an issue taken up on the final chapter.

Acknowledgments

I wish to thank the co-editors of this volume for their suggestions, both conceptual and editorial.

Section III

Perspectives on teaching GIS&T

12

The University of Minnesota master of geographic information science (MGIS) program: a decade of experience in professional education

Susanna A. McMaster and Robert B. McMaster

Department of Geography, University of Minnesota, Minneapolis, Minnesota, USA

12.1 Introduction

As the demand for a skilled geospatial workforce increases (Gewin, 2004; Phoenix, 2000), there are a series of challenges that educators face in meeting the needs of the professional GIS community. Research is needed to understand marketplace needs and demands better, in particular for those professional students who are now retooling into GIS&T from other areas such as engineering or urban planning. We also need a clearer understanding of the advantages and disadvantages of the different models of delivery for GIS&T education, and, after over a decade of operation, how well these models are fulfilling the needs of the professional community.

With the publication of the *Geographic Information Science and Technology Body of Knowledge* (DiBiase *et al.*, 2006), research is now needed on how different curricula fit into this model structure. Geographic information educators must also consider future trends, and how their own programs need to be consistent with the knowledge and techniques in industry and government. Some considerations include:

- What levels of computer science and mathematical education do our students need?
- What new areas, such as mobile technologies and web-based access methods, do we need to emphasize?
- What are the basic skills in business and management that might be needed?

Teaching Geographic Information Science and Technology in Higher Education, First Edition.
Edited by David J. Unwin, Kenneth E. Foote, Nicholas J. Tate and David DiBiase.

- How do programs prepare their students for GIS&T certification, and do we adjust our curricula to prepare students for such certification?
- Do we need explicit education in professional issues such as project management?

A growing question relates to the differences between a true professional versus an academic degree. The differences can be slight, but often focus on the practical 'training' balanced against the conceptual material. Does the degree prepare the student for a particular career, or is it more general in nature? In the United States educational system, there are standardized 'first-professional' degree programs, including those in medicine, dentistry and law. But other degrees such as planning, architecture and business also focus heavily on preparation for a particular profession. Professional degrees in geographic information science are mostly at the Master's level, and include coursework in the technology of the field such as database management, GPS, and spatial programming, as well as conceptual-level material. Students in non professional 'academic' programs normally are pursuing degrees in geography, natural resource assessment, and computer science, and coursework is only partly oriented towards GI Science. Additionally, each type of program often addresses the topic of ethics, however, the focus may be different. An academic program tends to focus on research ethics while professional programs address professional ethics and codes of conduct. Admissions criteria may also differ as professional programs will often take into account an applicant's work experience. Academic programs tend to focus more on theory and research with graduates who go onto academic careers, while graduates of professional programs seek careers as GIS&T practitioners. Finally, time to degree completion can vary in a professional program if students are working full time; most academic programs are completed in two years. Any such categorization, of course, is problematic as most degrees are an amalgam of both. For example, Frank and Raubal (2001) argue for a GIS&T education model that incorporates more training in business aspects of spatial data as well as a GI engineering approach in order to better meet market demands.

Given the growing demand for GIS&T education in the United States and other countries, effective models of access need to be developed. We also need to consider the possibility of internationalizing the curriculum to facilitate better discussions of GIS&T education approaches among countries (Solem *et al.*, 2006).

12.2 Models for delivery of professional GIS&T education

As identified by the UCGIS Education Committee, there is intense pressure on students, educators and employers to understand the myriad requirements for professional GIS&T education, and to develop appropriate and flexible models of delivery (UCGIS, 1997). Key models include distance learning, residential masters degrees in GI Science and certificate programs, short courses and workshops, degrees in other disciplines with GIS&T courses (e.g., forest resources) and combined approaches. Numerous databases of GIS&T academic programs now exist and assist prospective students in understanding the range of choices available to them. For example, the Esri online database of academic GIS programs allows a user to find GIS certificates, degree programs, graduate programs, distance learning and Esri Virtual Campus listings (http://edcommunity.esri.com/universityprograms/). A 27 January 2011 query of distance education programs in their database resulted in 85 matches, one for graduate programs resulted in 173 matches, and another for certificates resulted in 301 matches.

With the growth in programs and increasing number being developed each year, there are also guidelines that have been developed to assist institutions interested in starting GIS education programs. For example, Gaudet (2001) describes the development of a geospatial technology competency model that can be used by educational institutions as a way to ensure their curriculum is training students to meet market demands. Additionally, ESRI (2002) published a white paper that discusses guidelines for creating higher education GIS&T programs. A better understanding of different models of access, their advantages and disadvantages, and an institution's specific context, can contribute further to the development of effective guidelines and best practices. In what follows, we discuss three key models for delivering professional GIS&T education including on-campus GIS&T certificate programs, on-campus masters degree programs and distance education GIS&T programs (offering certificates and masters degrees) and their major advantages and disadvantages. Next, we provide an in-depth case study discussion of the MGIS program at the University of Minnesota that illustrates one approach using a residential delivery model.

Model 1: the GIS certificate

One of the earliest models for the delivery of professional GIS&T education was the certificate program where students would take a limited set of classes with a focus on a particular set of skills and/or application area. Such programs allow for students to take a more limited set of focused classes that enable a base-level exposure to the theory, applications and technology of geographic information science. Wikle (1999) and Wikle and Finchum (2003) provide an overview of GIS&T certificate programs in the US and Canada including both on-campus and distance based approaches. He suggests developing programs based on application areas (e.g., urban GIS) and position types (e.g., technician versus manager). Since the publication of his article, many more such programs have been created. Wayne (2002) estimated that there were 70–100 such certificate programs in North America in 2002.

The advantages of GIS certificate programs are that students can continue to focus on an applied area, such as forestry, urban planning and criminology and are able to bring their own unique expertise to the field, enabling a coupling of problems in their own areas with spatial solutions. A disadvantage of GIS certificates is that, since such programs can vary greatly in the depth and breadth of the coursework, many employers are uncertain about what a certificate represents in terms of a student's GIS&T capabilities.

Model 2: distance education

A growing number of students have migrated to the distance education model to earn both certificates and masters degrees in GIS&T. For older, non-traditional students often trying to balance career, family and other commitments, there are certainly significant advantages to the distance education approach. Given the recent growth in GIS&T, and the number of professionals wishing to redirect their careers in GIS&T but not being able to gain easy access to the 'in class on campus' experience, a distance-based approach is ideal. One of the primary models for distance education GIS&T programs is the Penn State World Campus that offers both an online GIS&T certificate and MGIS degree. DiBiase (2004) details the development of their online MGIS degree including its curriculum and a process for assessing and improving the program on a continuous basis. UNIGIS, founded in 1990, is the oldest GIS distance education model representing an international partnership of fourteen educational institutions (http://www.unigis.org, see Chapter 24). For academic institutions that are located in remote or more rural areas, the distance education model is often a more successful approach to establishing a professionally-oriented GIS&T program.

Such programs are advantageous because they allow for flexibility in terms of workload, time and budgets. Many distance education programs have successfully converted to an on-line environment and early indications are that, pedagogically, such a delivery format is successful, especially for motivated professional students. Innovations in sharing on-line courses among domestic and international campuses are also likely to add to the flexibility of distance-based programs although given the importance of tuition revenue this has been slow to happen. An example of this is the partnership between Leeds, Southampton and Penn State Universities described in Chapter 29. However, there are problems with these models. Students must have strong motivational, organizational and technical skills to complete them and they can experience high drop out rates if strong effort is not put into establishing effective mechanisms for advising and communication. Also, there tends to be less development of camaraderie and networking amongst the student body unless a program develops alternative online communication tools to facilitate group and one-on-one discussion. On the other hand, with effective communication tools in place, an even wider network can be established in comparison to traditional programs.

Model 3: residential masters programs

Over the past ten years, there has been steady growth in residential degrees, such that most regions of the United States now have several of these programs. Some of these include courses at the University of Minnesota, Clark University, San Diego State University, the University of Washington and many others. The residential model takes the traditional approach that assumes students will succeed through the experience of being on campus and in the classroom with a formal lecture, working in laboratories with other students, and benefiting from other types of formal and informal professional and social interaction. Students also have easy access to campus resources and more opportunities for face-to-face interaction with faculty, students, staff and local professionals (especially if the institution is located in a larger metropolitan area). Interaction with faculty expertise and advising is theoretically easier, students do not have to purchase their own software (although many do), and access to libraries and other resources is less complicated. On-campus programs can also be beneficial since many successful programs have a dedicated staff person who assists with the day-to-day management of the program including handling student needs in a timely manner. These students who are often returning for such professional degrees have high expectations for immediate and personalized service. A significant benefit for many professional students is the ability to work on campus in related positions, thus gaining practical and applied experience. Most of the Minnesota MGIS students have found employment working in a variety of positions such as in remote sensing laboratories, facilities management units, and as research assistants.

A major disadvantage is that students must commit to relocating to a campus. For professionals who may already have an established career along with other personal commitments it is extremely difficult to relocate. Also, courses are offered on a fixed schedule that provides little flexibility for the working professional GIS&T student, but some key courses

Table 12.1 Summary advantages and disadvantages of some course delivery approaches

	Residential certificate	Distance education	Residential masters
Advantages	Focus on application area and targeted skill set	Flexible, self-paced curriculum; especially good for non-traditional students with full-time jobs	Face-to-face interaction with faculty, students and staff
	Good for students who want to enhance existing skills quickly	Good option for providers in more remote locations; greater access to potential market	Access to campus resources and off-campus internship and employment opportunities depending on location
	Provider is able to develop program more easily than a full masters degree program	Provider can utilize skilled instructors beyond the local area through online courses	Provider has easy access to local market and professional GIS instructors in the local area
Disadvantages	Can have variation in depth and breadth of course work	Better for self-directed students comfortable with technology-enhanced learning	Less flexible course schedule
	Employer uncertainty about how to assess GIS capabilities (vs. Masters degree)	Potential high drop out rates if effective communication tools not in place	Relocation to campus necessary
	In many cases, credits cannot transfer into Masters programs if students wish to continue for advanced degree	Requires provider to have staff and infrastructure capable of delivering technology and support to run online program	Course offerings can be constrained depending on faculty expertise and access to local professionals as instructors

can accommodate professional students by being offered in the evening. It is clear that a major tradeoff from the distance-based curriculum is flexibility. Table 12.1 summarizes the advantages and disadvantages of these three approaches.

12.3 A case study: the University of Minnesota MGIS program

The initial justification

As a case study of a residential program we discuss the twelve-year-old masters in GIS (MGIS) program at the University of Minnesota. A number of factors and issues were considered in deciding whether or not, and in what form, to establish the program. By the middle of the 1990s, it was clear that geographic information systems and science was and would be a growth area in the private and public sectors. The Twin Cities remains a major focus for geographic information science development and application, home to a large number of GIS&T professionals who might form an initial market for an advanced degree program and serve as community faculty members to complement the core academic faculty. In particular, the location in the capital city (St Paul) results in a large demand for GIS&T specialists to help in the various state government agencies (Department of Natural Resources, Department of Transportation, Pollution Control Agency). Another factor was a strong geography department at the University of Minnesota whose faculty members, along with colleagues from allied disciplines such as forestry, soil science, public health, and computer science, could provide an instructional core. Finally, other factors that favored establishing a program included external pressure on the university to provide career education and, given its urban location, the relative dearth of professional masters in the university.

A brief history

The MGIS program at the University of Minnesota was one of the first professional masters degrees in GIS&T in the US and it remains one of the most comprehensive by offering a combination of substantive depth and cutting-edge technical training. The program is part of a larger movement in which GIS&T as a discipline is still in a growth period in both academia and beyond.

The MGIS program was conceived in 1995 as part of the geography department's ten-year plan. It had become clear that there was tremendous demand for professional GIS&T education/training in the Twin Cities metropolitan region and that the University of Minnesota needed to play a leadership role. Upon completion of the MGIS proposal, written by Robert McMaster and William Craig, it was approved in 1996 by the department, college, graduate school and University of Minnesota Board of Regents. The program was formally established in 1997 and housed in the geography department. Two years later, in 1999, the program received its first external review, and in response a series of changes were made, including renovation of space, creation of an Associate Program Director position, and curriculum refinement. A full external assessment of the MGIS program was completed in 2006 and is reported on later in this chapter.

The program was originally designed and continues to focus on an interdisciplinary approach to GIS&T education. Yet, based on the existing university structure, the program had to be housed on one unit, in this case the department of geography that already had a well-established academic program. For financial matters the MGIS program is located within the geography department in the College of Liberal Arts with a separate budget maintained within the department. The MGIS budget primarily includes the Associate Program Director salary, a small marketing line, software and other program needs, as well as teaching specialist salaries. The academic and professional programs were built on a shared faculty, shared resources (laboratories, space, technology), and, most importantly, a shared curriculum. For example, geography's GIS course offerings are part of the MGIS curriculum and the GIS faculty in the geography department are part of the MGIS faculty, but the MGIS program also maintains a separate set of technology courses and includes faculty outside of the department and the college on its graduate faculty roster.

We have graduate faculty from three other departments (computer science, forest resources, soil, water, and climate) and an on-campus research centre (Center for Urban and Regional Affairs). We also have informal relationships with other units such as conservation biology, the Minnesota Population Center, the School of Public Affairs, landscape architecture and biostatistics (public health). Many of the MGIS students are employed as research and teaching assistants in these areas. We would like to strengthen our relations with these and other programs on campus. The major hindrance to becoming more interdisciplinary has been institutional, with some faculty in other colleges having been hesitant to join a graduate faculty outside their home college. Additionally, for some of the GIS&T technology courses the program relies on highly qualified instructors referred to as teaching specialists. These are leading professionals working in the GIS industry and are ideal instructors for our students. These teaching specialists conduct the same course evaluations as academic faculty and also participate in an annual review process at the university.

The program also relies on the laboratory facilities in other units such as the Remote Sensing and Geospatial Analysis Lab, the Soil Landscape Analysis Laboratory and the Environmental Resources Spatial Analysis Center. While originally designed for the local GIS&T professional, the program quickly established a nationally and internationally diverse student body. Today, we have an interesting mix of domestic and international students with a range from novices to those with in-depth knowledge of computer science and previous GIS&T experience, a wide range of application areas, a range from full-time students to professionals in diverse fields, and students with varied financial needs.

The MGIS is a non-thesis, terminal (does not continue to a PhD within the geography department) degree with the following educational goals:

- To provide a balance between theory, applications and technology.
- To promote an interdisciplinary approach to GIS&T.
- To ensure our curriculum keeps pace with the ever-changing nature of the discipline.
- To enable students to have research and professional development experience.

The administrative and curricular structure that was implemented when the program was established in 1997 has undergone several modifications based on regular program assessment and evaluation. This includes formal external review of the program by outside examiners as well as self-imposed, internal assessment in the form of student and advisory board feedback.

Administrative structure

Although the MGIS program is housed in the geography department and many of the faculty members have their tenure base in geography, MGIS is an independent interdisciplinary graduate program. It has its own administrative structure including an Administrative Committee, Executive Committee, and Advisory Board. The Administrative Committee works on key programmatic issues (e.g., budget requests) and includes the following:

- A Program Director who oversees all aspects of the program's administration.
- An Associate Program Director (APD) who handles the day-to-day operations of the program, supports the Director of Graduate Studies (DGS) and is the primary

liaison between students and the program as well as the teaching specialists and the program.

- A DGS who supervises and coordinates the administration and governance of graduate studies within the program and serves as a liaison between the program faculty and the Graduate School.

The Executive Committee functions as the governing board for the program so that the entire faculty do not need to meet regularly for that purpose. However, they do vote on specified matters and the program schedules meetings of the MGIS faculty primarily to discuss curriculum matters. The current Executive Committee includes members of the Administrative Committee, MGIS faculty from other departments such as forest resources, and a student representative who is normally the student-elected president of the GIS Student Organization.

The MGIS program has had an Advisory Board since 2000 whose role is to provide advice and make recommendations on how well it is educating its students to meet market demands in a quickly changing profession. It is composed of highly experienced GIS&T professionals from government (e.g., Minnesota Department of Natural Resources and Metropolitan Council) and private (e.g., Best Buy Corp. ESRI and Rowekamp Associates) sectors and also includes a program graduate.

Curriculum design

The MGIS curriculum emphasizes three components—conceptual, technical and applied education in GIS&T. Courses for the program are divided into three broad categories. First, core courses provide the conceptual and theoretical underpinnings for a comprehensive, well-rounded knowledge of GIS, including an introductory seminar. These cover the fundamentals of GIS&T, including *Spatial Data Acquisition*, *Data Structures*, *Spatial Analysis*, *Remote Sensing* and *Cartography*. Some also include applied coursework in which students direct their knowledge towards solving a variety of social and biophysical problems. For example, in the *Urban GIS* course, students assist local community groups by using GIS&T to examine community issues. Second, a set of technology courses focuses on specific software and techniques in GIS&T and were developed using the budget provided by the student tuition fees. Technical courses include *Introductory* and *Advanced ArcGIS*, *Internet GIS*, *Desktop Mapping*, *Spatial Data Administration*, *Surveying and GPS* and *Spatial Programming*. Thirdly, elective courses provide additional breadth to the program by allowing students to take courses related to their area of interest. A detailed listing of the MGIS courses according to these three categories is available on the MGIS website at http://mgis.umn.edu/courses/courses.html which provides more information.

Students must complete thirty-five credits (one semester course $\approx$ three credits at the University of Minnesota; two credits of work outside of class is required for each credit hour in class) that include four required core classes: *Survey of GIScience*, *Principles of GIScience*, *Advanced GIScience* and an 8000 level *Seminar in GIScience*. Students must also complete six credits in technical courses including *ArcGIS I* and *ArcGIS II* with six credits outside the program and six credits of electives. All students must complete one research paper, one technology paper and one professional conference-style poster and an oral examination.

Key program modifications

We have found that in order to respond to market needs and the dynamic nature of the profession it is essential to assess and evaluate the program structure and curriculum on a regular basis. In response to external reviews and internal assessments from student and alumni surveys and administrative and advisory board recommendations the program has been modified better to serve our students and prepare them for a competitive job marketplace. Formal external reviews were held in 1999 and 2006 and included three external reviewers from other universities. In addition, we also seek feedback from our students, faculty and Advisory Board through the use of online surveys, student representative feedback, faculty meetings and Advisory Board meetings. Significant curriculum modifications took place in 2001 and 2005. We are considering additional refinements in 2010 based on internal program review and an upcoming advisory board meeting.

Some examples of significant administrative and curriculum modifications include:

- Appointment of an Associate Program Director in 2000.
- Renovation of space including updating laboratory facilities and student space.
- Changes in the required courses (e.g., addition of *ArcGIS II* as a required course).
- Addition of the three Plan B paper/project option more closely tied to course work.

Program statistics: admissions and other trends

Our students come from diverse undergraduate backgrounds ranging from the humanities to the natural sciences (including, for example, majors in geography, art, music, spanish, civil engineering, physics, computer science, agricultural engineering, mathematics, and environmental science). Many of them also have some additional graduate work in other disciplines such as environmental science, ecology, archaeology, information systems and public health. Admission to the MGIS program is handled by an admissions committee twice a year, as we have fall and spring entrants to the program. The program requires a personal statement, transcripts and three letters of recommendation. The committee examines applications holistically; we offer guidelines but do not set absolute standards for matters such as undergraduate grade point average (GPA) (we advertise a preferred GPA of 3.0) and english language test scores. The Graduate Record Examination (GRE) is not required since they prove to be a less useful diagnostic for our professionally-based program. Many of our students have been out of education for many years and GRE scores do not accurately reflect their capabilities. Additionally GRE are often used to make decisions about financial aid that we do not offer since we are a professional program. In the United States at the graduate level, financial aid tends to be focused on academic programs. Unlike more traditional academic programs, professional work experience is also considered in our admissions process. Once enrolled, however, MGIS students can often find employment as teaching and research assistants on campus. Finally, we are trying to attract more PhD students from other programs, such as forest resources or public health, that see the MGIS as a useful complement to research and a form of added security in the increasingly competitive academic job market.

Table 12.2 shows the number of applications received per year since 1997-98 as well as the number admitted, matriculated, dropped, graduated and continuing (i.e., still active in the

Table 12.2 Annual applications, admissions, matriculations, dropouts and graduations

Year	Applied	Admitted	Matriculated	Dropped out	Graduated	Continuing	Dismissed	Transferred
2009–10	33	22	14	1	8	5	0	0
2008–09	40	24	20	2	14	3	0	0
2007–08	34	22	15	2	11	2	0	0
2006–07	37	27	21	4	13	4	0	0
2005–06	33	22	14	4	7	0	0	0
2004–05	33	24	16	4	11	1	0	0
2003–04	42	24	20	6	14	0	0	0
2002–03	60	29	21	8	12	0	0	1
2001–02	74	35	27	10	15	0	1	1
2000–01	72	39	23	8	15	0	0	0
1999–00	46	25	17	9	8	0	0	0
1998–99	34	27	17	6	11	0	0	0
1997–98	29	21	16	8	8	0	0	0

program). We experienced a peak of admissions in 2001–02 and since then have decreased to a more stable steady state. Currently, we believe an annual class size of about fifteen to twenty students best fits our current capabilities.

Because we are a professional degree program, the length of time students are in our program varies greatly. Many of our students work full-time and thus take only one course per term and may take from four to six years to complete their degrees. In contrast our full-time international and domestic students take about two years to complete their programs. Overall, the number of students dropping out has decreased and we believe one factor influencing this trend has been a change from requiring a single capstone thesis-like project, that some students could not complete, to three more focused projects closely tied to courses specifically, a research paper, a technical project and a conference-style poster). Some other reasons students did not complete their degrees include work and family demands, personal issues, moving home and change in interests. Overall, we average about ten graduates per year and have now graduated over 140 MGIS students since the program started.

Program strengths and successes

The comparative advantages of the MGIS program include its focus on interdisciplinarity, the size and scale of the host university, and its curriculum. The program benefits from its position in a very large public university (with over 150 academic programs and 3000 faculty), a large metropolitan region that includes state and federal government agencies, many private companies (both large and small) that utilize GIS&T, and non-profit organizations. This cluster of private and governmental institutions means that we have a pool of students who wish to enhance their GIS&T education and skills through formal education, and provides the opportunity for internships and employment for students in the program. Often, private companies will pay the tuition for their employees who are pursuing the MGIS degree. More recently, we have an increasing number of undergraduates from the university who are interested in pursuing the MGIS degree. This may be related to a relatively new interdisciplinary undergraduate minor in *Geographic Information Science* that has been very well received by students majoring in a variety of fields. The size and scale of the University of Minnesota means that students have access to numerous academic departments, schools and centres, and tremendous resources in terms of libraries, facilities and other services. As a professional program, we are also able to draw upon an immense local pool of highly skilled and experienced GIS practitioners who can teach our technology courses and serve as advisers and mentors to our students.

Measures of success

A key to maintaining a successful program no matter what model it uses is to ensure that methods for assessing and evaluating the program are in place. We have used a combination of different approaches to ensure that we are satisfying our students' needs and meeting marketplace demands for GIS professionals. These methods include consultation with our Advisory Board, regular external reviews of the program, feedback from students, graduates and faculty (e.g., acquired from surveys and meetings), ensuring that effective advising occurs and students are given timely feedback on their progress in the program (e.g., regular student reviews).

As part of an external review of the department in autumn 2006 we completed a program evaluation that included a survey of our current students and graduates. First, we conducted an internal survey of our current active students (fifty-nine total) and an external survey of our then sixty-one graduates by developing an online survey form that could be easily and quickly completed. We asked our current students to provide us with feedback on four key topics: GIS interests, course evaluation, advising and program satisfaction, receiving forty-four responses out of a total of fifty-nine students. The results were as follows:

- GIS interests: The most frequently listed interest areas were natural resources and urban and regional planning. In addition to these areas, students cited interests in public health, archeology, software development and programming, Internet GIS, visualization and cartography, spatial analysis and geostatistics, transportation, business and mobile GIS.
- Course evaluation: The majority of our students found our technology courses and core courses to be either 'very valuable' or 'valuable'. Only a few courses had higher percentages in the neutral category but even among these more students found them to be valuable than not. All of our required courses were ranked in the very valuable or valuable categories.
- Advising: We were ranked 'average' to 'very good' in advising knowledge, providing feedback to students on their progress in the program, and for the utility of our graduate handbook. We received a 'good' in terms of availability; this is likely to happen as access to advisors can vary greatly. Overall, this reflects our efforts to develop a useful handbook to guide our students through the program and to revise our student annual review process.
- Program Satisfaction: thirty-five of the students participating in the survey (85%) felt 'very satisfied' (29%) or 'satisfied' (56%) with the MGIS program.

Students were also asked to inform us about their greatest obstacles to finishing. Many students cited work and family commitments, advisor and course availability, and personal issues.

Second, as part of the same review we asked our alumni to provide us feedback on two key topics: program satisfaction and course evaluation. The program satisfaction section included feedback on overall ranking of satisfaction, if the student would select the program again, and if the degree was useful in obtaining employment. We received thirty-six responses, with the following results:

- Program Satisfaction: twenty-nine of the respondents were 'very satisfied' (38.9%) or 'satisfied' (42.9%) with the program overall. Almost 81% said that they would select the MGIS program again and about 73% said that the degree was either 'very useful' or 'useful' in obtaining employment. Individual responses to our request for feedback on program improvements included suggestions on additional kinds of courses or training such as project management, enterprise GIS, and programming, as well as ways to improve current course offerings and the need for more faculty.

- Course evaluation: The majority of our students found our technology courses and foundational courses to be either 'very valuable' or 'valuable'. All of our required courses were ranked in the 'very valuable' or 'valuable' category.

Overall, the results of both surveys indicate that our students and graduates are very satisfied with our program and the curriculum. This fits well with our program being a national model for professional GIS education. Additionally, our recent external review highlighted the MGIS program as a success in the department. It is clear, however, that improvements and additions can be made to the program. We must also continue to meet the challenges that are specific to a professional degree program and meeting the needs of both traditional full-time students and non-traditional working students.

Future challenges for MGIS

As we continue to evaluate our program we are mindful of some key challenges that such programs may face, including:

- Maintaining stability in curriculum but also offering courses that reflect changes in technology.
- Maintaining stability in faculty in order to maintain a stable curriculum.
- Compromising between meeting student expectations for what they 'need' to get GIS jobs and maintaining the faculty vision of a stable and effective curriculum.
- Developing alternative opportunities for students to meet those expectations and support learning outside the formal curriculum.
- Streamlining the curriculum and making effective use of existing courses.
- Coordination amongst faculty on course content and transition.
- Meeting demands for an increased budget, e.g., additional courses and faculty.
- Making stronger linkages with additional university units and the private sector.
- Increasing committed financial support for students due to the program's professional status.

As a result of completing the external review in Fall 2006, we completed a ten-year plan that addresses a number of these challenges and how we can remain competitive and sustainable as a program (MGIS Program, 2007). Some key challenges we face are listed below.

First, GIS staff are stretched thin. A relatively small number of GIS faculty and professional staff administer the MGIS program, advise students and teach MGIS courses. This situation is exacerbated by a recent surge in the retirements of faculty who do quantitative teaching and research with MGIS students. While GIS faculty enjoy working with MGIS students, they face serious opportunity costs that include time not spent on research, service to their departments, and teaching in PhD and undergraduate programs. One indication of this need for more faculty is the fact that several areas are not taught or under-taught in when compared to a variety of peer institutions that also offer graduate GIS&T education, such as advanced spatial statistics, spatial programming, transportation, urban GIS, public health mapping and cartography. Similarly, UMN offers few undergraduate GIS courses. We instead are limited by and large to cross-listed graduate courses that are not pedagogically suited to many undergraduates. These restrictions in turn limit our ability to attract undergraduate majors to GIS, an area with obvious intellectual attraction and relevance to employment after graduation.

Second, attracting new faculty is a challenge. MGIS faculty have excellent personal and professional relationships but it is difficult to enlist new faculty. One key advantage of having the MGIS at the University of Minnesota is that our students work with an interdisciplinary faculty drawn from several colleges. Many professors report, however, that they face increasing administrative pressure to limit their involvement in MGIS in light of curricular needs in their home departments. One of the strongest pulls for faculty in other programs for working with the MGIS program is access to our students, whether as a member of our graduate faculty or agreeing to serve on student committees. This pull is offset by a major hindrance to developing formal MGIS relationships with new faculty, namely the university's current budget model, which gives the lion's share of tuition in a given course or program to the home college and department.

Third, we face growing competition from other GIS programs, particularly in the Twin Cities and Minnesota where MGIS has gone from being the only program of its kind to being one of several. Larger programs with far greater marketing budgets and online components, such as from Penn State and Redlands, increasingly command the national and international market. While we feel MGIS is sufficiently different from other programs to remain competitive at local to global scales, mounting competition and our limited marketing resources make it increasingly difficult to deliver our message to the best applicants.

Fourth, the University of Minnesota risks falling behind in GIS&T. GIS&T and cognate disciplines such as remote sensing, modeling and computer science are rapidly emerging as key intellectual foci across the information, natural and social sciences. Recent articles in *Nature* (Gewin, 2004) and the *Proceedings of the National Academy of Sciences* (Wachter, 2005) are but two examples of the growing number of articles that describe the importance of GIS&T to a wide array of disciplines as both an integrative approach and a research topic in and of itself. Key institutions such as Harvard University and the University of Pennsylvania are creating new graduate programs in GISc and spatial analysis.

In meeting these challenges, several issues are being considered. First, although growing the program is in some respects desirable it is very unlikely. Given that we have fixed costs, such as administration and technology, and spare capacity in classroom and laboratory seats for most GIS&T and related courses in most departments, increasing its size would make it more efficient overall. Growth is essentially impossible, however, without significantly diminishing the integrity of the MGIS degree or substantially increasing our marketing budget and faculty support for administration, teaching and advising.

Conversely, shrinking the program is also not the answer. The MGIS program was designed as a two-year masters program requiring thirty-five semester hours of coursework plus one to three 'Plan B' papers (non-thesis projects). This original design has been maintained over our first decade and continues to distinguish our program from many of our national competitors that have shorter or more narrowly defined programs. We therefore have little desire to reduce the total credits to create a one-year degree program. Nor can we move our courses on-line to foster distance-learning, as doing so is a very expensive proposition the online market is dominated by other programs, particularly ESRI and Penn State, and other entrants appear on a regular basis. However, even maintaining the program requires greater faculty support. Given that GIS faculty resources are stretched thin and that we have seen a decrease in non-GIS faculty who teach quantitative courses, we need several new faculty members in several key departments who have GISc as a research and teaching focus. Departments that are natural homes to these faculty include, but are not limited to, geography, forest resources, applied economics, public policy and computer science. At the same time, despite difficulties in doing so, we must broaden our efforts to encourage our students to work with faculty in other programs across campus. Our university is large and there is a diverse range of GIS-related interests and research occurring on our campus (at least fourteen units across several colleges offer some form of GIS). We consider interdisciplinarity as one of our major advantages over some of our national competitors, yet it is perhaps our greatest problem internally. As discussed throughout this chapter, the current tuition and funding structure within the university discourage faculty and departments from engaging in interdisciplinary teaching and research.

Finally, we must continue to innovate. We are actively exploring ways to streamline the program without sacrificing quality, as seen in our recent efforts to support advising through increased use of information technology and by offering greater structure in 'Plan B' paper formats. In order to maintain student quantity and quality, we must leverage our strengths: our reputation for excellence, being one of the few metropolitan MGIS programs, and having the local Twin Cities and on campus markets. Our ability to use these strengths is offset by growing competition and a marketing budget that is smaller by an order of magnitude than that of competing programs. A related concern, and one that is much harder to address, is the growing competition from other University of Minnesota graduate programs that incidentally have a GIS component.

12.4 Best practices for professional GIS education

Based on these experiences, we suggest some best practices to consider in the establishment and maintenance of a sustainable professionally-oriented masters level program (MGIS Program, 2007). These include:

- A need for a strong administrative home. The MGIS program has been fortunate to have a strong collegiate (College of Liberal Arts) and departmental (geography) home. Both have been supportive with budgets, curricular changes and staff support (initially, the geography department provided the support staff).
- A need for campus-wide partnerships. The MGIS program's success has been built, in part, on the strong relationships and faculty support from many other units. This has enabled us to offer expertise and coursework in computer science, remote sensing, biostatistics, planning and engineering.
- A need for pedagogical flexibility. Successful programs must undergo constant evaluation and curriculum modification where and when necessary. MGIS has modified degree requirements and coursework on several occasions based on feedback from students and the advisory board.
- A need to establish and utilize the expertise of an advisory board. The board should be composed of GIS professionals representative of different sectors in academia, private, non-profit and government.
- A need to conduct both external and internal assessments of the program to ensure high quality student services and a curriculum that meets market demands.
- A need to include highly qualified GIS practitioners/professionals as part of the teaching staff.

12.5 Conclusion: future challenges in GIS education

The future for professional education in geographic information science is bright. For instance, the Office of the President of the United States and the Department of Labor have identified geospatial technologies, alongside nanotechnology and biotechnology, as one of the three most important high-growth industries in the twenty-first century (Gerwin, 2004). For professionals seeking to enhance their GIS education, it is an opportune time to do so as there are many choices to meet their specific needs. It is important for existing programs as well as institutions interested in developing professional GIS programs to find their niche in the market in order to be successful and sustainable. To ensure that we can deliver the best possible GIS education to professionals, there are a number of broader challenges that need to be addressed. These can be posed in the form of potential discussion/research questions:

- What exactly is the nature of the distinction between academic and professional training for geospatial technologies?
- Do different educational institutions play different roles in providing academic vs. professional training?
- How may transfer of courses be best accomplished between institutions providing education from differing viewpoints?
- At what level should technology management be required for students in both tracks and where should the emphasis lie?
- Is a professional master's degree terminal (i.e., does not continue to the doctoral degree as with a traditional masters degree)?

References

DiBiase, D. (2004) Engaging stakeholders in program planning for an online master of GIS degree program. *Proceedings of the American Congress on Surveying and Mapping Annual Conference.* Nashville TN, 19 April.

DiBiase, D., M. DeMers, A. Johnson, K. K. Kemp, A. T. Luck, B. Plewe, and E. Wentz (eds) (2006) *Geographic Information Science and Technology Body of Knowledge.* Washington, DC: Association of American Geographers and University Consortium for Geographic Information Science.

Esri (2002) *Esri white paper on guidelines for developing a successful and sustainable higher education GIS program.* Redlands, CA: Esri Press.

Frank, A. U., and M. Raubal (2001) GIS education today: From GI science to GI engineering. *URISA Journal*, 13(2): 5–10.

Gaudet, C. (2001) *Workforce Development Models for Geospatial Technology.* The University of Southern Mississippi: Geospatial Workforce Development Center.

Gewin, V. (2004) Mapping opportunities. *Nature*, 427(22): 376–377.

MGIS Program (2007) *Master of GIS at ten years: Celebrating the past and planning for the future.* Department of Geography, University of Minnesota.

Phoenix, M. (2000) Geography and the demand for GIS education. *Association of American Geographers Newsletter*, June: 13.

Solem, M., L. Chalmers, D. DiBiase, K. Donert, and S. Hardwick (2006) Internationalizing professional development in geography through distance education. *Journal of Geography in Higher Education*, 30(1): 147–160.

UCGIS University Consortium for Geographic Information Science (1997) UCGIS education priorities white paper on professional education. http://www.ucgis.org/priorities/education/priorities/profeduc.htm (accessed August 12, 2011).

Wachter, K. W. (2005) Spatial demography special feature: Spatial demography. *Proceedings of the National Academy of Sciences*, 102: 15299–15300.

Wayne, C. (2002) GIS education for the working professional. *Geospatial Solutions*, July 2002.

Wikle, T. A. (1999) GIS education through certificate programs. *URISA Journal*, 11(2): 53–60.

Wikle, T. A., and G. A. Finchum (2003) The emerging GIS degree landscape. *Computers, Environment and Urban Systems*, 27(2003): 107–122.

13 Geospatial education at US community colleges

Ann Johnson
National Geospatial Technology Center, Del Mar College, Corpus Christi, Texas, USA

13.1 Introduction

This chapter documents the rise of geospatial programs at community colleges in the United States from the late 1980s to the present. It highlights key challenges that confront community colleges in the US as they work to respond effectively to current and future workforce needs. Community colleges are important to geospatial education because they account for such a large portion of the overall US higher education infrastructure. According to the Association of American Community Colleges (AACC, 2009), the nation's 1,175 two-year colleges enrolled 44 percent of US undergraduates – 11.7 million students – in 2009. In 2008 community colleges awarded 328,268 Certificates and 612,915 Associate of Arts or Science degrees. While many graduates enter or return to the workforce, many others transfer to four-year colleges and universities to pursue baccalaureate degrees.

The capacity of the community college sector to provide geospatial education and training has expanded dramatically since the mid-1990s, when the first attempts were made to identify providers in US higher education. In 1995, Jay Morgan and colleagues at Towson State University asked over 3,700 academic departments at higher education institutions worldwide if they offered or planned to develop GIS courses or certificate or degree programs (Morgan *et al.*, 1996). Only eight community colleges responded. Of those, only six indicated they had a GIS program, while two others stated that programs were planned. By contrast, more than 820 four-year colleges and universities responded that programs were in place or planned, or were presumed to have programs from data collected in earlier surveys.

It seems certain that many community colleges that offered geospatial education in 1995 did not respond to the Towson survey. In 1998, Johnson estimated that 150 community colleges offered GIS courses, certificates or degrees (Johnson, 1998). Her estimate was based on direct contact with community college clients in her role as a higher education manager at Esri (formerly Environmental Systems Research Institute, Inc.). The survey therefore overlooked institutions that offered GIS education and training but used other vendors' software

Teaching Geographic Information Science and Technology in Higher Education, First Edition.
Edited by David J. Unwin, Kenneth E. Foote, Nicholas J. Tate and David DiBiase.

products, such as Intergraph, AutoCAD or MapInfo. Even so, it's clear that the proportion of community colleges offering GIS education in the mid-1990s was much smaller than the proportion of four-year colleges and universities offering GIS education. Factors explaining the difference may include the distinctly different missions of two-year and four-year institutions, the availability of necessary computing hardware and software, and the differing levels of faculty expertise at the time.

More recently, a 2010 survey carried out by Lakeland Community College with support from the National Geospatial Technology Center of Excellence (a.k.a. GeoTech Center) suggests the extent to which geospatial education has caught on in US community colleges. Lakeland found that more than 445 community colleges – 38 percent of all 1,175 community colleges in the US – now offer at least one course related to geospatial technology (Rudibaugh and Ferguson, 2010). The survey also found that of the 445 colleges, 145 offer Certificate programs and 69 offer Associate Degrees (NGTC, 2011). The following section provides the background needed to understand this trend.

13.2 The community college mission

While some institutions are called junior or technical colleges, most in the US are referred to as community colleges. Recognizing that fact, the AACC changed its name from 'Junior' to 'Community' in 1991 (AACC, 2009). Community colleges may offer academic (for credit) programs or vocational (non-credit) programs – sometimes referred to as career and technical education (CTE) – or both. Some community college faculty members teach in both academic and vocational programs, while others specialize in one or the other. Some US states distinguish between 'community' colleges that confer academic credits that are transferrable to baccalaureate degree programs at four-year colleges and universities and 'technical' colleges that provide vocational training only. Some community colleges are expanding their reach beyond their local communities by offering online distance education. In general, however, the mission of these institutions is to provide low-cost, locally-accessible, lifelong learning opportunities that are responsive to the needs of the communities they serve (Sullivan, Brase and Johnson, 2008). Thanks in part to subsidies provided primarily by state governments, community colleges are relatively inexpensive – average annual tuition among US institutions was $2,402 in 2009, compared to $6,585 for four-year public institutions (AACC, 2009). California community colleges charge $26 per credit hour. Because of their relatively low cost, their non-competitive admissions requirements, and their emphasis on job placement, community colleges tend to attract more diverse student populations than four-year colleges and universities. Community college students include individuals who:

- did not complete high school;
- do not wish to matriculate in a four-year college or university directly after high school;
- only need a certificate or associate degree;
- want training and education to qualify to work in a particular occupation (e.g., respiratory therapist, emergency response, etc.);
- have not acquired an adequate educational background to enter a university;
- have a baccalaureate or higher degree but need to upgrade their education to meet the changing skills required to do their current job;
- need new skills to advance in their current career; or
- want to acquire entirely new skills to enter a new field.

Unlike most four-year colleges and universities, community colleges' first priority is to meet local workforce needs by preparing students for particular occupations. The research and development that gave rise to the geographic information science and technology (GIS&T) field therefore took place at universities (Longley *et al.*, 2005), not community colleges. Chrisman recounts meetings, conferences and training events at universities in the early 1960s (Chrisman, 2006), as well as early funding support from organizations like the National Science Foundation. University-based programs at Northwestern and Harvard Laboratory of Computer Graphics pioneered development of early software programs like SYMAP (Chrisman, 2006). Graduates went on to produce commercial software products including ERDAS in the mid 1970s and ArcInfo in 1982. MapInfo was commercialized in 1986 as a spin off from Rensselaer Polytechnic Institute, and IDRISI grew out of the geography program at Clark University in 1987.

Advances in computer capabilities, printers and printing techniques, programming languages and software all encouraged university departments to set up their own computer labs rather than be tied to a campus computing center (Chrisman, 2006). Graduate and upper division undergraduate programs were thus poised to create and advance GIS technology through graduate faculty and student research, as well as to teach emerging concepts and applications of the technology. In contrast, community colleges didn't participate in the development of the technology because their faculty members were focused on established workforce needs. For example, computer-aided drafting (CAD) was a popular topic at community colleges during the formative years of GIS&T. For some community colleges CAD served as a gateway to GIS due to their similar requirements for computing hardware and skills. CAD training programs that emphasized applications in environmental management or land use (e.g., forestry) more readily transitioned to GIS than those emphasizing engineering design. Their focus on undergraduate education, along with their lack of experience in attracting sponsored research projects, denied many community colleges the support needed to hire graduate teaching assistants or to equip their teaching labs adequately. Universities were more accustomed to applying for and managing grants and this helped many universities acquire the resources required to effectively use and advance GIS&T.

13.3 Early adopters of GIS at community colleges

Even into the early 1990s, computers at community colleges were most likely found in computer science departments, engineering programs or a few other technology-focused programs. Often these programs included computer-aided drafting (CAD) courses that used computers capable of running GIS software. Of the six community colleges that indicated they had a GIS program in the Towson survey of 1995, only one was in a geography department (Red Rocks Community College). Academic programs in forestry, agriculture, natural resources, science and engineering, and environmental technology accounted for the remaining five (at Central Oregon, Haywood, Spokane, Mesa and Lansing Community College, respectively). Two of the programs listed CAD software while others listed ArcInfo, MapInfo, IDRISI, AtlasGIS, ERDAS, GenaMap, GRASS and Intergraph MGE. Besides access to the necessary computing resources, what these early adopters may have had in common was local employers who sought graduates with GIS skills.

Faculty members who developed these programs shared their experiences through presentations at conferences like the national symposia on GIS in higher education (GISHE

'96 and '97). Abstracts from those events provide some insights. Unfortunately, few presenters submitted full papers for conference proceedings or subsequent journal publication, so most of their stories are undocumented. However, anecdotal evidence from personal communications with some of those involved is available to shed some light on how the early adopters got started. In general, the anecdotal evidence confirms that GIS programs at community colleges were most likely to appear where there was demand for workers from local businesses or government agencies.

Robert Welch created a GIS training program at Lansing (Michigan) Community College in 1986. This may have been the first such program at a two-year college in the US (Welch, 2001). Welch began working on the program curriculum in 1984 and received approval in 1985 (Welch, *personal communication* 2009). He was encouraged to start the program by local users including the Michigan Department of Transportation. Working with IBM on an early software program, he recalls crashing the college's computing system when his IT person tried to install it. Welch was advised that the software 'could be installed, but not to add data or do analysis as it would crash the system again.' His administration was skeptical about starting a GIS training program due to the few requests from industry for trained workers, but also because of the need for more hardware and software. Welch applied to Intergraph to become a Center of Excellence and was granted software (Intergraph MGE) and workstations to start the program. This innovative program has continued to grow and evolve. With a new focus on 'green' technology, Lansing's program increased from 60 student applicants in 2008 to 368 in 2009.

Throughout the 1980s most colleges were hampered by a lack of faculty development opportunities and curriculum resources. Then, in 1988, the National Science Foundation founded the National Center for Geographic Information and Analysis (NCGIA) at the University of California at Santa Barbara, the State University of New York at Buffalo, and the University of Maine. One of the Center's goals was to enhance the quality of GIS education (Kemp, 1996). It published a Core Curriculum in GIS in 1990. Consisting of course notes on seventy-five topics, the Core Curriculum was used by early adopters at community colleges as well as university faculty worldwide.

Other community colleges began to get involved. The Community College of Southern Nevada formed a GIS program in 1991 in a department called Computing and Information Technology in the College's Business and Industries Division. It was funded through the Yucca Mountain Project with support from the Clark County user community. Clark County government was itself an early adopter of GIS technology; in 1991 it implemented a GIS system for the Clark County Assessor office (Clark County, 2009). Meanwhile, Southern Nevada Community College taught ArcInfo on Sun platforms using SunOS (Phalke, *personal communication* 2009). In 1992, Central Oregon Community College's GIS program began with a chance hallway conversation between two faculty colleagues, Art Benefiel and John Schaeffer. They discovered a mutual interest in doing something with GIS (Benefiel, *personal communication* 2009; Schaeffer, *personal communication* 2009). Benefiel offering his first courses in 1993 using GeoSQL, built around AutoCAD and a SQL database. The program benefited from having a receptive clientele in Central Oregon's forestry program, and from a faculty familiar with CAD.

Few textbooks were available through the 1980s. Those that did exist were written with upper division students in mind, and were therefore not well suited to the lower division students at community colleges. By the early 1990s, however, the situation improved. *Geographic Information Systems: A Guide to the Technology* appeared in 1991 (Atenucci *et al.*, 1991) followed by the first English edition of Tor Bernhardsen's introductory text in 1992 (Bernhardsen, 1999). Soon Esri began publishing laboratory books that included data and step-by-step exercises, including the *Getting to Know* series that helped colleges offer hands-on GIS exercises without the faculty having to write every lab.

13.4 Later adopters

By the mid-1990s, many community colleges were better positioned to begin teaching GIS. Computer capabilities had increased, software products were beginning to mature, and access to geospatial data had improved, including via the World Wide Web (Bernhardsen, 1999). More and more working professionals sought to add or update their GIS skills at community colleges. As more community colleges got involved, there was a growing need for opportunities for faculty to learn the technology well enough to begin to teach it. The National Science Foundation (NSF) helped by funding several early initiatives to help develop curriculum and provide faculty development opportunities. NSF funding levels for community college programs increased from $7 million in 1993 to $35 million in 1995, $22 million of which was awarded through its Advanced Technology Education (ATE) program (NSF, 1996). ATE continues to be an important source of funding for GIS-related programs at community colleges. Three grants awarded in 1995 were specifically intended to build capacity to provide GIS and Global Positioning System (GPS) education at community colleges.

One of these three key grants was to Indiana State University (DUE 9553694). Its *GIS for the 21st Century* project aimed to introduce community college faculty to the principles of GIS in sufficient depth to prepare them to develop and teach classes about it (NSF, 1996). The grant proposed to train twenty-five two-year college faculty members in three week-long workshops over two consecutive summers in 1996 and 1997. University educators provided lectures about GIS concepts, while experienced community of college educators provided hands-on practical training. The community college educators included Schaeffer and Benefiel from Central Oregon Community College and J.R. Peay from Southern Nevada Community College. Although the grant provided support for only twenty-five participants, five additional community college educators were allowed to participate if they paid for their own housing and transportation. In all, approximately thirty individuals completed the two-summer program. The length of time, the hands on training in IDRISI, ArcView 3.0, ArcInfo, AutoCAD and MapInfo, and the networking opportunities provided by working together for six weeks over two summers succeeded in preparing the educators to begin teaching GIS. Some returned to their campuses and started programs or added courses immediately (e.g., at San Bernardino Valley and Pierce Community Colleges). Others applied for and were awarded NSF grants of their own (e.g., Houston, Hawkeye and Henry Ford Community Colleges). Some others took longer to get started, but eventually most introduced GIS into existing classes or created new classes. Most participants stayed in touch long after the workshops. While some have since retired from teaching, many remain active in community college GIS education. This includes several individuals involved in NSF Centers of Excellence, including Mike Rudibaugh at Lakeland Community College, Terry Brase at Kirkwood Community College, and the author.

In 1995, another consequential grant was awarded to the Northwest Center for Sustainable Resources for a project called *Visions for Natural Resource Education and Ecosystem Science for the 21st Century* (ATE DUE 9553760). This center focused on agriculture, wildlife, fisheries and forest management and included GIS and GPS as technology areas. GIS workshops were held for faculty. One of the most important outcomes, however, was a DACUM job analysis for the occupation of GIS Specialist. DACUM stands for Developing a Curriculum. It is a facilitated process in which expert workers representing a single occupation come together as a panel and define the duties and tasks, along with knowledge, skills and tools needed for that occupation (DACUM, 2009). The outcome is a DACUM chart that specifies what should be included in a curriculum designed to provide the practical skills for that occupation.

In 1995 NSF also funded the *Community College for Innovative Technology Transfer* (CCIT) project hosted by Prince George's Community College with twelve colleges linked to a local National Aeronautics and Space Administration (NASA) Center. This early grant enabled project staff and NASA education specialists to provide remote sensing-oriented workshops and curriculum development support for community college educators.

In 1996, the NCGIA embarked on a new initiative to develop a *GIS Core Curriculum for Technical Programs* (CCTP). This web-based resource began with a meeting in Santa Barbara in August 1996 and included faculty from several of the early community college GIS programs. Unlike the original NCGIA Core Curriculum in GIS, however, the CCTP did not succeed in attracting many volunteers to create content, nor did it gain wide acceptance. It may have been that the later CCTP was less sought after than the earlier Core Curriculum because other texts and materials had become available. The initiative did succeed in bringing educators together to talk about common needs and resources, and those conversations helped build a sense of community among GIS educators at community colleges.

Other sponsors, including state government agencies, the US Department of Labor (DOL) and NASA, began funding grants related to GIS at two-year colleges. In the summer of 1996, Joan Clemens, the Community College Affinity Group advisor for the Association of American Geographers (AAG), hosted a meeting at the University of California at Los Angeles for community college faculty interested in GIS. At the meeting, Marshall Gartenlaub volunteered to write a grant proposal to help fund workshops to help train faculty interested in learning GIS. Gartenlaub wrote the proposal in two weeks. Soon the California Community College Chancellors Office awarded the funds requested to start the *California Community College GIS* (C3GIS) support network. Over the next two years, more than 120 community college educators received hands on training in GIS during four-day workshops developed and conducted by C3GIS. This project helped California Community Colleges get a head start in GIS education. Today, more than half of California's 110 community colleges offer geospatial courses or programs. In 2007 the Chancellor's office awarded a second grant to sustain the C3GIS support network for eighteen more months. C3GIS became *California Community College Geospatial Technology Information Support*. It provided workshops, webinars and created resources for two-year programs (C3GIS, no date).

By 1997, more than 250 community colleges across the US had access to Esri software (Phoenix, 1997). By 1998, more than 150 schools listed geospatial courses, certificates and degrees (Johnson, 1998). Additional NSF- and DOL-funded projects help prepare faculty, create curriculum and develop programs. But major questions remained for community colleges that were considering developing GIS education and training programs, including:

- Which academic department(s) should host GIS courses?
- Should it be an academic program or a vocational (CTE) program?

- Should an institution develop a course or courses *about* GIS, or discipline-based courses taught *with* GIS?
- Should it be focused on concepts or hands on software use?
- Should field data collection and GPS be included?
- Should remote sensing be included?
- If it is a program, should it lead to a certificate or associate degree?
- What is the most effective method to teach GIS – a short course, distance, semester-long course?
- Which is the largest clientele for this type of program – current workers, job seekers, or traditional students?
- How do faculty learn the technology and how do they keep up to date?
- How should the software be managed – within a department or by the campus IT department?

There were also questions about how to determine the need for such programs in the absence of relevant occupational codes and employment estimates at DOL. While some programs did focus on software training, most included at least some fundamental concepts. However, there was little consistency in the number of courses or the number of semester credits that constituted a community college certificate program. Existing programs ranged from as few as 6 to more than 30 credits. There was also growing realization that teaching GIS effectively requires different teaching methods than most other subjects. Experience showed that disjointed lecture and lab sessions were less effective than formats in which lectures were shorter than usual and hands-on activities were emphasized. Typical college room arrangements and scheduling patterns didn't accommodate such integrated approaches, however.

Then in 1998 the NSF funded the *GIS Access* grant (Cypress College DUE ATE 9850306). This project provided workshops based on a problem-based, active learning pedagogy in which participants learned GIS by addressing authentic problems and data. By the second year of the grant, this semi-structured, open-ended methodology was modified to include short lectures on important concepts reinforced immediately by hands-on software use. Participant evaluations indicated that this revised approach was effective. The NSF ATE program continued to fund grants related to GIS and geospatial technology including many grants awarded in the late 1990s and early 2000s. A search of the NSF website for grants related to GIS and community colleges in the period from 1990 to 2001 found the following awards totaling $2,843,726. Hawkeye Community College DUE ATE 9553751, 1995; Henry Ford Community College DUE ATE 9752086, 1997; Alamo Community College DUE ATE 9750585, 1997; Houston Community College DUE ATE 9850344, 1998; Cypress College DUE ATE 9850306, 1998; Evergreen Valley College DUE ATE 0101576, 2001; Burlington County College, DUE 0088634, 2001.

13.5 Development of geospatial curricula and programs

Although they were designed and carried out primarily with four-year programs in mind, two national-scale initiatives were of interest to community college educators. One was the *Model Curricula* project organized by the University Consortium for Geographic Information Science (UCGIS). Another was NASA's Geospatial Workforce Development initiative.

UCGIS identified eight education challenges during its 1997 summer assembly. One challenge led to the formation of a Model Curricula Task Force in 1998, chaired by Duane Marble. The Task Force issued a draft *Strawman Report* in 2003. This outlined a comprehensive vision for GIS&T curricula reform at a national scale. Although progress slowed for a time, momentum was regained in 2005 within UCGIS Education Committee, chaired by David DiBiase. The committee sought support needed to advance the entire Model Curricula vision. However, encouraged by project sponsor Esri to complete a deliverable within a year, the Committee then focused on a core element of the vision – a *GIS&T Body of Knowledge* (*BoK*). Emulating similar efforts in computer science and related fields, the *BoK* presents 1,660 educational objectives in a hierarchical format of 10 'Knowledge Areas,' 73 'Units' (including 26 'Core' Units), and 329 'Topics.' (DiBiase, *et al.*, 2006).

Preceding the initial UCGIS project, NASA in 2001 funded a project to bring together workforce development specialists from the University of Southern Mississippi with representatives of geospatial professional organizations, education, business, industry and government to study the key roles and competencies needed by employees in the geospatial industry. A lasting impact of the original *Geospatial Technology Competency Model* the project produced was to identify four categories of core competencies, including not only technical and analytical competencies, but also business and interpersonal competencies needed by the geospatial workforce (Gaudet *et al.*, 2003). Nearly a decade later, DOL's new *Geospatial Technology Competency Model* would reflect a similar holistic approach to competency modeling (DiBiase *et al.*, 2010).

While the outcomes of the UCGIS *BoK* and NASA initiatives were of interest to community college educators, many felt that they were too high level for their programs and students, and that a different kind of support for community college curriculum development was needed (Sullivan *et al.*, 2008). This mismatch is partly explained by the fact that both the *BoK* and the original GTCM were developed primarily with four-year baccalaureate programs in mind. However, four-year baccalaureate programs in GIS or GIS&T were extremely rare; most educators at four-year colleges and universities concentrated on upper-division undergraduate and graduate curricula. Consequently, little guidance was available to community colleges offering lower-division undergraduate courses and programs. The lasting contributions of the *BoK* and original GTCM were to lay the groundwork for subsequent efforts like the new DOL GTCM (DiBiase *et al.*, 2010) and DACUM job analyses (Johnson, 2010), both carried out by the GeoTech Center.

Other projects around this time concentrated specifically on community colleges. In 2005 a group led by the National Council for Geographic Education (NCGE) – with support from NASA, the US Geological Survey (USGS), and Esri – focused on capacity building for remote sensing education at two-year schools. A workshop held at NSF in 2005 (NCGE, 2006) led to the *Integrated Geospatial Education and Technology Training* (iGETT, NSF DUE 0703185) project in 2008. Begun in 2008, this three-year project provided forty educators (mainly from community colleges, but also a few high schools and universities) with training in remote sensing concepts and techniques.

Many educators became convinced that community colleges could provide the capacity to train geospatial workers, but lacked an organization to provide the national-scale coordination to guide and support this effort. A report commissioned by DOL and prepared by the Geospatial Information & Technology Association (GITA), the Association of American Geographers (AAG), and the Wharton School of the University of Pennsylvania seconded this concern (GITA, 2006). The study's stated goals were to define the geospatial industry and its workforce needs. Its *Phase I Report* recommended that two-year colleges play a stronger role in geospatial education (GITA, 2006). These sentiments would soon coalesce into a major proposal to form the community college equivalent of NCGIA.

13.6 National Geospatial Technology Center of Excellence – the GeoTech Center

In the summer of 2005 a group of community college educators met at the Esri Education User Conference in San Diego to discuss support for GIS programs. The meeting led to a proposal to NSF to define the need and potential for national center to catalyze and coordinate capacity-building efforts at US two-year institutions. The proposal was funded in June 2006 (NSF DUE 0603424). To identify the needs, project personnel conducted online surveys of college instructors. Survey responses revealed ten key issues of concern to community college faculty members. The project sponsored a national forum in 2007 to discuss the issues and produce a set of draft recommendations. The process culminated in *A Plan for the National Coordination of Geospatial Technology Education from a Community College Perspective* (Sullivan *et al.*, 2008), a report that presented a prioritized list of goals for a national center. Forum participants began to discuss how to work together to enact the report's findings. In 2007, these efforts coalesced around a proposal to NSF's ATE program for a National Geospatial Technology Center of Excellence (NGTC), which has since become known as the GeoTech Center. This collaborative effort, lead by principal investigator Philip Davis at Del Mar College in Corpus Christi, Texas, was funded in September, 2008.

The GeoTech Center's aims are to provide:

- an online repository of resources for geospatial educators including sample curriculum, articulation agreements, course notes, lessons and best practices;
- a national voice for two-year colleges in geospatial technology education;
- a core competency model for GIS technicians;
- faculty development and training opportunities;
- networking and collaboration opportunities between colleges, schools, universities and industry; and
- capacity building to prepare community colleges to help meet demand for a growing geospatial workforce.

As of this writing, partners include eight community colleges – Del Mar, Cayuga, Central New Mexico, Central Piedmont, Gainesville State, Kentucky Community & Technical College System, Lakeland and Southwestern – and two universities – San Diego State University and The Pennsylvania State University. One kind of support many educators request is guidance about curricula and program planning. To this end the GeoTech Center sponsored a novel effort to consolidate multiple DACUM job analyses for GIS technicians (Johnson 2010). It also facilitated completion of a new GTCM for DOL's Employment and Training Administration (DiBiase *et al.*, 2010). For the first time, the higher education sector – including community colleges – has a holistic, comprehensive and industry-validated framework for assessing the alignment of curricula with workforce needs.

Another important goal of the center is to be a voice for community colleges to help foster communication and cooperation among community college educators, government agencies, professional organizations, other educational institutions, and employers. The GeoTech Center also offers a variety of ways for faculty and students to learn and share, including webinars, instructor-lead workshops and social networking.

The GeoTech Center website includes a resource repository that provides a way for educators to find or share resources. The kinds of resources that community college educators say they want include best practices for developing GIS&T programs, model course outlines, suggestions on what is needed to set up and maintain a lab facility, and best practices in working with information technology (IT) departments. Centralization and standardization of IT practices seems essential to maintain computers and Internet networks efficiently across a campus. Unfortunately, geospatial programs have different requirements than most other campus users. These include support for installation of software and software updates, data storage and access, student project storage, student administrative rights, network access and downloading data. Further challenges arise as college IT departments in the US implement increasingly stringent security measures in response, at least in part, to federal legislation concerned with privacy of student records. Colleges need the help of the GeoTech Center to work with their IT departments for the most effective ways to set up and support GIS and geospatial programs.

13.7 Future geospatial education at community colleges

The geospatial industry continues to evolve and spread into more occupational domains. New technologies and uses are emerging that require workers to update their knowledge and skills continuously. The GeoTech Center is actively working on ways to help colleges build, expand and update GIS programs to meet the lifelong learning needs of the geospatial workforce. The Center aims to help community colleges keep current with these trends by providing resources to meet their students and communities' needs. Specific focus areas include:

- Providing alternative ways for students to access the needed software and data less dependence on campus lab facilities and IT support;
- Providing faculty professional development opportunities through new online or web-based formats;
- Providing a one-stop location for help with program planning, development and self-assessment based on the new GTCM;
- Serving as a clearinghouse to help colleges learn about new or emerging geospatial technology trends; and
- Keeping community colleges aware of new trends in education including those that deal with infrastructure changes or mission changes such as allowing more units to be taken at colleges that count toward four-year degrees.

More and more college administrators are using GIS for the business side of higher education. One of the first such efforts was undertaken in 1998 by Henry Ford Community College in Michigan. The college used GIS to find out where their students lived and worked in relationship to the location of the college. They found that students were coming from unexpected places. This discovery helped them market programs to perspective student populations more effectively. Similar efforts by the City College of San Francisco to map their campus and Tacoma Community College to improve marketing and outreach are providing administration with new information and tools to manage their campuses.

Because public awareness of geospatial technology and associated career opportunities remains low, some colleges still struggle to build or sustain enrollments. The GeoTech Center promotes several tactics to help colleges with recruit and retain students. One is to work more closely with local high schools to market career-oriented college programs and develop articulation agreements with high schools and universities. More colleges are also working to have GIS courses approved as general education offerings as a way to bolster enrollment.

Miller (1988) defines general education as '. . . the conscience of higher education, the part of a university that is concerned most directly with the individual student's responsibility to society at large.' Community colleges have an opportunity to promote geospatial awareness courses that foster spatial literacy and spatial thinking abilities that enrich students' understanding of the local, national and global communities in which they live. A survey undertaken by the Association of American Colleges and Universities (AACU, 2009) suggests that more colleges are turning to general education courses that emphasize integrative approaches to contemporary societal challenges like sustainability. Characterized as a green technology by DOL, GIS and the spatial thinking abilities it supports can enrich general education curricula and create new opportunities for easing transfers of academic credit between two-year and four-year institutions. For example, Southwestern College, among others, has developed courses that qualify for general education as well as articulate with San Diego State University, a four-year undergraduate and graduate research institution (Tsou and Yanow, 2010).

Anthony Salcito, Microsoft's Vice President for Worldwide Education states in his blog *Education Trends to Watch for in 2010* that educational institutions will need to be more focused on directly connecting education programs and the workforce (Salcito, 2010). Salcito also suggests that blended learning (hybrids of online and classroom instruction), new devices including mobile technology, and cloud computing are becoming increasingly important. His predications span the entire formal education enterprise, not just geospatial education and training at community colleges. Still, it is trends like these that the GeoTech Center needs to help community college educators who specialize in GIS and related technologies to keep on top of.

Unprecedented numbers of students at all educational levels are becoming aware of spatial thinking and visualization through tools like Google Earth. The challenge and opportunity for educators is to reveal to students the potential for more sophisticated problem solving and insights provided by more sophisticated software tools and analyses. And as students flock to new education and training programs such as 'green' energy and sustainability studies, community college educators are challenged anew to demonstrate the relevance of geospatial technologies and methods. Lifelong learning is as much a necessity for GIS educators as it is for GIS professionals. Supporting this need is the GeoTech Center's core mission.

US president Barack Obama has emphasized the central role of community colleges in stimulating and sustaining economic recovery (Obama, 2009). Obama announced a \$12 billion community-college initiative to help colleges update and upgrade facilities, develop new technologies and increase graduation rates. Community colleges interested in

GIS need to make the most of this and other opportunities to help increase the capacity and quality of GIS education across their campuses. The GeoTech Center may be the community college sector's best hope for fulfilling its potential to help prepare the nation's geospatial workforce for continuing success.

References

American Association of Community Colleges (2009) Community college facts at a glance. http://www.aacc.nche.edu/AboutCC/Documents/fastfacts2009.pdf (accessed March 4, 2011).

Antenucci, J. C., K. Brown, P. L. Croswell, and M. J. Kevany (1991) *Geographic Information Systems: A Guide to the Technology*. New York: Van Nostrand Reinhold.

Association of American Colleges and Universities (2009) New survey finds colleges moving away from pure 'cafeteria-style' general education requirements, with only fifteen percent now using distributions requirements alone. http://www.aacu.org/press_room/press_releases/2009/membersurvey092.cfm/ (accessed March 4, 2011).

Bernhardsen, T. (1992) *Geographic Information Systems*. Arendal, Norway: Viak.

California Community College Geospatial Technology Information Support (n.d.) http://c3gis.net (accessed February 20, 2011).

Chrisman, N. (2006) *Charting the Unknown: How Computer Mapping at Harvard Became GIS*. Redlands, CA: Esri Press.

Clark County, Nevada (2009) *Clark County Assessor's History*. http://www.accessclarkcounty.com/100/Pages/History_Assessor.aspx.

DACUM.org (2009) An Online Resource for Occupational Analysis. http://www.dacum.org/ (accessed August 14, 2011).

DiBiase, D., M. DeMers, A. Johnson, K. Kemp, A. Luck, B. Plewe, and E. Wentz (eds) (2006) *The Geographic Information Science and Technology Body of Knowledge*. Washington, DC: Association of American Geographers and University Consortium for Geographic Information Science.

DiBiase, D., T. Corbin, T. Fox, J. Francica, K. Green, J. Jackson, G. Jeffress, B. Jones, B. Jones, J. Mennis, K. Schuckman, C. Smith, and J. Van Sickle (2010) The new geospatial technology competency model: Bringing workforce needs into focus. *URISA Journal*, 22(2): 55–72. http://www.urisa.org/files/URISA%20Journal%20Vol.22%20Issue%202.pdf (accessed March 4, 2011).

Gaudet, C., H. Annulis, and J. Carr (2003) Building the geospatial workforce. *URISA Journal* 15(1): 21–30. http://www.urisa.org/files/Gaudetvol15no1.pdf (accessed August 14, 2011).

Geospatial Information Technology Association and Association of American Geographers (2006) *Phase I Report: Defining and Communicating Geospatial Industry Workforce Demand*.

GeoTech Center (2010) GeoTech – Empowering colleges, expanding the geospatial workforce. http://www.geotechcenter.org/ (accessed March 4, 2011).

Johnson, A. B. (1998) Community college GIS programs – filling a vital niche in GIS professional education. GISED Conference. Ypsilanti, MI.

Johnson, J. (2010) What GIS technicians do: A synthesis of DACUM job analyses. *URISA Journal* 22(2): 31–39. http://www.urisa.org/files/URISA%20Journal%20Vol.22%20Issue%202.pdf (accessed March 4, 2011).

Kemp, K. (1996) The original NCGIA Core Curriculum in GIS. Second International Symposium of GIS in Higher Education, GISHE '96. Columbia, MD. http://www.ncgia.ucsb.edu/conf/gishe96/program/ncgiacores.html (accessed September 10, 2009).

Longley, P. A., M. F. Goodchild, D. J. Maguire, and D. W. Rhind (2005) *Geographic Information Systems and Science*, 2nd ed. Chichester: John Wiley & Sons.

Miller, G. (1988) *The Meaning of General Education: The Emergence of a Curriculum Paradigm*. New York: Columbia University, Teachers College Press.

National Research Council (2006) *Learning to Think Spatially: GIS as a Support System in the K-12 Curriculum.* Washington, DC: The National Academies Press.

National Science Foundation (1996) *NSF 96-83 Activities in Support of Two-Year College Science Mathematics, Engineering, and Technology Education – FY 1995 Highlights.* http://www.nsf.gov/pubs/stis1996/nsf9683/nsf9683.txt (accessed August 14, 2011).

National Center for Geographic Education (2006) *Integrating Geographic Information Systems and Remote Sensing for Technical Workforce Training at Two-Year Colleges.* National Council for Geographic Education Geospatial Education Workshop Report. http://www.aag.org/roundtable/resources/ncge-report-2-23-2006.pdf.

Morgan, J. M., B. Fleury, and R. A. Becker (1996) *1996 Directory of Academic GIS Education.* Dubuque, Iowa: Kendall/Hunt Publishing Company.

Obama, B. (2009) *The White House excerpts based on the president's speech at Macomb Community College, Warren, Michigan, July 14, 2009.* http://www.whitehouse.gov/the_press_office/Excerpts-of-the-Presidents-remarks-in-Warren-Michigan-and-fact-sheet-on-the-American-Graduation-Initiative/ (accessed August 14, 2011).

Phoenix, M. (1997) GIS in community colleges – Where we stand. GISHE '97, Third International Symposium on GIS and Higher Education. Chantilly, VA.

Rudibaugh, M., and B. Ferguson (2010) Mapping geospatial education at US community and technical colleges. *URISA Journal* 22(2): 41–44. http://www.urisa.org/files/URISA%20Journal%20Vol.22%20Issue%202.pdf (accessed August 14, 2011).

Salcito, A. (2009) *Education trends to watch for in 2010.* (Blog). http://blogs.msdn.com/microsoftuseducation/archive/2009/12/28/education-trends-to-watch-for-in-2010.aspx (accessed August 14, 2011).

Sullivan, D., T. Brase, and A. B. Johnson (eds) (2008) *A Plan for the National Coordination of Geospatial Technology Education from a Community College Perspective.* http://www.marinetech.org/workforce/geospatial/GSTreport7_17_Final.pdf (accessed September 1, 2009).

Tsou, M.-H., and K. Yanow (2010) Enhancing general education with geographic information science and spatial literacy. *URISA Journal* 22(2): 45–54. http://www.urisa.org/files/URISA%20Journal%20Vol.22%20Issue%202.pdf (accessed August 14, 2011).

Welch, R. I. (2001) Community colleges: the front line in GIS education. *Proceedings of the Twenty-First Annual Esri User Conference*, San Diego, CA.

14

The GIS Professional Ethics project: practical ethics for GIS professionals

David DiBiase[1], Francis Harvey[2], Christopher Goranson[3] and Dawn Wright[4]

[1]*John A. Dutton e-Education Institute, Penn State University, University Park, Pennsylvania, USA*
[2]*University of Minnesota, Minneapolis, USA*
[3]*GIS Center, Department of Health and Mental Hygiene, City of New York, New York, USA*
[4]*Department of Geosciences, Oregon State University, Corvallis, Oregon, USA*

14.1 Introduction

Engagements with 'GIS ethics'

Governments, militaries, commercial enterprises and other interests have relied on maps and mapping for centuries and upon aerial surveillance technologies since World War I. Computerized geographic information systems, digital remote sensing, and satellite navigation systems are relatively recent developments. As these automated technologies matured and their applications became widespread in the late 1980s and early 1990s, scholars and practitioners began to express concerns about the ethical implications of their use. Brian Harley (1988) was in the vanguard of scholars who questioned the assumption that maps are impartial and value-neutral depictions. By 1991, he challenged map makers to consider whether there could be 'an ethically informed cartography, and if so, what should be its agenda?' (Harley, 1991, 13)

At the same time, Pickles (1991) highlighted the use of GIS as a surveillance technology, while Smith (1992) alleged that the makers and users of geospatial technologies were complicit in the killings associated with what he considered to be a morally questionable Gulf War. By 1995, a substantial literature focused on ethical and epistemological critiques of GIS and related technologies had appeared (e.g., Pickles, 1995), and a widening gulf of misunderstanding and mistrust had separated critical scholars from proponents and practitioners of GIS and related technologies (Schuurman, 2000).

Meanwhile, GIS began to show signs of coalescence as a distinct (if heterogeneous) profession. Among the earliest considerations of professional ethics in cartography and GIS

Teaching Geographic Information Science and Technology in Higher Education, First Edition.
Edited by David J. Unwin, Kenneth E. Foote, Nicholas J. Tate and David DiBiase.

was an 'ethics roundtable' published in 1990 (McHaffie *et al.* 1990). Contributors identified implications of inaccurate maps and data, intellectual property issues, and conflicts of interest as important ethical issues. Soon thereafter, Monmonier (1991, 1996) pointed out ways in which maps can be used to mislead decision-makers and the public, and proposed design guidelines to foster ethical practice by cartographers. Speaking to the challenges for the nascent profession, by 1993, Craig had laid the groundwork for a GIS Code of Ethics (Craig, 1993). Onsrud (1995) recommended a survey of the moral reasoning of GIS professionals in response to a set of 'ethical conflict scenarios' as a way to gauge the extent to which moral consensus about GIS practice exists. A code of conduct derived from 'observation and analysis of current practice contexts and moral conditions,' Onsrud (1995, 94) wrote, should be 'evaluated and honed by continuously reassessing (the Code's) conformance with theory.'

Compliance with a GIS Code of Ethics (written primarily by Craig) is included among the requirements for certification as a GIS Professional (GISP) by the GIS Certification Institute (http://www.gisci.org), a program that began accepting applications in 2004. GISCI's Ethics Committee now seeks to compile a collection of authentic case studies to help GISPs develop moral reasoning skills (Craig, 2006). Meanwhile, following the lead of more established fields like computer science, the University Consortium for Geographic Information Science (UCGIS) developed a *Geographic Information Science and Technology Body of Knowledge* (*BoK*, DiBiase *et al.*, 2006). Published by the AAG, the *BoK* highlights such ethical and legal issues as privacy, access, intellectual property, and others, among the 329 topics that comprise the GIS&T knowledge domain. 'Ethical aspects' is included among the 'core' units that UCGIS recommends as part of every geospatial certificate and degree program curriculum.

As GIS&T continues to cohere into a distinct field, emerging technologies introduce increasingly worrisome ethical challenges for academics and professionals, including such location-based services as 'human tracking.' Dobson and Fisher (2003, 47) challenge society to 'contemplate a new form of slavery characterized by location control', arguing that '. . . the countless benefits of [location-based services] are countered by social hazards unparalleled in human history.' Clearly, there is an urgent need for practical ethics education that bridges the gap between academic engagement and professional practice in GIS&T.

Lessons learned from computer ethics education

The Association of Computing Machinery (in collaboration with the IEEE and other organizations) has issued a series of recommended curricula for undergraduate degree programs in computer science since 1968. From the outset, the recommendations included topics related to ethics and social issues. By 1994, however, observers concluded that the recommended curricula 'fell short of providing sufficient detail and guidelines about how to implement [understanding of the social and ethical context of computing] within the curriculum' (Martin and Weltz, 1999, 7) In response, NSF funded a project called *ImpactCS* to 'define the core content and methodology for integrating social impact and ethics topics across the computer science curriculum' (Martin and Weltz, 1999, 7) Following this effort, the *Computing Curricula 2001* (ACM/IEEE, 2001) recommended 'Social and Professional Issues' as one of fourteen knowledge areas, including the core unit 'SP4 Social and Ethical Responsibilities.'

To qualify for accreditation by ABET's Computing Accreditation Commission, undergraduate degree programs in computer science must demonstrate that there is 'sufficient coverage of social and ethical implications of computing to give students an understanding of a broad range of issues in this area' (Computing Accreditation Commission, 2004, 4). A 2005 survey of a quarter of the ABET-accredited computer science programs revealed that 55 percent of them meet this standard by teaching their own computer ethics course, 30 percent meet the standard by incorporating discussions of social and ethical implications of computing into other computer science courses, and only 15 percent of them outsource ethics to other academic departments, typically philosophy (Quinn, 2006a, b). In other words, addressing ethical issues 'in house' is the norm for accredited undergraduate computer science programs. We believe that the same should be true of in GIS&T education. As Davis (1990, 37) points out, 'being concerned about the ethics of one's profession is a professional virtue. A professional [GIS&T] teacher can teach that virtue by example in a way a philosopher cannot.'

Articles that describe, prescribe and evaluate computer ethics education are plentiful. Many describe teaching strategies (e.g., Wahl, 1999); face-to-face and computer mediated discussion (e.g., Sanders, 2005; Grodzinsky *et al.*, 2004), interactive tutorial and decision support software (e.g., Goldin, Ashley, and Pinkus, 2001; Robbins, Wallace, and Puka, 2004), and student evaluation methods including rubrics (Moskal, Miller, and King, 2002) and peer review (e.g., Grodzinsky *et al.*, 2004). A common theme in the computer ethics education literature is the goal of giving the students the opportunity to transform themselves. As Martin and Holz put it: 'Our belief is that ethics cannot be taught; rather what can be taught is a framework for evaluating ethical dilemmas and making decisions. In accepting the premise that technology is value-laden, we stress the need to teach a methodology of explicit ethical analysis in all decision-making related to technology' (Martin and Holz, 2005, section 1.2).

14.2 GIS Professional Ethics project

Unlike computer science, most of the academic disciplines that offer degree programs with specializations in GIS&T are not specifically accredited in the US, even though an estimated 50,000 students enroll in GIS classes yearly in US higher education institutions. For this reason GIS&T curricula vary widely, and few require formal ethics training. Consequently, research publications on ethics education in the context of GIS&T are rare. Rarer still are case studies that illuminate the distinctive ethical problems that confront geospatial professionals. The GIS Professional Ethics project (http://gisprofessionalethics.org) promotes practical ethics training for current and aspiring GIS&T professionals by creating and sharing a collection of open educational resources, especially case studies, and evaluating their effectiveness within three graduate programs in the US. These materials are available for use in GIS professional education in the US and internationally. To launch the project and take the diversity of US higher education into account, in 2007 the US National Science Foundation sponsored design, development and evaluation of graduate GIS ethics seminars within Departments of Geography and Geosciences at the University of Minnesota, Oregon State University, and Penn State University.

At the University of Minnesota, material developed in the project is included in two courses offered in the Professional Master's of GIS degree program (Chapter 12). Taught each fall, students in GIS 8501 spend one week on an introduction to applied ethical approaches and the seven-step method (Table 14.1). Small groups of students apply this method to a case from the gisprofessionalethics.org website. In GEOG 5563, *Advanced Geographic Information Science*, a required course for students in the program, the seven-step method is revisited and students complete a take-home assignment using the method with a case from the gisprofessionalethics.org website. Approximately twenty-four students are introduced to ethical issues each year through these two classes. At Oregon State University, the online course GEO 567, *Responsible GIS Practice*, includes a review of ethical theories and processes of moral reasoning, extensive discussion of papers and case studies, and required interviews with practitioners and development of additional original case studies based on those interviews. Around thirty students take this course each year by way of offerings in the fall and winter terms. The central activities of the courses are an interview with a GIS professional in the student's local area and the development of an original GIS ethical case study. The interview involves recruiting a willing and suitable GIS professional, obtaining informed consent, conducting the interview, and preparation of post-interview documentation. The purpose of the interview is to learn about the 'moral ecology' of the organization within which the professional works. The case studies follow the structure of existing case study documents (also from gisprofessionalethics.org) Ultimately, the goal of preparing your own case studies is to increase:

- Ethical sensitivity with the ability to identify and discriminate among ethical issues.
- Ethical knowledge via familiarity with rules of conduct.
- Ethical creativity via the ability to see 'beyond the dilemma.'
- Judgment, given the increased likelihood that students will act appropriately.

Penn State's ethics education takes place within its Online Master of Geographic Information Systems (MGIS) program (http://gis.e-education.psu.edu). The Penn State MGIS program attracts working adults who are only able to participate part-time and at a distance via the instructor-led online classes. Formal training in responsible scholarship and professional practice is mandated for all Penn State graduate students. MGIS students fulfill this mandate either by completing an online workshop or a two-credit online class called *Professionalism in GIS&T*. Both were developed as part of the GIS Professional Ethics project. Thirty to fifty adult learners complete the Penn State's ethics courses each year.

14.3 Teaching practical ethics by the case method

In the context of professional ethics, case studies are realistic workplace scenarios that challenge students to analyse ethical problems rationally and to identify reasoned solutions (see example in Table 14.2 below). Helping students develop stronger moral reasoning skills is an overarching goal of ethics education (Dark and Winstead, 2005). The 'case method' is a common pedagogical technique for strengthening the moral reasoning skills of students in business, medicine, law, engineering, and computer and information science (Davis, 1999; Keefer and Ashley, 2001; Quinn, 2006c). Professional ethicists recommend that students be provided with frameworks to guide their case analyses. Students at Penn State, Oregon State, and Minnesota are required to analyse cases using Davis' (1999) 'seven-step guide to ethical decision making' (outlined in Table 14.1 below) or similar models suggested by Keefer and Ashley (2001) and others.

Table 14.1 Davis' (1999) seven-step guide to ethical decision making

Step 1. State problem. For example, 'there's something about this decision that makes me uncomfortable' or 'do I have a conflict of interest?'
Step 2. Check facts. Many problems disappear upon closer examination of situation, while others change radically.
Step 3: Identify relevant factors. For example, persons involved, laws, professional code, other practical constraints.
Step 4: Develop list of options. Be imaginative, try to avoid 'dilemma'; not 'yes' or 'no' but whom to go to, what to say.
Step 5: Test options. Use such tests as the following: *Harm test:* does this option do less harm than alternatives? *Publicity test:* would I want my choice of this option published in the newspaper? *Defensibility test:* could I defend choice of option before Congressional committee or committee of peers? *Reversibility test:* would I still think choice of this option good if I were adversely affected by it? *Colleague test:* what do my colleagues say when I describe my problem and suggest this option as my solution? *Professional test:* what might my profession's governing body or ethics committee say about this option? *Organization test:* what does the company's ethics officer or legal counsel say about this?
Step 6: Make a choice based on steps 1–5.
Step 7: Review steps 1–6. What could you do to make it less likely that you would have to make such a decision again? Are there any precautions can you take as individual (announce your policy on question, change job, etc.)? Is there any way to have more support next time? Is there any way to change the organization (for example, suggest policy change at next departmental meeting)?

14.4 Example case analysis

Table 14.3 below contains an edited version of a Penn State student's analysis of the example case study (Table 14.2) following Davis' (1999) seven-step guide.

14.5 Conclusions

In the process of academic engagement with ethical issues arising from the increased use of GIS and corresponding professionalization of the field in the US, educators involved in professional GIS ethics education have recognized the importance of pragmatically engaging students with related ethical and moral issues. The GIS Professional Ethics project contributes an important foundation for teaching ethical and moral thinking as a distinct engagement with the use, applications and responsibilities of GIS professionals. This chapter reviews these developments and demonstrates a promising method that prepares students to analyse ethical problems rationally and to respond thoughtfully. As in allied fields the case method is a key pedagogical strategy for strengthening the moral reasoning abilities of current and aspiring GIS professionals. The GIS Professional Ethics project provides a collection of case studies online for educators who wish to incorporate ethics in geospatial education and training. Comments and additions to the collection are welcome.

Preparing these materials was an important outcome of the project. We also have some reflections on the lessons we have learned. First, there is the important observation that the

Table 14.2 Example case study produced as part of the GIS Professional Ethics project

Case Study: Mapping Muslim Neighborhoods

A GIS Professional employed as director of the Center for Risk and Economic Analysis of Terrorism Events at the University of Southern California receives an inquiry from an officer of the Los Angeles Police Department (LAPD) The officer, Commander Michael P. Downing, seeks the laboratory's assistance in a 'community mapping' project whose purpose is to 'lay out the geographic locations of the many different Muslim population groups around Los Angeles,' and to 'take a deeper look at their history, demographics, language, culture, ethnic breakdown, socio-economic status, and social interactions.' The community mapping project is to be one component of a counter-terrorism initiative that aims to 'identify communities, within the larger Muslim community, which may be susceptible to violent ideologically-based extremism . . .' (Downing 2007, 7) The director invites Downing to send the laboratory a Request for Proposal (RFP)

Soon after the telephone contact, Commander Downing is invited to Washington DC to explain the LAPD plan to the US Senate Committee on Homeland Security and Governmental Affairs. Committee chairperson Sen. Joseph Lieberman cites it, among other similar projects, as an example of effective local-level counter-terrorism strategy.

News of the Senate Hearing and the LAPD plan is reported by the major media outlets including the New York Times, KNBC Los Angeles, and National Public Radio. Within days, representatives of three local Muslim groups along with the American Civil Liberties Union sent a letter to Commander Downing expressing 'grave concerns about efforts by the Los Angeles Police Department ('LAPD') to map Muslim communities in the Los Angeles area as part of its counter-terrorism program.' The signatories argued that the community mapping project

> . . . seems to be premised on the faulty notion that Muslims are more likely to commit violent acts than people of other faiths. Singling out individuals for investigation, surveillance, and data-gathering based on their religion constitutes religious profiling that is just as unlawful, ill-advised, and deeply offensive as racial profiling' (Natarajan *et al.* 2007, 1)

Meanwhile, the LAPD's RFP arrives at the Center. The well-funded project will involve considerable GIS work and will provide support for both student interns and professional staff. However, the director worries about the unfavorable publicity and possible legal action that might attend the project given the allegations of racial profiling. How should the director respond to the RFP?

---

integration of ethics in GIS&T programs must respond to circumstances for the programs and the context of the programs. Second, the applied ethics approach aims to prepare students for the value-based decisions their professional careers will involve. Instructors can draw on these experiences and discuss their ideas and experiences with others at the gisprofessionalethics.org website. Third, students are very open to the applied approach as it holds more direct relevance for students with a strong application interest than abstract engagements through ethical theories.

Indeed, this last point may be somewhat disconcerting for educators with a strong grounding in theoretical ethics or theological studies who see this approach as mere application of a check list. Indeed, it seems apt to conclude with the remark, signaled in the title of this chapter, that the focus of these courses and project was and is on practical ethics. There are still important needs for theoretical ethic engagements, above all, to help inform the evolving role of ethics and morals in GIS&T. That said, we feel the thousands of computer

Table 14.3 Analysis of example case study by graduate student in Penn State's ethics seminar

Step 1: State Problem.

There are a number of potential ethical issues with this case, among them: Would work provided in response to the LAPD's RFP align with the mission of the Center for Risk and Economic Analysis of Terrorism Events? Does the Center's mission conflict with the University's? Should possible legal action or negative publicity influence the director's decision to respond to the RFP? Will the Center alienate its sponsors if it declines to submit a proposal? Will responding to the RFP alienate the University, the public, and more specifically the Muslim community? Could this project be considered racial profiling?

Step 2: Check Facts.

Fact: The Center's mission is 'improve our Nation's security through the development of advanced models and tools for the evaluation of the risks, costs and consequences of terrorism and to guide economically viable investments in homeland security.'

Fact: University prides itself as 'pluralistic, welcoming outstanding men and women of every race, creed and background' and 'private, unfettered by political control, strongly committed to academic freedom.'

Fact: The LAPD has specifically invited the Center director, a certified GIS Professional, to submit a proposal in response to the RFP.

Fact: The LAPD hopes to identify Muslim neighborhoods within the city's Muslim community that may 'be susceptible to violent ideologically-based extremism.' (Downing 2007, 7)

Fact: Representatives from three local Muslim groups and the ACLU object to the mapping project, claiming racial profiling (Natarajan *et al.*, 2007)

Fact: LAPD portrays the mapping project as a 'Community Engagement Plan' and specifically rejects charges of profiling.

Fact: 'Racial profiling,' according to one definition, 'occurs whenever police routinely use race as a factor that, along with an accumulation of other factors, causes an officer to react with suspicion and take action' (Cleary, 2000) A dictionary definition is 'the act or process of extrapolating information about a person based on known traits or tendencies.'

Who might be able to offer guidance to the director for further direction? The University's legal counsel, ethics officer or a conflicts of interest board, and the University's own IRB can provide guidance. Professional societies may offer additional suggestions.

Step 3: Identify relevant factors.

Do the mission statements of the Center and the University conflict? An organization concerned with terrorism assessment seems an awkward fit within a university that prides itself in pluralism and independence from government influence, notwithstanding its commitment to academic freedom. The missions may in fact conflict, though on the surface they appear to be simply driven by separate objectives.

Is the project really profiling, and in what sense does mapping constitute profiling? Given the definitions cited above, the proposed project may at least be unethical, and at the worst, illegal and unconstitutional. The project would be particularly problematic if, unlike public Census surveys, it involves identification of individuals or small groups suspected of potential terrorist activities.

(*continued*)

Table 14.3 (*Continued*)

Is even the appearance of profiling more damaging to the University and the Center than the benefit of receiving the funding? If the funding primarily provides a short-term gain for what becomes a longer-term conflict with the media, public, and student body, the rationale for the Center's involvement is questionable.

Which of the GISCI Rules of Conduct pertain to this case? The GISCI Rules of Conduct represent a 'set of objectives toward which [GISPs] must continually strive' (GISCI, 2008) The Center director affirmed compliance in his application for GISP certification.

Under Section I. Obligations to Society

Rule 1: '. . . Some applications of GIS products and services may harm individuals (directly or indirectly) while advancing government policies that some citizens regard as morally questionable'

Rule 6, Rule 10: A GISP shall not accept an assignment that may be 'in violation of GISCI Rules of Conduct' or 'violate the law.'

Under Section II. Obligations to Employers and Funders

Rule 6: A GISP will not assist a client who in conduct is 'illegal' or 'unethical.'

Under Section IV. Obligations to Individuals in Society

Rule 3: 'We shall allow people to know whether they are included in a database and to see the information listed about themselves. We shall encourage them to correct any inaccurate information about themselves . . .'

Rule 5: 'If there is a potential for harm to the individual, spurious and questionable data about him or her will be validated, removed, or denoted.'

How would RFP deliverables differ from publicly available Census data? Downing described some of the information the community mapping project was to collect, including data on Muslim population groups, detailing their 'history, demographics, language, culture, ethnic breakdown, socio-economic status, and social interaction' (Downing 2007, 7) Some but not all of these data are publicly available from the U.S. Census Bureau. It can be assumed that the community mapping project would go beyond simply aggregating existing public data sets.

Step 4: Develop list of options.

Option #1: Submit a proposal in response to the RFP.

Option #2: Don't submit a proposal.

Option #3: Request a modification to the existing RFP, or respond in a manner that removes any suggestion that profiling will occur.

Option #4: Spin off the Center from the University.

Step 5. Test options.

Option #1: Submit a proposal in response to the RFP.

Harm test: If awarded, the results of the project might alienate a community, cause irreparable harm to the University, and contribute to profiling.

Publicity test: Negative publicity is likely, though clients and sponsors will approve.

(*continued*)

Table 14.3 *(Continued)*

Defensibility test: The Center can defend applying from within it's own mission statement, but may struggle to do so within the confines of the University.

Reversibility test: If a member of the Muslim community, the director might have natural reservations about the focus and outcome of the project.

Colleague test: Center staff would likely support the decision. Other university colleagues might oppose it.

Professional test: There are numerous potential conflicts with the GISCI's Rules of Conduct.

Organization test: The University's ethics officer/legal counsel may have serious reservations.

Option #3: Request a modification to the existing RFP, or respond in a manner that removes any suggestion that profiling will occur.

Harm test: Causes LAPD to reconsider end-goals of RFP.

Publicity test: Public may not recognize the distinction, or care.

Defensibility test: Defensible but calls the Center's existence somewhat into question.

Reversibility test: University's concerns would likely be alleviated if the RFP itself were modified.

Colleague test: Colleagues would likely support and respect such a decision.

Professional test: Avoids potential conflicts with the GISCI's Rules of Conduct.

Organization test: This choice would probably be supported by legal counsel/ethics officer.

Option #4: Spin off the Center from the University.

Harm test: Could undermine the Center's viability and decrease revenue to University. Precedent could harm other University-based DHS Centers of Excellence.

Publicity test: Some may wonder why was the Center established in the University in the first place.

Defensibility test: Defensible but calls the Center's existence into question. Also questions DHS in providing the funding, and the University for pursuing the creation of the Center in the first place.

Reversibility test: DHS and University officials might be hesitant to back Center in the future.

Colleague test: Colleagues would likely question such a decision.

Professional test: Avoids potential conflicts with the GISCI's Rules of Conduct but might call into question the original intent and approval of the Center.

Organization test: This choice might be supported by legal counsel/ethics officer, if done correctly

Step 6: Select choice based on steps 1–5.

Option #3: Request a modification to the existing RFP, or respond in a manner that removes any suggestion that profiling will occur.

Step 7: Review steps 1–6.

The Center director should reflect on potential conflicts between the Center's mission and the University's. An ethics statement may help avoid future conflicts. The key issue is whether or not the project does indeed involve profiling. If I were the director I would not submit a proposal unless I was confident that the project would not infringe on the rights and privacy to U.S. citizens are entitled science programs adopting applied ethics alone speaks to the relevance it holds for many GIS&T programs.

Acknowledgments

The National Science Foundation (NSF) supported the research and development reported here (award #0734888). However, opinions expressed do not necessarily reflect those of NSF. Thanks to project consultants Michael Davis, Chuck Huff, Michael Keefer and Michael Solem.

References

ACM/IEEE (Association for Computing Machinery / IEEE Computing Society) (2001) *Computing Curricula 2001*. http://www.computer.org/education/cc2001/ (accessed March 23, 2005).

Cleary, J. (2000) *Racial Profiling Studies in Law Enforcement: Issues and Methodology*. Minnesota House of Representatives, Research Department. http://www.house.leg.state.mn.us/hrd/pubs/raceprof.pdf (accessed March 3, 2009).

Computing Accreditation Commission (2004) *Criteria for Accrediting Computing Programs*. Accreditation Board for Engineering and Technology (ABET), http://www.abet.org (accessed March 23, 2006).

Craig, W. J. (1993) A GIS code of ethics: What can we learn from other organizations? URISA Journal, 5(2): 13–16. http://www.urisa.org/certification/craigeth.pdf (accessed March 23, 2005).

Craig, W. J. (2006) *Ethics for GIS professionals* (Panel discussion), Association of American Geographers Annual Meeting, Chicago IL.

Dark, M. J., and J. Winstead (2005) Using educational theory and moral psychology to inform the teaching of ethics in computing. *Information Security Curriculum Development Conference '05, September 23–24*, Kennesaw, Georgia, Association of Computing Machinery.

Davis, M. (1990) Who can teach workplace ethics? *Teaching Philosophy*, 13(1): 321–38.

Davis, M. (1999) *Ethics and the University*. London: Routledge.

DiBiase, D., M. DeMers, A. Johnson, K. Kemp, A. Luck, B. Plewe, and E. Wentz (eds) (2006) *The Geographic Information Science and Technology Body of Knowledge*. Washington, DC: Association of American Geographers and University Consortium for Geographic Information Science.

Dobson, J. E., and P. F. Fisher (2003) Geoslavery. *IEEE Technology and Society Magazine*, 22(Spring): 47–52.

Downing, M. P. (2007) *Statement before the Committee on Homeland Security and Governmental Affairs, United States Senate*. Washington DC, October 30. http://hsgac.senate.gov/public/index.cfm?Fuseaction=Hearings.Detail&HearingID=483590e6-9f4e-4aa6-b595-8ca3791e4acb (accessed June 12, 2008).

GIS Certification Institute (2008) *Rules of Conduct for Certified GIS Professionals (GISPs)*. http://www.gisci.org/Ethics_and_Conduct/rules_of_conduct.aspx (accessed July 6, 2009).

Goldin, I. M., K. D. Ashley, and R. L. Pinkus (2001) Introducing PETE: Computer support for teaching ethics. *Proceedings of ICAIL-2001*, Association for Computing Machinery, pages 94–98.

Grodzinsky, F., S. Gehringer, L. S. King, and H. Tavani (2004) Panel: Responding to the challenges of teaching computer ethics. *SIGSCE '04 Proceedings, Association for Computing Machinery*, pages 280–281.

Harley, J. B. (1988) Maps, knowledge, and power, IN: D. Cosgrove, and S. Daniels (eds) *The Iconography of Landscape*. Cambridge: Cambridge University Press, pages 277–312.

Harley, J. B. (1991) Can there be a cartographic ethics? *Cartographic Perspectives*, 10: 9–16.

Keefer, M., and K. D. Ashley (2001) Case-based approaches to professional ethics: A systematic comparison of students' and ethicists' moral reasoning. *Journal of Moral Education*, 30(4): 377–398.

McHaffie, P., S. Andrews, and M. Dobson and two anonymous employees of a federal mapping agency (1990) Ethical problems in cartography: A roundtable commentary. *Cartographic Perspectives*, 7: 3–13.

Martin, C. D., and H. J. Holz (2005) *Non-apologetic computer ethics education: A strategy for integrating social impact and ethics into the computer science curriculum*. http://www.southernct.edu/organizations/rccs/resources/teaching/teaching_mono/martin_holz/martin_holz_intro.html (accessed July 5, 2009).

Martin, C. D., and E. Y. Weltz (1999) From awareness to action: Integrating ethics and social responsibility into the computer science curriculum. *Computers and Society*, 29(2): 6–14.

Monmonier, M. S. (1991) Ethics and map design: Six strategies for confronting the traditional one-map solution. *Cartographic Perspectives*, 10: 3–8.

Monmonier, M. S. (1996) *How to Lie with Maps*. Chicago: University of Chicago Press.

Moskal, B., K. Miller, and L. A. Smith-King (2002) Grading essays in computer ethics: Rubrics considered helpful. *Proceedings of SIGCSE'02, Association for Computing Machinery*, 27 February–3 March, pages 280–281.

Natarajan, R., P. Bibring, F. Y. Khera, S. Syed, H. Ayloush (2007) Letter to Commander Downing, LAPD. http://www.aclu-sc.org/attach/m/muslim_mapping_letter.pdf (accessed 5 July 2011).

Onsrud, H. J. (1995) Identifying unethical conduct in the use of GIS. *Cartography and Geographic Information Systems*, 22(1): 90–97.

Pickles, J. (1991) Geography, GIS, and the surveillant society. *Papers and Proceedings of Applied Geography Conferences*, 14: 80–91.

Pickles, J. (ed) (1995) *Ground Truth: The Social Implications of Geographic Information Systems*. New York: Guilford.

Quinn, M. J. (2006a) *Ethics for the Information Age*. 2nd ed. Reading, MA: Pearson Addison-Wesley, Longman.

Quinn, M. J. (2006b) On teaching ethics inside a computer science department. *Science and Engineering Ethics*, 12(2): 335–343.

Quinn, M. J. (2006c) Case-based analysis: A practical tool for teaching computer ethics. *Proceedings of SIGCSE '06, Association for Computing Machinery, March 1–5*, pages 520–524.

Robbins, R. W., W. A. Wallace, and B. Puka (2004) Supporting ethical problem solving: An exploratory investigation. *Proceedings of SIGMIS '04, April 22–24*, pages 134–143.

Sanders, A. F. (2005) A discussion format for computer ethics. *Proceedings of SIGCSE '05, Association for Computing Machinery*, pages 352–355.

Schuurman, N. (2000) Trouble in the heartland: GIS and its critics in the 1990s. *Progress in Human Geography*, 24(4): 569–590.

Smith, N. (1992) History and philosophy of geography: Real wars, theory wars. *Progress in Human Geography*, 16: 257–271.

Wahl, N. J. (1999) YAATCE – Yet another approach to teaching computer ethics. *Proceedings of SIGCSE '99, Association for Computing Machinery*, March, pages 22–26.

15

An exploration of spatial thinking in introductory GIS courses

Injeong Jo, Andrew Klein, Robert S. Bednarz and Sarah W. Bednarz

Department of Geography, Texas A&M University, College Station, Texas, USA

15.1 Introduction

This chapter explores the degree to which spatial thinking is embedded in the curriculum of introductory GIS courses taught in the U.S. This is accomplished by examining two key components of GIS courses: (1) syllabi, which represent the way in which geospatial subject matter is organized, presented and assessed; and (2) the textbooks assigned in courses, specifically, the spatiality of questions presented in frequently used GIS textbooks. It is our central premise that spatial thinking is a key component of GIS education but that it is often not explicitly taught in a way that helps students become more competent and spatially literate users of geospatial technologies. First, we define spatial thinking and the complex relationship between spatial thinking and GIS. Next we investigate the syllabi and textbooks used in GIS courses in the United States to assess the status of spatial thinking. Our goal is to provide insights on the design and development of university-level spatially enhanced courses.

15.2 GIS and spatial thinking

The publication of *Learning to Think Spatially* by the National Research Council (National Research Council, 2006) increased an already growing interest in spatial thinking. The authors of the study (geographers, astronomers, psychologists and educationists) argued that spatial thinking is an amalgam that involves knowledge about three components—the nature of space, methods of representation of spatial information, and processes of spatial reasoning. The authors went on to assert that '[s]patial thinking uses representations to help us remember, understand, reason, and communicate about the properties of and relations

Teaching Geographic Information Science and Technology in Higher Education, First Edition.
Edited by David J. Unwin, Kenneth E. Foote, Nicholas J. Tate and David DiBiase.

between objects represented in space' (National Research Council, 2006, 27) and, although a wide variety of representation tools support spatial thinking, that 'GIS had a clearly demonstrated potential as a support system for spatial thinking' (National Research Council, 2006, 221).

The notion that representation generally, and GIS specifically, might support spatial thinking was proposed before the publication of *Learning to Think Spatially*. Several geographers, probably most notably Golledge and Self, claimed that using GIS could help students develop spatial abilities (cf. Albert and Golledge, 1999; Audet and Abegg, 1996; Kerski, 2000; Self, Gopal, Golledge and Fenstermaker, 1992). The research results supporting the positive effect of GIS on spatial thinking, however, are inconsistent. Whereas many researchers found a positive correlation between GIS learning and spatial thinking, others did not.

Lee and Bednarz (2009, 194) argued that '[t]he impact of GIS learning varied widely among students and the spatial skills assessed,' but demonstrated that, in an experimental setting, GIS learning helps students think spatially. They also argued that gains in students' spatial thinking were more highly correlated with some aspects of GIS coursework than others, namely, that stronger correlations existed between spatial thinking gains and GIS laboratory-based practicum scores than between spatial thinking and mid-term or final exam scores. A major question that must be addressed is: What is the source of this improvement? In the next section we examine questions in GIS textbooks for information to answer this question.

15.3 Spatiality of questions in GIS textbooks

Questions, either those asked verbally, or presented in textbooks, are used to stimulate students' thinking and problem-solving skills (Myers and Savage, 2005; Pizzini, Shepardson and Abell, 1992; Vogler, 2005; Wilen, 2001). Not all questions, however, cause students to think (Costa, 2001; Hamaker, 1986; Mills, Rice, Berliner and Rosseau, 1980; Nosich, 2005; Wilen, 2001). Researchers have found that questions which 'ask a learner to repeat or recognize information exactly as it was presented in instruction' (Andre, 1979, 282) elicit only low level cognitive processes whereas questions that 'ask the student to mentally manipulate bits of information previously learned to create an answer, or to support an answer' (Winne, 1979, 14) facilitate higher-level thinking (Hamaker, 1986). The relationship between the cognitive level of questions and the level of student thinking makes it imperative that students be asked higher-order questions.

As discussed in the previous section, the relationship between spatial thinking and GIS is complex and unclear. Thinking spatially requires knowing, understanding and remembering spatial information and concepts. It also requires the flexible use of such knowledge whenever needed through the use of tools of representation to understand, remember and communicate the knowledge effectively. Reasoning with spatial information to solve problems and make decisions is critical as well. Questions designed to stimulate students' spatial thinking therefore should require students to use spatial concepts, spatial representations like maps and appropriate reasoning skills. Thus, it is relevant to our purpose to examine how questions in GIS textbooks support the development of spatial thinking skills.

Major textbooks used for undergraduate introductory geography GIS courses in the United States were selected for inclusion in this study following a protocol designed to identify the most-widely used introductory GIS textbooks. Schools with undergraduate geography programs that offered specializations in GIS or GIS certificate programs were identified by referring to the *Guide to Geography Programs in the Americas* (Association of American Geographers, 2009). From the 140 schools that met the first criterion, the 36 schools that had 100 or more undergraduate geography majors were selected to identify the textbooks used by the largest number of students. Textbook and general information about these schools' introductory GIS courses were determined from online materials. For schools lacking online information, the GIS course instructor or academic advisor was contacted. This process generated textbook information for 32 introductory GIS courses. One program did not assign a textbook, and we were unable to obtain the relevant information from three additional schools. Eight different textbooks were used by the 32 undergraduate programs. The four most frequently used books (those authored by Bolstad, Longley *et al.*, Chang, and Harvey) were determined based on both the number of schools and the number of students that used each text. With the exception of the Harvey textbook, the textbooks selected for this analysis correspond to the three most commonly used textbooks in a recent syllabi analysis by Vincent (2010).

In analysing all chapters in the texts, five common themes emerged: (1) data models, (2) cartography, map projections and coordinate systems, (3) GIS data collection, (4) spatial analysis, and (5) spatial modeling. We judged that the themes most directly related to spatial thinking skills were cartography; map projections and coordinate systems; and spatial analysis. In total the four textbooks contained 318 questions related to these two topics which constituted the sample we analyzed.

The number of questions contained in the chapters of the four textbooks varied. Bolstad's (2008) text yielded 87 questions, Longley *et al.* (2005) 37 questions, Chang's (2010) 123 questions, and Harvey's (2008) 71 questions. Table 15.1 lists the number of questions both by textbook and by theme.

The spatiality of questions was evaluated using a taxonomy of spatial thinking developed by Jo and Bednarz (2009) (Figure 15.1). The main components of spatial thinking—concepts of space, tools of representation and processes of reasoning (National Research Council, 2006) are captured by the model's three-dimensional structure. These concepts and cognitive processes are based on previous research and a comprehensive review of relevant theories.

To define 'spatial concepts,' the first of the components of spatial thinking, Golledge's (1995, 2002) scheme, which categorizes spatial concepts at three levels: spatial primitives, simple-spatial concepts and complex-spatial concepts, proved most useful. Spatial primitives describe characteristics such as location or place-specific identity, the fundamental building blocks of space. Simple-spatial concepts are developed or constructed from spatial primitives. Distance is a simple spatial concept because it represents the interval between two locations. Complex-spatial concepts are derived either by combining simple-spatial concepts or from combinations of spatial primitives and simple-spatial concepts. For instance, 'network' is formed by sets of connected locations; 'hierarchy' is developed by

Table 15.1 The number of questions by textbook and by theme

	Bolstad		Longley *et al.*		Chang		Harvey	
Theme*	1	2	1	2	1	2	1	2
Number	21	66	24	13	39	84	50	21
Total	87		37		123		71	

*Theme 1: Cartography, map projections, and coordinate system; Theme 2: Spatial analysis

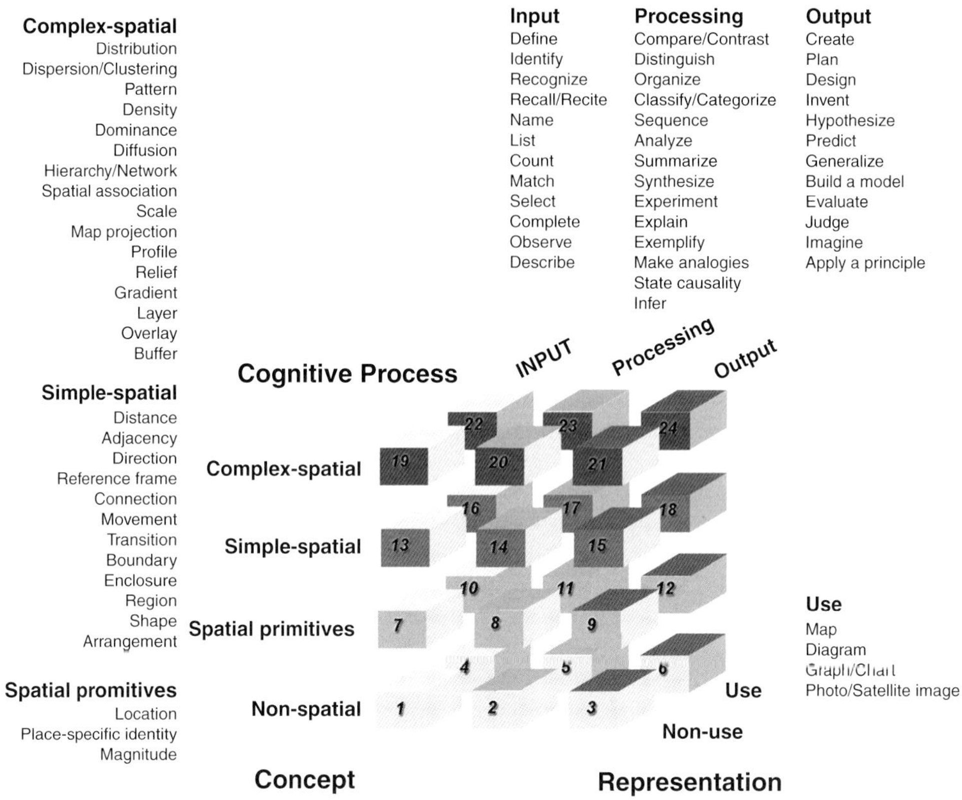

Figure 15.1 A taxonomy of spatial thinking. Each of the 24 cells represents a unique characteristic in terms of three components of spatial thinking involved. For example, cell 24 stands for a complex-spatial concept, use of a representation, and the output level cognitive process. Numbers were assigned for referential convenience. Reproduced, with permission, from Jo, I., and Bednarz, S. W. (2009). Evaluating geography textbook questions from a spatial perspective: Using concepts of space, tools of representation, and cognitive processes to evaluate spatiality. Journal of Geography, 108, 4–13. © Taylor and Francis

Table 15.2 Question coding examples

		Coding Scheme		
Question	Concept	Representation	Cognitive process	Categorization
'Define the RGB and CMYK color models'	RGB and CMYK color models (Non-spatial)	None (Non-use)	Define (Input)	Cell 1
'Use the county data and descriptive statistics from Question 2 to draw a boxplot. What kind of data distribution does the boxplot show?'	Data distribution (Complex-spatial)	Boxplot (Use)	Create and generalize (Output)	Cell 24

combining location and magnitude (both spatial primitives) with 'connectivity' (a simple spatial concept). All together, thirty-one concepts essential to spatial thinking were identified for inclusion in the taxonomy and categorized into three levels representing the concepts' level of abstractness and complexity.

To conceptualize 'cognitive processes,' the second spatial thinking component, a dozen taxonomies of thinking skills (e.g., Bloom *et al.*, 1956; Costa, 2001; Krathwohl, 2002; Marzano, 2001; Moseley *et al.*, 2005) were reviewed, and the framework suggested by Costa (2001) was adapted to our purposes. The resulting hierarchy of cognitive processes divided the nature of thinking into three subcategories: (1) input level, (2) processing level, and (3) output level. Input-level thinking is characterized by the recall or acquisition of information such as defining, observing, identifying, recalling and listing. Cognitive processes at the input level may not lead to reasoning, but they do ensure acquisition of knowledge required for reasoning. At the processing level, learners manipulate and act on the information acquired via the input level. Cognition at this level may be defined as reasoning in that students make sense of, and go beyond, the information gleaned during input. Finally, at the output level students produce new knowledge or create products derived from the information gained through the lower-level processes. Cognition at this level is considered reasoning and represents the highest level of difficulty and complexity.

While the taxonomy classifies spatial concepts and cognitive processes by both complexity and abstractness, this was not the case for the third component, 'tools of representation.' Because the nature of a learning activity using representations relies upon the learning context more than the nature of the representations used, it did not seem appropriate to include a measure of the complexity of representations. Thus, the taxonomy differentiates only on whether or not a representation is part of a question.

The taxonomy can be used to analyse questions from the four textbooks by locating each in one of the 24 cells of the three-dimensional array representing the taxonomy (Figure 15.1). Each cell represents a unique combination of the three components of spatial thinking. For example, a question posed about a non-spatial concept (e.g., RGB and CMYK color models) is irrelevant to any spatial representation like a map or diagram, and also requires the lowest level of cognitive processes (e.g., define) and would fall in cell 1 in Figure 15.1. Alternatively, a question that requires an understanding of a complex-spatial concept (e.g., overlay), asks the students to use or create a spatial representation, and requires the highest level of cognitive processes (e.g., create) would be located in cell 24.

Thus the taxonomy classifies questions according to all three spatial cognitive elements and allows for easy identification of questions requiring simple concepts, lower-level cognitive processes, and no spatial representation (front cells at the lower left) versus those employing complex concepts, higher-level reasoning, and the use of representations (rear cells on the upper right). The location also enables one quickly to detect questions that, for example, require higher-level concepts but low-level reasoning (cells on the upper left) or any other combination of the components of spatial thinking.

The taxonomy was used to categorize each of the 318 questions from the four textbooks. Examples of question coding are presented in Table 15.2. This exercise allowed us to determine both the level of each component of spatial thinking required by the question and the extent to which the three components were integrated.

The results of the analysis of the nature of spatial concepts employed in the textbook questions are presented in Table 15.3. Approximately 40 percent of the 318 questions did

Table 15.3 Concepts required, by textbook

	Bolstad		Longley *et al.*		Chang		Harvey		Total	
	Frequency	Percent	Frequency	Percent	Frequency	Percent	Frequency	Percent	Frequency	Percent
Non-spatial	28	32.18	14	37.84	47	38.21	39	54.93	128	40.25
Spatial primitive	9	10.34	2	5.41	4	3.25	0	0	15	4.72
Simple-spatial	22	25.29	7	18.92	30	24.39	12	16.90	71	22.33
Complex-spatial	28	32.18	14	37.84	42	34.15	20	28.17	104	32.70
Total	87	100.00	37	100.00	123	100.00	71	100.00	318	100.00

Table 15.4 Concepts required, by theme

	Theme 1 Cartography, map projections, and coordinate systems		Theme 2 Spatial analysis	
	Frequency	Percent	Frequency	Percent
Non-spatial	62	46.27	66	35.87
Spatial primitive	13	9.70	2	1.09
Simple-spatial	32	23.88	39	21.20
Complex-spatial	27	20.15	77	41.85
Total	134	100.00	184	100.00

not require the understanding or use of a spatial concept. Questions that asked about spatial primitives, such as location, place-specific identity, and magnitude, comprised about 5 percent while questions that asked about simple-spatial concepts, including distance, reference frame and shape, comprised about 22 percent. The remaining 33 percent of questions dealt with complex-spatial concepts, such as map projection, overlay and spatial pattern, probably reflecting the importance of understanding these higher-level spatial concepts in learning GIS.

The percentage of concepts falling into each of these categories was broken down by textbook. In all the texts studied, non-spatial concepts were the focus of approximately one-third of the questions. Harvey's book contained the most questions concerning non-spatial concepts (about 55 percent). Of the questions classified as spatial, simple-spatial-concept questions were most prevalent in Bolstad's book while Longley's book contained the highest percentage of complex-spatial-concept questions (Table 15.3). Overall Bolstad's book contained the most questions associated with spatial concepts at all levels (spatial primitive, simple-spatial, and complex-spatial in total, but only slightly).

Differences in the level of spatial concepts required were also analysed by theme (Table 15.4). As noted previously, questions concerning theme 1 (cartography, map projections and coordinate systems) accounted for 134 of the 318 sample questions (42.14 percent); the remaining 184 questions (57.86 percent) addressed theme 2 (spatial analysis). Table 15.4 reveals that the nature of the questions varies significantly by theme. Approximately 46 percent of questions that asked about cartography, map projections and coordinate systems focused on non-spatial concepts whereas for the spatial analysis theme, 42 percent were complex-spatial-concept questions. Perhaps not surprisingly, many questions posed in the spatial analysis chapters required higher-level spatial concepts than those presented in cartography, map projections and coordinate systems chapters.

Table 15.5 reveals that significant variation in the use of spatial representations exists among the questions from the four textbooks. Approximately 77 percent of 318 questions asked students neither to use nor create representations. The Bolstad and Chang textbooks were similar in that 27 to 30 percent of the questions required the use of representations. In contrast, 38 percent of questions in Longley's book required students to use or create a spatial representation while no questions did in Harvey's book.

Table 15.6 reports the variation by theme in the questions' use of spatial representations. A slight difference between themes exists. More than 80 percent of the questions in theme 1 and about 70 percent in theme 2 do not require using or creating spatial representations.

Table 15.5 Tools of representation required, by textbook

	Bolstad		Longley *et al.*		Chang		Harvey		Total	
	Frequency	Percent	Frequency	Percent	Frequency	Percent	Frequency	Percent	Frequency	Percent
Non-use	61	70.11	23	62.16	89	72.36	71	100	244	76.73
Use	26	29.89	14	37.84	34	27.64	0	0	74	23.27
Total	87	100.00	37	100.00	123	100.00	71	100.00	318	100.00

Table 15.6 Tools of representation required, by theme

	Theme 1 Cartography, map projections, and coordinate systems		Theme 2 Spatial analysis	
	Frequency	Percent	Frequency	Percent
Non-use	115	85.82	129	70.11
Use	19	14.18	55	29.89
Total	134	100.00	184	100.00

The level of cognitive processes required to answer the questions is reported in Table 15.7. Across all texts, almost 47 percent of the questions required input-level thinking, about 34 percent of the questions required processing-level, and the remaining 20 percent demanded output-level cognition. As is apparent from Table 15.7, a greater percentage of the questions in Longley *et al.*'s book demanded processing- and output-level of thinking. Bolstad's text contained the fewest processing-level questions (about 26 percent), while Harvey's book asked the fewest output-level questions (about 6 percent).

Table 15.8 reveals little variation by theme in the cognitive processes required. Questions posed about both themes 1 and 2 frequently required input-level thinking (about 51 percent and 43 percent, respectively). Output-level cognitive processes were required the least for questions addressing both themes (about 16 percent and 23 percent, respectively).

The final analysis evaluated the extent to which questions integrated the three components of spatial thinking: concepts of space, representations and processes of reasoning. Questions located in cells 1, 2, 3, 4, 5, 6, 7, 8, 9, 13, 14, 15, 19, 20 and 21 were judged to be questions that did not integrate the three components, whereas questions placed in cells 10, 11, 12, 16, 17, 18, 22, 23 and 24 incorporate all three components. Questions integrating the three components were differentiated from simplest level to intermediate level and to the highest level by the complexity of concepts and cognitive processes involved: simple 10, 11 and 16; intermediate 12, 17 and 22; and complex 18, 23 and 24.

Overall, approximately 86 percent of 318 questions failed to integrate the three components. Of the 14 percent of questions in which all three components were incorporated, 3 percent required the simplest level spatial thinking, focusing on spatial primitives or input-level thinking; approximately 2 percent demanded intermediate level spatial thinking; and about 8 percent required the highest level spatial thinking (Table 15.9). The four textbooks were similar in that over 75 percent of their questions did not integrate the three components, despite some variation in degree (Table 15.9). For the questions integrating the three components, Bolstad's and Longley *et al.*'s books included relatively more questions calling for higher level of spatial thinking (cells 12, 17, 18, 22, 23 and 24).

Figure 15.2 provides a few examples of questions requiring higher level spatial thinking. These questions demand knowledge of complex-spatial concepts such as buffer, overlay, distribution, spatial interpolation and density. In addition students are instructed to use a variety of representations such as diagrams, maps and graphs as a tool. The cognitive processes required to answer these questions are at the output level.

Table 15.10 shows that theme 2 questions integrated the three components more than theme 1. In addition, more questions concerning theme 2 were associated with highest level spatial thinking (cells 18, 23 and 24).

Table 15.7 Cognitive processes required, by textbook

	Bolstad		Longley *et al.*		Chang		Harvey		Total	
	Frequency	Percent	Frequency	Percent	Frequency	Percent	Frequency	Percent	Frequency	Percent
Input	45	51.72	6	16.22	57	46.34	40	56.34	148	46.54
Processing	23	26.44	16	43.24	41	33.33	27	38.03	107	33.65
Output	19	21.84	15	40.54	25	20.33	4	5.63	63	19.81
Total	87	100.00	37	100.00	123	100.00	71	100.00	318	100.00

Table 15.8 Cognitive processes required, by theme

	Theme 1 Cartography, map projections, and coordinate systems		Theme 2 Spatial analysis	
	Frequency	Percent	Frequency	Percent
Input	68	50.75	80	43.48
Processing	45	33.58	62	33.70
Output	21	15.67	42	22.83
Total	134	100.00	184	100.00

15.4 Spatiality in course syllabi

Recognizing that courses are much more than textbooks, we further explored the extent to which spatial thinking is embedded in the curriculum by examining the syllabi of 20 introductory GIS courses. As with the question analysis, the largest geography programs were targeted. Although we recognize that a syllabus cannot fully capture the totality of any course, we feel it is reasonable to assume that faculty construct their syllabi faithfully to represent course content. Syllabi were collected at the same time that we acquired information about textbook use. Syllabus analysis examined: (1) the extent to which the content of the courses emphasized those topics most related to spatial thinking; (2) whether the syllabi indicated a greater emphasis on spatial thinking than is evidenced by the analysis of questions in course textbooks; (3) whether courses with a greater emphasis on spatial thinking tend to use textbooks whose questions score higher in spatial thinking; and (4) if courses emphasizing lab practicums or a final project employ a textbook emphasizing higher-level spatial thinking skills.

In 2006, the University Consortium for Geographic Information Science (UCGIS) in collaboration with the Association of American Geographers (AAG) published the *Geographic Information Science & Technology Body of Knowledge* (*BoK*) (DiBiase, *et al.*, 2006) which attempts to codify the field in a systematic manner. This framework is employed in our preliminary examination of the alignment between the content of introductory GIS courses and the major knowledge areas of the *BoK*. The knowledge areas and core topical areas used in the evaluation of syllabi are presented in Table 15.11.

The syllabi of the 20 selected courses were analysed to compare topics, grading schemes and assigned textbooks. This information is summarized in Table 15.12. Key to the analysis was the assignment of each syllabus topic to the appropriate *BoK* area after a careful review of each of the ten *BoK* areas. Only topics that could be appropriately assigned to a particular *BoK* area were included. For example, class topics such as 'lab work' or 'time to work on project' were not assigned. Some syllabi identified topics only on a weekly basis while others were more specific. In all cases, the most specific information concerning the topics available on the syllabi was used.

Some overlap in the topics covered between *BoK* areas exists (Table 15.11). In general, when a choice had to be made, topics were allocated to the broadest appropriate *BoK* area. For example, attribute queries are a common topic when discussing databases. While attribute queries are specifically contained within the Analytical Methods (AM) knowledge

Table 15.9 Integration of the three components, by textbook

	Bolstad		Longley *et al.*		Chang		Harvey		Total	
	Frequency	Percent	Frequency	Percent	Frequency	Percent	Frequency	Percent	Frequency	Percent
Not integrated	69	79.31	29	78.38	105	85.36	71	100	274	86.16
Cells 10, 11, & 16	6	6.90	1	2.70	3	2.44	0	0	10	3.14
Cells 12, 17, & 22	2	2.30	3	8.11	3	2.44	0	0	8	2.52
Cells 18, 23, & 24	10	11.49	4	10.81	12	9.76	0	0	26	8.18
Total	87	100.00	37	100.00	123	100.00	71	100.00	318	100.00

Q1. Sketch out the output from a variable distance buffer applied to the set of points shown below. Draw output buffers that dissolve the boundaries between areas that fall within multiple buffers.

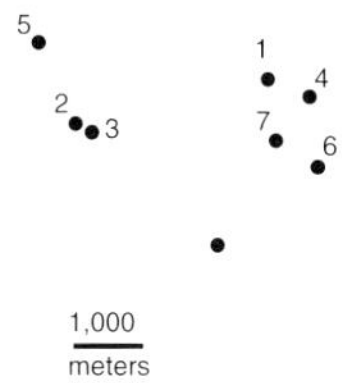

ID	Distance
1	250
2	0
3	1500
4	500
5	0
6	500
7	250
8	250

(Bolstad, 2008, 376)

Q2. Sketch both the output polygons and the resultant attribute table from the overlay shown below:

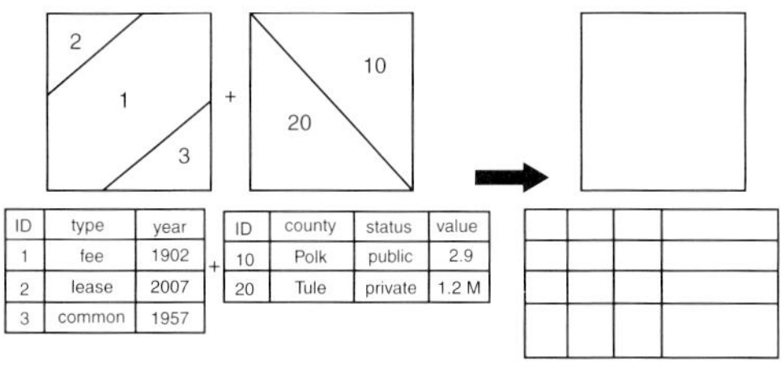

(Bolstad, 2008, 376)

Q3. Use the county data and descriptive statistics from Question 2 (percent population change by county for your state between 1990 and 2000) to draw a boxplot. What kind of data distribution does the boxplot show?

(Chang, 2010, 215)

Q4. You are given two digital maps of New York City: one shows landmarks, and the other shows restaurants. One of the attributes of the restaurant layer lists the type of food (e.g., Japanese, Italian, etc.). Suppose you want to find a Japanese restaurant within 2 miles of Times Square. Describe the steps you will follow to complete this task.

(Chang, 2010, 215)

Q5. You are given a map showing the home locations of the customers of an insurance agent and asked to construct a map showing the agent's market area. Would spatial interpolation or density estimation be more appropriate, and why?

(Longley *et al.*, 2005, 339)

Figure 15.2 Example questions associated with high level spatial thinking

area, most syllabi do not specifically mention this topic but rather allude more generally to relational databases which are encompassed within Data Modeling (DM). Therefore for most courses, attribute queries were included under DM and not AM where it is explicitly mentioned. Of the *BoK* areas, the Conceptual Foundations (CF) category was the most difficult to define as it is probable that many courses include discussion of these fundamental topics within a discussion of other areas, although only a few syllabi clearly identified topics falling exclusively into this category. Table 15.12 summarizes the distribution of topics by *BoK* areas and illustrates the percent of topics in each category, arranged alphabetically.

Table 15.10 Integration of the three components, by theme

	Theme 1 Cartography, map projections, and coordinate systems		Theme 2 Spatial analysis	
	Frequency	Percent	Frequency	Percent
Not integrated	123	91.79	150	81.52
Cells 10, 11, & 16	5	3.73	5	2.72
Cells 12, 17, & 22	4	2.99	4	2.17
Cells 18, 23, & 24	2	1.49	25	13.59
Total	134	100.00	184	100.00

Table 15.11 Major knowledge areas of the GIS&T and their relevant core units

Analytical Methods (AM) Geometrical measures, Basic analytical operations, Basic analytical methods	**Data Manipulation (DN)** Representation transformation, Generalization and aggregation
Conceptual Foundations (CF) Domains of geographic information, Elements of geographic information	**Geocomputation (GC)**
Cartography and Visualization (CV) Data considerations, Principles of map design, Map use and evaluation	**Geospatial Data (GD)** Earth geometry, Georeferencing systems, Datums, Map projections, Data quality, Land surveying and GPS, Aerial imaging and photogrammetry, Satellite and shipboard remote sensing, Metadata, Standards and infrastructures
Design Aspects (DA) Database design	**GIS&T and Society (GS)** Ethical aspects of geospatial information and technology
Data Modeling (DM) Database management systems, Tessellation data models, Vector and object data models	**Organizational and Instructional Aspects (OI)** Institutional and inter-institutional aspects, Coordinating organizations

Our analysis also investigated the percentage of the class grade assigned to lecture, labs and projects, and these findings are also included in Table 15.12. The uneven length of the grading bars indicates that for some courses a portion of the grade was not determined by these three areas, for example, credit for class participation. The textbooks used in the course are also identified in Table 15.12 with optional textbooks italicized.

The analysis revealed that four of the surveyed courses differ somewhat from the rest. For one (Ohio University), the syllabus was not sufficiently detailed to analyse the topical areas, so is not included in the topical analysis. Three other courses (Arizona State University, Central Michigan and Montana State) are more general introductions to geographic information science and technologies in which GIS is discussed along with cartography and remote sensing. The books selected for these courses reflect this different approach and place greater emphasis on cartography.

Overall, the syllabi of these introductory GIS courses were similar although some courses also introduced other geospatial technologies such as GPS and remote sensing. The lab practicum component typically represented 30 to 50 percent of the total grade. Eight of the courses (38 percent) required final projects which have been found by Vincent (2010) to be used increasingly for student evaluation over the past decade. Moreover, Lee and Bednarz (2009) demonstrated both through statistical analysis of results of a spatial thinking test and from student interviews that final projects improved students' spatial thinking skills.

The order in which topics were presented was strikingly similar and consistently focused on five of the *BoK* areas (Table 15.11). Most, if not all, of the topics listed fit easily within Analytical Methods (AM), Conceptual Foundations (CF), Cartography and Visualization (CV), Data Modeling (DM) and Geospatial Data (GD). The courses we examined placed less emphasis on two additional areas, GIS&T and Society (GS) and Organizational &

Table 15.12 Syllabus analysis. The second column indicates the *BoK* knowledge areas included in the syllabus; the third column the weighting of assignments in grading, and the fourth column the textbooks used (optional textbooks are italicized)

University	GIS&T breakdown	Grading breakdown	Textbook(s)
	AM | CF | CV | DM | GD | GS&OI	Lecture | Lab | Project	
Arizona State University			Campbell
Brigham Young University			Chang
California State – Long Beach			Ormsby
California State – Northridge			Bolstad Ormsby
Central Michigan University			Campbell
Florida State University			Longley *et al.*
Montana State University			Dent Ormsby
Ohio University			Ormsby
Salisbury State University			Longley *et al.* Ormsby
Southern Illinois University – Edwardsville			Chang
St. Cloud State University			none listed
Texas State University			Longley *et al.* Price
University of California – Santa Barbara			Longley *et al.*
University of Colorado – Boulder			Bolstad *Wade and Sommer*
University of Colorado – Colorado Springs			Clarke Gorr and Kurland
University of Iowa			Bolstad
University of Kansas			Bolstad Ormsby
University of South Florida			Gorr and Kurland *Bolstad*
University of Wisconsin			Longley *et al.*
West Virginia University			Bolstad Gorr and Kurland

Institutional Aspects (OI). The inclusion of these two knowledge areas mainly relates to the history, future and application of GIS&T.

Most important to our analysis is that the courses do vary in their emphasis on topical areas that could be assumed most related to spatial thinking. Of the five major knowledge areas, it is reasonable to assume that the AM, CV and GD knowledge areas place greater emphasis on knowledge and skills related to the three aspects of spatial thinking that are components of the taxonomy used to analyse textbook questions. The AM knowledge area encompasses basic GIS operations such as buffering and cartographic overlay. Because understanding and use of these spatial operations requires a fundamental knowledge of many spatial concepts, inclusion of such topics could indicate that students are instructed in these spatial concepts. In nearly 60 percent (11 of 19) of the courses, AM comprised 30 percent or more of the identified topics. Similarly, the Geospatial Data knowledge area includes topics requiring fundamental knowledge of spatial concepts including Earth geometry, referencing systems, datums and map projections. In over one-fourth (5 of 19) of the courses, GD comprised 30 percent of the topics covered.

The Bolstad, Chang and Longley *et al.* books were adopted by 21 of the courses we reviewed along with a number of laboratory manuals or other materials (Gorr and Kurland, 2008; Ormsby, 2010; Price, 2010; Wade and Sommer, 2006). It is difficult to ascertain a clear relationship between courses with a greater percentage of spatial-thinking topics and the assigned textbook. Wide variation in the relationship between the textbooks used and the courses' *BoK* topical areas exists. This is perhaps not surprising. Ideally, faculty construct courses based on their own perceptions of relative importance of topics which ideally are drawn from multiple sources, not simply the table of contents of a selected text.

Nevertheless, a few interesting relationships between the course content and adopted text are evident. Those courses adopting the Longley *et al.* and Chang texts tend to place greater than average emphasis on the AM knowledge area. The five courses that use Bolstad's text (found to have the most spatial-concept questions) are courses that have strong emphasis on GD.

Finally, we examine whether courses that emphasize labs or include a final project tend to adopt a textbook that emphasizes higher level spatial thinking skills. Based on anecdotal evidence and student feedback (e.g., Lee and Bednarz, 2009), a common belief amongst GIS students and instructors is that undertaking a GIS project improves a student's spatial thinking skills. Approximately half of the courses required only a lab-type instruction manual. However, with the exception of West Virginia University, courses adopting a lab-oriented book (e.g., Ormsby or Price) tended to have a higher proportion of the final grade assigned to the lab component of the course or a summative project. Interestingly, of the eight courses with a lab practicum, seven required both a textbook and a lab manual. Two courses we reviewed adopted the Chang textbook and in both, the labs and projects comprised over half the grade. The case is much less clear for the Bolstad and Longley *et al.* textbooks which were used in classes where the labs and/or projects varied widely as a percentage of the total grade.

15.5 Conclusions

This chapter has explored the degree to which spatial thinking is embedded in introductory GIS courses by analyzing the spatiality of textbook questions and course syllabi. Considerable variety with respect to support for spatial thinking exists in both the textbooks and the course syllabi. In general, however, we found little evidence of explicit instruction in, or opportunities for students to develop, spatial thinking skills.

The analysis of textbook questions revealed that nearly half did not require the knowledge of spatial concepts. Over 75 percent did not ask students to use or create spatial representations, and only 20 percent called for output-level cognitive processes. Finally, 86 percent of textbook questions failed to integrate the three components of spatial thinking.

The course syllabi we examined seldom, if ever, explicitly mentioned spatial thinking or gave any evidence of instruction in this essential component of geospatial analysis, although we realize that the representation of the course in the syllabus and the course itself are not necessarily isomorphic. We found little relationship to exist between the spatiality of the textbook questions and the emphasis placed on spatial concepts in the syllabi.

Based on these results, we conclude with two observations and several recommendations. First, instructors should realize that significant differences exist between textbooks in their support of spatial thinking. Given the relatively low level of spatial thinking we found embedded in the questions of several current popular textbooks, instructors should be aware they cannot necessarily count on textbooks to stimulate spatial thinking skills in their students. It is important that instructors choose their textbooks with that awareness. For those GIS instructors who intend to develop spatial thinking skills, some texts offer more support than others, but by themselves are unlikely to produce students who possess effective spatial thinking skills. If spatial thinking is to be developed, it must be embedded in the classroom and laboratory. Thus, we recommend that more attention be paid to explicit instruction in spatial thinking.

Instructors seeking to develop spatial thinking in their students should consider ways to encourage this early in the course. Our textbook analysis reveals that questions covering topics which typically occur at the beginning of the course (e.g., cartography, map projections and coordinate systems) do not require as much spatial thinking as topics such as spatial analysis which often appear later in the course.

We believe this exploratory work could be expanded to assess the philosophy of and pedagogy employed by GIS instructors to better understand how, if at all, spatial thinking is embedded in the teaching of introductory GIS courses. We encourage textbook authors and publishers to consider revising end-of-chapter questions to better support the three identified components of spatial thinking; to create more questions that offer students opportunities to integrate concepts of space, representations, and processes of reasoning; and to use questions that require output-level cognitive skills.

References

Albert, W. S., and R. G. Golledge (1999) The use of spatial cognitive abilities in geographic information systems: The map overlay operation. *Transactions in GIS*, 3(1): 7–21.

Andre, T. (1979) Does answering higher-level questions while reading facilitate productive learning? *Review of Educational Research*, 49: 280–318.

Association of American Geographers (2009) *Guide to Geography Programs in the Americas*. Washington, DC: Association of American Geographers.

Audet, R., and G. L. Abegg (1996) Geographic information systems: Implications for problem solving. *Journal of Research in Science Teaching*, 33(1): 21–45.

Bloom, B. S., M. D. Engelhart, E. J. Furst, W. H. Hill, and D. R. Krathwohl (1956) *Taxonomy of Educational Objectives: The Classification of Educational Goals.* New York, NY: David McKay Company, Inc.

Bolstad, P. (2008) *GIS Fundamentals: A First Text on Geographic Information Systems*, 3rd ed. White Bear Lake, MN: Eider Press.

Chang, K.-T. (2010) *Introduction to Geographic Information Systems*, 5th ed. New York, NY: McGraw-Hill.

Costa, A. L. (2001) Teacher behaviors that enable student thinking, IN: A. L. Costa (ed.), *Developing Minds: A Resource Book for Teaching Thinking*, Vol. 3. Alexandria, VA: Association for Supervision and Curriculum Development, pages 359–369.

DiBiase, D., M. DeMers, A. Johnson, K. Kemp, A. Luck, B. Plewe, and E. Wentz (eds) (2006) *The Geographic Information Science and Technology Body of Knowledge.* Washington, DC: Association of American Geographers and University Consortium for Geographic Information Science.

Golledge, R. G. (1995) Primitives of spatial knowledge, IN: T. L. Nyerges and M. J. Egenhofer (eds), *Cognitive Aspects of Human-Computer Interaction for Geographic Information Systems*). Dordrecht, The Netherlands: Kluwer Academic Publishers, pages 29–44.

Golledge, R. G. (2002) The nature of geographic knowledge. *Annals of the Association of American Geographers*, 92: 1–14.

Gorr, W. L., and K. S. Kurland (2008) *GIS Tutorial: Workbook for Arcview 9*, 3rd ed. Redlands, CA: Esri Press.

Hamaker, C. (1986) The effects of adjunct questions on prose learning. *Review of Educational Research*, 56: 212–242.

Harvey, F. (2008) *A Primer of GIS: Fundamental Geographic and Cartographic Concepts.* New York, NY: The Guilford Press.

Jo, I., and S. W. Bednarz (2009) Evaluating geography textbook questions from a spatial perspective: Using concepts of space, tools of representation, and cognitive processes to evaluate spatiality. *Journal of Geography*, 108: 4–13.

Kerski, J. J. (2000) *The implementation and effectiveness of geographic information systems technology and methods in secondary education.* PhD diss. University of Colorado, Boulder.

Krathwohl, D. R. (2002) A revision of Bloom's Taxonomy: An overview. *Theory into Practice*, 41: 212–218.

Lee, J., and R. S. Bednarz (2009) Effect of GIS learning on spatial thinking. *Journal of Geography in Higher Education*, 33(2): 183–198.

Longley, P., M. F. Goodchild, D. J. Maquire, and D. W. Rhind (2005) *Geographic Information Systems and Science.* Chichester: Wiley

Marzano, R. J. (2001) A new taxonomy of educational objectives, IN: A. L. Costa (ed), *Developing Minds: A Resource Book for Teaching Thinking*, Vol. 3. Alexandria, VA: Association for Supervision and Curriculum Development, pages 181–189.

Mills, S. R., C. T. Rice, D. C. Berliner, and E. W. Rosseau (1980) The correspondence between teacher questions and student answers in classroom discourse. *Journal of Experimental Education*, 48: 194–204.

Moseley, D. M., J. Elliott, M. Gregson, and S. Higgins (2005) Thinking skills frameworks for use in education and training. *British Educational Research Journal*, 31: 367–390.

Myers, M. P., and T. Savage (2005) Enhancing student comprehension of social studies material. *The Social Studies*, 96: 18–23.

National Research Council (2006) *Learning to Think Spatially.* Washington, DC: National Academy Press.

Nosich, G. M. (2005) Problems with two standard models for teaching critical thinking. *New Directions for Community Colleges*, 130: 59–67.

Ormsby, T. (2010) *Getting to Know ArcGIS Desktop*, 3rd ed. Redlands, CA: Esri Press.

Pizzini, E. I., D. P. Shepardson, and S. K. Abell (1992) The questioning level of select middle school science textbooks. *School Science and Mathematics*, 92: 74–79.

Price, M. H. (2010) *Mastering ArcGIS*, 4th ed. Dubuque, IA: McGraw-Hill.

Self, C. M., S. Gopal, R. G. Golledge, and S. Fenstermaker (1992) Gender-related differences in spatial abilities. *Progress in Human Geography*, 16, 315–342.

Vincent, P. (2010) Structure and Content of Introductory GIS Courses. Paper presented at the 2010 National Conference on Geographic Education.

Vogler, K. E. (2005) Improve your verbal questioning. *The Clearing House*, 79: 98–103.

Wade, T., and S. Sommer (2006) *A to Z GIS: An Illustrated Dictionary of Geographic Information Systems*, 2nd ed. Redlands, CA: Esri Press.

Wilen, W. W. (2001) Exploring myths about teacher questioning in the social studies classroom. *The Social Studies*, 92: 26–32.

Winne, P. M. (1979) Experiments relating teachers' use of higher cognitive questions to student achievement. *Review of Educational Research*, 49: 13–50.

16

Teaching spatial literacy and spatial technologies in the digital humanities

David J. Bodenhamer[1] **and Ian N. Gregory**[2]

[1]*The Polis Center, Indianapolis, Illinois, USA*
[2]*Department of History, Lancaster University, Lancaster, UK*

16.1 Spatial literacy in the humanities: an introduction

In August 2008 almost 150 scholars from four continents gathered at the University of Essex for a conference on the uses of geographic information systems (GIS) in history and other humanities disciplines. The sessions suggest the wide reach of GIS within the humanities. Some papers could have been anticipated – the heart of GIS is its ability to manage geographic space and analyse the spatial configuration of a problem, usually with a map, so it was not surprising to witness the technology used to examine territorial and boundary changes, population migrations, wars, or environmental changes, all topics that relate to the geography of human behaviour. But other presentations would have surprised anyone who thinks of GIS primarily as mapping software:

- Can GIS help us understand the Bible?
- Unfolding the landscape drawing method of Japanese historical folding-screens in a GIS environment.
- Mapping the performance culture of 19th-century Nottingham.
- Geomedical analysis of disease records in early 20th-century Kyoto.
- Mapping the Lake District: Towards a literary GIS.
- The financial geography of the US oil industry during the Civil War era.
- Historical GIS and carbon emissions reduction programs.

The truly surprising realization was not how many topics were open to the use of GIS, but rather how common it is now to see these applications at professional conferences. The Social Science History Association and its sister organization the European Social

Teaching Geographic Information Science and Technology in Higher Education, First Edition.
Edited by David J. Unwin, Kenneth E. Foote, Nicholas J. Tate and David DiBiase.

Science History Conference, American Historical Association, Association of American Geographers, Modern Language Association, and numerous other scholarly organizations report a surge in the number of presentations employing this increasingly ubiquitous technology. Professional journals and university presses also are experiencing the increased presence of humanities GIS, with Indiana University Press launching a series on the spatial humanities in 2010.

The use of spatial technologies in the humanities is part of a broader movement in what is increasingly known as the digital humanities. This term covers generally the use of digital and computing technologies to create and disseminate knowledge about human society and culture, past and present, but it takes many different forms (see Greengrass and Hughes, 2008; Schreibman *et al.*, 2004). Textual analysis, data mining, quantification, visualization, serious gaming and multi-media explorations are all part of this new approach to knowledge, both as tools and methods of discovery as well as ways of knowing or perceiving society and culture. Students at all levels are aware of the explosion of digital materials in libraries or across the internet, but less well recognized is the influence of digital humanities on the development of knowledge itself. Computing technologies structure and present information differently than the ways that humanists have traditionally encountered it. Painstaking research in hundreds of archival documents, often scattered across different locations, is rapidly giving way to Google searches, online retrieval and software-specific parsing of electronic materials. Literary scholars quickly discover patterns of word usage, many of them unknowable by traditional means; religionists link text and images via the web to find associations that likely would have escaped them otherwise; linguists and anthropologists use speech analysis software to trace linguistic and cultural connections among groups that might not be possible to discern by the human ear alone. In all of these instances, the act of making information digital opened it to possibilities for the creation of new knowledge.

The same potential is true with the new approach to space made possible by the application of spatial technologies, especially GIS and its allied tools. Doing this effectively requires the integration of three distinct skill sets. The first is the technical skills of using and understanding GIS software and other related spatial technologies. The second is understanding the importance of space and geography to the discipline concerned. The third involves bringing these together to make a contribution to knowledge within humanities disciplines in ways that are sympathetic to the traditions, approaches and sources used within that discipline. This is a process of evolution not revolution (Bodenhamer, 2010; Churchill and Hillier, 2008; Gregory and Ell, 2007). The ability to bring these together is what we term spatial literacy. It is important to note however that an understanding of the importance of space has a long tradition that predates spatial technologies and it is to this that we now turn.

16.2 Spatial thinking

Over the past two decades, the humanities and social sciences especially have advanced a more complex and nuanced understanding of space, or, as David N. Livingstone (1999, 7) has written, 'there has been a remarkable "spatial turn" among students of society and culture.' Subject matter once organized largely by periods of time, with names such as the Great Depression or the Age of Discovery, increasingly embraces themes of region, disapora, colonial territory, and contact zones and rubrics such as 'border' and 'boundary.' The shift has been accompanied by, and reinforced through, an equivalent concern with material culture and built environment, in observations of local representation in dress, architecture, eating, music and other cultural markers of space and place. Climate, topography, hydrology and landscapes likewise have reemerged as important considerations in the investigation of literatures, histories, and social and political life. As a result, our sense of space and place has become more complex and problematic, but in the process it has assumed a more interesting and active role in how we understand history and culture (Cosgrove, 2004; Olwig, 2002).

Today, humanists are acutely aware of the social and political construction of space and its particular expression as place. Spaces are not simply the setting for historical action but are a significant product and determinant of change. They are not passive settings but the medium for the development of culture: 'space is not an empty dimension along which social groupings become structured,' sociologist Anthony Giddens (1984, 364) notes, 'but has to be considered in terms of its involvement in the constitution of systems of interaction.' All spaces contain embedded stories based on what has happened there. These stories are both individual and collective, and each of them link geography (space) and history (time). More importantly, they all reflect the values and cultural codes present in the various political and social arrangements that provide structure to society. In this sense, then, the meaning of space, especially as place or landscape, is always being constructed through the various contests that occur over power. There is nothing new in this development – the earliest maps reveal the power arrangements of past societies – but humanities scholarship increasingly reflects what may in fact be the greatest legacy of postmodernism, the acknowledgement that our understanding of the world itself is socially constructed.

At its core, the spatial turn rejects the universal truths, grand narratives and structural explanations that dominated the social sciences and humanities during much of the last century. The spatial turn, above all, is about the particular and the local, without any supposition that one form of culture is better than another. Its claim is straightforward: to understand human society and culture we must understand how it developed in certain circumstances and in certain times and at certain places. From this knowledge, we can appreciate that the world is not flat but incredibly complicated and diverse. This view no longer seems new because humanists have embraced it eagerly; now, we all recognize the particularity of space, the importance of place. It is, in fact, a postmodern view. But for all the uses we make of this insight – and for all its explanatory power – the concepts of space and place employed by humanists frequently are metaphorical and not geographical. Far less often have we grappled with how the physical world has shaped us or how in turn we have shaped perceptions of our material environment.

Although the *Annales* school, most notably through its chief practitioner Ferdinand Braudel, had, since the 1930s, urged scholars to pay attention to *geohistoire* – the linkage of geography and history – most humanists paid much less attention to the environmental context for human behaviour and much more to the actions, associations and attitudes that made a space particular or, in short, a place. These places could even exist in imagined space or in memory. They also could be personal – emotional space or the body in space – and even metaphorical or fictional, a woman's place, for example, as in Virginia Wolfs' story, *A Room of One's Own*. The spaces of interest to humanists bore little relationship to GIS, with its emphasis on physical or geographical space. But in one area of history, scholars have begun to apply the new spatial technology and, in the process, discover its potential and limits for their work.

16.3 Spatial technologies and spatial literacy in the humanities

History and archaeology are the two disciplines within the humanities in which spatial literacy has made the most progress, however it is interesting to note that these have occurred in parallel rather than together. In archaeology interest in GIS came from its ability to help deal with the fact that archaeological data are inevitably spatially referenced as they are found at specific locations (Connolly and Lake, 2006; Wheatley and Gillings, 2002). In history, by contrast, the use of GIS originated from the quantitative social science end of historical geography, frequently based on the use of census data, from where it has spread into other areas of the discipline.

Archaeologists have been using GIS for over 20 years with several major texts appearing in the 1990s (for example, Allen *et al.*, 1990; Lock and Stancic, 1995). The use of GIS in history developed later and is probably little over a decade old. As late as 1997 a survey of historical geography made no mention of historical GIS; few historians in the mid 1990s had even heard of GIS, much less its application to their discipline. Anne Knowles (2008a, 7), an early US advocate, has defined historical GIS as having the 'elements of *geohistoire*, historical geography, and spatial and digital history' and identified more by its characteristics than any theoretical approach or body of scholarship. Among these characteristics are the dominance of geographical questions and geographical information in framing inquiries, usually fashioned as patterns of change over time, and the use of maps to present its results. The growth of historical GIS, as it has become known, has been impressive. The first dedicated publication came in 2000 when *Social Science History* dedicated a special issue to historical GIS. Since then there have been three more special journal issues – *History and Computing* (2001), *Historical Geography* (2005) and *Social Science Computer Review* (2009) – a good practice guide (Gregory, 2003), three collections of essays (Knowles, 2002; Knowles, 2008b; Boonstra and Shuurman, 2009), and a university press book (Gregory and Ell, 2007). The key to proving the success of the field, however, lies in the fact that research that uses spatial approaches is increasingly appearing in mainstream books and journals as well as these more specialist publications, and that this research covers a wide range of topics within the discipline of history. Examples include work on topics as diverse as Medieval England (Campbell, 2000; Campbell and Bartley, 2006; Lloyd and Lilley, 2009), the causes of the Dust Bowl on the US Great Plains (Cunfer, 2005), urban decline in US cities (Gordon, 2008), long-term mortality change in nineteenth and twentieth century Britain (Gregory, 2008 and 2009), the US iron industry in the antebellum period (Knowles and Healey, 2006), and slavery before the American Civil War (Thomas and Ayres, 2003). Work of this kind is increasingly referred to as 'spatial history' rather than historical GIS, stressing the fact that it has moved beyond the application of GIS technology, to make use of the approaches the technology has to offer to make an applied contribution to a wide range of topics within the discipline.

At the same time as the full potential of spatial history begins to be realized, there has also been a shift towards using spatial technologies in a wide range of other humanities disciplines including religious studies, classics, literary studies and linguistics. This is increasingly referred to as 'humanities GIS' or, more commonly, 'spatial humanities,' a topic on which a university press has set up a series which has produced one edited volume (Bodenhamer *et al.*, 2010a) and has an additional five titles under contract.

Despite its success, especially among younger scholars, most historians – indeed, most humanists – have not adopted GIS or, more fundamentally, found it helpful. What is becoming increasingly frustrating to its practitioners is that the technology is not finding its way into the toolkit of these scholars. After all, human activity is about time and space, and GIS provides a way to manage, relate and query events, as well as to visualize them, that should be attractive to researchers.

Currently, the problems with GIS as a platform for humanities research are well recognized. Spatial technologies in general, and especially GIS, are expensive, complex and cumbersome, despite recent advances that have driven down costs and simplified the user experience. They require significant investments of time to learn both the language and techniques of the toolsets they employ. GIS and its cousins are literal technologies: they favor precise data that can be managed and parsed within a highly structured tabular database. Ambiguity, uncertainty, nuance and uniqueness, all embedded in the evidence typically available to humanists, do not admit readily to such routinization. GIS also has difficulty managing time, which is a major problem in disciplines that orient their study to periods and epochs. Time is merely an attribute of location within a GIS, but it is a much more complicated concept for humanists, who view it at multiple scales – days, centuries and epochs – and often inexactly, as in the Middle Kingdom or the Age of Industrialization, and who may not know either time or location with any certainty.

More important, the use of GIS requires humanists to be alert to issues that are not part of their training or culture. Humanists, for instance, are logo-centric – they traditionally have relied upon words, with their halos of meaning, to parse complexity, ambiguity and uncertainty – yet GIS relies heavily on visualization to display its results. The use of GIS requires collaboration between and among technical and domain experts, thereby putting humanists, who work in isolation and often are inept in the lingo, at a two-fold disadvantage. Finally, for many humanities scholars GIS appears reductionist in its epistemology. It forces data into categories; it defines space in limited and literal ways instead of the metaphorical frames that are equally reflective of human experience; and, while managing complexity within its data structures, it too often simplifies its mapped results in ways that obscure rather than illuminate (Bodenhamer, 2008).

Above all, GIS demands the use of spatial questions in its applications, whereas most humanists think rarely about geographical space. The fact that humanists typically do not employ geographical concepts in their analyses, however, begs several questions: Do humanists discern a connection between their methods and the methods of geography? Do they understand how spatial competencies can affect their scholarship? Do humanists perceive themselves to be spatially literate?

16.4 Charting spatial literacy in the humanities: a survey

In 2009, a team of three researchers, one from the UK (Ian Gregory, Lancaster University) and two from the US (David Bodenhamer, Indiana University Purdue University Indianapolis), and James Wilson (James Madison University) received a SPLINT (Spatial Literacy in Teaching) grant to investigate spatial literacy in the humanities. The project, 'GIS and the Humanities: Towards an Educational Strategy in Britain and America,' aimed to map core spatial competencies onto the themes and methods embraced by historians and other humanists in their work, based on consultation with user communities. The humanities community of interest was defined broadly and included the disciplines of history, linguistics and literary studies, cultural studies, religious studies, and archaeology, among others.

The team used three approaches – key informant interviews, web-based survey, and an expert workshop – to develop its understanding of the core competencies as defined by the target communities. Specifically, the project sought to:

1. Determine how much humanities scholars know about spatial literacy and the use of GIS, as well as the extent of their interest in spatial concepts, spatial thinking and spatial technologies.
2. Determine what knowledge, training and skill development they required or desired and in what form this content should be delivered.
3. Recommend strategies and actions to address the need for spatial literacy in the humanities.

Invitations to participate distributed through a host of list-servers, established networks and professional associations generated 201 respondents to the survey, which was open for six weeks. Two-thirds were male, a figure that generally matches the gender ratio in the humanities for established professionals. As shown in Figure 16.1, a regular distribution among faculty ranks was evident with, at the junior end of the scale 37 percent describing themselves as researchers and, at the senior end, 21 percent who were professors. Forty-nine percent of respondents came from North America (US, 37 percent; Canada, 11 percent), 42 percent from Europe (Great Britain, 20 percent), and the remainder from elsewhere, with Japan (3 percent) registering the most respondents from this group. Almost 62 percent came from institutions that offer graduate programs (58 percent doctoral, 3.5 percent

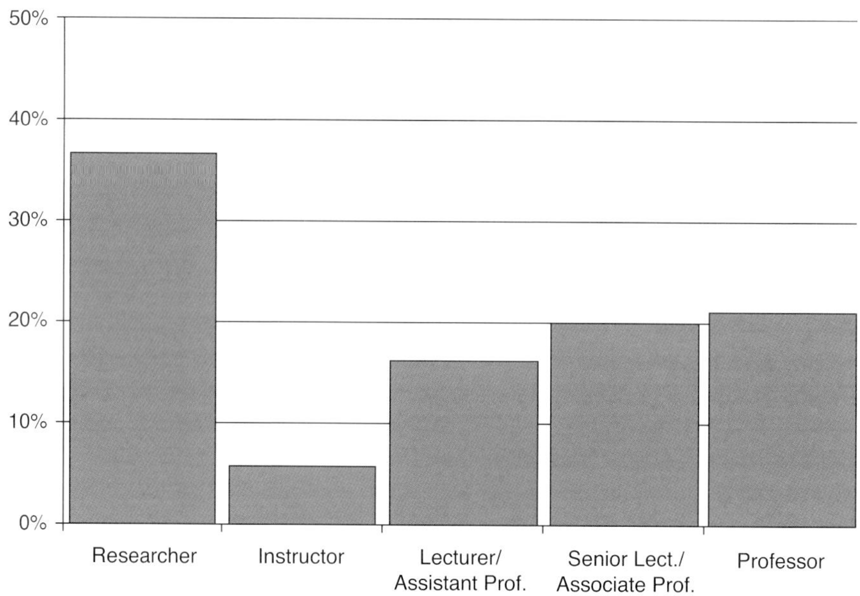

Figure 16.1 Rank of respondents to the survey

master's), whereas 16 percent taught only at the baccalaureate level. The remaining 22 percent were recorded only as 'other.' Although historians and geographers were the most frequent respondents (31 and 29 percent, respectively), many other disciplines were represented, including archaeology (7 percent), literary studies (5 percent), and religious studies (4 percent), among six others.

Table 16.1 shows how respondents rated their level of knowledge about GIS. In the first of the three questions respondents were asked to rank their expertise of software and web applications as either *expert* ('I can perform most functions'), *moderate* ('I can use many functions'), *average* ('I can perform some functions') or *beginner* ('I can perform simple functions'). Fifty-two percent described themselves as expert or moderate users of desk-top GIS, whereas 27 percent were beginners. Similar results were found for digital globes (e.g., Google Earth, 49 percent vs. 22 percent), whereas fewer respondents classed themselves as expert in GPS (31 percent). Familiarity with spatial databases ranked lowest with only

Table 16.1 Responses to questions about respondents' spatial literacy

What is your level of knowledge about the following software and Web applications?					
	Expert	Moderate	Average	Beginner	No response
Desktop or Web GIS	23.8	28.1	14.1	27.0	7.0
'Spatial databases (e.g., Oracle Spatial)'	6.0	14.1	16.3	39.1	24.5
'Digital Globes (e.g., Google Earth)'	13.4	35.5	25.3	22.0	3.8
Global Positioning Systems	10.8	20.5	24.9	30.8	13.0
What is your level of expertise with the following analytical functions in GIS?					
	Expert	Proficient	Familiar	Beginner	No response
I can create a spatial database	17.6	23.5	14.7	27.1	17.1
I can run spatial autocorrelations	8.8	11.8	20.0	35.3	24.1
I can perform relational joins	24.9	17.2	15.4	21.3	21.3
I can perform an overlay analysis	24.7	12.4	14.7	30.6	17.6
I can do cluster analysis	17.1	12.4	12.9	34.7	22.9
I can create a publishable map for a journal	32.1	16.1	16.1	22.6	13.1
I can place a map on the Internet	21.3	18.3	17.8	29.6	13.0
What is your knowledge of literature of the following debates and/or developments regarding GIS?					
	Knowledgeable	Aware	No knowledge		
Critical GIS	19.2	37.7	43.1		
GIS and Society	27.1	47.0	25.9		
Neogeography	7.2	24.7	68.1		
Volunteered Geog. Information (VGI)	9.6	32.9	57.5		
Text mapping	8.4	38.9	52.7		

20 percent in the expert and moderate category; a high proportion, 24 percent, did not respond, suggesting perhaps that they either did not use these resources or had never heard of them. Questions about proficiency revealed a split among respondents who used many advanced GIS functions (e.g., relational joins or cluster analysis) and an equally large group who reported use only at the beginning level (e.g., create publishable maps). Although not included in the table, removing geographers from the responses increased the percentage of beginners significantly in all categories, implying an acquaintance by humanists with spatial technologies but not expertise in them. Asked to rank their knowledge of the literature on Critical GIS, GIS and Society, neogeography, volunteered geographical information (VGI), and text mapping, the respondents said they had no knowledge or only general awareness on each topic, with over 60 percent citing no knowledge of such prominent concepts such as neogeography. Removing geographers from these responses reduced the number of 'knowledgeable' responses slightly and led to a corresponding increase in the other proportions but did not change the overall pattern significantly.

Survey participants reported a high degree of unstructured learning when asked about their training in GIS and other spatial technologies. Sixty percent learned from a mentor or colleagues, professional literature and websites. Formal training in the form of courses and workshops were the least common responses, with fewer than half of the respondents reporting these avenues for learning. This outcome remained the same when geographers were excluded. The lack of formal instruction may explain the general lack of familiarity with topics associated with geographical information.

Respondents confirmed reports in the literature when asked about barriers to the use of spatial technologies in the humanities (see Bodenhamer, 2008; Bodenhamer *et al.*, 2010b; Jessop, 2008). By far the most important concern was the lack of required technical skills, with 89 percent noting it as a serious (48 percent) or somewhat serious issue. The costs of developing data, lack of technical support, and lack of methodological skills were also seen as problems by over 75 percent of respondents. Perhaps because of the nature of the survey audience, which doubtless had more response from scholars interested in technology or in Digital Humanities more broadly, few respondents believed GIS and other spatial technologies were inappropriate tools for humanities research. But most respondents believed that knowledge of the literature was either a very serious or somewhat serious concern (59 percent). Costs of developing the data and lack of appropriate data also were cited as issues. When geographers were removed from the respondents, all of these issues were viewed as significantly more of a problem.

Given these results, what solutions did respondents endorse? Seventy percent cited the need for more training in methods, best practice, software, web applications, data development and visualization; less than 40 percent thought more exposure to theory was important. But respondents shied away from a tools/methods-oriented approach when asked about undergraduate and graduate curricula, overwhelmingly citing the need for coursework in spatial thinking (94 percent), cartographic representation, and spatial analysis, along with training in GIS software, spatial database construction, spatial data development, critical theory and literature. Instruction in 'other spatial visualization' and development of spatial databases were seen as the least important (>30 percent).

When asked what skills students currently have, the order of the categories remained the same – spatial thinking, cartographic representation, etc. – but fewer than one in four respondents believed students currently were proficient in any of these areas. Overwhelmingly, respondents believed spatial skills need to be improved across the board, with spatial thinking being seen as particularly important and spatial analytic skills notably lacking among current students. The number of respondents who cited the importance of these skills increased somewhat when asked about the graduate level but the order remained the same. All the various means for providing skills – coursework, workshops, online instruction and labs, among others – received endorsement from three of four respondents, with 90 percent agreeing strongly that the best method of learning was with project work.

16.5 Teaching spatial literacy in the humanities: challenges

The survey on the use of spatial technologies in the humanities revealed two basic themes of importance to scholars interested in the spatial humanities and historical GIS. First, it is important to develop an effective way to acquire the technical skills in computer-based tools, such as GIS. More important, it is necessary to help scholars with an interest in space to enhance their spatial thinking and spatial analytical skills.

The first problem is in some ways the easiest to solve, although perhaps the most difficult to finance. But the survey and conversations with key informants led to the same proposals – workshops, online training, seminars, conferences and courses available within researchers' host universities. The issue, however, is not one of software skill development alone. More important in many ways is an introduction to the types of digital resources available for humanities research. Many historians, for instance, are aware of such datasets as the census or ships' manifests but often think of them only in tabular form, useful for manipulating with programs such as SPSS or SAS. Yet these resources are eminently spatial, as are many textual repositories in digital form, such as the Early English and Early American imprint series. It also is necessary to teach scholars about documentation practices (metadata), preservation and migration of data, and use of non-proprietary file formats. At one point in the development of digital humanities, any of these issues – software, metadata, preservation of data, and the like – might have been large obstacles but rarely is this the case any longer. Not only has software of all sorts become more intuitive and easier to use but most now provide a means for capturing essential metadata automatically, leaving the researcher primarily the task of describing limits of the data or manipulations performed in preparing it.

Several other issues must be addressed in resolving the problem of skill development. First, the training must be affordable. The survey results were clear on this point. Affordability did not mean simply a reasonably priced training session, but also readily obtainable and supported software and both financial support and time to build the necessary skill sets. Simply providing these skill development opportunities will not be sufficient, however, unless scholars have a way to attend them inexpensively. Second, the researcher's home institution must provide ways to encourage the use of these skills through such means as IT professionals who understand both the technology and the research needs of humanists. Finally, the effort put forward by researchers must find recognition within the home institution and within the discipline. The standards of peer review are important for professional growth and reputation yet few people who begin in the spatial humanities will have immediate results to submit to this process. Institutions and disciplines must find the means to evaluate what is being learned and how it is being applied and not simply wait until a result is available for peer review. In brief, the spatial humanities, like the digital humanities more generally, involve risk-taking and experimentation – and sometimes risks don't pay off and experiments fail. Few researchers will venture in this direction unless they have some assurance that temporary failure will not harm their careers.

Even if these issues are addressed successfully, another problem remains – and, in fact, it is the larger obstacle. How can the humanities cultivate spatial literacy, which is the first step toward making the humanities truly spatial? Scholars at undergraduate, postgraduate and professional level must become aware of the importance of space and how it affects them and their discipline. This circumstance rarely exists in any humanities discipline. Closely related to it is how to help scholars understand the different ways in which space can be conceptualized, as well as how to integrate time and space. In this matter, the history of geographical thought is important, particularly issues associated with the way that space was conceptualized in a quantitative manner during the quantitative revolution and the reaction against this and towards a more humanistic representation of place during the so-called cultural turn, of which the spatial turn is a variant.

16.6 Teaching spatial literacy in the humanities: delivery strategies

Educating humanists in spatial literacy to provide them with both technical skills and spatial thinking requires a strategy that will work across the range of professional and student levels. Given the relative newness of these technologies and approaches within the humanities we believe this requires a five-prong strategy, namely: (1) that the research base be deepened to demonstrate how effective spatial approaches can be in delivering new knowledge across a range of humanities disciplines; (2) that efforts be made to include senior academics, junior academics and PhD students in using spatial technologies; (3) that spatial literacy be taught formally at postgraduate level in suitable humanities courses; (4) that spatial literacy be introduced at undergraduate level; and (5) that access to the limited and disparate resources relevant to spatial literacy become more accessible to humanists. These will be discussed in more detail below.

Deepening the research base

For spatial approaches to be truly accepted across the range of humanities disciplines, they must prove their worth in each of the many subjects that make up the humanities. To date this has only happened within history and archaeology, nevertheless, the potential of spatial approaches is leading to a growing enthusiasm for them in many other humanities disciplines, with literary and cultural studies, in particular, emerging as key areas. To convert this enthusiasm into acceptance, early adopters of the technologies must deliver high-quality research of the kind that makes other researchers, including those with little interest in technology and perhaps little interest in geography, realize how the use of spatial approaches is making a difference to the broader understanding of their discipline. This means that research has to be published that goes beyond describing database construction, methodologies, and research potential, and instead illustrates research results of significance to the broader discipline. An example of this is Gregory and Cooper's (2010) paper which describes the development of a database of texts on early writing about the English Lake District. This was subsequently developed in a second paper (Cooper and Gregory, 2011) which explores the new knowledge that this offered by comparing and contrasting the geographies inherent in the writing of two early Lake District tourists: Thomas Gray and Samuel Taylor Coleridge. In many cases moving to this stage is difficult and involves long lead times. It requires researchers to focus on the substantive research questions that the technology can answer as well as the more technically focused questions. This may benefit from inter-disciplinary research between humanities researchers and GIScientists where each partner can concentrate on their area of interest. This is, however, frequently problematic owing to the lack of availability of suitable GIS experts who are prepared to be involved as there must be intellectual benefits that accrue from their involvement in addition to some of the other barriers to inter-disciplinary research described above.

In many ways the most important stage in developing spatial literacy this is also perhaps the most difficult one because much depends on the ability of early adopters to drive an agenda that is inclusive of a wide range of research interests and not simply focused on their own research subjects. However, it also offers an advantage to early adopters because the field is so new that pioneering research is often more easily recognized and rewarded. Effective strategies for advancing nascent fields such as spatial history and spatial humanities include expert workshops, specialist meetings, tightly focused conferences and networks where scholars can share ideas and approaches with each other.

Widening the pool of researchers

Success in establishing the spatial humanities more completely will ultimately require enlisting new practitioners. Three levels of audience are likely to be relevant here: graduate students, junior academics who have the time to learn and apply spatial approaches, and senior scholars who may want to include a spatial component within research projects but will often employ staff to do this rather than learn the technology for themselves. Conference presentations and published research are traditional ways to encourage the participation of individuals who are first entering the field. Both methods expose the value of spatially-oriented research and allow novices to begin the essential process of developing a necessary grounding in the literature, as well as learning 'who is doing what' with spatial technologies. This process, although well-tested, is slow, so more focused approaches are likely to be required.

Two established venues for introducing scholars at all levels to the potential of spatial humanities are outreach seminars and short courses, both of which are especially suitable for periods when budgetary restraints prevent more comprehensive approaches. Outreach seminars are short meetings that take a day or less. Their aim is to provide an overview of spatial approaches, what they have to offer to the humanities, and what a humanist needs to do to go further. These could be suitable for all three levels of academic (senior, junior and graduate student) or could be tailored specifically to one particular level. This format of meeting is unlikely to be able to provide any significant software training. Introducing the technology and building skills requires short courses offered over a period of two to five days. Based on extensive experience, we believe that two days is perhaps the minimum time required to give a beginner enough knowledge of a GIS software package for them to build on themselves. Longer courses, preferably four or five days long, have the advantage of allowing the instructor to provide more in-depth software training and combine this with the theory of spatial data handling, exemplar projects, and allowing students to bring their own data to work on in the later stages of the course.

Developing post-graduate training

Providing more in-depth training in spatial literacy requires formal training for post-graduate students on master's and PhD courses. Currently, there is only one fully-fledged master's course in this field, the MA in Historical Resources Management at Idaho State University. Given the economic constraints present at most Anglo-American universities, it is unlikely that similar programs will emerge quickly, if at all. However, in many ways, what is most needed is not a specialized degree but a more intentional embrace of spatial literacy in humanities postgraduate work. Digital humanities programs, which are growing in number, offer one such venue, and numerous disciplines already have IT courses in which the introduction of spatial technologies can be incorporated. Training of this type, notably at the PhD level, should combine teaching of what spatial approaches can and have offered to their discipline at an academic level, while providing technical skills to allow students to employ a spatial approach, if appropriate, in their own research.

This solution cannot be implemented immediately but must evolve over time. Problems to be overcome include a lack of suitably qualified instructors and an absence of suitably equipped labs. One strategy may be to develop collaborations with geography departments. Another approach would be for this type of training to be provided by a different institution, perhaps using distance learning approaches. Software-specific training, such as is offered by Esri – the best established GIS software vendor in the academic market – is also an option, but will not address humanities examples and will require the student to reason by analogy from exercises based on planning or environmental management to the issues presented by his or her own research.

Undergraduate training

Undergraduate courses can be less ambitious in that they need to focus more on what spatial literacy is and how it has made an impact on their disciplines rather than on technical skills. In other words, undergraduate education needs to focus on exemplars of why space is important within their discipline. This is potentially much easier to teach than post-graduate level courses as it requires neither the technical expertise in the instructors or the IT resources for large groups. Numerous resources exist for the development of this understanding, including but not limited to materials offered by the Spatial Literacy in Teaching program (http://www.spatial-literacy.org/) and the ThinkSpatial resources from Spatial@UCSB.

Developing the knowledge base

A further step that could assist at this stage is construction of a knowledge base that draws what is currently a very fragmented and inter-disciplinary literature together in a single location. Currently the Historical GIS Research Network website (http://www.hgis.org.uk)

goes some way towards doing this for historical GIS. Expanding this for the humanities more generally would provide a portal that allows interested parties with a starting point from which to explore the current state of the arts within spatial humanities generally and their discipline more specifically. There is also the need for specialist publications that will help educate people about the uses of spatial approaches in the humanities. In particular, there is a clear niche for textbooks that provide software training aimed specifically at humanists, and a primer that provides a higher-level overview of how to implement spatial approaches in the humanities.

16.7 Spatial technologies for the spatial humanities

The issue of which spatial technology to use in teaching humanities is a somewhat problematic one. At one extreme there are two large, well developed packages that have been commercially available for a number of years: ArcGIS (or ArcView or ArcInfo) and MapInfo. Of these ArcGIS, developed by Esri, has succeeded in dominating the academic market in Europe and North America for a number of years. These packages have a number of advantages: they come from well-established commercial companies and have set the industry standard; they provide a wide-range of functionality, far wider than most humanists are likely to use; they have well laid-out used interfaces; and professional support and updates are readily available. The disadvantages are: they are relatively expensive, although generous academic discounts are sometimes available; they were never designed for humanities use, in particular that they are based on tabular data rather than unstructured texts; and while they are the industry standard they rely on proprietary software and data formats. They are also not very easy to use, partly because of the range of functionality offered, and are not easy to learn without some formal – and often expensive – training.

At the opposite extreme, virtual globes such as Google Earth have gained rapid, widespread acceptance. These are not GIS software packages but are instead viewers that allow users to quickly and easily explore spatial data located against a background image of the Earth's surface. The advantages of virtual globes are: they are usually free; they are very simple and intuitive to use; they are based on open-source data formats such as Keyhole Markup Language (KML), an XML-based implementation which makes them relatively easy to use with unstructured textual information such as Text Encoding Initiative (TEI) encoded texts; and they are very easily used to disseminate data over the internet, a major limitation with commercial GIS packages. The disadvantages are: they are very limited packages that only allow basic mapping with very limited querying ability; they possess awkward database-creation functions; and they have limited analytical capabilities. In short, they are effective for introducing people to spatial technologies and as a way of disseminating the outputs from a GIS database over the internet, but they do not allow the majority of essential spatio-analytical operations to be conducted (Rakshit and Ogneva-Himmelberger, 2008; Tuttle *et al.*, 2008).

A serious issue is the dearth of options between these two extremes. A variety of open source and free GIS software packages are available including GRASS, Quantum GIS and TimeMap, the last of which was developed specifically for humanities use. Although popular among some archaeologists, none of these has, however, gained widespread acceptance among the mainstream GIS community and there are always questions of stability and long-term sustainability in the use of open-source packages that lack the backing of a major commercial company.

What, then, is the best package to teach? The answer is not simple: virtual globes require little more than a basic introduction but do not take the students very far. In many ways ArcGIS seems to be the best package to teach, perhaps in collaboration with Google Earth as an internet dissemination tool. Despite its problems for humanists, ArcGIS is still the most comprehensive GIS package, and it is the most widely used in all markets, including the academic market. Its software formats such as Shapefiles have also become the *de facto* standard in an area sadly lacking in accepted open standards. But ArcGIS should not be offered as the only option. It is not always suitable for humanities data and must be taught with knowledge of the open source and free software also available for use in the spatial humanities.

16.8 Conclusion

Spatial thinking is valuable for humanists because we are drawn to issues of meaning, and space offers a way to understand fundamentally how we order our world. We already have seen that contemporary notions of space are myriad: what was once a reference primarily to geographical space, with its longstanding categories of landscape and place, is now modified by class, capital, gender and race, among other concepts, as an intellectual framework for understanding power and society in times near and distant. We recognize our representations of space as value-laden guides to the world as we perceive it, and we understand how they exist in constant tension with other representations from different places, at different times, and even at the same time. We acknowledge how past, present and future conceptions of the world compete simultaneously within real and imagined spaces. We see space as the platform for multiplicity, a realm where all perspectives are particular and dependent upon experiences unique to an individual, a community, or a period of time (Massey, 2005, 9–15). This complex and culturally relativistic view of space, the product of the last several decades, has reinvigorated geography as a discipline, just as it has engaged scholars within the humanities, with one key difference: humanists, unlike geographers, often do not grapple with the effect of physical space or geographical influences on human behavior and culture.

Ultimately, humanities scholarship must incorporate both geographical and metaphorical meanings of space and place, otherwise we will not be able to take advantage of the vast potential of GIS and other spatial technologies to stimulate and re-vitalize our fields of study. Fortunately, GIS has the ability to help us revisit our understanding of traditional problems, as well as to raise new questions worthy of study. Spatial technologies are powerful analytical engines, but like all tools, they must be properly understood more completely before they can be used effectively. High quality research is already being conducted and disseminated that will demonstrate the utility of spatial approaches across the disciplines of the humanities. This development will continue, but it should be assisted by a variety of strategies to speed its uptake.

Acknowledgments

Work on this chapter was funded by SPLINT under a fellowship awarded to David Bodenhamer, Ian Gregory and James Wilson ‘GIS and the Humanities: Towards an Educational

Strategy in Britain and America.' Construction of the Historical GIS Research Network website was funded by the Economic and Social Research Council under grant RES-451-25-4307.

References

Allen, K. M. S., S. W. Green, and E. B. W. Zubrow (1990) *Interpreting Space: GIS and Archaeology.* New York, NY: Taylor and Francis.

Bodenhamer, D. J. (2008) History and GIS: Implications for the discipline, IN: A. K. Knowles (ed) *Mapping the Past: How Maps, Spatial Data and GIS are Changing Historical Scholarship.* Redlands, CA: Esri Press, pages 220–233.

Bodenhamer, D. J. (2010) The potential of spatial humanities, IN: D. J. Bodenhamer, J. Corrigan, and T. M. Harris (eds) *The Spatial Humanities: GIS and the Future of Humanities Scholarship.* Bloomington, IN: Indiana University Press, pages 14–30.

Bodenhamer, D. J., J. Corrigan, and T. M. Harris (eds) (2010a) *The Spatial Humanities: GIS and the Future of Humanities Scholarship.* Bloomington, IN: Indiana University Press.

Bodenhamer, D. J., J. Corrigan, and T. M. Harris (2010b) Introduction, IN: D. J. Bodenhamer, J. Corrigan, and T. M. Harris (eds) *The Spatial Humanities.* Bloomington, IN: Indiana University Press, pages vii–xv.

Boonstra, O., and A. Schuurman (eds) (2009) *Tijd en Ruimte: Nieuwe Toepassingen van GIS in de Alfawetenschappen.* Utrecht, Netherlands: DANS.

Campbell, B. M. S. (2000) *English Seigniorial Agriculture 1250–1450.* Cambridge, UK: Cambridge University Press.

Campbell, B. M. S., and K. Bartley (2006) *England on the Eve of the Black Death: An Atlas of Lay Lordship, Land and Wealth, 1300–49.* Manchester, UK: Manchester University Press.

Churchill, R., and A. Hillier (2008) Teaching with GIS, IN: A. K. Knowles (ed) *Placing History: How Maps, Spatial Data, and GIS are Changing Historical Scholarship.* Redlands, CA: Esri Press, pages 61–94.

Connolly, J., and M. Lake (2006) *Geographical Information Systems in Archaeology.* Cambridge, UK: Cambridge University Press.

Cooper, D., and I. N. Gregory (2011) Mapping the English Lake District: A literary GIS. *Transactions of the Institute of British Geographers,* 36: 89–108.

Cosgrove, D. (2004) Landscape and Landschaft. *GHI Bulletin,* 35: 57–71.

Cunfer, G. (2005) *On the Great Plains: Agriculture and Environment.* College Station, TX: Texas A&M University Press.

Giddens, A. (1984) *The Constitution of Society: Outline of the Theory of Structuration,* Oxford, UK: Oxford University Press.

Gordon, G. (2008) *Mapping Decline: St. Louis and the Fate of the American City.* Philadelphia, PA: University of Pennsylvania Press.

Greengrass, M., and L. Hughes (eds) (2008) *The Virtual Representation of the Past.* Farnham, UK: Ashgate.

Gregory, I. N. (2003) *A Place in History: A Guide to Using GIS in Historical Research.* Oxford, UK: Oxbow.

Gregory, I. N., and P. S. Ell (2007) *Historical GIS: Technologies, Methodologies and Scholarship.* Cambridge, UK: Cambridge University Press.

Gregory, I. N., and R. G. Healey (2007) Historical GIS: Structuring, mapping and analysing geographies of the past. *Progress in Human Geography,* 31: 638–653.

Gregory, I. N. (2008) Different places, different stories: Infant mortality decline in England and Wales, 1851–1911. *Annals of the Association of American Geographers,* 98: 773–794.

Gregory, I. N. (2009) Comparisons between the geographies of mortality and deprivation from the 1900s to 2001: Spatial analysis of census and mortality statistics. *British Medical Journal*, 339(b3454): 676–679.

Gregory, I. N., and D. Cooper (2010) Thomas Gray, Samuel Taylor Coleridge and geographical information systems. *International Journal of Humanities and Arts Computing*, 3: 61–84.

Jessop, M. (2008) The inhibition of geographical information in digital humanities scholarship. *Literary and Linguistic Computing*, 23: 39–50.

Knowles, A. K. (ed) (2002) *Past Time, Past Place: GIS for History*. Redlands, CA: Esri Press.

Knowles, A. K. (2008a) GIS and History, IN: A. K. Knowles (ed) *Mapping the Past: How Maps, Spatial Data and GIS are changing Historical Scholarship*. Redlands, CA: Esri Press, pages 1–25.

Knowles, A. K. (ed) (2008b) *Mapping the Past: How Maps, Spatial Data and GIS are Changing Historical Scholarship*. Redlands, CA: Esri Press.

Knowles, A. K., and R. G. Healey (2006) Geography, timing, and technology: A GIS-based analysis of Pennsylvania's iron industry, 1825–1875. *Journal of Economic History*, 66: 608–634.

Livingston, D. (1999) Science, region, and religion: The reception of Darwin in Princeton, Belfast, and Edinburgh, IN: R. L. Numbers and J. Stenhouse (eds) *Disseminating Darwinism: The Role of Place, Race, Religion, and Gender*. Cambridge, UK: Cambridge University Press, pages 7–38.

Lloyd, C. D., and K. D. Lilley (2009) Cartographic veracity in medieval mapping: Analyzing geographical variation in the Gough Map of Great Britain. *Annals of the Association of American Geographers*, 99: 27–48.

Lock, G. R., and Z. Stancic (1995) *Archaeology and Geographic Information Systems: A European Perspective*. London, UK: Taylor and Francis.

Massey, D. (2005) *For Space*. London, UK: Sage.

Rakshit, R., and Y. Ogneva-Himmelberger (2008) Application of virtual globes in education. *Geography Compass*, 2: 1995–2010.

Schreibman, S., R. Siemens, and J. Unsworth (eds) (2004) *A Companion to Digital Humanities*. Oxford, UK: Blackwell.

Olwig, K. (2002) *Landscape, Nature and the Body Politic: From Britain's Renaissance to America's New World*. Madison, WI: University of Wisconsin Press.

Thomas, W. G., and E. L. Ayers (2003) An overview: The differences slavery made: A close analysis of two American communities. *American Historical Review*, 108: 1298–1307.

Tuttle, B. T., S. Anderson, and R. Huff (2008) Virtual globes: An overview of their history, uses and future challenges. *Geography Compass*, 2: 1478–1505.

Wheatley, D. W., and M. Gillings (2002) *Spatial Technology and Archaeology: The Archaeological Application of GIS*. London, UK: Taylor and Francis.

17 Discussion and prospect

David J. Unwin

School of Geography, Birkbeck College, University of London, London, UK

17.1 Introduction

At first sight the five chapters in this section seem to have little in common, but it will be seen they all share a concern to extend teaching and learning in GIS&T out beyond the walls of the academy in which it started, typically departments of geography in established universities as an advanced or master's level program. Two concerns are evident. The first, illustrated in the chapters by Susanna and Robert McMaster, David DiBiase *et al.* and Ann Johnson, is for the education of students with what is primarily a professional motivation. The second, illustrated in the chapters by Injeong Jo *et al.* and by David Bodenhamer and Ian Gregory, arises when the GIS&T is taken out of its box and transferred into contexts that have no past connection with academic geography. This involves isolating and teaching the spatial concepts and ideas that underpin the work by people who, insofar as their way into geography was through the use of a proprietary GIS software rather than systematic geographical study, I have in the past called *accidental geographers* (Unwin, 2005).

17.2 Developing a GIS profession

In their accounts Susanna and Robert McMaster, David DiBiase *et al.* and Ann Johnson all illustrate some of the very real problems that any discipline with a strong industrial and technological drive is likely to meet in providing programs for professionals, either already in, or attempting to enter into, that industry. The educational need here is to provide the essential training required by anyone entering the GIS&T workforce at almost any level to enable them to 'push the right buttons' in whatever GIS system they employ, understand 'which buttons should be pressed', and be able to explain 'why these are the right buttons'. Understandably, purely academic programs tend to concentrate on the second of these questions, whereas a technical education will rightly concentrate on the first. This education/training dilemma is one that has been with the GIS&T educator community for many years (see, for example Unwin, 1991; Forer and Unwin, 1999). Introducing the

Teaching Geographic Information Science and Technology in Higher Education, First Edition.
Edited by David J. Unwin, Kenneth E. Foote, Nicholas J. Tate and David DiBiase.

third 'why' question brings in issues to do with so-called transferable skills and appropriate professional behavior.

A decade and a half or so ago, there was a lively debate in both the US and UK about whether or not GIS&T should be considered as a profession (AGI, 1999; Barr, 1995), a debate that in both countries seems to have been settled by the GIS Certification Institute in the US (see http://www.gisci.org/) and the Royal Geographical Society's Chartered Geographer status in the UK (see http://www.rgs.org/OurWork/CharteredGeographer/Chartered+Geographer.htm). It is clear that routine use of GIS&T by individuals has most, if not all, of the characteristics that are usually listed as properties of a profession, such as a relationship to a specialist body of knowledge, a need for specialized training, some sense of shared 'mission,' and common languages, customs and lore. What the certification/chartered status schemes do for individuals who recognize the need is to provide a supporting infrastructure containing some form of accreditation/certification status when joining, a clear commitment to formal continuing professional development (CPD), and acceptance of a mandatory code of behavior that has a strong ethical component (see for example Dale, 1991; Huxhold, 1991; Goodchild and Kemp, 1992; Obermeyer, 1993; Rix and Markham, 1994).

As the chapters in this section show, taking GIS&T education out of its native habitat in academia to serve a would-be professional audience poses some new challenges for which those involved may not necessarily have been prepared and are of a type that almost any discipline confronting the same education/training/professional dilemmas will have to address. Direct GIS&T skills are important, but it is clear that education for a profession needs to add a whole raft of transferable skills as well as clear guidance on correct professional practice. Three comments are in order.

Teaching resources

First, there are the inevitable issues of resources for teaching and learning in a professional context. As Susanna and Robert McMaster show, even for a well-established program in a major university, the pace of development of the 'technology', as distinct from the 'science' parts of GIS&T creates a continuing tension between a need for some necessary curriculum stability while at the same time offering opportunities to work with the very latest tools. In the early days it was relatively easy, necessitating perhaps knowledge of a well-known computer operating system, a command line driven GIS and a so-called 'high level' programming language (such as FORTRAN, BASIC or C) to do things that were not in the toolbox that came with the GIS. Nowadays the number of computing environments that could be taught is very large indeed and at any moment in time it is seldom clear which of these would be most appropriate. As Ann Johnson shows, without the stimulus provided by generous grant aid or donations of software and hardware it is unlikely that technically-oriented two-year colleges in the USA could even enter the field. In the past, some form of grant-in-aid seems to have been a necessary condition for development of these professional programs.

Professional education

Second, in using GIS&T teaching to address the needs of professionals in the industry, it would obviously be useful to have some idea of the educational aims the industry itself values.

GIS&T educators in USA, UK and Australia have been very active in trying to establish the professional need using a variety of approaches. An early attempt by the UK Association for Geographic Information (AGI) reported by Bennion *et al.* (1996) used a focus group composed of representatives of both the industry and academia to create a model of the geospatial industry based very largely on the very well-developed *industry structure model* of the British Computer Society (see www.bcs.org). In essence this model was a matrix resulting from a cross-tabulation of levels of responsibility for different sectors of the industry (data capture, analysis, mapping etc), each cell of which was provided with an agreed description of the necessary competencies to operate in that sector and at that level. In Australia an early analysis of the required competencies was provided by the Australia and New Zealand Land Information Council (ANZLIC, 1996). In the US a similar approach has been taken for graduates from geography programs by Solem *et al.* (2006) and for the specifics of the geospatial industry by Gaudet *et al.* (2001, 2003), DiBiase *et al.* (2010) and Johnson (2010). Although the numbers of roles and competencies differ, in recognizing 12 roles within the industry and 39 relevant competencies the geospatial technology competency matrix assembled by Gaudet *et al.* (2001) is very similar to the AGI's model for the UK. The work for the US Department of Labor, Employment and Training Administration (DOLETA) reported by DiBiase *et al.* (2010) provides both an update and an expansion of their results. Johnson (2010) provides an analysis of the competencies required by a GIS&T technician. It would seem that the industrial needs are well established and the messages are remarkably consistent in that what employees look for are so-called transferable skills, with analytical ability valued more than theoretical and factual knowledge, together with a mixture of GIS&T specific skills such as in GIS, cartography, fieldwork and conventional data analysis which are also valued highly. A worry I have is that there may be a gap between the claims that result from these kinds of analysis and the actual behavior of both employers and employees in work. Without a sustained survey it is impossible to be certain about this, but regular scanning of advertisements for GIS&T professionals gives the impression that it is often technical skills that are asked for, with transferable skills seldom mentioned. This suggests that there may be a difference between what employers say that they need and what in fact they ask for when they hire staff. Almost exactly the same difference has been noted in the IT industry at large (see for example Lethbridge, 1999).

Teaching issues

Third, it seems to me that courses targeted towards the professional already in work will almost certainly require a different mixture of materials and skills from those delivered to full time undergraduate students possibly emphasizing so-called generic or transferable skills, formal project management methods such as PRINCE II in UK, and business practice. Teaching such generally unfamiliar material in a professional context will almost certainly imply using equally unfamiliar methods, but, with the exception of teaching the ethical implications discussed by DiBiase *et al.*, there is as yet very little GIS&T literature on this. An underlying theme of much of this volume is our contention that GIS&T educators have pioneered in many areas of teaching but I strongly suspect that, in methods appropriate for professionals in work, we have much to learn from colleagues elsewhere in higher education. In his account of his teaching GIS&T to undergraduate students in business studies, Ifan Shepherd (2009) addresses not only issues to do with the organizational framework for such teaching, but also the requisite changes in emphasis in the content of what is taught and the associated learning objectives. He notes, for example, the obvious but important truth that business students are unlikely to be as familiar with, or as interested in, geographical ways of thinking as their contemporaries in departments of geography.

17.3 Spatial thinking and literacy

Shepherd (2009) makes a second, equally fundamental, point in his observation that 'The geographical knowledge and skills that are perhaps most difficult for business students to acquire are ones that trained geographers tend to take for granted' and illustrates this with the observation that 'These include our obvious habit of referring to 'west' or 'north-east', rather than the student's habit of referring to 'the left of London', or 'the top right of the USA' (Shepherd, 2009, S36). Far more important than knowing how to 'drive' a GIS is a group of qualities that have been called *spatial thinking*. This is defined by the US National Research Council in its publication *Learning to Think Spatially* as 'a collection of cognitive skills comprised of knowing concepts of space, using tools of representation, and reasoning processes' (National Research Council, 2006, 12).

In different ways, the chapters by Injeong Jo *et al.* and Bodenhamer and Gregory in this volume address some of the issues that arise when GIS&T is taken out to a wider world. In his forward to this volume, Michael Goodchild observes that the development of what has been called *neogeography* creates further educational challenges for the GIS&T community. It is one thing to have immediately to hand high resolution aerial photography of the area surrounding you, courtesy of Google Earth, displayed on some location-aware device such as your mobile (cell) telephone, together with, for example, a 'pin map' display of the coffee shops within walking distance, but it is quite another to make best use of this information and understand its limitations. Such knowledge involves some degree of what has been called *spatial literacy*, the ability to understand, represent, and reason about concepts such as location, distance and direction that are intrinsically spatial in nature. There is no geographical monopoly of these concepts, nor are those trained as geographers necessarily as literate in them as might be imagined but, equally, many professional geographers are surprised to discover that much of what is part of their taken for granted world is not shared by others. As Bodenhammer and Gregory observe 'More important, the use of GIS requires humanists to be alert to issues that are not part of their training or culture' and, a little later, that 'Above all, GIS demands the use of spatial questions in its applications, whereas most humanists think rarely about geographical space.' It is important to note that an understanding of the importance of space has a long tradition that predates spatial technologies and that much the same comments about contemporary spatial literacy might have been made from the very first moment that our species represented their world in map form. The rising popularity of the spatial humanities and historical GIS only make these issues more compelling (Bodenhamer *et al.*, 2010).

The ease with which a modern GIS can be used has created a problem for GIS&T educators, since the vast majority of the new users will have discovered the utility of a spatial perspective through working within the software they happen to have and have no contact, formal or otherwise, with academic GIS&T. To such users it may well be that the technology and the concepts it expresses are unproblematic, with GIS use being only slightly more complex than a conventional database or spreadsheet. The dangers of this should be apparent and can be seen by the practiced eye at virtually any GIS trade meeting in the form of analytical results that are simply wrong, or studies that do not appreciate the subtleties of the ways in which the underlying true geography of the problem has been distorted by its representation in digital data. 'Driving' a GIS without any appreciation of the spatial concepts that it implements 'under the bonnet/hood' can be a very dangerous activity.

All of this begs a series of questions such as:

- What is a spatial concept?
- Can they be taught?
- If so do GIS&T courses and textbook actually deliver?

Over the years there have been a number of attempts to define the set of essential spatial concepts. Many years ago John Nystuen published a short paper with the title *Identification of some fundamental spatial concepts* (Nystuen, 1963) in which he attempted, somewhat in the spirit of dimensional analysis, to reduce the number of such concepts to the minimum number of independent ones necessary to the spatial point of view of the geographer. The primitives to emerge were *direction or orientation, distance* and *connection or relative position*, with *boundary* as an extra 'notion not considered.' In a presidential address to the Association of American Geographers Reginald Golledge (2002) presented a list of some nineteen concepts whose understanding would constitute the process he at the time called *Geographic Thinking and Reasoning*. More recent work by Injeong Jo and Sarah Bednarz (2009) has attempted to group these elements into an hierarchy and Don Janelle (n.d.) provides a list in which eight such concepts are listed, but with the important addition of a claim that these same concepts are demonstrable at all spatial and temporal scales from the galactic to the sub-atomic. This type of claim is testable, and the team at the University of California at Santa Barbara that created the website http://www.teachspatial.org use a simple empirical approach, similar to that taken in the chapter by Jo *et al.*, by examining relevant disciplinary texts and listing the terms used to describe spatial concepts. As might be expected, the emphasis given to each term varies strongly with discipline, whether geography, design, psychology, science education, linguistics, geosciences or social science, but there is also considerable agreement on some concepts. Finally, in the introduction to a workbook *Numbers aren't nasty: A workbook of spatial concepts* (Unwin, 2011) whose creation was funded by the UK Spatial Literacy in Teaching (SPLINT) Initiative, I develop a schema that I hope helps clarify educational aims and intended learning outcomes that are appropriate for the explicit teaching of spatial concepts but that at the same time enables them to be mapped into ideas that will be familiar to most GIS&T practitioners.

The second question, whether or not spatial concepts can be taught has an obviously positive answer and the workbook itself utilizes at least one way by use of simple numerical and related 'thought exercises.' *Learning to Think Spatially* examined the same issue and asserted that 'GIS had a clearly demonstrated potential as a support system for spatial thinking' (National Research Council, 2006, 221). Despite the ideas presented by Tsou and Yanow (2010), I suspect that most of this potential has yet to be realized, since as Jo *et al.* show in their chapter, the textbooks used by the GIS&T community do not seem to address spatial concepts as directly as perhaps they should and there is a similar deficiency in the course syllabuses they examined.

Personally, as a geographer who has taught methods of spatial analysis for over forty years, I find this a humbling yet challenging note on which to end this overview: despite all that has been accomplished, not all these efforts have as yet yielded expected dividends and much more work remains in deciding how to teach GIS&T to professionals in work and spatial thinking to just about the entire planet outside the walls of academic geography.

References

AGI (Association for Geographic Information) (1999) *Continuing Professional Development: Manual & Log Book*, London: Association for Geographic Information.

ANZLIC (Australia and New Zealand Land Information Council) (1996) *Land Information Management Training Needs Analysis: National Report*. Canberra: ANZLIC Secretariat.

Barr, R. (1995) Excuse me, do you have a license to drive that GIS? *GIS Europe*, February: 16–17.

Bennion, F., B. Capper, and D. J. Unwin (1996) *Professional Development for the Geographic Information Industry*. London: Association for Geographic Information.

Bodenhamer, D. J., J. Corrigan, and T. M. Harris (2010) *The Spatial Humanities: GIS and the Future of Humanities Scholarship*. Bloomington: Indiana University Press.

Dale, P. (1994) *Professionalism and Ethics in GIS*. London: Association for Geographic Information, GIS Notes.

Forer, P., and D. J. Unwin (1999) Enabling progress in GIS and education, IN: P. Longley, M. Goodchild, D. Maguire, and D. W. Rhind (eds) *Geographical Information Systems. Volume 2: Management Issues and Applications*, Chichester: John Wiley & Sons, pages 747–756.

Gaudet, C., H. Annulis, and J. Carr (2001) *Workforce Development Models for Geospatial Technology*, Hattiesburg, MS: University of Southern Mississippi.

Gaudet, C., H. Annulis, and J. Carr (2003) Building the geospatial workforce. *URISA Journal*, 15(1): 21–30.

Golledge, R. G. (2002) The nature of geographic knowledge. *Annals of the Association of American Geographers*, 92(1): 1–14.

Goodchild, M. F., and K. K. Kemp (1992) GIS accreditation: What are the options? *ACSM Bulletin* (Nov/Dec.): 44–47.

Huxhold, W. E. (1991) The GIS profession: Titles, pay, qualifications. *Geo Info Systems*, March: 12–22.

Janelle, D. G. (n.d.) *Spatial concepts and spatial reasoning in the social sciences: an agenda for undergraduate education*. http://spatial.ucsb.edu (accessed March 21, 2011).

Jo, I., and S. W. Bednarz (2009) Evaluating geography textbook questions from a spatial perspective: Using concepts of space, tools of representation, and cognitive processes to evaluate spatiality. *Journal of Geography*, 108: 4–13.

Johnson, J. (2010) What GIS technicians do: a synthesis of DACUM job analyses. *URISA Journal*, 22(2): 31–40.

Lethbridge, T. C. (1999) *The Relevance of Education to Software Provision: Data from the 1998 Survey*. Computer Science Technical Report, TR-99-06. Ottawa: University of Ottawa.

National Research Council (NRC) (2006) *Learning to Think Spatially*, Washington DC: National Academies Press.

Nystuen, J. D. (1963). Identification of some fundamental spatial concepts. *Papers of the Michigan Academy of Science, Arts, Letters*, 48: 373–384. Anthologized at least twice, of which the most influential was that in B. J. L. Berry and D. F. Marble (1968) *Spatial Analysis: A Reader in Statistical Geography* (Englewood Cliffs, NJ: Prentice Hall). The most accessible is probably now J. Agnew, D. N. Livingstone, and A. Rogers (1996) *Human Geography: An Essential Anthology*. Oxford: Blackwell.

Obermeyer, N. J. (1993) Certifying GIS professionals: challenges and alternatives. *Journal of the Urban and Regional Information Systems Association*, 5: 67–76.

Rix, D., and R. Markham (1994) GIS certification, ethics and professionalism. *Proceedings AGI'94*, 6.2.1–6.2.5.

Shepherd, I. D. H. (2009) From geography department to business school: Strategies for transplanting GIS courses between disciplines, *Journal of Geography in Higher Education*, 33(Supplement 1): S28–S45.

Solem, M., I. Cheung, and M. B. Schlemper (2008) Skills in professional geography: An assessment of workforce needs and expectations. *The Professional Geographer*, 60(3): 356–337.

Tsou, M.-H., and K. Yanow (2010) Enhancing general education within geographic information sciences and spatial literacy. *URISA Journal*, 22(2): 45–54.

Unwin, D. J. (1991) The academic setting of GIS. IN: D. J. Maguire, M. F. Goodchild, and D. W. Rhind (eds) *Geographic Information Systems: Principles and Applications*. London: Longman.

Unwin, D. J. (2005) Fiddling on a different planet, *Geoforum*, 36(5), 681–684.

Unwin, D. J. (2011) *Numbers Aren't Nasty: A Workbook of Spatial Concepts*. Leicester, UK: University of Leicester, SPLINT Initiative. Available from http://spatial.ucsb.edu.

Section IV

Digital worlds and teaching GIS&T

18
Virtual geographic environments

Gary Priestnall[1], Claire Jarvis[2], Andy Burton[1,4,3], Martin J. Smith[4] and Nick J. Mount[1]

[1] *School of Geography, University of Nottingham, Nottingham, UK*
[2] *Department of Geography, University of Leicester, Leicester, UK*
[3] *Now at Computing and Technology Team, Nottingham Trent University, Nottingham, UK*
[4] *Department of Civil Engineering, University of Nottingham, Nottingham, UK*

18.1 Introduction and context

Virtual Geographic Environments

In this chapter, we use the term virtual geographic environment (VGE) in a broad sense, to refer to a form of reconstruction of a landscape, either real or imaginary, that gives the impression of a three-dimensional environment that allows varying types and degrees of interaction with geographic information and other human beings, often from a 'first person' perspective. Three-dimensional *virtual* models are now used by millions owing to the widespread use of computer generated imagery (CGI) in movies, computer games, and increasing uptake of virtual globes such as Google Earth. As a result of this greater familiarity with sophisticated visual environments, more students should be comfortable with the basic modes of interaction with VGEs, albeit through entertainment rather than teaching and learning. In combination with the more accessible authoring tools now available, we argue that there is considerable potential for the more widespread and creative use of visual techniques in teaching and learning.

We present here a variety of technologies that offer opportunities for use in teaching and learning contexts, in an order that reflects, generally speaking, increasing levels of visual immersion and in most cases increasing levels of interaction. Discussion starts with the use of static and animated 3D visualization, followed by the work flows necessary to build 3D models of the type seen in more interactive VGEs. We then present a range of technologies for interacting with VGEs, from the opportunities presented by virtual globes such as Google Earth, moving through to a consideration of augmented and virtual realities. In addition to providing a technical overview of these different types of VGE, we also explore the capabilities and challenges associated with each, to help educators new to this field make an assessment of the various possibilities for adoption within the curriculum. Further,

Teaching Geographic Information Science and Technology in Higher Education, First Edition.
Edited by David J. Unwin, Kenneth E. Foote, Nicholas J. Tate and David DiBiase.

we reflect on some of the pedagogic benefits and also the practicalities of utilizing these technologies in teaching, and suggest a phased curriculum relevant to incorporating VGEs at various levels of the higher education curriculum.

Progression in the use of 3D environments in teaching and learning

One of the most common ways to direct the learning experience remains the use of slide shows containing text, static imagery and occasionally embedded videos. In the context of VGEs the incorporation of static images can still be useful to place objects or phenomena into their spatial context. With no interaction possible, annotations for scale and orientation are important, and may need to be added if the image is a screenshot from a more interactive or visually immersive experience. Static visualization can be of value in controlling the message being given to the viewer, although with this goes the risk of misinterpretation owing to a lack of understanding of the quality of data or processing lineage that have resulted in the end product.

Static images are often extracted during analytical procedures carried out within geographical information system (GIS) software such as ArcGIS. Such systems include components for manipulating, analysing and visualizing layers of geographic information in 3D (ArcScene) and also for visualizing such data on a digital globe surface (ArcGlobe). Where there is a desire to convey more photo-realistic visualizations of such data, or to use it as the basis for a more interactive or visually immersive environment, other packages designed for landscape modeling and visualization are often used. Some software packages have high fidelity static representations as perhaps their main focus, for example Bryce and Terragen, and the more generic 3D modeling package 3D Studio Max. These are powerful and flexible packages, but as a consequence the learning curve can be steep. There are particular challenges from a GIScience perspective in that most operate around local or arbitrary coordinate systems and so importing, integrating and exploring geospatial data can be problematic, although Visual Nature Studio is an example of a landscape visualization package designed to use GIS data layers in a more integrated fashion. The Virtual Terrain Project website (http://www.vterrain.org/) offers more information on software options along with data resources related to landscape modeling.

The use of animated sequences generated from 3D environments adds a small amount of interaction, as they can be paused for discussion, or replayed. Many packages (such as Bryce, 3D Studio Max, and Maya) use the concept of *keyframe animation*, where the user defines the characteristics of the environment at certain key points along a timeline, and the system interpolates all frames of an animation between those points. Such animated sequences can include the fly-through, where the camera's position is animated along a timeline (see Figure 18.1), but also include the animation of objects, textures and environmental variables in the scene. Some animation formats such as QuickTime can preserve hyperlinks present in the package generating them, and therefore offer a slightly more interactive experience. Animations can be effective in demonstrating the spatial context of features, and several more interactive environments are adding authoring tools to make the creation of virtual tours easier, such as Google Earth.

As with static visualization, a challenge is to communicate the messages contained in the animation carefully, and the addition of captions and annotation using a simple video editing package like Movie Maker or iMovie can be very effective. Many modes of teaching

Figure 18.1 The flight path of a camera (left), with the position of one key frame (right) highlighted by an arrow

delivery and also assessment formats have not traditionally supported animated, let alone interactive, graphical components. Animations are, however, likely to become easier to create, and the ease of sharing and embedding movie files through YouTube and similar sites should encourage their wider use within a teaching and learning context.

While the animated sequence can have value, the ability to break away from a predetermined route and explore a virtual environment at will, opens up many possibilities, but also presents design challenges. The way in which novice users are allowed to interact can vary greatly depending upon how the rendering interface software is designed, and also how interface devices such as keyboard, mouse or joypad are programmed (Ware and Osborne, 1990). Effective interaction is a cornerstone of the computer games industry, and the number of degrees of freedom a user is offered, and how intuitively the user can navigate an environment, are just some of the design challenges that are relevant to implementing VGEs (Shepherd and Bleasdale-Shepherd, 2008).

An early example of real-time interaction on the web was virtual reality modeling language (VRML) using web browser plugins such as Cortona 3D, and later Extensible 3D (X3D), an ISO standard XML-based successor to VRML, incorporating extensions to VRML like GeoVRML. There are also open source options such as Java 3D which benefit from large user communities, and also non-open source options including Shockwave 3D, which benefits from the ability to use the multi-media development environment Adobe Director and the robust browser plug-in Shockwave Player. Figure 18.2 shows an example of a teaching application developed in Shockwave 3D called *Locata*, designed to explore the ability of students to associate various 2D representations with perspective views generated within the same landscape (Priestnall and March, 2008). When developing interactive applications a skill set combining programming and graphical user interface design is vital, or at least an acceptance of the potentially steep learning curve.

Many browser-based interactive environments suffer from rendering issues when attempting to re-create realistic geographic landscapes involving a terrain surface and objects on that terrain. Another issue is that in a teaching environment, interactive browser-based applications need careful checking before being deployed on multiple machines in teaching laboratories due to variations in local systems configurations and security settings. Initiatives such as OpenGL and Microsoft's ActiveX have for some time allowed programmers to create bespoke 3D graphics applications and more recent developments aim to help

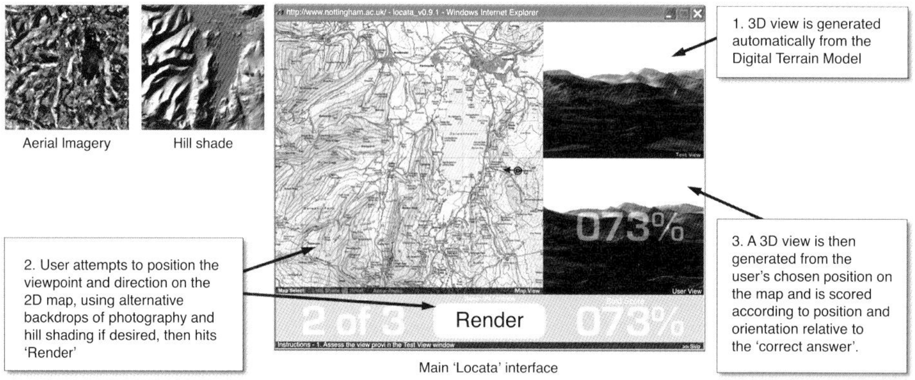

Figure 18.2 Locata, an example of a web-based Shockwave 3D application (http://www.nottingham.ac.uk/~lgzgp/locata). Reproduced by permission of Harvey Maps

programmers access the power of the computer's graphical capabilities as efficiently as possible, including Google's web-graphics API O3D and the Unity development environment. The more widespread availability of standard, powerful, 3D graphics across all browsers and devices, should ensure that 3D graphics via the web becomes an increasingly significant tool for deploying interactive applications.

VGEs are considered to have a high degree of interaction but specifically in terms of controlling the users positioning and orientation within a 3D reconstruction of a landscape. Reference to 3D is used here to represent systems which allow objects and the observer to be placed in three-dimensional space, rather than systems which give the impression of 3D but do not allow interaction in 3D space (x,y,z), sometimes referred to as 2.5D systems.

There are a range of techniques for allowing 3D interaction with virtual models, offering a wide variety of capabilities. The degree to which any one technique is 'fit for purpose' depends upon the main goals of a particular activity. In the context of VGEs these goals can include some combination of the following.

- *Engaging the user in some activity, often task-related, game-like, or collaborative.* This is clearly an attribute of successful computer games, but is also a goal of teaching and learning. There can be various levels of engagement from being 'wrapped up' in the game or set of tasks, right through to suspending disbelief in all details of the narrative of the game (Qin *et al.*, 2009). Many activities can involve collaboration with others, and some VGEs offer a means through which people can communicate remotely in real-time.
- *Enhancing spatial context when sharing and exploring geographical information.* Careful use of geographical information systems (GIS) can present the spatial interrelationships between features quite effectively, but the portrayal of spatial context can be enhanced further by use of 3D perspective scenes (Hubona and Shirah, 2005; Tavanti and Lind, 2001). Some techniques can offer powerful mechanisms for placing volunteered geographic information (VGI) into a broader landscape context, and alongside existing authored content, and can provide flexible tools to help explore both data and environment.

- *Conveying a sense of visual realism, where the virtual models would be considered photorealistic.*
  Virtual models can be made to offer different levels of realism from quite abstract or schematic representations through to the very photo-realistic. The degree to which photorealism is actually required has been considered by Appleton and Lovett (2003), in this case in a planning context, assessing the point at which models become acceptable to the public as representations of real scenes. In a teaching and learning context, levels of realism need to be applied to the extent they are needed to emphasize the principal learning goals; in a biogeographical context, for example, tree and plant realism would be more important than for planning.
- *Attempting sensory immersion, where the user is given a sense of being there.*
  Some techniques aim to make the user feel like they are inside the virtual environment by attempting to dominate their field of vision with the virtual model but also let other sensory actions control their interactions. This sense of telepresence or of 'being there' is a hallmark of the classic virtual reality (VR) concept (Steuer, 1992; Witmer and Singer, 1998; Heim, 1993). As for realism, the degree to which this sensory immersion is important will relate to the learning task.

Before considering a range of techniques and technologies for interacting with VGEs, we consider some of the issues related to the creation of the 3D models that underpin such environments.

18.2 Building virtual geographic environments

Data components of a VGE

VGEs often contain the same basic components each with a number of options for data sources, processing lineage and rendering, these components include:

- Ground or terrain model.
- Ground texture (often provided by images draped on the ground model).
- Discrete objects, usually including models of buildings and vegetation, but sometimes also including models of cars and people.
- Models of physical dynamic phenomena such as water bodies.
- Environmental conditions (lighting, haze).
- Other sensory elements such as sounds emitted from objects.

A full description of the options for building every component of a VGE is beyond the scope of this chapter, however we offer an overview of the basic workflows for building the ground model and surface objects, which form the core elements of most VGEs. Further recommended reading can be found in Ervin and Hasbrouck (2001) and Mach and Petschek (2007).

Options for creating the base terrain model

Digital terrain data can be obtained relatively easily at a range of resolutions. Airborne radar and laser-scanning (LiDAR – Light Detection And Ranging) most commonly provide two types of dataset. Firstly the digital surface model (DSM), also known as the canopy model, represents the tops of buildings and trees as well as the ground surface in open areas. The second dataset is derived by a semi-automated procedure of editing out surface features from the DSM and interpolating across the gaps, the result being referred to variously as digital terrain model (DTM), digital elevation model (DEM), or a bare earth model. These datasets are typically supplied in regular grid form with spatial resolutions in the order of 5m for radar and 50cm for LiDAR.

An alternative to radar and LiDAR is stereo photogrammetry. This technology uses the concept of stereoscopy to produce the 3D coordinates of points on the landscape. Photogrammetry has been extensively used for mapping, spatial information gathering and to determine heights on the landscape using overlapping aerial photography (film) or digital images. The photographs can be captured at a wide range of scales, and the digital images captured at a range of ground sample distances (GSD), either way resulting in ground coordinates at a range of qualities. Today, the use of specially made digital aerial cameras has produced an efficient workflow and high quality results. Digital surface models generated by photogrammetric processes are largely automated in the measurement of corresponding image points which have been identified by image matching. The use of perspective geometry to represent the camera geometry enables the 3D coordinates of points on the ground to be determined. Once a point cloud is generated representing a DSM it can be handled in the same way as LiDAR and radar point clouds. Point clouds of a resolution appropriate for the application should be produced.

The DTM from any of these techniques can form the base for a VGE, upon which models of buildings, trees and other components described in section 2.1 can be added. Challenges exist however, in that raw data volumes can be large. To be used in virtual models where real-time rendering is required, the data need to be simplified greatly while attempting to preserve the shape of the surface. Some systems render surfaces using a *mesh* data structure, which can represent a grid of height posts but may also use non-regular triangular mesh structures as in the GIS TIN (triangular irregular network) data structure (Tate *et al.*, 2007). These can have the advantage of representing finer ground detail or critical breaks of slope using many smaller triangular facets, and using larger triangles where detail is less important or the landscape simply contains fewer undulations.

Options for creating surface objects

The detail contained within 3D models of surface objects, and therefore the whole virtual scene, can vary according to the required end purpose, and the type of rendering environment that will be used. Other considerations are the time, skill sets and data resolutions available for a particular project. Many 3D rendering packages allow multiple levels of detail (LOD) to be modeled, such that when objects are far away from the observer the system does not render, or renders a very simplified version of the object, and more detailed versions are loaded as the user becomes closer to the object centre. An example of this is given in Figure 18.3. The LODs may vary in both complexity of modeling, and resolution of textures. Even if varying LOD are not applied, the choice of a single LOD requires thought and should be appropriate to the task in hand. For example, web-based interactive models involving distance views of objects would require much lower detail than models intended to be explored in a first person perspective in a semi-immersive VR theatre. Kolbe *et al.* (2005)

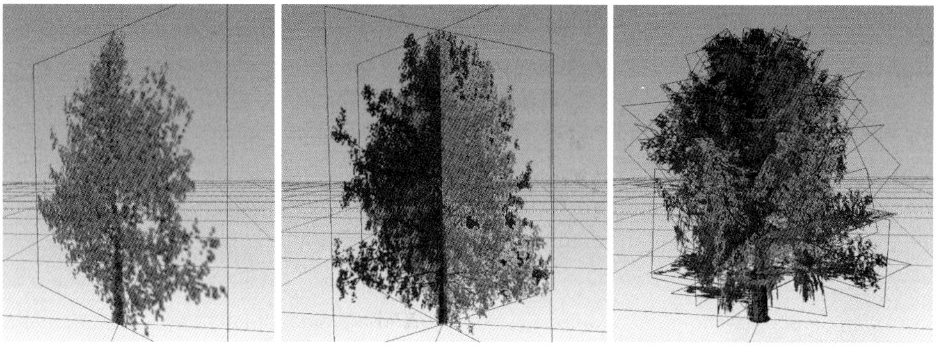

Figure 18.3 An oak tree model at 3 LODs – (left) billboard with a low resolution alpha texture, (middle) cross-board with higher resolution alpha texture and (right) full modeled trunk and branches. From Smith et al (2010)

provide a discussion of the LOD concept in the context of urban modeling, with particular emphasis on the CityGML information model.

Digital photogrammetry can offer a route for modeling buildings as the DSM generated includes the roof tops, and is therefore particularly useful for modeling roof detail and texture (Smith *et al.*, 2010). Alternatively, 3D modeling packages can be used. Sophisticated 3D packages like 3DS Max and Presagis Creator, which allow varying LODs to be created, have substantial learning curves and are relatively costly; however *entry level* routes to 3D model building and sharing have emerged. Of particular note here is Google SketchUp, which has a free version, and uses 3D model interchange formats such as Collada which integrate well with Google Earth and other systems.

Creating 3D building models by matching ground-based photographs using software like Photomodeler or Photosketch offers an alternative way of creating representations of real buildings, where model geometry is derived from overlapping photos which also provide the textures for facets of the 3D model. Higher end 3D modeling packages are continually being refined and more advanced tools to support the creation of realistic models, textures and movements, including humans, will have an impact where the skills are available to use them. The capabilities of entry level modeling packages are increasing rapidly, as exemplified by Google SketchUp, and plug-ins for it, such as Light-Up for realistic modeling of light sources and reflections.

Arguably the most accurate method for generating a point cloud representation of a building model is to use a 3D terrestrial laser scanner (Schwarz, 2010), as illustrated in Figure 18.4. The laser scanner uses reflected laser light to measure the distance between the scanner and an opaque object (e.g. building surface). It shines the laser in a pre-defined grid pattern onto the object measuring the distance at each point, and is able to create 3D coordinates of a 'cloud' of points representing the surface of the object, for example a building façade (see Figure 18.4). The time taken for the scan depends on the make and type of scanner as well as the density of points in the point cloud. The accuracy of the coordinates can be affected by a number of factors including: the angle of incidence on the surface, the material's reflectivity, obstructions, atmospheric conditions, and the measurement quality of the scanner. Software may then be used to optimize the point cloud to create an optimized model, reducing the complexity of what can be very large data sets.

Figure 18.4 The Leica HDS3000 laser scanner and an example point 'cloud' extract

Vegetation and particularly tree models can be created using plant library software, such as RealNat. This software allows the creation of particular randomly generated samples of plants with specific ages at particular times of year. Once a library of usable models has been created from the software, they can be placed individually or if software allows, scattered over the terrain using vector or raster land use maps. Examples of free vegetation libraries can be found in the Google 3D Warehouse (http://sketchup.google.com/3dwh) or via the Virtual Terrain Project website (http://www.vterrain.org/Implementation/Plants).

A generalized work flow

Work flows for building VGEs can vary greatly depending on end user requirements, but a general example is given in Figure 18.5 involving the following broad stages:

- *Acquire source data for the terrain surface (Figure 18.5, left).*
  Radar, LiDAR or photogrammetrically derived DSMs can have surface features removed to leave the bare earth model, or spot heights can be manually lifted from the DSM at critical points or breaks of slope, a practice common in photogrammetry. The DSM is often used to determine the heights and positions of 3D buildings and trees to be added later. Alternatively a bare earth model can be acquired from a data provider, or can be surveyed manually if detailed terrain in a small area was important.
- *Derive bare earth terrain along with textures (Figure 18.5, center).*
  Whatever the source of the bare earth model, a common practice is to acquire colour aerial images and drape a georeferenced image over the terrain. Problems with this approach include the presence of leaning buildings and trees which can be particularly confusing once 3D surface objects are added. However this problem can be minimized by increasing the overlap of successive images and using only the central area, which gives a vertical view, to create the ortho-image. Also when VGEs are explored in first person from close to the ground, the photo drape can appear pixelated or fuzzy. Some systems will allow alternative textures to be displayed when the terrain is viewed from nearby.

Figure 18.5 Building VGEs, digital surface models (DSMs) from LiDAR (left). Reproduced by permission of Intermap Technologies Corporation, 'bare earth' digital terrain model (DTM) with aerial photography (center) © 2011 Google Earth, and with the addition of 3D building models and trees (right). © 2011 Google Earth

- *Create 3D surface objects and register to the terrain (Figure 18.5, right).*
  Using any of the techniques mentioned in this section, surface objects can be built and placed on the surface. Some techniques allow a building footprint from aerial imagery to be used as a guide during the modeling process, allowing automatic registering of the model to the terrain (e.g. SketchUp when used with Google Earth). Other techniques may require the model to be created independently and scaled, positioned and orientated later to fit with the terrain model (e.g.: ground based photogrammetric techniques and 3DS Max).

For more information about 3D model building work flows in a teaching and learning context see Smith *et al.* (2008, 2010).

18.3 Technologies for interacting with Virtual Geographic Environments

This section presents a summary of some of the techniques and technologies which allow users to interact with VGEs. The approaches have very different capabilities and the order in which they appear here has, for convenience, been broadly aligned to increasing sensory immersion in the classic VR sense (Steuer, 1992), using greater sophistication in interactivity and the power to manipulate content as further discriminating factors. Interactivity is used here to include the ability to communicate with computer-generated content and other users via their computer representation of themselves (their avatar). The following list offers a brief description of these techniques and a few examples, and includes static visualization, animation and interactive models from section 2 for context.

Examples of techniques for interaction with Virtual Geographic Environments:

a. **Static Visualization**
   Snap-shot images rendered from either 3D models or 2.5D extruded surfaces to give the impression of a sense of depth. Fly-throughs are also possible.
   ArcGIS (http://www.esri.com/software/arcgis)
   Genesis (http://www.geomantics.com/genesis4.htm)
   Terragen (http://www.planetside.co.uk)

b. **Computer Animation**
Movie sequences derived from 3D modeling packages designed to introduce dynamic visual stimulus. This includes the traditional fly-throughs, but also the animation of objects, textures and environmental variables in the scene.
Bryce (http://www.daz3d.com/i/software/bryce7/download?_m=d)
3D Studio MAX (http://usa.autodesk.com/3ds-max)
Maya (http://usa.autodesk.com/maya)

c. **Interactive Models**
Tools allowing real-time navigation of 3D models, offering various degrees of freedom Interactions possible. Can be standalone but usually web-based.
VRML/Cortona (http://www.parallelgraphics.com/products/cortona)
X3D (http://www.web3d.org/about/overview)
Java 3D (http://java.sun.com/javase/technologies/desktop/java3d/downloads/index.html)
Shockwave (http://www.adobe.com/products/shockwaveplayer)
O3D (http://code.google.com/apis/o3d)

d. **Virtual Globes**
Typically complete online 'mirror worlds' serving geographic information, often of a 'volunteered' nature. Interactions are designed to preserve spatial context when browsing.
Google Earth (http://www.google.com/earth/index.html)
Biosphere3D (http://www.biosphere3d.org)
NASA World Wind (http://worldwind.arc.nasa.gov)
Historically many **others** (http://geography2.blogspot.com/2005/11/listing-of-virtual-globes.html)

e. **Online Virtual Worlds**
Typically fantasy environments constructed largely from user-generated content. These introduce the presence of avatars and more advanced communications channels. Some are accessed via games clients or consoles.
Second Life (http://secondlife.com)
Blue Mars (http://www.avatar-reality.com)
Via games clients or consoles:
World of Warcraft (http://www.wow-europe.com/en/index.xml)
PlayStation Home (http://uk.playstation.com/psn/pshome)

f. **Games Engines**
Development environments allowing the creation of virtual worlds with sophisticated modes of interaction, more realistic physics and behaviours of virtual objects and avatars.
Panda 3D (http://panda3d.org/)
CryENGINE (http://www.crytek.com/cryengine)
Unity (http://unity3d.com)

g. **Semi-immersive VR**
Hardware-Software configurations to offer greater visual immersion, often used for landscape visualization and photo-realistic simulation, typically rendering in stereo. Includes large screens, curved screens, domes and caves.
Vega Prime (http://www.presagis.com)
StereoServer (http://www.virtalis.com)
Vizard (http://www.worldviz.com/products/vizard)

h. **Fully Immersive VR**
   Fully immersive Virtual Reality usually involves a head mounted displays (HMD), an inertial device, and sometimes haptic sensors to attempt multi-sensory immersion.
   Head-Mounted Displays (http://www.3dvisor.com)
   Data Gloves (http://www.vrealities.com/glove.html)
   Full Haptic Workstation (http://www.inition.co.uk)
   Augmented Reality utilizes similar technology but fuses elements of the virtual model with a video stream of the real scene played back through the HMD or projected into a translucent visor.
   Full outdoor AR system (http://www.tinmith.net)

Virtual globes

The coupling of interactive 3D web graphics with the efficient delivery of geographic information has resulted in 3D interactive models of the Earth with consistent and intuitive navigation control. These act as web-based portals allowing the uploading and sharing of 3D models and other user generated content. The potential applications of such technologies in geography education has been widely recognized (Rakshit and Ogneva-Himmelberger, 2008; Tuttle *et al.*, 2008). While sometimes referred to as *mirror worlds*, the term *virtual globes* will be used here to refer to such developments.

While there were many early examples of virtual globes, most notably NASA's World Wind, the launch of Google Earth in 2005 was particularly significant, achieving rapid dominance helped by the widespread acceptance of the Google brand. The interface is designed to offer an intuitive interactive platform through which to explore and share geospatial information. The base terrain data is relatively coarse (including for example the 90m resolution shuttle radar topographic mission data) but with high resolution photography for much of the globe it can prove useful as a teaching resource for showing the spatial context of various physical landscapes (Tooth, 2006) and by using map overlays, for example to assist in teaching geology (De Paor and Whitmeyer, 2011).

From a GIScience perspective there are several ways in which the use of Google Earth can be incorporated into the curriculum. One focus can be on the use of the technology itself, for example the appropriate use of the touring facility and encouraging good practice in the design of tours which preserve the viewer's sense of orientation by the careful use of additional placemarks to control the flight path. The issues involved in using Google Earth as a tool for interactive landscape visualization, as discussed by Sheppard and Cizek (2009), have formed the focus for a teaching exercise at the University of Nottingham to reconstruct a wind farm, shown in Figure 18.6. Students register an image overlay of the proposed site plan and place individual 3D models of wind turbines, in Collada format, as indicated on the plan, and assess the likely visibility for the wind farm from key points around the landscape. They reflect on the extent to which this technique can replicate real world visibility by repeating the exercise for an existing wind farm and comparing 3D views with actual photographic evidence from the field.

This exercise also raises awareness of the lack of detail in surface objects modeled in Google Earth, particularly in rural environments, which reduces its usefulness for landscape visualization, but also where first person exploration on the ground is required, as with other virtual environments. Even where broad landscape views are considered, as

Figure 18.6 Google Earth™ exercise to reconstruct a wind farm on the edge of a national park. Placing Collada models in Google Earth™ using image overlay of planning application (left), wind farm model with orientation device placed at a known height above ground (center), and perspective view from a ground level point of interest looking towards the wind farm (right). © 2011 Google Earth

in Figure 18.7, the difference is marked, although this is not surprising as photo-realistic landscape visualization is not the primary goal of virtual globes. Data capture efforts are being concentrated initially on urban areas, including the modeling of trees in version 6 of Google Earth realized in 2010, but it is quite feasible that the representational fidelity will also improve in rural areas in the near future. Biosphere3D represents a virtual globe which may not have as broad a user community as Google Earth but was designed more specifically for landscape visualization.

There are clearly many opportunities for using Google Earth as a platform to gather VGI in class exercises, and to share this using keyhole markup language (KML) files. Due to the integration with 3D modeling package SketchUp, it also offers a useful platform for teaching the model building process itself. It is likely that the incorporation of GIS data will become easier via more widespread KML exporters, and the wider use of Google Earth's API will allow the use of the 3D rendering engine in teaching related web pages. Also new developments in interactive hardware such as interactive tables as in Microsoft's Surface technology could allow even more intuitive interaction with virtual environments.

(a) (b)

Figure 18.7 Photograph looking towards Derwent Water (left) and equivalent view in Google Earth™ (right). © 2011 Google Earth

Online virtual worlds

Online virtual worlds allow users to explore and communicate via their avatars but unlike virtual globes they are typically fantasy landscapes. Some are accessed via desktop client game software, often described as massively multiplayer online role-play games (MMORGs), for example World of Warcraft. Online virtual worlds which are constructed almost entirely from user-generated content, such as Second Life, offer more interest to educators in terms of the potential flexibility to explore the design of different learning spaces (DeFreitas, 2006).

The use of online virtual worlds to showcase geographical data has received some attention, following the idea of using scale models as in the 'Virtual GIS room' project (Neves *et al.*, 1999). The important addition however is the presence of avatars, not only as a means of communicating to other users, but as a scaled point of reference. The incorporation of geographic data can be achieved via the in-world builder, creating 3D primitives (or 'prims') from imported surface model data, or using an offline development environment such as OpenSimulator (http://opensimulator.org). Geographic data can also be processed in third party 3D modeling packages such as Blender or 3D Studio MAX and exported as 'Sculpt Maps'. For more information on the incorporation of geographic data in Second Life see Mount *et al.* (2009a,b).

Figure 18.8 shows three examples of the use of scale models, beginning with the use of live web data feeds (from www.fboweb.com) to position aircraft around Los Angeles. The central image shows a scale model of Birmingham, UK incorporating live web links and photographic panoramas using the Google Earth API, importing KML data to generate 2D and 3D models of the city centre (http://www.daden.co.uk). Finally on the right is a scale model of the English Lake District as used in the DELVE project for collaborative placement of wind farms, see Mount and Priestnall, Chapter 20.

Challenges associated with the use of geographic data in Second Life include restrictions on the units of land available and the numbers of primitive objects in each region, owing to the way space is organized across a grid of servers. This prevents the creation of large geographic spaces at 1:1 scale to the avatar, and scale models emerge as the most common solution. The graphical appearance of Second Life is relatively crude compared to computer games, as is the control over the avatar's view upon the virtual environment; interactions lack the degrees of freedom offered in other visualization and games environments. The use of in-world communication tools such as voice chat, can be unreliable and can prove a barrier to engaging new users in discussions, which is a problem if environments like Second

Figure 18.8 Examples of using scale models from geographic data in Second Life™. © Daden Limited 2011. Images courtesy of Daden Limited

Life are to gain acceptance as platforms for collaboration. Ongoing interest in online virtual worlds is demonstrated by the release of Blue Mars (http://www.bluemars.com), which offers many of the attributes of Second Life but underpinned by computer games engine technology and different approaches to organizing space within the virtual environment.

Games engines

The computer games industry is having an increasing impact upon developments in 3D graphics technologies, interaction devices, interface design and in popularizing 3D virtual environments. The high financial returns from producing a successful game often mean large investments in development time are possible, leading to very detailed environments and robust and intuitive modes of interaction. Teaching and learning activities have long been influenced by interactive game-based approaches, and the concept of playfulness in web-based interactive learning packages was explored by Chou (2003). Some examples of games-based learning include: simulation games explored by Leutner (1993), engineering themed Java-based games described by Crown (1999), and multi-player games in Shang *et al.* (2006).

Often game-like environments contain some form of computer simulation, for example the visual simulation of landform evolution in Lou *et al.* (2005) which enabled students to view the results of their actions on a model of process – form relationships of eroding hill slopes. Figure 18.9 shows an example using the Shockwave 3D web-based games platform, where the concept of quick response (QR) codes (2D bar codes) was illustrated through

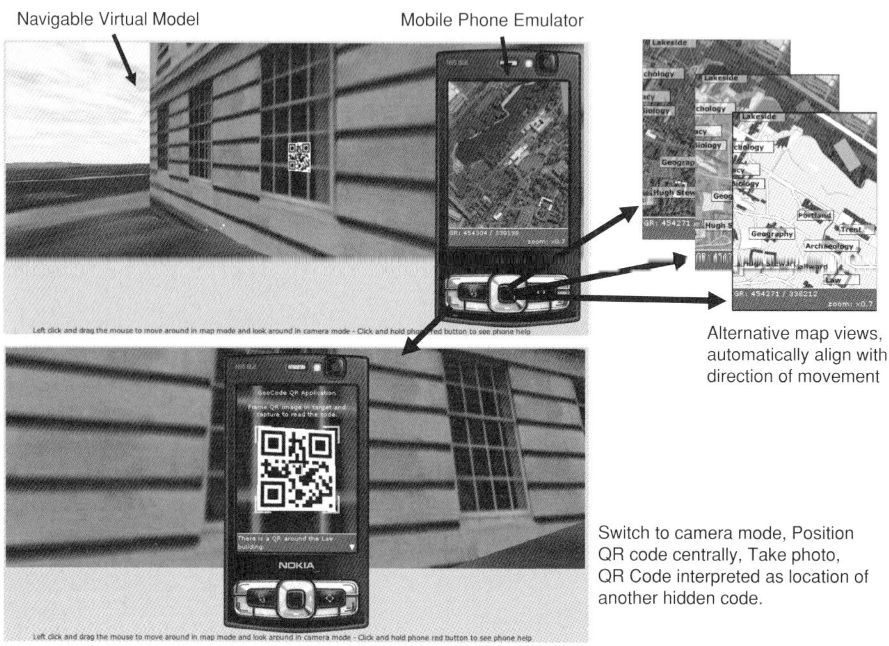

Figure 18.9 A games-based emulator of a mobile mapping application (http://www.nottingham.ac.uk/~lgzgp/lbs)

a high-level emulation of the functions of a mobile phone QR code reader. Here students navigate a 3D model of a university campus searching for hidden QR codes, each of which can be decoded via the mobile phone to reveal a text-based clue as to the location of the next target. The application also emulates the functions of a phone-based mapping system allowing the user to switch between map layers whilst navigating the campus model, all user interactions with the interface being logged for future analysis.

Some computer games offer very effective mechanisms for aiding user navigation and orientation, such as on-screen context maps and floating orientation symbols, techniques which have been adopted respectively in the Google Earth exercise described earlier (see Figure 18.6) and in the semi immersive laboratory work described later in this section. A characteristic of next generation games consoles has been online multi-user gaming, and games design involving multiple users is clearly of interest for the future development of collaborative VGEs. The computer games industry has also been a driver for innovative and intuitive interaction devices such as the *wiimote*, and there are continued explorations into controller-free interaction using image recognition techniques as in Microsoft Xbox *Kinect*.

In addition to the design ideas and interaction techniques emerging from the games industry, the actual software development environments could potentially offer platforms on which to build virtual geographic environments. Several computer games developers have made available the underlying games engines upon which their games are based, which allows users to benefit from the significant investments made in development, these gains including:

- Photo-realistic visualization of detailed objects, textures and atmospheric effects.
- Fast, efficient rendering techniques.
- Physics engines allowing realistic simulation of human movement, vehicles etc.
- Controlled interactions, variety of degrees of freedom possible.
- Import capabilities for geographic data via 'sandbox' type development environment.
- Large support networks of developers and resources.

One of the most popular games engines has been CryENGINE, with examples of its use with geospatial data emerging in fields such as landscape design (Herrlich, 2007). Another example is shown in Figure 18.10, where students are required to measure dimensions of tree trunks in a woodland but must develop a strategy for taking a sub-sample rather than visiting every tree (Ashfield *et al.*, 2010).

Although games engines can offer the software infrastructure and development tools for building a VGE, this does not guarantee that the resulting environment will be as intuitive and effective as most computer games are, in terms of supporting users in undertaking various tasks. Many games have benefitted from meticulous storyline design and usability testing, some of the relevant factors involved in usability research being summarized by Isbister (2008). A challenge facing the development of VGEs using games engines is the steep learning curve, particularly when making the transition from working with geographic data and coordinate systems.

Where computer games, virtual globes, virtual worlds and Web 2.0 technologies in general converge is often termed the *metaverse*, the Metaverse Road Map being an example of an initiative attempting to summarize these developments. Hudson-Smith and Crooks (2008) also provide a discussion focusing on the implications of these technologies for the use

Figure 18.10 Forest sampling tool implemented in the Cryengine games engine. Courtesy of Crytek GmbH © 2011. All rights reserved. Crytek and CryEngine are registered trademarks or trademarks of Crytek GmbH in Europe, the USA and other territories

of geographic information, and the Digital Urban blog (http://www.digitalurban.org) is particularly useful in the context of modeling and visualizing urban environments.

The techniques for exploring virtual environments discussed so far have sometimes been referred to as *fishtank VR*, taking VR in the most general sense of being technology which allows a user to interact with a computer-simulated environment, be it a real or imagined one. Perhaps a more common understanding of VR is when hardware and software is used in an attempt to make the user feel more visually immersed in the virtual scene. Advances in refresh rates of monitors are seeing an emergence of stereo displays for gaming but going beyond the desktop there are a number of options which attempt to combine stereo vision with display devices which dominate the user's field of vision, and the first step towards this would be referred to as semi-immersive virtual reality.

Semi-immersive virtual reality

Technologies aimed at increasing measurable visual immersion can be used to allow the virtual scene to become dominant over the user's more immediate physical environment, blocking out the physical world to varying degrees (Biocca, 1992). In the context of Virtual Reality these techniques are often said to increase the sense of *presence* (or *telepresence*) that a user feels in the virtual environment (Azuma *et al.*, 2001; Heim, 1993; Riva, 2007; Steuer, 1992; Witmer and Singer, 1998). This usually involves stereo display technologies which attempt to exploit one of the major advantages of binocular vision, that of depth perception or *stereopsis*.

The use of stereoscopic projection systems aims to trick the brain into believing that there is a depth, and therefore volume, to the objects appearing on a flat 2D Screen. This is done by providing each eye of the viewer with a slightly different viewing position of the same scene. In order to view a stereoscopic scene therefore, the graphics card must render each of the views seen by two virtual cameras in the scene, and the stereo presentation equipment must deliver these individually to the user's eyes. Clearly, therefore, rendering a stereoscopic virtual geographic environment causes twice the rendering load on the system, unless the frame rate of the system is decreased.

There are several options for rendering in stereo, firstly *active* systems, which use shutter glasses which alternately display frames from the left and right virtual cameras, to the left and right eye of the user. This switching occurring in the glasses is synchronized with the refresh rate of the monitor, which must be sufficiently high for the human eye not to notice the flicker. The sense of depth can be strong and these systems are often used in digital photogrammetry, although the glasses can be expensive.

Cheaper *passive* stereo solutions which can be deployed on larger groups of users include the use of two-colour anaglyph systems, which typically use red and blue filters set in plastic or card glasses. These filters allow the human eye to fuse the left and right images which are displaced slightly on the display screen. Although a cheap system, the colours of the stereo scene can be poor, and somewhat dark, although a more expensive blue/green filter can improve this situation.

Another passive solution involves the use of low cost polarized glasses where each lens filters either horizontally or vertically polarized light. Similarly filters on the data projectors create two separate images one horizontally polarized, the other vertically. A more expensive silver coated screen is required to prevent depolarization of light on the screen surface. The user sees only one of these images through each eye, achieved without the need for active shuttering of the lenses.

In addition to the use of stereo display technology there is often an attempt to fill more of the user's field of vision than a typical monitor does. True VR might be considered to involve an individual using a head mounted display (HMD) to achieve this, as described in the next section, but another option is the use of large screens, which are also suitable for more than one viewer. Here, we refer to such techniques as semi immersive VR.

The Geowall project (http://www.geowall.org) explored such techniques in the context of Earth Science education using configurations involving a pair of data projectors to produce stereo visualization using low cost eye wear. Large stereo displays driven by stereo data projectors, or one or more pairs of regular data projectors, are perhaps the first step towards VR in that they aim to provide more visual immersion than a desktop monitor. Figure 18.11 shows a large flat screen at the University of Nottingham, and a large curved screen at the University of Leicester. Both installations utilized the Vega Prime rendering software and were part of SPLINT (SPatial Literacy IN Teaching), a Centre for Excellence in Teaching and Learning (CETL) funded by the Higher Education Funding Council for England (HEFCE), see the project website (www.splint-cetl.ac.uk) for more details.

One use of large screen semi-immersive displays is for landscape visualization, and the example on the left in Figure 18.11 shows an exercise where students present the locations of wind farms resulting from a GIS-based site suitability analysis in their landscape context. Master's level students in GIScience and environmental management used the lab space to justify their choice of locations and to respond to ad-hoc queries about the visibility of the proposals from various points around the landscape. After the first year of running the exercise it emerged through focus group meetings that many students experienced great difficulty re-orientating themselves within the virtual model, especially when confronted with the task of flying to a new position when requested. As a result, a game-like context map was implemented, using a simulated GPS signal being transmitted from the ceiling of

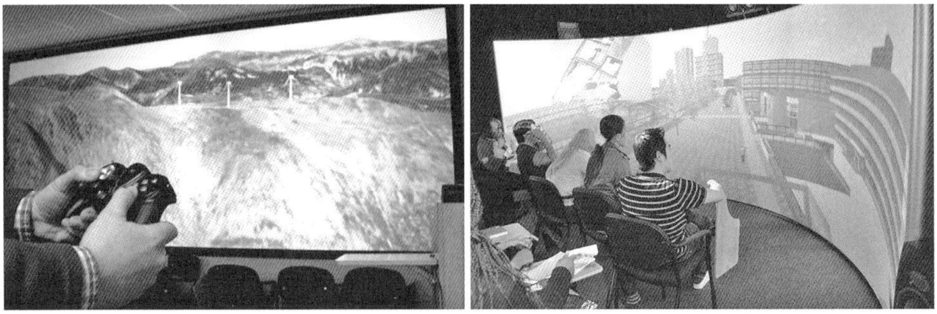

Figure 18.11 Semi-immersive virtual reality using flat screen (left) and curved screen (right). Base data for visualisation derived from OS MasterMap. Ordnance Survey © Crown Copyright. All rights reserved

the lab via Bluetooth, previously developed to test mobile mapping applications in the lab (Priestnall and Polmear, 2007). The context map application was written in Visual Basic and ran on a Bluetooth-enabled laptop connected to a large LCD monitor, and displayed the position and cone of vision of the observer over a range of different base maps.

Another role of landscape visualization can be to familiarize students with field sites they will subsequently visit, which was also a role performed by various virtual field course emerging throughout the 1990s (Williams *et al.*, 1997). Figure 18.12 illustrates one such example, using semi-immersive VR in advance of a field course to the Almeria region of SE Spain.

This use of VR arose from observations that students were taking the majority of an eight-day field course to orient themselves adequately within the relatively large field area. It was felt that there was a need to increase their spatial awareness in order to encourage more critical consideration of individual field locations in a broader landscape context. In part, the issue appeared to relate to the use of a coach for transport between sites, which resulted in a lack of active navigation on the part of the students (e.g. Hahm *et al.*, 2007).

The 3D immersive environment was used shortly prior to the field course to guide a number of the students along the main routes that would be travelled during the field course, pointing out the main features from which students could take their bearings in the

Figure 18.12 Semi-immersive virtual reality for landscape orientation in Almeria, Spain, showing broad landscape view (left) and ground level detail (right). Visualisation derived from the Digital Terrain Model product for Andalucia & Land Use Data product for Andalucia. Copyright Junta de Andalucia. All rights reserved

field. The significant characteristics of the basin and range topography were highlighted, and environmental contextualization was provided through discussions of water availability, degrees of landscape dissection and typical vegetation assemblages. During the field course an orientation-focused questionnaire was used at different stops during the first day to evaluate whether use of the VR exercise had a positive learning effect on orientation ability in the field for those who viewed it. Half of the stops where questions were asked had been explicitly viewed during the orientation, and half had not been directly alluded to or directly visited. Twenty-nine out of forty-two students took part in the orientation, and after randomly removing thirteen questionnaires from this cohort there remained twenty-six respondents, half of whom had experienced the VR orientation and half who had not. The results suggested, with 99 percent confidence, that a significant and improved difference in ability to orientate in the field occurred amongst students who had undertaken the semi-immersive virtual tour versus those who had not. This improvement occurred particularly at the stops explicitly visited in VR; some broader navigational learning is suggested in addition to site specific benefits, but this will require further investigation. Sun *et al.* (2004) for example suggest that orientation-free learning is particularly associated with active rather than passive learning, but equally the multiple angles from which the terrain is viewed may mitigate against this.

The use of stereo in semi-immersive VR is most effective when objects in the foreground of the scene predominate, as opposed to open landscape views. The example shown in Figure 18.13 illustrates the use of a large curved stereo display for exploring the accuracy and availability of global positioning system (GPS) signals in urban environments. Positional uncertainty in GPS results in part from the effect of surface objects interfering with the lines

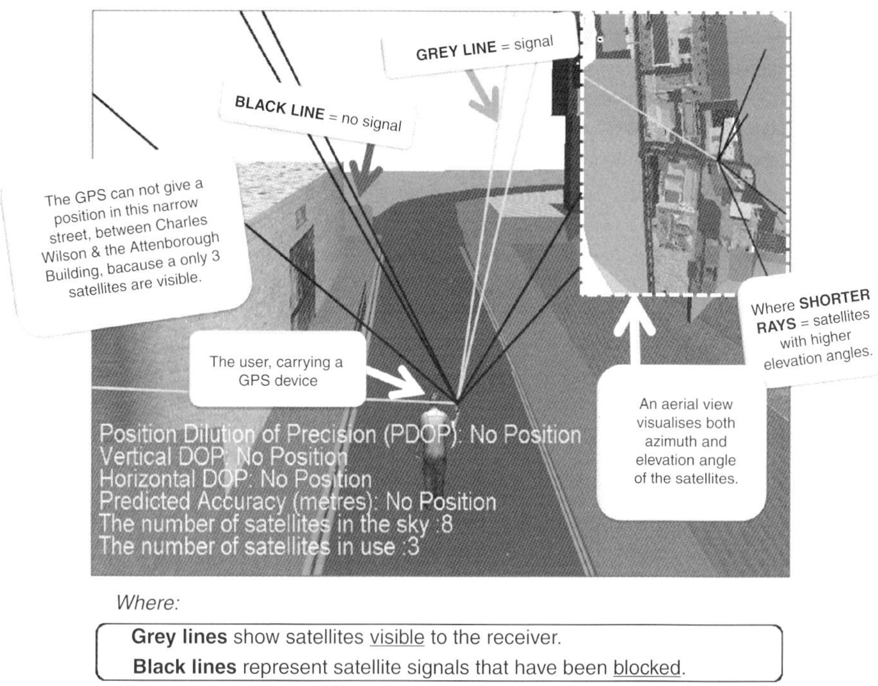

Figure 18.13 Semi-immersive virtual reality for explaining GPS accuracy issues. Base data for visualisation derived from OS MasterMap. Ordnance Survey © Crown Copyright. All rights reserved

of sight from the satellites to the receiving device, and this example uses stereoscopic VR to represent these 3D vectors in the context of those surface objects.

Those using handheld GPS equipment need to be able to use it judiciously; aware of the nature and extent of positional uncertainty in comparison with the scale of the phenomenon they are using it to measure. For example, we have observed undergraduate students considering the use of handheld GPS to measure the relative heights of river terraces when the error of vertical measurement given by the equipment exceeds the change in height of the feature, or relying on GPS to provide positional information in dense rainforest or within urban canyons. These situations have arisen even when GPS variation has been explained in 2D, within a lecture context, but also through active learning in the field, where students using relatively unfamiliar equipment can be unclear whether operational difficulties relate to signal occlusion, equipment breakdown or software misunderstandings.

In this visualization example, variations in the accuracy and availability of GPS are explained visually within the context of a virtual urban university campus environment. Generally, a wider spacing between satellites and a receiver produces smaller errors (Langley 1991), while signal outage is caused by nearby buildings and trees; the contribution of receiver-satellite geometry to positional errors is quantified by various dilution of precision (DOP) values, which are computed using a least squares solution involving a matrix of line of sight vectors. This real-time 3D visualization tool illustrates both changes in occlusion and DOP as an avatar moves around the campus (Figure 18.13), allowing students at lower levels to engage with basic numerical variations that occur in PDOP under different spatial configurations and those at higher levels to interactively and intuitively explore the mathematics behind different GPS signals. Students who viewed the 3D virtual reality campus tool before undertaking active learning in the equivalent learning space outside were considerably better at diagnosing the reasons for GPS signal variation of outage when undertaking the subsequent practical exercise (Li *et al.*, 2010); further, their ability to make judgements regarding the appropriate use of GPS in other contexts later was enhanced.

Setting up a large screen stereo configuration can be time consuming and costly. The system hardware is complex and there can be steep learning curves associated with the software to build and render virtual models across a stereo display. For example the configurations shown earlier in Figure 18.11 required four and six data projectors respectively, each with a high specification PC to drive it, and separate hardware and software to blend the individual stereo images into a larger one. Prospects for more affordable stereo VR may be boosted by the re-emergence of stereo in cinemas, television and gaming, which may encourage the development of cheaper more powerful stereo monitors, but also the emergence of single stereo data projectors for large screen applications as described here.

The process of integrating data from multiple sources within a 3D rendering environment is not trivial, for example the terrain requires time consuming pre-processing involving re-sampling and tiling to ensure acceptably smooth rendering. Similarly aerial photography drapes needed to be optimized and land cover maps produced from which the software can render 3D vegetation at user-defined densities. The aforementioned issues are particularly significant when considering the creation of virtual models for large upland areas where vegetation and reasonably high fidelity terrain definition is important. Data volumes can be unwieldy and the challenge is handling the trade-off between the quality of terrain representation and the speed and smoothness of rendering. Over and above these challenges are the efforts involved with creating individual buildings to place on the terrain base, as described in section 2, and also the manipulation of parameters relating to the various levels of detail (LOD) of objects. The experiences of building the landscape models and setting up the stereo rendering environment for the wind farm exercise, and related teaching and learning activities, are described in more detail in Burton *et al.* (2008). Adding humans to such scenes, for example to facilitate collaborative learning, in anything but a crude way presents further challenges (Sanchez-Vives and Slater, 2005) and is the subject of ongoing research (e.g., Peters, 2010).

Many of the examples discussed have emphasized quite detailed replications of physical reality, which in many domains remain important to enable learners to believe they are present in a simulation, and behave accordingly, for example in medical simulation (Gutierrez *et al.*, 2007). In the context of VGEs attempts can be made to increase the level of visual immersion further by using displays which fill most or all of the viewer's peripheral vision, including hemispherical domes and cave systems (Li and Longley, 2006). The feeling of being present in a virtual environment, termed telepresence, could be seen as equally reliant upon the degree and sophistication of interactivity offered by a particular technology or technique in the classic VR sense (Steuer, 1992). Many of the devices used to achieve this attempt to complement full visual immersion with techniques to allow a user control their view within the scene and manipulate objects in the virtual environment as intuitively as possible, and this results in an individual experience referred to here as fully immersive VR.

Fully immersive virtual reality

Fully immersive VR might be seen as offering an increased *breadth of information* (Steuer, 1992) characterized by attempts to utilize a greater number of senses over and above visual immersion, leading to more haptic and multi-sensory experiences (Pausch *et al.*, 1997; Sanchez-Vives and Slater, 2005). The first characteristic is to ensure the user's field of view is completely dominated by the virtual scene, usually through the use of a head-mounted display (HMD), either in the form of a helmet or glasses. These feature two small liquid crystal display (LCD) or organic light emitting diode (OLED) displays with magnifying lenses, one for each eye. HMDs may also be coupled with head-tracking devices, allowing the user to 'look around' the virtual world by moving their head, eliminating the need for a separate controller. Performing this update quickly enough to avoid inducing nausea in the user requires high graphics processing speeds to render images at acceptable frame rates. If six axis position and orientation sensing is used then the wearer may move about within the limitations of the equipment used.

The addition of a real-time positioning device such as GPS allows the development of outdoor VR, where alternative virtual scenes can be viewed in the context of the real world position. Figure 18.14 shows a configuration used in the SPLINT project based upon the GeoVisionary visualisation system developed by Virtalis in collaboration with the British Geological Survey described in more detail in Jarvis *et al.* (2008). A Bluetooth GPS centres the virtual model on the user's current position, with an inertial head-tracking device controlling where the user is looking within the virtual scene. At any point the user can flip up the visor to reveal the real scene. If the user wanted an alternative viewpoint onto the scene a joypad can be used to override the GPS and initial unit and enter a free flight mode as in lab-based VR. The system can also allow them to load and unload particular models from the environment, or change image drapes on the terrain surface.

Figure 18.15a shows a screenshot from the system, here using part of the University of Nottingham campus model for testing registration issues, and Figure 18.15b a screenshot from the system being used on a field course in Cumbria, NW England, visualizing retreating

Figure 18.14 A schematic of the SPLINT head-mounted display configuration

glacial ice in a valley. This configuration was one of a range of techniques for augmenting the visitor experience evaluated by students during the fieldtrip. Other techniques included a PDA-based field assistant and a phone-based mediascape which delivered media according to user-defined trigger regions in the landscape. The exercise formed part of a level 3 GI Science field trip and encouraged a critical evaluation of the effectiveness of various technologies in the field, including the degree to which various forms of digital geographic information helped the user construct knowledge about the landscape around them. The in-field VR configuration proved popular and worked well as a proof of concept, with very responsive head-driven movements in the VGE, and the HMD offering clear graphical display even in bright sunlight, something which proved problematic for other techniques. The combination of hardware used was, however, extremely cumbersome and with high cost for a single user experience it remains impractical for anything other than specialist teaching of the techniques and technologies themselves. A more detailed discussion of the field exercise can be found in Priestnall *et al.* (2009).

Figure 18.15 Screenshots from the outdoor VR system based upon GeoVisionary

The final steps towards full VR include the use of data gloves to allow users to point to, grab and manipulate objects within the virtual space, usually involving visual representations of their virtual hands and fingers in the scene. Hardware and software configurations which incorporate a sense of feeling by allowing *haptic* force feedback are also under development, which, along with the other components, form what could be referred to as *full haptic workstations*. From a teaching and learning perspective their use is currently limited by system complexity and high costs, and it is usually only specialist training where simulation of real world experience is paramount, such as in military training, where such investments are made.

Augmented reality

Fully immersive VR replaces the user's view of the real world completely with a view of a virtual model, however if computer generated information is used to augment the real scene in some way then it would be referred to as augmented reality (AR). A typical hardware configuration would involve similar components to the system described in Figure 18.14, but instead of rendering only the virtual environment, elements of that virtual model are rendered on top of the user's view of the real scene. This can be achieved by either projecting the computer generated information onto a translucent visor, or by merging the computer generated information with a video stream of the scene in front of the user. Figure 18.16 illustrates the latter option, here allowing a user to create and edit 3D geometries in an outdoor environment using data gloves (Piekarski, 2006).

The cost and system complexity of full AR typically prohibits adoption in a teaching and learning context apart from occasional use in the context of technical evaluations as with the in-field VR discussed previously. Full AR involves even more technical challenges as it requires high accuracy positioning and inertial measurements in attempting to match the 3D geometry with the view of the real scene. Perhaps a more achievable goal in the near future is to augment the real scene with more schematic information which does not require

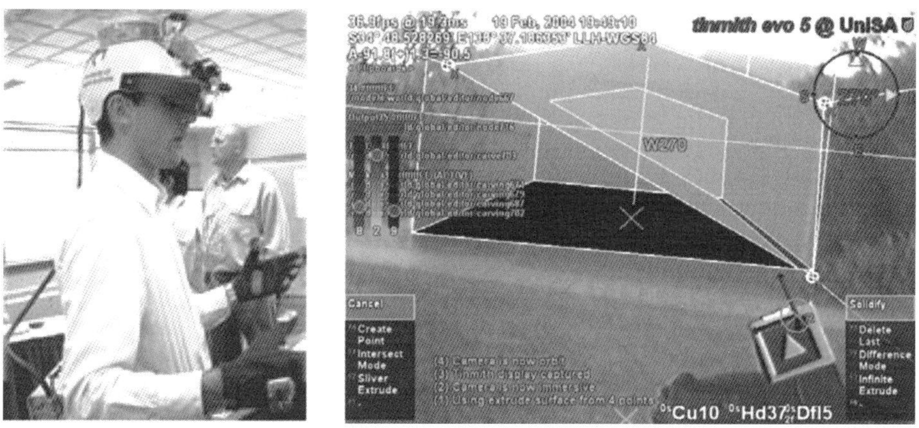

Figure 18.16 True augmented reality: The Tinmith system including data gloves (left), merging computer generated geo-registered 3D geometries with a video stream of an outdoor scene (right). Reproduced by permission of Wayne Piekarski. http://www.tinmith.net/, 9 March 2011

Figure 18.17 The LAYAR augmented reality browser. Reproduced by permission of Layar B.V.

such accurate placement on the video stream of the real scene. The arrival of mobile devices with camera, positioning capabilities, integral digital compass and powerful development environments has resulted in great interest in what could be termed *handheld AR*.

As such technologies become embedded in personal mobile devices then the broader adoption of handheld AR in a teaching and learning context seems more likely. In a field-work context there are possibilities for augmenting the real geographic environment with elements of a virtual model and information from geographic databases. Applications such as Wikitude (http://www.wikitude.org/) and Layar (http://www.layar.com/), released on phones with GPS and digital compass, have begun to popularize the concept of augmenting the real scene with information from a spatial database. Figure 18.17 illustrates the Layar application being used in fieldwork, here the names of mountain peaks being displayed as icons broadly aligned with the real features in the horizontal axis, but using an optional vertical displacement to suggest proximity to the viewer. Free authoring environments such as Hoppala! Augmentation (http://www.hoppala.eu/, which offers easy map-based creation of Layar points of interest and 3D models, should open up handheld AR to the broader teaching and learning community.

18.4 Discussion

Practical reflections and futures

As outlined above, there is a rich set of possibilities for the use of virtual geographic environments in the teaching of geography and related subjects. A technological approach has been used to organize the range of options according to the degree of immersion they afford; the classification could be seen as following the classic VR notion of progressive sensory immersion (Steuer, 1992), using sophistication and richness of interaction as a discriminant at each stage.

The first set of techniques introduced were essentially desktop-based, or *fishtank virtual reality*, and offer the user progressively more sophisticated interaction with elements of the virtual world and other people. At this level, aspects of social networking within virtual worlds and games environments are likely to be particularly strong in future as developments in the representation of people and their emotive facial responses and gestures develops (e.g. Peters, 2010). Further, interaction devices emerging within the games industry are expected to assist learners in controlling VGE interfaces in a more intuitive manner, and at more reasonable cost, in the future. The reality is that many of the techniques traditionally used on the desktop could be made to work using larger or more visually immersive technologies, albeit perhaps not easily in stereo; but they tend not to be. It is also the case that static image-based representation, a form of graphical output which offers someone a snapshot view into a 3D world in a very controlled fashion, still predominates as a way to direct learning.

Stepping beyond this, *semi-immersive virtual reality* goes beyond the desktop and attempts to make the virtual model more dominant in the user's field of vision. Increasingly, we expect to see convergence between the technologies associated currently with virtual globes, virtual worlds and games engines; additionally, with burgeoning 3D games and film industries, the lines between desktop and semi-immersive technologies are already blurring. The capital outlay required to engage with developing semi-immersive teaching tools, both in terms of hardware and software, is decreasing in real terms. However, the time to develop teaching tools in semi-immersive virtual reality using proprietary software currently available should not be underestimated.

The section on *fully immersive virtual reality* introduced the idea of devices which totally replace the user's view with the virtual model and then introduce additional modes of interaction, such as haptic control, which bring in additional senses. Technologically, handheld augmented reality offers exciting possibilities for enhanced in-situ content at affordable cost, particularly with the increasingly wide availability of mobile phones carrying a camera, GPS, accelerometer and compass.

Computer-based visualization has been central to teaching and learning across a wide range of disciplines, and some initiatives such as the Visual Learning Lab at the University of Nottingham (visuallearninglab.ac.uk) have demonstrated the continued interest in exploring more interactive and collaborative visual techniques. While understanding the role of visual literacy in a teaching context (Stokes, 2001), including the merits of more interactive approaches (Najjar, 1998), we should also acknowledge the importance of usability and design issues (Ainsworth *et al.*, 2002; Goldman, 2003) when considering the effectiveness of such approaches. Interactive visual learning tools may not stand alone as effective devices for learning, and students may not necessarily know how to get the best from them (Yeo *et al.*, 2004).

Technical learning: a VGE curriculum

From the perspective of teaching geographical information science in particular, VGEs can be used to facilitate teaching regarding GI principles that involve inherently 3D perspectives such as TIN structures, contours, GPS signal availability, viewsheds and error in interpolated surfaces. They have been particularly effective in visualizing the unseen (e.g. historical landscape, future potential landscape or GPS signal) and to reveal

Phase 1: Task & intended audience
Who is the VGE for? What is its purpose?
Are VGE interactivity and 3D characteristics relevant to the task and audience context?

Phase 2: Design issues
Use of colour & symbolism, texture & light
Degree of realism (abstract to photorealistic) & levels of detail
Degree of interaction
Degree of immersion required (factual or emotive effect? Stereo?)
Principles of human-computer interaction

Phase 3: General technical approach
Model building (data sources; terrain; surface objects; workflow)
Choice of technology (e.g. Virtual Globe, virtual world, games engine, VR modeling software)
Model rendering (e.g. VR or games engine; environmental effects; sensory elements. Bespoke or proprietary?)
Type of interaction (e.g. accelerometer, GPS input, gyro mouse, wiimote, tracking system)
Technical options for enhancing immersion

Phase 4: Evaluation approach
Check to reality (where appropriate)
Speed of processing/interaction fit for technical context?
Interface and content usability – is it navigable, and does it perform for the purpose intended?

Figure 18.18 Phases within the VGE curriculum

mathematical form in a visual manner to those without a background in mathematical notation.

Additionally however, the development and evaluation of VGEs and their associated 3D models and interaction methods is in itself a highly current aspect of visualization in GIScience. In reflecting on the practicalities of utilizing VGE technologies in teaching, we also move towards expressing a curriculum flow considered relevant to incorporating the design and building of VGEs at various levels of the HE curriculum. The relevance of this approach reflects the fact that VGEs are increasingly being constructed in many different settings, involving surveyors and photogrammetrists, graphics programmers, games developers and software engineers in addition to geographical information scientists.

Figure 18.18 highlights the different phases involved in developing a VGE, whether for a teaching purpose or otherwise. The curriculum flow pulls together and develops concepts from the Geographic Information Science & Technology *Body of Knowledge* (DiBiase *et al.*, 2006) for this interdisciplinary setting and fast moving technical arena. The model flow applies regardless of expertise or complexity of technology used.

Firstly, in Phase 1, we draw on the cartographic tradition within which geographical visualization is rooted to highlight the continued importance of considering message focus, audience type and task when designing communication media. In general, cartographic theory (e.g. the visualization cube of MacEachren, 1994) suggests that high levels of interactivity (inherent in VGEs) are unnecessary for public communication. More recent cartographic work however also emphasizes the potential of 3D (also inherent in VGEs) in communicating information at a more naive perspective to the user, without the rotation of perspective required when interpreting a 2D map. Overall, it remains the case that particular care needs to be taken to avoid cognitive overload when using VGEs for a specific purpose, and in this sense structured games environments or guided VGE tours offer advantages over initial free-movement explorations. Importantly, the developer also needs to consider how many people they are expecting to present their visualizations to, and what equipment and technical skills they anticipate that they might have. While Google Earth is relatively universal, ethical restrictions may apply to the use of virtual worlds such as Second Life, and the computer graphics power required by games engines is not insignificant. Further, and specific to this group of visualization approaches, given that VGEs can be time intensive to develop depending on approach, is the task they are communicating about inherently 3D in nature? Would 2D be equally appropriate, if not more so? Before diving into development aspects of VGEs, we consider that it is important for students to have a good grounding on these fundamental issues that might be forgotten in pursuit of a highly technical teaching agenda.

In Phase 2, we draw on visual tradition more broadly, for example as exemplified by the work of Tufte (1990). The complex interactions between colour and symbolism when interpreting data, for example, are only compounded in 3D (Jobst and Germanchis, 2007) and extended in regard to light and texture. Kriz (2009, 59) argues '... it seems to be unfortunately evident that our society has difficulty dealing with multimedia representations for means of geo-communication – at least from a cartographic perspective.' It follows that empowering future VGE designers to make careful and considered representation of abstract items within a landscape is a key component of a rounded VGE curriculum. Further, the importance of simplicity in design, emphasized by Tufte, can be translated in a VGE setting into a need to maximize developmental effort on communicating the intended theme to the level of realism or abstraction required for a particular task. Thus, for example, stylised vegetation may not be appropriate for biogeographical learning where careful rendering of vegetation assemblages may be important, but where the emphasis is on landscape form the simpler photo-drape over a more detailed terrain model may be highly effective; Phase 1 of the curriculum directly informs Phase 2. Debates regarding the nature of virtuality and the perceived need for realism in VGEs are best sparked within a VGE learning context (see Section 18.3), even where the level of further technical engagement in VGE development does not go beyond Phase 2. For circumstances where the curriculum is intended to develop advanced designers, the timetable might go on to consider methodological approaches to levels of detail and to consider, practically speaking, how high levels of realism might be obtained in Phase 3. Similarly, stereo enablement in a VGE can be an important and effective interpretation aid in some circumstances, for example the angle of GPS in Section 18.3, but less so in others e.g. orientation within a broad landscape (also Section 18.3). Stereo can also assist more generally in enhancing a sense of immersion and place within a VGE, where this is key to a task. The need for stereo is foremost a design issue, subsequently requiring expertise in rendering engines and advanced programming skills using 3D libraries such as OpenGL within Phase 3 of a curriculum. The degree to which immersion and interaction are required of a task are Phase 2 design issues that can be debated from a variety of angles and with varying depth; technical options for enabling these characteristics are considered only after a good grounding in fitness for use and design has been gained. Cartographic and graphic protocols and human-computer interaction theories merge within Phase 2, which we argue is incomplete without consideration of both traditions regardless of the disciplinary home within which the VGE curriculum sits in any one instance.

Sections 2 and 3 of this chapter have largely focused on matters that fall within Phase 3 of our suggested curriculum, relating to the technical implementation of VGEs with an emphasis on tools and approaches within a teaching and learning context. In this discussion of a VGE curriculum, we emphasize the wide range of levels at which the Phase 3 curriculum might be approached according to discipline and learning goal. At one level, both for the teacher and student alike, Google Earth offers an accessible, easy to use medium to introduce VGEs, perhaps in conjunction with other software tools such as Sketch-Up for more senior students. As Weber *et al.* (2010) outline too, gamepads and joysticks can offer a means of introducing different approaches to interacting with VGEs at relatively early stages in the curriculum. Online virtual worlds and XML frameworks such as Geovisionary offer a range of intermediate options, while those emphasizing the development of technical skills might move students on to programming in Vega Prime, Open GL or a games engine. Similarly, Phase 4 curriculum activities can be explored in a stepped manner, from the evaluation of VGE tools to support fieldwork by undergraduates (e.g. Section 18.3) to the more sophisticated consideration of human computer interaction (HCI) evaluation protocols. Thus, our suggested curriculum flow operates linearly as a progression but also affords the revisiting of themes at different levels over a period of time as understanding grows, according to the pedagogic notions of scaffolding.

18.5 Conclusion

In technological terms, this chapter has outlined many positive developments that will enhance the use of VGEs as a teaching tool. Packages for 3D model creation and visualization, either static or animated, are getting easier to use and more widely available. There is much activity, and competition, in the area of web-based 3D graphics standards, and we can expect powerful development environments which will allow novice programmers to harness the power of real-time 3D previously only available to expert games developers. We expect the base data within virtual globes such as Google Earth to become ever more detailed and faithful to reality, and that the tools for more easily importing models and GIS data into such environments will get better. In the case of online virtual worlds there may be less reliance on in-world development tools and more support for import from specialist modeling environments. Having said this, new online virtual worlds are appearing, such as Blue Mars, based on games engine technology and addressing many of the technical restrictions in existing virtual worlds which have influenced the design of certain spaces where many concurrent users might be required. The computer games industry could well be an even more significant driver than it has been in the past. Not only are powerful games engines being opened up, and seen as viable alternatives to high end visualization systems, but an interest in novel interaction devices and cheap stereo solutions may soon offer cost effective Virtual Reality which has so far resided in the domain of more specialist research and development groupings. The increasing graphical power and higher bandwidths in mobile devices, coupled with the rapid developments in browser based games, should see virtual spaces being accessed in many different ways.

There are a lot of technical challenges remaining, but it is clear that many of the component parts already exist to provide more powerful and more accessible virtual environments for teaching and learning. Exciting times lie ahead. In pursuing these themes, we also note the need to develop visualizations and learning spaces fit for their purpose, that structure and support learning. Knowing when, and where, to step back to simplicity in either technological approach, interface or content is paramount when using digital environments for teaching and learning.

Acknowledgements

The images used in the Locata application reproduced in Figure 18.2 featured aerial photography from Getmapping, a hillshaded version of the NEXTmap Digital Surface Model from Intermap, and a section of the Harvey's British Mountain Map: The Lake District (Reproduced with permission from scanned paper mapping, © Harvey 2008). Figure 18.5 contains an image derived from LiDAR Digital Surface Model data from Infoterra. Building models in Figure 18.5 and 18.15 have been partly created using Blom Aerofilms Ltd aerial photography.

References

Ainsworth, S. E., P. Bibby, and D. Wood (2002) Examining the effects of different multiple representational systems in learning primary mathematics. *Journal of the Learning Sciences*, 11: 25–61.

Appleton, K., and A. Lovett (2003) GIS-based visualisation of rural landscapes: Defining 'sufficient' realism for environmental decision-making. *Landscape & Urban Planning*, 65: 117–131.

Ashfield, S., C. H. Jarvis, and J. Kaduk (2010) Serious games for geographical field skills: An exploration. *Proceedings of IEEE VS-Games 2010*, pages 71–74.

Azuma, R., Y. Balliot, R. Behringer, S. Feiner, S. Julier, and B. MacIntyre (2001) Recent advances in augmented reality. *IEEE Transactions in Computer Graphics and Applications*, 21: 34–47.

Biocca, F. (1992) Virtual reality technology: A tutorial. *Journal of Communication*, 42: 23–72.

Burton, A., G. Priestnall, G. Polmear, and N. Mount (2008) Real-time landscape visualisation: Experiences in a teaching and learning context *Geographical Information Science Research UK (GISRUK) Conference*, 2–4 April 2008, Manchester Metropolitan University.

Chou, C. (2003) Interactivity and interactive functions in web-based learning systems: A technical framework for designers. *British Journal of Educational Technology*, 34: 265–279.

Crown, S. W. (1999) Web-based learning: Enhancing the teaching of engineering graphics. http://www.imej.wfu.edu/articles/1999/2/02/index.asp (Accessed October 1, 2008).

DeFreitas, S. (2006) *Learning in Immersive Worlds: A Review of Game-Based Learning*. Bristol: JISC.

De Paor, D. G., and S. J. Whitmeyer (2011) Geological and geophysical modelling on virtual globes using KML, COLLADA and Javascript. *Computers & Geosciences*, 37(1): 100–110.

DiBiase, D., M. DeMers, A. Johnson, K. Kemp, A. Luck, B. Plewe, and E. Wentz (eds) (2006) *The Geographic Information Science and Technology Body of Knowledge*. Washington, DC: Association of American Geographers and University Consortium for Geographic Information Science.

Ervin, S., and H. Hasbrouck (2001) *Landscape Modelling*, New York: McGraw-Hill.

Goldman, S. R. (2003). Learning in complex domains: When and why do multiple representations help? *Learning and Instruction*, 13: 239–244.

Gutierrez, F., J. Pierce, V. Vergara, R. Coulter, L. Saland, T. P. Caudell, , T. E. Goldsmith, and D. C. Alverson (2007) The effect of degree of immersion upon learning performance in virtual reality simulations for medical education, IN: J. D. Westwood, R. S. Haluck, H. M. Hoffman, G. T. Mogel, R. Phillips, R. A. Robb, and K. G. Vosburgh (eds), *Medicine Meets Virtual Reality 15*, Amsterdam: IOS Press, pages 155–160.

Hahm, J., K. A. Lee, S.-L. Lim, S.-Y. Kim, H. T. Kim, and J.-H. Lee (2007) Effects of active navigation on object recognition in virtual environments. *CyberPsychology & Behaviour*, 10: 305–308.

Heim, M. (1993) *The Metaphysics of Virtual Reality*. New York: Oxford University Press.

Herrlich, M. (2007) A tool for landscape architecture based on computer game technology. *17th International Conference on Artificial Reality and Telexistence*, November 28–30, pages 264–268.

Hubona, G. S., and G. W. Shirah (2005) Spatial cues in 3D visualization. *Lecture Notes in Computer Science.* Springer: Berlin/Heidelberg, pages 104–128.

Hudson-Smith, A., and A. T. Crooks (2008) Building space in the machine: Digital tool kits for mirror worlds. *AGORA: Dutch Journal for Social-Spatial Questions*, 24(4): 27–29.

Isbister, K. (2008) *Game Usability.* San Francisco: Morgan Kaufmann.

Jarvis, C., G. Priestnall, G. Polmear, and J. Li. (2008) Geo-contextualised visualisation for teaching and learning in the field. *Geographical Information Science Research UK (GISRUK) Conference*, 2–4 April 2008, Manchester Metropolitan University.

Jobst, M., and T. Germanchis (2007) The employment of 3D in cartography – An overview. IN: W. Cartright (eds) *Multimedia Cartography.* Berlin / Heidelberg: Springer pages 217–228.

Kolbe, T., G. Groger, and L. Plumer (2005) CityGML – interoperable access to 3D city models. IN: *Proceedings of the International Symposium on Geoinformation for Disaster Management,* pages 21–23.

Kriz, K. (2009) Are we living in a cartographic illiterate society? *Cartography and Art.* Berlin/ Heidelberg: Springer, pages 59–68.

Langley, R. B. (1991) The mathematics of GPS. *GPS World*, 2: 45–50.

Leutner, D. (1993) Guided discovery learning with computer-based simulation games: Effects of Adaptive and non-adaptive instructional support. *Learning and Instruction*, 3: 113–132.

Li, C., and P. Longley (2006) A test environment for location-based services applications. *Transactions in GIS*, 10: 43–61.

Li, J., C. H. Jarvis, and C. Brunsdon (2010) The use of immersive real time 3D computer graphics for visualisation of Dilution of Precision (DOP) in virtual environments. *International Journal of Geographical Information Science*, 24: 591–605.

Lou, W., J. A. Stravers, and K. L. Duffin (2005) Lessons learned from using a web-based interactive landform simulation model (WILSIM) in a general education physical geography course. *Journal of Geoscience Education*, 53: 489–493.

MacEachren, A. M. (1994) *Some Truth with Maps: A Primer on Symbolization and Design.* Washington D. C.: Association of American Geographers.

Mach, R., and P. Petschek (2007) *Visualization of Digital Terrain and Landscape Data.* New York: Springer.

Mount, N., G. Priestnall, D. Weaver, and A. Burton (2009a) Wading through Derwent Water: Taking digital terrains from the real world to Second Life. *Geographical Information Science Research UK (GISRUK) Conference*, 1–3 April 2009, Durham University.

Mount, N. J., C. Chambers, D. Weaver, and G. Priestnall (2009b) Learner immersion engagement in the 3D virtual world: Principles emerging from the DELVE project. *ITALICS*, 8: 40–56.

Najjar, L. (1998). Principles of educational multi-media user interface design. *Human Factors*, 40: 311–323.

Neves, J. N., P. Goncalves, J. Muchaxo, and J. P. Silva (1999) A virtual GIS room: Interfacing spatial information in virtual environments, IN: A. S. Camara and J. Raper (eds) *Spatial Multimedia and Virtual Reality (Research Monographs in GIS).* London: Taylor and Francis, pages 147–156.

Pausch, R., D. Proffitt, and G. Williams (1997) Quantifying immersion in virtual reality. *Proceedings of 24th Conference on Computer Graphics and Interactive Techniques*, ACM Press/Addison-Wesley, pages 13–18.

Peters, C. (2010) Animating gaze shifts for virtual characters based on head movement propensity. In *Proceedings of IEEE VS-Games 2010*, pages 11–18.

Piekarski, W. (2006) Simple collaborative indoor-outdoor modelling using mobile augmented reality, *Proceedings of the First International workshop on mobile geospatial augmented reality*, Banff, Alberta, Canada, 29-20th May 2006.

Priestnall, G., and J. March (2008) Learning to read the landscape: Interactive visual tools for exploring spatial thinking. *Geographical Information Science Research UK (GISRUK) Conference 2–4 April 2008*, Manchester Metropolitan University.

Priestnall, G., and G. Polmear (2007) A synchronised virtual environment for developing location-aware mobile applications. *Proceedings of the 15th Annual Geographical Information Science Research UK Conference (GISRUK '07)*, National Centre for Geocomputation, Maynooth, Ireland.

Priestnall, G., E. Brown, M. Sharples, and G. Polmear (2009) A student-led comparison of techniques for augmenting the field experience. *Proceedings of Mlearn 2009*, Orlando, Florida, October 28–30 2009.

Qin, H., P. R. Pei-Luen, and G. Salvendy (2009) Measuring player immersion in the computer game narrative. *International Journal of Human-Computer Interaction*, 25: 107–133.

Rakshit, R., and Y. Ogneva-Himmelberger (2008) Application of virtual globes in education. *Geography Compass* 2(6): 1995–2010.

Riva, G. (2007) Virtual reality and telepresence. *Science*, 318: 1241–1242.

Sanchez-Vives, M. V., and M. Slater (2005) From presence to consciousness through virtual reality, *Nature Reviews Neuroscience*, 6: 322–339.

Schwarz, B. (2010) LIDAR: Mapping the world in 3D. *Nature Photonics*, 4: 429–430.

Shang, J., M. S. Y. Jong, F. L. Lee, and J. H. M. Lee (2006) VISOLE: A new game-based situated learning paradigm. *Proceedings of the Sixth IEEE International Conference on Advanced Learning Technologies (ICALT'06)*, pages 1082–1083.

Shepherd, I. D. H., and I. D. Bleasdale-Shepherd (2008) Multi-sensory data realization in virtual worlds: Abstraction or pragmatism. Paper presented at the *International Conference on Virtual Geographic Worlds*, Hong Kong, 7–8 January 2008.

Sheppard, S. R. J., and P. Cizek (2009) The ethics of Google Earth: Crossing thresholds from spatial data to landscape visualization. *Journal of Environmental Management*, 90: 2102–2117.

Smith, M. J., A. Burton, N. Kokkas, G. Priestnall, and G. Polmear (2008) Raising spatial and visualisation awareness in a teaching environment. *The XXI Congress of the International Society for Photogrammetry and Remote Sensing*, 3–11 July 2008, Beijing, China.

Smith, M. J., A. Burton, and N. Kokkas (2010) Teaching and learning strategies for 3D urban and landscape modelling, *The XXIV International Federation of Surveyors Congress*, Sydney, Australia, 11–16 April 2010.

Steuer, J. (1992) Defining virtual reality: Dimensions determining telepresence. *Journal of Communication*, 42: 73–93.

Stokes, S. P. (2001) Visual literacy in teaching and learning: A literature perspective. *Electronic Journal for the Integration of Technology in Education*, 1. http://ejite.isu.edu/Volume1No1/Stokes.html (accessed October 1, 2008).

Sun, H.-J., G. S. W. Chan, and J. L. Campos (2004) Active navigation and orientation-free spatial representations. *Memory & Cognition*, 32: 51–71.

Tate, N. J., P. F. Fisher, and D. Martin (2007) Geographic Information Science and Surfaces. IN: J. P. Wilson and A. S. Fotheringham (eds) *The Handbook of Geographic Information Science*. Oxford: Blackwell, pages 239–258.

Tavanti, M., and M. Lind (2001) 2D vs 3D, Implications on spatial memory. *Proceedings of the IEEE Symposium on Information Visualization 2001 (INFOVIS'01)*, IEEE Computer Society.

Tooth, S. (2006) Virtual globes: A catalyst for the re-enchantment of geomorphology? *Earth Surface Processes and Landforms*, 31: 1192–1194.

Tufte, E. (1990) *Envisioning Information*. Cheshire CT: Graphics Press.

Tuttle, B. T., S. Anderson, and R. Huff (2008) Virtual globes: An overview of their history, uses, and future challenges. *Geography Compass*, 2(5): 1478–1505.

Ware, C., and S. Osborne (1990) Exploration and virtual camera control in virtual three dimensional environments. *Computer Graphics*, 24: 175–183.

Weber, A., B. Jenny, M. Wanner, J. Cron, P. Marty, and L. Hurni (2010) Cartography meets gaming: Navigating globes, block diagrams and 2D maps with gamepads and joysticks. *The Cartographic Journal*, 47: 92–100.

Williams, N., A. Jenkins, D. Unwin, J. Raper, T. McCarthy, P. Fisher, J. Wood, K. Moore, and J. Dykes (1997) What should be the educational functions of a virtual field course? *Computer Aided Learning 1997*. Exeter, UK, University of Exeter School of Education, pages 351–354.

Witmer B. G., and M. J. Singer (1998) Measuring presence in virtual environments: A presence questionnaire. *Presence: Teleoperators and Virtual Environments*, 7(3): 225–240.

Yeo, S.,R. Loss, M. Zadnik, A. Harrison, and D. Treagust (2004) What do students really learn from interactive multimedia? A physics case study. *American Journal of Physics*, 72: 1351–1358.

19

Using web-based GIS and virtual globes in undergraduate education

Lynn Songer

Department of Social Science, Lane Community College, Eugene, Oregon, USA

19.1 Introduction

It is difficult to find a computer user who has not experienced the 'wow' factor of Google Earth, digital maps and other geospatial technologies. These ubiquitous internet tools are used daily to locate features. For a geography educator that is both the good and bad news. People are aware and excited about these geographic innovations yet, for the most part, are unaware of their power to better understand the world in which we live (Goodchild, 2009; Schöning *et al.*, 2008). These tools are well suited for use in education to teach the why of where and to help develop and test hypotheses. Digital maps, web-based GIS, and virtual globes are much more than on-line atlases. They are tools for spatial analysis and problem solving. I am writing this section as an instructor who uses web-based GIS and virtual globes to enhance geography teaching, not as a developer or programmer of web-based GIS and virtual globe applications. While I teach GIS and have created many maps that are hosted on a web-based GIS site, my focus here is to discuss how instructors can use existing geospatial applications to teach important geospatial and problem-solving skills to undergraduate students, particularly those in geography and environmental science. The beauty of these applications is that one need not know how to create them to use them effectively in education.

I begin this chapter with a broad framework for thinking spatially in undergraduate classes and will argue for the importance of including digital geography in education. Then I will provide a working definition of digital maps, web-based GIS, and virtual globes and briefly discuss three web-based GIS applications – ArcIMS, ArcServer and MapServer – and three virtual globes – NASA World Wind, Google Earth, and ArcExplorer. This is followed by specific examples of these technologies being employed to teach geospatial concepts in undergraduate geography and environmental science.

Teaching Geographic Information Science and Technology in Higher Education, First Edition.
Edited by David J. Unwin, Kenneth E. Foote, Nicholas J. Tate and David DiBiase.

19.2 A framework for geospatial thinking in undergraduate education

Spatial thinking is an important life skill. The ability for college students to learn to think spatially is important for success in a number of careers and is a fundamental life skill. Spatial thinking involves: 'recognizing a shape or pattern amid a cluttered or noisy background, visualizing a three-dimensional object, describing the position and orientation of objects encountered in the real world, remembering the location and appearance of previously seen items, and envisioning the processes by which objects change shape' (NRC, 2006, 108). For example, assembling a bicycle, properly installing a child's car seat, packing for a move, finding your way to a new doctor's office using a map or verbal directions, or giving a delivery person directions to your house all require the ability to problem-solve using spatial thinking (NRC, 2006; Wegner and Girasek, 2003; Gerber 2003).

In the workplace, spatial skills are increasingly needed in business, government, resource management and social services (SEAC, 2007; USDL, 2005). For example, banking, retail and insurance industries employ spatial thinking for market analysis and to understand service needs. National, state and local governments use the power of spatial thinking to establish and regulate policy, assess and collect revenue, and plan for economic development. Spatial thinking is critical in the area of public health and safety where it is used to manage health care services, monitor public health, identify disease clusters, explore sites of environmental risk, plan emergency response for fire and other natural hazards, plan tactics and strategies for law enforcement, and analyse crime incidences. Management of resources – forests, fisheries, parks and agricultural lands depend on the spatial thinking concepts and skills. Finally, workers involved in managing economic infrastructures – such as transportation networks and utilities – rely heavily on the ability to employ and understand spatial thinking.

Geospatial thinking can encompass concepts of distance, direction, scale and arrangement viewed in ways to understand spatial relationships (NRC, 2006). When conveying these concepts to students, it is helpful to have a more clearly defined framework for spatial thinking. Project GeoSTART (The Association of American Geographers *et al.*, 2008), provides a practical guide that defines and describes eleven geospatial thinking concepts (Table 19.1). These concepts go far beyond simple location and observation and can be taught and reinforced by using geospatial applications such as web-based GIS or virtual globes.

In addition to teaching and reinforcing these eleven geospatial skills, web-based GIS and virtual globes can be used to teach concepts of scale and spatial exception. Identifying spatial exceptions combines sequential thinking skills to identify what is different than expected or to predict events. Changing map scale changes the level of detail on the map and impacts how that information is evaluated. In addition, researchers believe that the visual components and hands-on interactivity of GIS help develop spatial reasoning and problem-solving skills (Songer, 2010; Lee and Bednarz, 2009; Kerski, 2003). General geography skills are also thought to benefit from using GIS applications (Drennon, 2005). Educators and researchers also report increased student motivation at all grade levels when GIS is taught within the curricula (Songer, 2010; Baker and White, 2003; West, 2003; Kerski, 2003; Keiper, 1999; Meyer, 1999).

Table 19.1 Geospatial concepts from GeoSTART teachers guide

Concept	Description
Analogy	Describing relationships between distant places. i.e. Climate of Los Angeles and Casablanca to predict an event at a little-known location based on criteria at a known location
Association/Correlation	Identifying the relationship between phenomena. i.e. High birth rate and short life span are negatively correlated, high evaporation rate and high temperature are positively correlated
Aura – Influence	Understanding the impact of a phenomenon on an area. i.e. Potential market areas, the impact of sea-level rise
Change/Movement	Describing spatial-temporal changes, i.e. gentrification, dune migration, diffusion of building type
Comparison	Analysis of two or more events at different locations. i.e. population, tornado frequency
Condition/Connection	Describing or evaluating natural and human features and their connections. i.e. river, railroad
Hierarchy	Identify systems. i.e. Census units, economic sectors, scale
Location	Using coordinates, describing relative location, direction, distance
Pattern	Identifying clusters, alignment, and dispersion of spatial phenomena i.e. landslide patterns, crime occurrence, tornado events
Region	Classify and organize spatial information. i.e. identify critical habitat, demarcate hazard zones or cultural groupings
Transition	Change over space and time. i.e. rainfall, crops, fluvial geomorphology, immigration

19.3 The potential impact of recent developments in online GIS

Despite the promise of GIS&T to benefit education, prior GIS technologies – notably desktop GIS – have been underutilized due to the overly technical nature of the software, the high cost of deploying it in the classroom and the steep learning curve associated with implementing a GIS curriculum (Bednarz, 2004; Kerski, 2003; Summerby-Murray, 2001; Lloyd, 2001; Walsh, 1992).

These historic impediments to educational use of geospatial technologies no longer block their application in the classroom. The availability and accessibility of geospatial information and geospatial technology is exploding with advances in internet technology. Access to a web-based GIS or a virtual globe requires only a computer, an internet browser, and a relatively robust internet connection. Today internet-based geospatial applications such as Google Earth, World Wind and many web-based GIS applications are used by hundreds of millions of people worldwide (Craglia *et al.*, 2008; Tuttle *et al.*, 2008). The imagery and maps are of good quality, often based on recent satellite imagery and/or current vector data. Web-based GIS and virtual globes provide analysis tools that can support problem-solving and data analysis in a number of disciplines and for a wide range of learning levels from early elementary to graduate (Carver, *personal communication* 2009; Baker and White, 2003;

Bodzin and Anastasio, 2006). The learning curve is less steep because menu options are limited and students need not manipulate data, making it less likely they will introduce errors. Many geospatial applications are free, or have free versions, so there is little or no software cost.

Innovations in geospatial and internet technology make the distinctions between digital maps, web-based GIS, and virtual globes less discernable. These applications occupy a geospatial technology continuum ranging from limited interactivity to full user access, including the ability to upload data. A key characteristic of an interactive digital map is that it allows the user to make map layers visible or not visible, and the map may be scalable. A web-based GIS application extends the interactivity of digital maps to provide additional geospatial tools such as: access to data tables, query and selection capabilities, ability to develop buffers, and dynamic coordinate information. A virtual globe is an application that provides a 3D perspective of the Earth or other body with satellite imagery. It allows the viewer to change viewing angle and perspective. Virtual globes provide some of the features seen in a web-based GIS, plus the ability to add local data – which can be viewed in geospatial context on a local computer without being uploaded to a server or other shared site. Because most digital maps, as defined here, have limited interactivity and fewer geospatial tools, I will focus this discussion on web-based GIS and virtual globe applications, which have more potential to impact student learning.

19.4 Web-based GIS

Web-based GIS sites provide access to existing interactive maps hosted on the internet. Each application offers a suite of tools and has specific advantages and disadvantages for educational use. Typically, the maps have been developed for a specific purpose or agency. For example: Tax lot maps for Klamath County, Oregon (http://klamathcounty.houstoneng.com/klamathmap/klamathmap.html); Community Crime View for the San Bernardino Police Department (http://sanbernardino.ca.crimeviewcommunity.com); US Ancestry Map by Lane Community College for education use (http://arcgis.lanecc.edu/website/ancestry/viewer.htm). In general, the process of creating a web-based GIS starts with the design of a map using a desktop GIS application. Once the map is created it is published or hosted to the internet using a web-map server. Creating the maps and hosting them on the internet requires considerable desktop GIS knowledge and server technology expertise.

However, once the maps are available on the internet, they can be accessed by any number of users with relative ease and at minimal expense. While the data layers available with each web-based GIS application are determined by the map developer, these technologies provide a wide range of tools that allow users to explore the data, select layers, develop spatial queries, view data tables and develop and test spatial hypotheses. Three such web-mapping applications are ArcIMS, ArcServer and MapServer.

ArcIMS (Arc Internet Map Server) is an application developed by Environmental Systems Research Institute (ESRI) that hosts dynamic maps on the internet via a server. The tools available with an ArcIMS web-based GIS are considerable and the out-of-the-box application has the look and feel of ArcGIS desktop. Standard tools allow the user to move about the map, 'zoom' in and out and 'pan.' The 'find' tool facilitates locating specific text from layer attribute tables and the 'identify' tool provides specific attribute information about a map feature by allowing the user to view the attribute table. ArcIMS also allows the user to measure linear map features and to change measurement units. Line, point and polygon features can be selected for analysis using interactive selection tools. Dynamic coordinate information is provided as the mouse cursor moves about the map.

ArcIMS provides high-level analysis capabilities with the 'buffer' and 'query' tools. Once a feature is selected, a linear or circular buffer of a specific distance can be created around the feature. The buffer can be generated around the selected feature or it can be set to identify another feature that intersects the buffer. Boolean logic queries can be created using the logic tool to select features that meet specific parameters. Users can define simple queries with one variable and one operator, or they can set up a complex query with multiple operators. Finally, ArcIMS provides data sharing with the ability to extract and download shapefiles for off-line analyses.

ArcGIS Server is a newer internet map service by Esri. Maps created with ArcServer provide a sharper map image that maintains the quality and color of the original map document. Additionally, images render more quickly than in ArcIMS maps. Tools are available for navigation and identifying features. The basic ArcGIS Server tools do not have the familiar icons used in ArcIMS. For example, the zoom in and out feature uses a more modern slider bar, not the plus and negative magnifying glass icons. One limitation of the maps served by ArcGIS Server is that queries are predefined by the map developer. With ArcIMS, the user is free to develop his or her own Boolean logic queries using fields, operators and variables from the data table but with the basic ArcGIS Server the user chooses from a list of predefined query statements.

MapServer is open-source software that allows the user to publish interactive maps over the internet. Web-based maps created in MapServer have the same tools and spatial analysis capabilities of ArcIMS and ArcGIS Server.

The tools and level of interaction on web-based GIS maps vary depending on the map publisher's purpose. Some map applications provide the ability to customize a site by adding or manipulating data. Some sites are designed to be data viewers with tools for spatial analysis of existing map data.

19.5 Virtual globes

Virtual globes offer a three-dimensional view of the earth's surface using mosaic areal photography and Landsat images. The three virtual globes discussed in this chapter – Google Earth, NASA World Wind, ArcGIS Explorer – are offered as free downloads and strong support for educational use. Additionally, virtual globes may require a 3D graphics card or other computer enhancements to render the images. The globe-like interface often helps new users get started quickly since it provides standard navigation tools to zoom in and out, rotate the axis and viewing angle, and determine latitude and longitude. Because virtual globes provide considerable imagery as well as additional layers of data – roads, political boundaries and points of interest – it is easier, in some cases, for users to get started in displaying their own data overlays and maps since most of the latest virtual globes allow users to add local data – points, line, polygons – to the global image using open application programming interfaces.

ArcGIS Explorer, by Esri, is a free application to view global maps and data. Linking to the ArcGIS Explorer resource page allows the user to view other data layers such as: streets, shaded relief, topographic maps with physical features, and world protected areas, and an 1812 Pinkerton map view. The layer manager in ArcGIS Explorer allows the user to change the order of layer and change the layer transparency. Edited map sessions can be saved for later use.

ArcGIS Explorer offers a measurement tool with several options. Users can choose to measure linear or area features and choose the measurement units. Measurements can be saved as map layers – lines or polygons – and can be added or removed from the map as NMC files (Esri map content files). ArcGIS Explorer allows users to add local data to create customized maps. One method is to import a text file or a comma-delimited file that has an attribute field and longitude and latitude fields. The user can assign a symbol and save the new layer. Another method to add local data is to open stand-alone shapefiles, KML/KMZ, NMC and GPX files, raster files, or features in a geodatabase.

NASA World Wind was developed at NASA Ames Research Center, and is an open-source application that provides high resolution Landsat images with additional data layers for topographic maps, political borders, a graticule and cloud bands. An exceptional feature available with World Wind is to access a number of Landsat imagery products: NLT Landsat 7 visible color (3, 2, 1), pseudo color images at 30 and 15 meter resolution, Geocover pseudo color at 15 meter resolution for early 2000 and 30 meter resolution for early 1990. World Wind tools allow the user to measure distance, determine bearing and load and view shapefiles and KML files.

Google Earth is by far the most well-known and well-used virtual globe. Google Earth has a free version and more sophisticated commercial versions. Google Earth provides Landsat imagery with overlays for roads, borders, weather, traffic, places of interest (for example, hotels, restaurants, museums and parks), and global awareness locations (for example: Greenpeace, the United Nations Environmental Program and UNICEF Water and Sanitation program). Google Earth also provides layers to view 3D terrains and 3D buildings. The free version of Google Earth provides tools to measure linear distance, change the coordinate units for latitude and longitude, add KML/KZM files, create folders, place marks, polygons, and lines that can be saved as a KML/KMZ files. Users can change the symbology of place marks from a provided list, or the user can import a customized image.

19.6 Examples of web-based GIS and virtual globes in undergraduate education

Web-based GIS is being successfully employed in the undergraduate classroom to teach and reinforce important geospatial skills – *analogy, association/correlation, aura/influence, change/movement, comparison, condition/connection, hierarchy, location, pattern, region and transition* (Table 19.1). Carver (2004) employs web-based GIS to teach spatial decision-making in a problem-based curriculum for undergraduate and graduate students. Students use query and selection tools to evaluate key attributes for site selections. Students compare *locations*, look for *spatial hierarchy*, evaluate *influence*, and test hypotheses. Carver reports that students showed improved understanding of geospatial concepts and were enthusiastic about using GIS. Web-based GIS allowed students to focus on geographic principles without distraction from complicated software or the cost of 'tooling up' (Carver, 2004, 436). Carver currently uses this and other web-based lessons developed at Leeds to provide an early introduction to GIS. Carver states that web-based GIS maps are 'great resources that allow us to lead our students into GIS gently' (Carver, 2009).

Bodzin and Anastasio (2006) used web-based GIS in an undergraduate earth and environmental science class to support watershed analysis and real-world problem solving related to land use and the environment. Students were introduced to spatial concepts – *location, influence, region, change* – as well as data analysis skills. GIS was needed for students to be able to handle large amounts of data, but a desktop GIS application would have made the learning too time consuming. Web-based GIS made this data-rich, inquiry-based activity possible. Students evaluated various data layers – geology, population, land use, transportation and forest type – to propose the location of a new home. Bodzin and Anastasio suggest that the map can be used to develop a geomorphology-focused lesson. 'The instructional use of this activity's materials and resources may be customized to emphasize the geologic occurrence of sinkholes and its implications for geoenvironmental engineering' (Bodzin and Anastasio, 2006, 96).

At Lane Community College in Oregon I have worked with colleague Eric Sproles on the Mapping Analyzing and Problem Solving with GIS (MAPS-GIS), a project to develop web-based GIS lessons – using ArcIMS – for undergraduate community college students. Working with faculty across the curriculum at Lane, we have developed lessons that teach geospatial thinking and embedded them within a number of disciplines: human services, economics, geography, psychology, environmental science, forestry, geology, chemistry, maths and computer science. In the lessons, students were taught to create spatial buffers to explore *aura* or *influence*, use Boolean logic queries to evaluate *correlation* and look for *analogs* in spatial data, discuss spatial *patterns*, evaluate scale, and search for *exceptions*. For example, lessons cover concepts such as tornado and earthquake location and intensity, global demographic patterns, ancestry and poverty correlations in the United States, age and religious adherence patterns in the United States, whale migration and correlation to coastal ecology, watershed analysis and learning to use latitude and longitude. The lessons are accompanied by a tutorial, instructor guides and student handouts. The lessons have proven to be portable and have been adopted by instructors at Atlantic Cape Community College in New Jersey and as the focus for high school teacher professional development workshops at Lakeland Community College, Ohio – as part of the outreach mission of the National Geospatial Technology Center in Corpus Christi, Texas.

Ming-Hsiang Tsou employs web-based GIS and virtual globe exercises in his undergraduate class, Geographic Information Science and Spatial Reasoning, at San Diego State University. In one exercise students go beyond simple *location* tasks to describe disease *patterns* of bird flu in the Middle East, exploring the geographic *distribution* and *diffusion* of the disease. Students predict or hypothesize the spread of the disease to understand the potential global impacts (Tsou, 2009). Not all the web-based GIS sites used by students in the class were developed at San Diego State. Tsou uses existing web-based GIS sites such as the FundRace 2008 web-mapping site at the Huffington Post (http://fundrace.huffingtonpost.com/) to engage students. In this activity students are asked to describe *patterns* of political funding and to posit uses for such information in developing a political campaign. Another existing web-based GIS site used by Tsou's students is the San Diego Crime Map. Using this map (http://mapping.arjis.org/) as a basis for their lesson, students create queries to explore *connections* between neighborhoods and types of crimes and to compare neighbourhood crime data. Tsou's work shows how adaptable existing web-based GIS sites are for use in education to teach spatial thinking at the undergraduate level.

At San Francisco State University, Jerry Davis (2009) developed several lessons using NASA World Wind for an upper division geomorphology course. After a classroom demonstration of the software, students are given the assignments to complete outside of class. 'I find that it brings them to a part of geography that I consider as under-emphasized in recent years in our discipline – exploration. I encourage students to explore the area to find additional features, and they must provide evidence from what they can see in the image to interpret the landforms and processes' (Davis, *personal communication* 2009). In each of the four activities, students are given the latitude and longitude of specific features – braided stream, point bar, cinder cone, caldera – then asked to *locate* that feature and provide supporting information about the landform's related geomorphology, or evidence of process. Students use measurements to evaluate feature characteristics, describe evidence of *change* and *movement*, discuss *hierarchy* and *connection*, and provide evidence of *transition*. To accommodate students whose computers won't run World Wind, Davis has provided directions to complete the activity in Google Earth.

GIS for the Humanities (2009), a curriculum development project at San Antonio College in Texas has to date developed fourteen undergraduate level web-based GIS lessons. Modules were developed across the college curriculum and include participation from history, foreign language, sociology and geography instructors. This diverse group of web-based GIS maps covers topics such as: globalization, the conflict in Palestine, crisis in Sudan, geography of Mexico, geography of Bulgaria. In the globalization lesson, students can evaluate the economic and the cultural aspects of globalization. Students are lead through the Boolean logic process to select data and are asked to describe distribution then hypothesize about *patterns* of global trade, identify economic *regions*, and look at cultural *hierarchy* and *influences* as they study *diffusion*.

Thomas Allen (2007) has used virtual globes to teach digital terrain visualization in an advanced geomorphology class. The use of virtual globes – NASA World Wind and Google Earth—to augment lecture materials and for online activities allowed students to visualize geomorphic processes at multiple scales which promoted an understanding of *regions* and *hierarchy*. Allen used the 3D view in conjunction with 'world tour' and 'fly through' features available in Google Earth to create an interactive component that aligned with representative site features in the text.

Rakshit and Ogneva-Himmelberger (2009) have developed four undergraduate level activities that highlight the use of three virtual globes: Skyline Globe, Google Earth and Virtual Earth. The lessons focus on evaluating urban sprawl, visualizing global internet use over time, toxic release inventory mapping and image classification. The lessons have not been implemented in the classroom, but Rakshit states that the activities are based on current research techniques and technologies (Rakshit, 2009). One advanced skill-level multi-part lesson developed by Rakshit and Ogneva-Himmelberger teaches students how to use Google Earth, Excel, GE Graph and Batch Coder to examine the *connection* between point source toxic waste and income in Worcester County, Massachusetts, USA.

It is important to note that while some of the examples presented here use geospatial technology specifically developed for student learning activities (Carver *et al.*, 2006; Bodzin and Anastasio, 2006, MAPS-GIS, 2009) every web-based GIS and virtual globe has the potential to be used in a variety of disciplines to teach important geospatial skills and concepts. For example, the web-based GIS tutorial developed by MAPS-GIS for general student use was employed in an undergraduate level discrete maths class to teach Boolean logic and demonstrate the use of Boolean logic for real-world problem-solving.

19.7 The future of web-based GIS, virtual globes and APIs in undergraduate education

Geospatial technologies are evolving and converging at a dramatic rate (Jackson *et al.*, 2009; Tuttle *et al.*, 2008). There are at least two likely evolutions in the geospatial technology revolution that will directly impact educational use. The first is the continued interactivity of users and spatial data in the form of neogeography – the use of cartographic tools and techniques by non-expert users – and volunteered geographic information (VGI) – user generated spatial data – (Goodchild, 2007). The second is the merging of the geospatial analysis capabilities seen in web-based GIS and virtual globes.

Increased user involvement in maps created by non-professionals and with user generated content, neogeography and VGI, is already underway. The Chesapeake Bay Field Scope Project, under development by the National Geographic Society (2009), combines the concepts of citizen science and VGI. Students living in the Chesapeake Bay watershed will collect and upload water-quality data to a web-based GIS site. This will provide a robust dataset that students can analyse to understand the health of the Chesapeake Bay watershed. The highly interactive project will make students active participants in GIS data collection, visualization and analysis.

The VGI potential of web-based GIS applications such as OpenStreet Map (http://www.openstreetmap.org/) and Wikimapia (http://wikimapia.org/) will allow instructors to create custom lessons and direct students to local real-world data. For example, a geomorphology text shows examples of fluvial geomorphology typically from a well known river system – the Mississippi River or the Nile River – to reach a national or regional audience. However, VGI applications offer instructors the opportunity to design lessons that are more relevant to their students. An instructor could upload to a VGI-enabled web-based GIS point and polygon markers that exemplify specific examples of fluvial geomorphology from a local river system. Students could be directed to the site and, as in the lesson examples developed by Jerry Davis, asked to describe and provide supporting information about the landform's related geomorphology and process. An extension of this activity for a more advanced class would be to reverse the process. Here students could be asked to find and mark examples of specific features and offer the same supporting discussion.

Merging the capabilities of web-based GIS and virtual globes to perform operations like, select, query and buffer will make geospatial application more powerful to teach geospatial thinking and analysis. Already the difference between the two types of geospatial applications is less distinct. Maps can be customized using an application programming interface (API) by integrating multiple geographic technologies. Google was among the first to make their APIs open and easy to use, both for Google Earth and Google Maps and others followed its lead. This means that Google Earth images and Google Earth maps can be customized and embedded into course web pages for easy student access without the need to download the Google Earth software. It also allows instructors to use customize geospatial technologies to create high-quality teaching materials without as much training as was once required to integrated GIS&T into higher education.

Geography and environmental science instructors can have a huge impact on how students think about current digital geography. Is Google Earth just a cool map to show where they live? Or is it a powerful tool to assist spatial decision-making? Once students are taught how to use the tools available in a web-based GIS, virtual globes and other emerging technologies to view the world from a geographic perspective and explore the world using geographic concepts, they will be in a better position to understand the why of where, to make better decisions and to ask better questions.

References

Association of American Geographers, George Mason University, Hunter College, and Howard University (2008) *Project GeoSTART: Geo-spatial Thinking Activities and Resources for Teachers of Geography and Earth Science.* http://www.aag.org/galleries/education-files/GeoStartComplete.pdf (accessed February 2011).

Allen, R. A. (2007) Digital terrain visualization and virtual globes for teaching geomorphology. *Journal of Geography*, 106: 253–266.

Baker, T., and S. White (2003) The effects of GIS on student's attitudes, self-efficacy, and achievement in middle school science classrooms. *Journal of Geography*, 102: 243–254.

Bednarz, S. (2004) Geographic information systems: A tool to support geography and environmental education. *GeoJournal*, 60: 191–199.

Bodzin, A., and D. Anastasio (2006) Using Web-based GIS for earth and environmental systems education. *Journal of Geoscience Education*, 54(3): 297–300.

Carver, S., A. Evans, and R. Kingston (2004) Developing and testing an online tool for teaching GIS concepts applied to spatial decision-making. *Journal of Geography in Higher Education*, 28(3): 425–438.

Craglia, M., M. Goodchild, M. A. Annonil, G. Camara, M. Gould, W. Kuhn, D. Mark, I. Masser, D. Maguire, S. Liang, and E. Parsons (2008) Next-generation digital earth: A position paper from the Vespucci initiative for the advancement of geographic information science. *International Journal of Spatial Data Infrastructures Research*, 3: 146–167.

Davis, J. (2009) *NASA World Wind education – Lessons.* http://worldwindcentral.com/wiki/Education (accessed September 2009).

Drennon, C. (2005) Teaching geographic information systems in a problem-based learning environment. *Journal of Geography in Higher Education*, 29(3): 385–402.

Gerber, R. (2003) The global scene for geographical education, IN: R. Gerber (ed) *International Handbook on Geographical Education.* Dordrecht: Kluwer Academic Publishers, pages 3–18.

GIS for the Humanities (2009) San Antonio College: Bringing internet mapping to the humanities. http://sacarcims.sac.alamo.edu/default.htm (accessed September 2009).

Goodchild, M. (2009) Neogeography and the nature of geographic expertise. *Journal of Location Based Services*, 3(2): 82–96.

Goodchild, M. (2007) Citizens as sensors: The world of volunteered geography. http://www.ncgia.ucsb.edu/projects/vgi/docs/position/Goodchild_VGI2007.pdf (accessed September 2009).

Jackson, M., D. Schell, and D. R. F. Taylor (2009) The evolution of geospatial technology calls for changes in geospatial research, education and government management. *Directions Magazine*, April. http://www.directionsmag.com/articles/the-evolution-of-geospatial-technology-calls-for-changes-in-geospatial-rese/122571 (accessed February 2011).

Kemp, K., M. Goodchild, and R. Dodson (1992) Teaching GIS in geography. *Professional Geographer*, 44(2): 181–191.

Keiper, T. (1999) GIS for elementary students: An inquiry into a new approach to learning geography. *Journal of Geography*, 98(2): 47–59.

Kerski, J. (2003) The implementation and effectiveness of geographic information systems technology and methods in secondary education. *Journal of Geography*, 102(3): 128–137.

Lee, J., and R. Bednarz (2009) Effect of GIS learning on spatial thinking. *Journal of Geography in Higher Education*, 33(2): 183–198.

Lloyd, W. (2001) Integrating GIS into the undergraduate learning environment. *Journal of Geography*, 100(5): 158–163.

MAPS-GIS (2009) Mapping, analyzing and problem solving with GIS. NSF-ATE Grant # 603492. http://gis.lanecc.edu (accessed September 2009).

Meyer, J., J. Buttrick, M. Olkin, and G. Zac (1999) GIS in the K-12 curriculum: A cautionary note. *Professional Geographer*, 51(4): 571–578.

NGS (National Geographic Society) (2000) *The Chesapeake Bay Fieldscope.* http://www.nationalgeographic.com/field/projects/cbfieldscope.html (accessed September 2009).

NRC (National Research Council) (2006) *Learning to Think Spatially: GIS as a Support System in the K-12 Curriculum.* Washington, DC: The National Academy Press.

Rakshit R., and Y. Ogneva-Himmelberger (2008) Application of virtual globes in education. *Geography Compass*, 2(6): 1995–2010.

Rakshit, R., and Y. Ogneva-Himmelberger (2009) Teaching and learning guide: Application of virtual globes in education. *Geography Compass*, 3(10): 1749–8198.

SEAC (2007) The spatial information industry workforce plan: Developing a 'people' industry. http://www.crcsi.ecampus.com.au/file.php/88/Workforce_Plan_PRESS_RELEASE.pdf (accessed September 2009).

Schöning, J., B. Hecht, M. Raubal, A. Krüger, M. Marsh, and M. Rohs (2008) Improving interaction with visual globes through spatial thinking: Helping users ask 'why? *Proceedings of the 13th International Conference on Intelligent User Interfaces.* Gran Canaria, Spain, January 13–16, pages 129–138.

Songer, L. (2010) Using web-based GIS in introductory human geography. *Journal of Geography in Higher Education*, 34(3): 401–417.

Summerby-Murray, R. (2001) Analyzing heritage landscapes with historical GIS: Contribution from problem-based inquiry and constructivist pedagogy. *Journal of Geography in Higher Education*, 25(1): 37–52.

Tsou, M. (2009) Geographic information science and spatial reasoning: Lessons for GEOG 104, a general education [GE] course. http://map.sdsu.edu/geog104/lab.htm (accessed September 2009).

Tuttle, B., S. Anderson, and R. Huff (2008) Virtual globes: An overview of their history, uses, and future challenges. *Geography Compass*, 2(5): 1478–1505.

US Department of Labor (2005) President's high growth job training initiative: Identifying and addressing workforce challenges in America's geospatial technology sector. http://www.doleta.gov/brg/pdf/Geospatial%20Final%20Report_08212007.pdf (accessed September 2009).

Walsh, S. (1992) Spatial education an integrated hands-on training: Essential foundations of GIS instruction, *Journal of Geography*, 91(2): 54–61.

Wegner, M., and D. Girasek (2003) How readable are child safety seat installation instructions? *Pediatrics*, 111(3): 588–559.

West, B. (2003) Students' attitudes and the impacts of GIS on thinking skills and motivation. *Journal of Geography*, 102(6): 267–274.

20

Trying to build a wind farm in a national park: experiences of a geocollaboration experiment in Second Life

Nick J. Mount and Gary Priestnall

School of Geography, University of Nottingham, Nottingham, UK

20.1 Introduction

In this chapter we present an example of the use of the 3D virtual world Second Life (www.secondlife.com) as a geocollaborative tool in a teaching and learning setting. We begin by reviewing some of the key developments in geocollaborative technologies of the last decade and the emergence of Web 2.0 and the possibilities it offers with respect to digital Earths and virtual worlds. We consider the general limitations of the Second Life environment for the creation, display and manipulation of geographic information alongside the technical difficulties associated with transferring geographic data from a GIS to Second Life. We identify 3D virtual worlds as a rapidly emerging frontier in geocollaborative technology and, therefore highlight the need for inclusion of material on the subject within GIScience curricula. We, therefore, present an example of a geocollaborative teaching and learning activity undertaken in Second Life in which synchronous multi-user interaction with spatial data sets formed the foundation of the learning task. Finally, we analyse results from focus groups formed of learner participants in the Second Life learning task to highlight some of the key challenges that educators considering integrating Second Life in their GIScience teaching and learning activities should be aware of.

Teaching Geographic Information Science and Technology in Higher Education, First Edition.
Edited by David J. Unwin, Kenneth E. Foote, Nicholas J. Tate and David DiBiase.

20.2 The emergence of 3D virtual worlds as geocollaborative environments

The acceptance of a stand-alone working model driven by the predominance of desktop GIS, as experienced throughout the early 1990s has been consistently challenged by user's increasing desire for synchronous, distributed geocollaboration. Initial responses to the challenge in the late 1990s focused on the development of distributed databases to allow remote, asynchronous geocollaboration through newly-developed web-based tools for analysis and visualization (Doyle *et al.*, 1998), and the participation of the public to support spatial decision making (Kingston *et al.*, 2000). Alongside these developments, opportunities for synchronous geocollaboration were also being considered (MacEachren, 2000). Here efforts were centred on the extensions of the single-user data flow model via the conceptualization of system architectures to provide visualization toolkits to multiple, synchronous users (Wood *et al.*, 1997) and the coupling of data visualization and immersive virtual reality environments (Wheless *et al.*, 1996). Some hints of what would be to come are evident during this period, with the integration of chat room functionality and shared windows containing geospatial display tools at the Upper Atmospheric Research Collaboratory project being particularly developmental (Olson *et al.*, 1998; Plale *et al.*, 1998). However, the majority of synchronous geocollaboration remained same-place (MacEachren, 2000) due to the difficulties associated with implementing visual geospatial data analysis in a distributed computing environment, the development and implementation of interaction concepts and the challenges associated with understanding human cognition of collaborative geographical analyses through distributed, visual interfaces (MacEachren, 2001).

The development of Web 2.0 technologies (O'Reilly, 2005) invoked massive change and has taken geocollaboration from the domain of small, specialist research and development projects driven by academics and discipline specialists, to the public at large in a way that was largely unforeseen in the late 1990s and early 2000s. Whilst the term Web 2.0 remains difficult to define, its core themes of supporting the bi-directional creation and sharing of user-based content through lightweight application programming interfaces (APIs) and distributed client-server architectures (Rinner *et al.*, 2008) has resulted in a computing environment that 'supports large numbers of users to interact with and supply information to central sites, and to see that information collated and made available to others' (Goodchild, 2007, 27). Indeed, many Web 2.0 mapping technologies have made great leaps towards overcoming the difficulties of visualizing geospatial data in distributed environments and have formalized the approaches to multi-user interaction with, and generation of that data. The result has been the emergence of a 'neogeography' (Eisnor, 2006; Haklay *et al.*, 2008), in which the profile of the geographic information user together with the scope of audience that can be reached by them, has changed from geographic information science professionals to members of the public. This has resulted in a renaissance in geographic information (Hudson-Smith and Crooks, 2008), with huge public engagement in collaborative geospatial Web 2.0 platforms such as Google Earth, NASA's World Wind and Microsoft's Virtual Earth, as both users and creators of spatial information. These virtual Earth platforms have built on the earlier, two-dimensional web mapping services such as Google Maps, but crucially offer the user the ability to navigate, tag and build geographic objects in 3D geographical space; thereby taking a step towards the existence of a mirror world (Gelernter, 1991) and a more immersive visualization experience. However, the extent to which they incorporate the necessary affordances for true geocollaboration remains limited by a lack of tools for integrated, real-time synchronous interaction and communication between users and their data. As a result, virtual Earths represent a hugely popular set of distributed geospatial visualization tools, but users remain firmly in the real-world looking in, and the desire for a more immersive, interactive experience (both in terms of interaction with objects and other users) remains.

To achieve online collaboration, technologies that offer tools to support the key components of social interaction and group working; namely generation (ideas/options), negotiation, choice, execution (McGrath, 1984), are required. Hence, tools that mediate understanding and support negotiation amongst participants, as well as tools for spatial data creation, manipulation and visualization are essential components of a truly immersive geocollaborative technology (MacEachren and Brewer, 2004). To this end, recent developments in online computer gaming are particularly relevant (Hudson-Smith and Crooks, 2008), with the technologies behind 3D virtual worlds representing a Web 2.0 frontier in which both the visual and social aspects of multi-user collaboration are supported (Mount *et al.*, 2009).

3D virtual worlds are social and visual three-dimensional online environments and, as with mirror worlds, they support advanced visualization of the 3D space afforded to the virtual environment (Minocha and Mount, 2009). However, unlike mirror worlds the visualization is experienced in the second person through a virtual representation of the user known as an avatar, and crucially the avatar has a real-time presence within the virtual environment itself and is able to interact with the entities in the environment. Multiple users are also supported with the result that avatars are able to see one another and interact with each other and the environment from within the virtual world in real-time. Integrated in the virtual world API are tools for instant messaging based communication among avatars, with some virtual worlds also offering voice-based chat. Arguably, then, virtual worlds offer far more of the tools necessary for social interaction, group working and geocollaboration than has existed previously and take us closer to the conceptual framework of geocollaboration technology described by MacEachren and Brewer (2004). Moreover, cases of professional scientific collaboration from within 3D virtual worlds are emerging (Huang *et al.*, 2008) together with technologies for the embedding of geographic information (Boulos *et al.*, 2008). Consequently, they are becoming recognized as one of the most exciting, distributed, synchronous computing environments for geocollaboration (Hudson-Smith and Crooks, 2008).

Recognition of the geocollaboration potential of virtual worlds has been paralleled by recognition that the integration of GIS data sets within them is crucial and to this end significant research efforts have been made. Forward-looking examples of GIS integration dating back more than a decade are well documented (e.g. Huang *et al.*, 2001). These early efforts are commonly centered on the conversion of 2.5D spatial data sets of extruded urban features to the virtual reality modeling language (VRML) for rendering (Dodge *et al.*, 1998) by compatible web-services and offer only a 'one-way' link between the GIS and the VRML with no means of feedback (Doyle *et al.*, 1998). From an analysis perspective, such early models generated a very limited spatial representation of geographical entities and very limited opportunities to do more than simply view rendered 3D representations. However, they did make it possible to quickly and efficiently convert GIS data to a format that could be integrated into emerging virtual worlds and distributed widely. They certainly contributed to GIS researcher's increasing interest in the integration opportunities that might emerge. However, early VRML approaches did little to support those wishing to integrate terrain data in virtual worlds and it was not until Active Worlds (www.activeworlds.com) supported the import of DEMs in a tiled fashion (a feature that is common to many contemporary virtual worlds) that significant progress was made here. However, the limitations of the proprietary 3D rendering engines used by many virtual worlds remains a key challenge to the successful integration of spatial data sets.

20.3 User creation of geographic information in Second Life

Second Life is a virtual world, with in excess of 15 million registered users. At any one time an average of 70,000 users are active within the world. It is maintained through a combination of client software and hosting servers with the client available as a free download to registered users. In accordance with the majority 3D virtual worlds users access the world through avatars in a way that is similar to various multiplayer online games and users are able to purchase digital space within which terrains can be created and objects built using the authoring tools, physics engine and object embedding capabilities provided by the client software.

Like the majority of 3D virtual worlds, Second Life supports much of the fundamental functionality of digital Earths with respect to its support for creating, viewing and sharing three-dimensional graphical entities in a virtual geographic space. However, it differs markedly in the ways in which the digital landscape, and entities that exist within it, are created and the extent to which notions and constructs of the real-world are allowed to constrain them. While users of mirror worlds can create terrains and geographical entities with considerable flexibility, constraints on what is possible occur due to the fundamental geodetic, cartographic and geographical principles that underpin the architecture of their APIs, and the baseline terrain and imagery provided to the user. These constraints are, of course, unsurprising given the primary purpose of a digital Earth is to replicate the geographic space of the real one.

By contrast, in Second Life traditional geographical constraints hold little sway. Spatial referencing is not constrained by the fixed three-dimensional figure of an oblate sphere, rather the entire world is created on a planar surface referenced not by proximity to a universally-recognized datum, but by the servers used to store the 3D entities which make up the Second Life world (Figure 20.1). There is, therefore, no horizon – rather perspective view is limited solely by the capabilities of the rendering engine coupled with the graphical capabilities of the hardware being used to access Second Life. Similarly, there is no true concept of diurnal cycles nor, indeed, a universal time. Rather, time is set by each individual user meaning that different users can visualize their environment under different illumination conditions concurrently.

The surface of Second Life is divided into squares, known as regions, each 256 m × 256 m and extending to 1000 m elevation. Within each region location is given relative to a set of x,y,z Cartesian coordinates with the origin located at the bottom left of region. Regions are identified by a textual name which refers the SL client software to the server on which a region's entities are stored rather than geographic location. Together the region name and x,y,z coordinates provide a SLURL (Second Life URL); a unique reference to space within the SL world. However, because regions are textually referenced by region name, rather than by a continuous spatial reference system, Tobler's Law (Tobler, 1970) breaks down at the region scale and traditional geographical relations between regions cease to exist. Indeed,

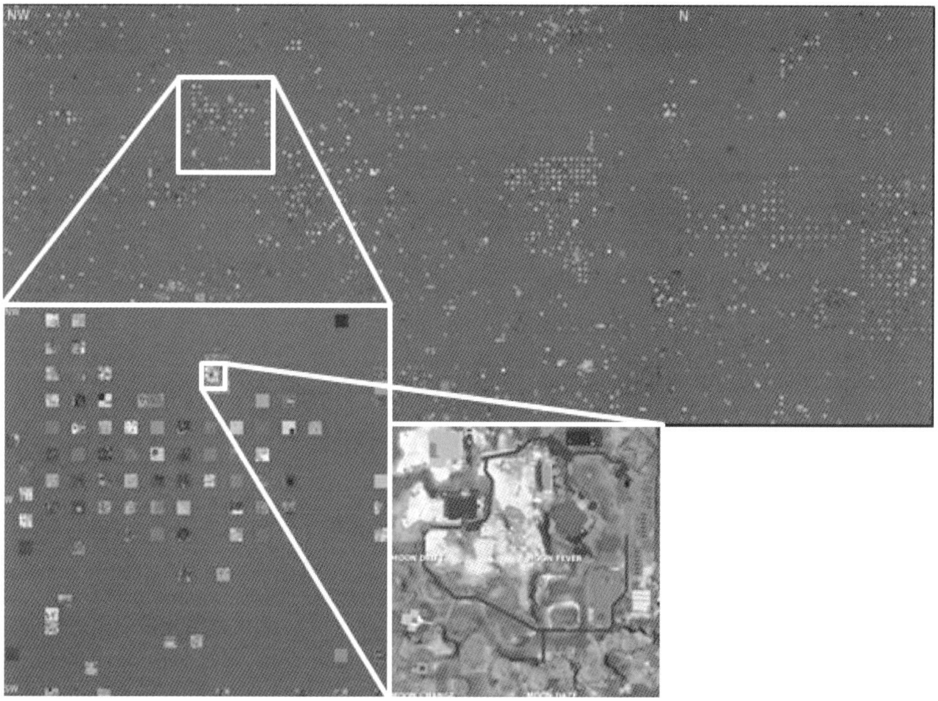

Figure 20.1 A subarea of the Second Life™ planar world at successive scales. Regions are visible as the myriad of small squares located within the blue 'ocean' of the world. MOONACY is an 8 sim estate based on the 4 seasons, featuring 16 music venues, owned by SL musician, Moondoggirl Moomintoog

the use of teleportation as the basic method of moving from region to region highlights how spatially disjunct Second Life can be. As a consequence, the majority of examples of geographic information in Second Life involve the projection of geo-information feeds onto in-world objects created to display and control them (see Boulos *et al.*, 2008 for an overview of existing technologies and examples). However, this approach to visualizing geographic data prevents the user's avatar from interacting directly with the data and restricts user's data manipulation options to only those allowed by the functionality of the display/control objects. The result is an environment for geocollaboration in which the key requirement of data manipulation can be highly constrained by the limitations of the techniques used to feed and project the geographic entities onto the 3D virtual world.

There are, however, opportunities to import geographical data directly into Second Life so long as the data remain within an individual region where the consistent Cartesian referencing system makes it possible to translate the spatial units of the geographic data from the real world to the virtual one. The process then becomes one of converting the spatial entities encoded in the geographic data into graphical primitives (known as prims) that can be rendered by the Second Life client, and the use of the associated attribute data to determine attributes of the prim (e.g. their colour, width, height etc.). Most objects within Second Life are composed of multiple prims and the construction of highly complex spaces can be achieved wholly within the in-world environment, or via import from other software packages. The development environment also allows interactivity to be built into objects

Figure 20.2 Spatially-referenced point data, extruded in the z-dimension to show population and integrated onto a projected base map of Sweden within Second Life™. Image © Binder Magnuz admin@magnuz-se.com

which can be useful in a geocollaboration context where certain actions, and their impacts, can be associated with in world objects.

One such example of this approach for point data is the Sweden 3D map of Magnuz (2008) (Figure 20.2) in which Swedish cities are represented as circular prims located on a base map, with the city population attribute used to assign the prim height; thereby producing a 3D graduate symbol map. This example highlights the relative ease with which point data can be integrated directly into the Second Life environment. However, the conversion of more complex spatial entities is limited by several factors including a maximum number of 255 prims that can be used in the creation of any object, the lack of an explicit encoding of the topological relationships among prims and the limitation of each prim's extents by the spatial extents of the region (256 m × 256 m × 1000 m). Consequently, examples in which large line and polygon data sets have been imported directly into Second Life from geographical information systems (GIS) remain elusive and the ability of the Second Life environment to support even basic GIS analytical functionality in-world is highly limited.

The one area in which the import of data from a GIS directly to Second Life has become more established is that of the import of continuous raster data as surfaces; especially digital terrains. This is due to Second Life's support for the creation of complex surfaces onto which other objects or image drapes can be placed. To achieve this, the properties of a variant of the standard prim called a sculpted prim (commonly known as a sculptie) have been exploited (Mount and Weaver, 2009). A sculptie is a three-dimensional mesh, no greater than 10 m × 10 m, whose surface morphology can be determined by image-based texture files and to which drapes can be added. The X,Y,Z coordinates of the mesh exist as a grid whose location is defined according to the spatial reference within the region in which it is to be located. The values ascribed to the X,Y,Z coordinates at each node in the mesh are defined

Table 20.1 Restrictions of the use of sculpties for rendering raster data and their impacts

Fundamental restrictions	Impact
Spatial Extent	A single sculptie cannot exceed 10 m × 10 m in area, although several, overlapping sculpties can be stitched together to make larger surfaces Mosaiced sculpties cannot exceed 256 m × 256 m × 1000 m Sculpties must be contained entirely within a single region
Resolution	A sculptie's resolution is restricted to 64 × 64 vertices and can be draped with an image that does not exceed 1024 × 1024 pixels The level of detail with which a sculptie is rendered depends on distance from the viewing location. Proximal to the avatar sculpties will render at 64 × 64 vertices, decreasing to 32 × 32 and 16 × 16 vertices with increasing distance Beyond about 100 m in SL space, the sculptie will not render
Interpolation	Interpolation between vertices uses a variation of the b-spline algorithm, with no additional options available
Mosaicing	Where multiple sculpties are used, an overlap of at least 1 vertex must be maintained to avoid artefacts in the rendered model

using RGB values from a supported image file, where the R value defines the location of a node's X component, the G value defines the Y component and the B values defines the Z component. Consequently, taking a raster file from a GIS to Second Life requires conversion of the raster to an RGB image coupled with any resampling necessary to ensure that the raster's spatial extents can be accommodated within the constraints imposed by both the sculptie and the Second Life region. The result is that any standard raster data layer from a GIS can be converted to a Second Life sculptie in theory, but a number of significant restrictions exist (Table 20.1). These restrictions mean that for raster layers whose spatial extents exceed the spatial extents of a single sculptie it will be necessary to mosaic several overlapping sculpties together; thereby necessitating the discretization of the raster data in the GIS prior to conversion to image files. Similarly, where mosaics of sculpties will result in surfaces whose spatial extents exceed those of a Second Life region (256 m × 256 m × 1000 m), a scaled representation of the data will be required; thereby necessitating data resampling in the GIS. It is clear then, that whilst theoretically possible, conversion of GIS-based raster data to Second Life sculpties is far from straightforward and this complexity is highlighted by Figure 20.3, in which a flow chart that documents the main steps involved in creating a digital terrain sculptie from a raster data layer is presented.

So, despite Second Life offering support for complex surfaces and image drapes it is problematic to create anything close to 1:1 scale representations of raster-based surfaces for anything but very small areas whose vertical extent does not exceed 1000 m. Even if large numbers of sculpties are stitched together, the constraints of the regional architecture make it difficult to build 1:1 terrains greater than 0.125 km^2 and for the majority of geocollaborative activities, this will prove unacceptable. Coupled with this, rendering restrictions mean that only proximal locations within the terrain are rendered, resulting in a highly restricted visualization in which perspective is highly compromised. Consequently, scaled terrains represent the most feasible solutions for geocollaborative activities, particularly where large digital surface models are required that can be rendered over large distances.

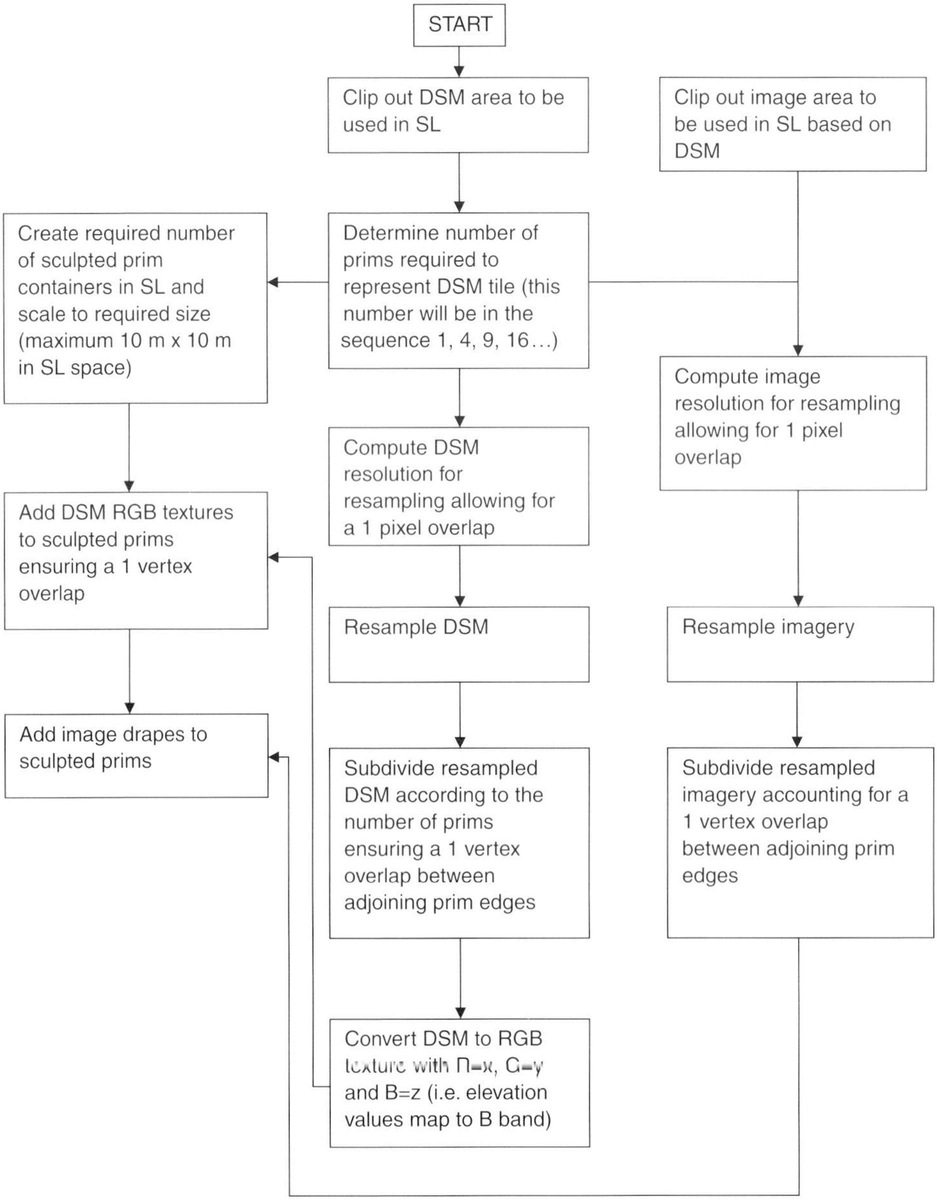

Figure 20.3 The workflow for converting a draped digital terrain model from a GIS to Second Life™

20.4 A teaching and learning example

There is little doubt that contemporary students of GIScience need to be made aware of the development in 3D Virtual Worlds such as Second Life. Indeed, degree and MSc programme curricula should include material on technologies for geocollaboration and future directions as these have been recognized as being of vital importance to the future of GIScience for in excess of five years (CIBGIIT, 2003). Where possible, this should include hands-on experience of integrating and interacting with geographic information from within virtual worlds as part of geocollaborative tasks because, as the geographical information science professionals of the future, current developments within 3D virtual worlds are likely to influence the ways in which they perform geocollaborative tasks during their careers. Indeed, courses that fail to encourage students to gain a general appreciation of the technical opportunities and limitations of the technologies and develop a critical view of their affordances, are arguably failing to prepare students for the changes in which they will participate as professionals.

As a part of the master of science programme in geographical information science at the University of Nottingham, UK, students undertake a geocollaborative learning task within Second Life centered on the development of a small wind farm within the English Lake District National Park. This learning task was originally developed as part of the JISC-funded Design of Learning Spaces in 3D Virtual Worlds (DELVE) project (Minocha *et al.*, 2009) and has since been integrated into the postgraduate programme in GIScience.

The task takes the format of a mock planning enquiry, held entirely within Second Life, to assess the impact of a set of proposed wind turbines in the Cumbrian landscape and the potential of a number of sites for development. The enquiry is undertaken in groups of three students plus a facilitating staff member, with each participant assigned the role of either a turbine developer, a wind farm opposition activist or a planning inspector according to the responsibilities outlined in Table 20.2. Participants are required to research and develop arguments to support their role prior to the planning enquiry with the basis for these arguments arising from the student's practical experiences of participating in GIS-based multi-criteria analyses for wind turbine location in other elements of the master's programme.

Prior to the enquiry, several possible locations for the wind turbines are identified using GIS-based site selection approaches (e.g. Baban and Parry, 2001; Ramachandra and Shruthi, 2005; Bishop and Miller, 2007) and planning policy guidance documentation (ODPM, 2004) and, for each potential site identified, scaled virtual turbines are placed into a photorealistic model of the Lake District terrain developed within Second Life, with the attributes outlined in Table 20.3 and according to the workflows described above. In many regards, the result is a Second Life equivalent of a Google Earth Geo-Mashup (Wood *et al.*, 2007), but with the major advantage that students can interact with the terrain data and image drapes, the turbine models and one another in-world and in real time (Figure 20.4). This allows participants to discuss and evaluate issues such as turbine visibility, dwelling proximity and site exposure in-world supported by the flexibility of moving turbines through the landscape and assessing alternative planning scenarios raised.

Each in-world planning enquiry is conducted synchronously over a period of 30 minutes with students located remotely from one another. Students are required to communicate their arguments to each another via Second Life's chat-based communication facilities, and to interact with and manipulate the terrain and wind turbine models to show the impacts of wind turbine development from their perspectives. Consequently, the spatial data display manipulation and interaction affordances of Second Life are coupled with its communication tools so that ideas and options for turbine placement can be presented and visualized, negotiated, chosen and executed through the placement of the turbine models within the digital terrain. In this respect, a truly geocollaborative teaching and learning exercise is experienced in which all of the major collaborative components cited by McGrath (1984)

Table 20.2 The roles and responsibilities of participants in the Second Life™ Lake District planning enquiry

Role	Responsibilities
Developer	• prepare by revisiting your wind farm choices from last learning task and determining how you intend to justify your wind farms • welcome the opposition activists and inspectorate when they arrive in Second Life and ensure introductions are made • explain your wind farm choices in terms of the criteria you identified as important and how these were used to arrive at your three options • provide a guided tour of the wind farm locations including some locations from which you would like other participants to visualise your wind farm • respond to any questions and queries participants may have • keep an eye on the time to ensure that your activities do not exceed 30 minutes • save and record the chat from your session
Opposition activist	• prepare by considering what the range of negative impacts a wind farm in the Lake District might impose • decide which of these will most impact upon Lake District residents • arrive in SL at the appointed SLURL at the appointed time and identify yourself to the developer • consider each wind farm suggested by the developer and identify which settlements might be most affected and the impacts that are likely for nearby residents • challenge the developer about the negative impacts you expect each wind farm to have on the residents and ask them to justify how these impacts will be minimised for residents
Planning Inspector	• prepare by considering the balance that needs to be struck between the protection of sensitive natural environments and the development of renewable wind energy, such that government targets of 20% renewable energy by 2010 may be met • arrive in SL at the appointed SLURL at the appointed time and identify yourself to the developer • listen to arguments and justifications offered by both developers and residents • request clarification and further explanation where discussions are unclear or where you need more information to evaluate the validity of the arguments being made • in light of the discussions and arguments, rank the three wind farms suggested by the developer according to the best (1) and worst (3) and explain this ranking to the participants • Indicate whether you consider either side has made a sufficient case to build/prevent building of the wind farm

Table 20.3 Summary specification of the in-world digital elevation model created to support the Lake District planning enquiry

Geographical extent of digital terrain model	30 km × 30 km
In-world extent of digital terrain model	0.06 km × 0.06 km
Scale of in-world digital terrain model	1:500
Number of mosaiced sculpties used	36
Sculptie mesh resolution (at maximum viewing resolution)	Approx. 140 m
Image drape resolution (at maximum viewing resolution)	Approx. 5 m

and MacEachren and Brewer (2004) are present. Moreover, students are also encouraged to supplement the Second Life environment with additional documentation and resources to support their role. To this end, large in-world posters (Huang *et al.*, 2008) outlining the GIS analytical procedures used (Figure 20.5) and in-world presentations highlighting environmental benefits and costs associated with the proposed turbines (Figure 20.6) supplement the geographical data.

Figure 20.4 Student avatars debating a turbine location scenario in Second Life™. It should be noted that the avatars appear to 'wade' through the landscape; a problem that relates to the way that Second Life's physics engine drives the interaction between the avatar and a terrain which has been imported without a suitable bounding box. Imagery supplied by Intermap Technologies Inc.

Figure 20.5 An example of a student-created poster outlining the methodological issues surrounding their multi-criteria analysis to identify potential wind turbine locations. Imagery supplied by Intermap Technologies Inc.

Figure 20.6 An in-world presentation highlighting negative impacts of wind turbine development. Imagery supplied by Intermap Technologies Inc.

20.5 Geocollaboration in Second Life: the learner perspective

The use, by educators, of 3D virtual worlds such as Second Life as teaching and learning tools in GIScience remains very much in its infancy and this, perhaps, reflects the extent to which the technology represents a current frontier in geocollaboration. However, the potential that Second Life offer is enormous; particularly in distance learning settings, or as a distributed tool for scenario testing in multi-user spatial decision making and group working or a tool for dissemination. But, early adoption of such a new technology in the classroom is fraught with potential difficulties for the learner who may have no prior knowledge or experience of 3D virtual worlds, and for the educator who may have little understanding of the specific challenges that the technology poses to the learners. For example, a learner's motivation to engage is an essential precursor for any learning activity (Keller and Suzuki, 2004; Keller, 1997), and the degree of success with which any geocollaborative learning activity within a 3D virtual world engages learners will depend partly on the different motivational factors underpinning that engagement (Yee, 2007). However, these factors may be poorly understood by learner and educator alike. Similarly, once engaged in the learning task, a range of factors will operate to promote or limit the geocollaborative experience of the learner; with these factors being associated with both the specific geocollaborative activity and the environment in which the activity is undertaken (Ritzema and Harris, 2008). Hence, it is important to evaluate these factors from the learner's perspective at the early adoption stage and incorporate the knowledge gained into future teaching and learning activities.

With this in mind, all students undertaking the Second Life learning activity described in the previous section also agreed to participate in a comprehensive evaluation of their learning experiences. A total of 18 students participated in the learning task, which was duplicated six times. A qualitative, inductive approach was then used to collect and analyse data relating to each learner's experience of the learning task, with a particular focus on geocollaboration. To this end, each duplicate of the learning task was followed by a semi-structured focus group conducted to elicit learner's perspectives on the key themes surrounding the opportunities and limitations of Second Life as a teaching and learning platform for geocollaboration and the factors underpinning these themes. The structure of the focus group prompts centred on three central topics:

1. The factors that affected the degree to which participants were able to engage in the geocollaboration.
2. The factors promoting the geocollaborative experience:
    a. The factors associated with the learning activity.
    b. The factors associated with the Second Life environment.
3. The factors limiting the geocollaborative experience:
    a. The factors associated with the learning activity.
    b. The factors associated with the Second Life environment.

An audio recording of each focus group was collected and transcribed and saved as a text file. The transcription was then split up into sections according to the specific question/prompt preceding a given block of text. The sections of text from the different focus group transcripts were then combined into separate files, with each file labelled according to the specific prompt to which the responses map. The result, therefore, was a set of files, each relating to a specific prompt and containing data of relevance to that prompt. Themes were then identified in all prompt files by two researchers working independently and quotes supporting each identified theme were coded in the transcripts. Themes from the researchers were then cross-referenced to ensure agreement before being tabulated.

Engagement in the in-world geocollaboration

Four themes emerged from the data that impacted the extent to which students were able to engage in the geocollaboration (Table 20.4). Unsurprisingly, the relevance of the task to each learner and the degree to which students had the necessary knowledge to fully participate in it were of particular importance. Indeed, this was one of the most strongly emerging themes and serves to highlight the extent to which the success of any geo-collaborative exercise hinges on engaging the right participants, both in terms of the background knowledge necessary to make valid and informed contributions to the collaboration, and those for whom the activity has greatest relevance. This knowledge may be subject-specific, but the theme also highlights that knowledge and preparation may also extend to those skills necessary to properly use the geocollaboration technology; in this case Second Life. In this regard, the effort needed to learn how to use Second Life for basic activities such as creating avatars, communicating via chat logs or moving spatial entities acted as a major barrier to many and meant that effective geocollaboration could only be achieved following extensive training in the Second Life environment. Even with several weeks of preparative training, several participants failed to engage in the geocollaboration fully due to their poor Second Life skills and this problem poses a major barrier to the wide-spread use of Second Life for geocollaboration. There is little doubt that part of the success of mirror worlds such as Google Earth, is the ease with which they can be accessed and operated and the fact that their effective use requires little training. Indeed, even the integration of user-created data is easily achieved. This is in contrast to the complexities of using Second Life which, while offering advanced geocollaborative functionality, requires high levels of skills to operate.

For many, and in accordance with the general findings of Ryan *et al.* (2006) and Yee (2007), a parallel was drawn between collaborating in Second Life and collaborating whilst playing multi-user computer games, and this parallel could be both motivating and demotivating. Certainly, those participants who were regular gamers were the most comfortable collaborating within Second Life and many have found the process of skills acquisition discussed above to be engaging; promoting their collaborative experience. However, for others Second Life lacked the opportunity for face-to-face communication that was considered essential for effective collaboration and this resulted in the virtual world being seen as a place to hide – arguably the complete antithesis of what a space for geocollaboration should be about. Similarly, the representation of oneself and others as avatars, and the necessary adoption of roles had both positive and negative effects. The willingness of individuals to project themselves onto virtual representations and roles is central to the effectiveness of avatar-based representation in a 3D virtual world, but this then replaces direct collaboration among individuals with a sense of being one step removed. For some, this increased their willingness to collaborate by promoting personal confidence, but it also comes with the risk that the collaboration becomes driven by roles and ideas founded in a virtual world

Table 20.4 Factors affecting the degree of engagement in the in-world geocollaboration

Theme	Key factors	Example quotes
Relevance and Knowledge	Relevance of activity	*'Yes, it's when you spend an hour like dressing the character, then you're thinking, what's this actually got to do with my course?'*
	Preparation required	*'But I guess like we obviously needed the first few sessions just to be able to understand what we were actually doing.'* *'Building objects, but I found that quite difficult, how to grasp the whole concept of how you can build. It's like you need to put quite a lot of effort into it.'*
	Preparation undertaken	*'I just think it was because the controls were new. We had only used them once before. It's like anything, if you used it a lot, it comes second nature.'* *'. . . Because obviously we haven't prepared like a great deal for it, then we don't really know what we're expecting.'*
Gaming Culture	Engagement with computer gaming	*'I just think it's weird. You're like combining learning with like kind of a computer game situation.'*
	Perceptions of geekiness	*'It seems a bit anti-social, it seems you are a bit like a geek in your bedroom and you can't go out and talk to real people.'* *'Don't hide in a virtual world.'*
Environmental Representation	Scale	*'It was hard to comprehend that turbines were so small and like and you talk about something that's so big and I know, if you change your logic it's easier to do, but if you can't it's tough.'*
	Accuracy and Level of Detail	*'Its limited by the detail that's been modeled. If all the towns were modelled accurately then you would be able to work much better.'* *'If the detail was much more life-like – if you could see the forest, if the trees were built, if the towns were built, if all the houses were built and you knew the area – then you would be able to go, 'That's that house, I can orientate myself over there.''*
	Believability	*'Having a tiny wind turbine and a massive person next to it is hard to believe.'*
Personal Representation	Avatars	*'I was sort of immersed in Second Life because I was a character in Second Life and the character was in the location and I was able to move the character around the data'*
	Roles	*'And then you would be thinking about the response to that as your character would be thinking about it, as role-playing a resident and that's when you'd let yourself think about just being in that world.'*
	Confidence	*'Because for someone who is shy as I am can't really talk to people that well. In Second Life you feel in a comfortable environment and you can say anything and it gives you time to think as well.'*

rather than the real one. This could impact on the extent to which the outcomes of the collaboration truly represent the views and actions of the participants and the extent to which the outcomes of activities in the virtual world are applicable in the real one.

The notion that certain actions and outcomes in the virtual world may not be applicable in the real world, and that this limits the potential of virtual worlds for geocollaboration is perhaps most clearly apparent in the environmental representation theme. Indeed, the limitations of Second Life, in respect of the lack of a proper spatial architecture and the extent to which spatial data can be integrated, have significant impact on the effectiveness with which the geocollaboration can be achieved because the visualization experience of the virtual world can be very different than that of the real one. Particularly significant was the need to use scale models of the terrain and turbines, which required participants to *'change their logic'* when interacting with the data and resulted in a lack of spatial detail and the perception of an inaccurate representation. It also resulted in the juxtaposition of avatars whose height far exceeded that of the turbines or, indeed, the terrain model and resulted in an experience that some found difficult to relate back to the real world. Similarly, restrictions on the level of detail that could be rendered meant that important spatial elements were missing and this made orientation difficult. This resulted in a visualization that at times lacked believability and for participants who found it difficult to suspend their disbelief, acted to reduce the extent of their engagement with the exercise.

In summary, one of the most essential aspects of any geocollaboration is the engagement of the participants in the collaboration and this is unlikely to be achieved if the interests of the participants and geocollaboration activity are poorly matched, irrespective of the environment supporting the activity. For many, Second Life is an engaging environment in which to collaborate and one in which the freedom to create avatars and develop and inhabit roles releases one from the constraints of the real world and promotes an enhanced desire to engage in collaborative activities. However, participants recognize a clear parallel between Second Life and computer games that can be both engaging and obstructive; the latter being particularly true when coupled with Second Life's complex controls and the resultant need for extensive skills acquisition prior to any geocollaboration. Second Life has its roots in a vision of a fantasy virtual world in which anything is possible and this flexibility is key to its potential for geocollaboration. However, it is also one of the factors limiting collaborator engagement as the lack of in-built support for spatial data can result in an environment that lacks believability and limits the extent to which participants perceive Second Life as a space capable of supporting formal, authoritative geocollaboration.

Factors promoting the geocollaboration experience

Whereas the above themes concentrated on the extent to which learners engaged with geocollaboration within Second Life, the following themes are focused more specifically on the factors and affordances of Second Life that promoted the geocollaboration experienced by the learners. To this end, five main themes emerged as being particularly supportive of geocollaboration (Table 20.5).

Geocollaboration is, by definition, a group-based experience and it is of little surprise therefore that the aspects of Second Life that support group interaction are highlighted as being important. Of particular interest in this respect was the ability of Second Life to alter power within the group by offering the same affordances and access to them to all

Table 20.5 Factors promoting the geocollaborative experience

Theme	Key factors	Example quotes
Group Working	Presence of others	*'I was engaged in the activities because I was communicating with everybody, so if the people weren't there, then I wouldn't probably be as engaged in the activity if it was just the landscape then.'* *'I think that the Second Life environment might be helpful with the people coming into contact with each other.'* *'You were interacting with other people. And by interacting with other people you felt, well I didn't feel I was in a class room at all.'*
	Issue identification	*'You could try different things with regard to this project. You could try putting it in a SSSI and see who gets annoyed with it. And you could experiment with getting participation out of people to see what they think about it. So it helps in that way.'* *'And like if there was something you missed, someone else would probably show it up, so it is sort of like, a safety net maybe.'*
	Decision making	*'And you can all make judgements whether that's important or not, and do it together.'* *'. . . and we were talking about the effects on the landscape on animals, habitats, those kind of things and how the accesses to power grids and to roads and how local communities are concerned about that and locations.'*
	Power differentials	*'I was asked proper questions, it wasn't simple questions, from higher people and I was giving the information, the knowledge, I knew about it.'* *'In Second Life I felt like I didn't have to be nervous, or I didn't always have to have that, you know, that look that formal look, I didn't have that worry, I can just say it how I want to say it.'*
Communication	Synchronous chat	*'I felt immersed due to the conversation, the debate which we were having between each other.'* *'Being able to chat with other people in, like, real-time made it possible to collaborate far more effectively.'*
	Speed of response	*'They could probably send me a report and say, 'This is what I think you should provide do', but in Second Life straight away I could tell them and it makes it much quicker.'*
	Logged chat	*'If you weren't sure what had been said, you could go back through the chat record and check.'*
Information Integration	Information integration	*'Being able to do everything in one place, I guess, because I could show the poster that we did, explain it and then show the locations and why they were chosen and their relationship to the landscape, all in one environment.'* *'If you were doing it in real life you would have to give a presentation in Power Point and then possibly go into the Visualisation Lab to do something else and then a discussion face-to-face.'*

(*continued*)

Table 20.5 (*Continued*)

Theme	Key factors	Example quotes
Information Integration	Information presentation	*'I put a presentation on screen actually into Second Life and attached images and I said, 'You know, you want to put a wind turbine on here?' As in, you know, 'You want to destroy this beautiful landscape? Destruction to nature?'*
	Interaction with in-world artefacts	*'I think probably like it all comes down to interaction, and the more someone interacts with the data the more they will get from it.'* *'It's really hard to explain but, compared to the working off paper that we used to do, assessing the criteria was much better. I mean asking questions and interacting with the terrain and stuff like that, it makes it completely different.'*
Scenario Testing	Test-bed	*'If you like wanted to refer back to anything its really, really easy to go and look at it from different perspectives'.* *'If there is a major flaw that you have missed, like there's a turbine right on the cusp of the buffer, you could notice that in Second Life. Whereas, show them a picture from a map, you're not really going to notice.'* *'[It gave us the opportunity to] test whether the sites chosen were sufficient with the real world.'*
Ownership	Ownership of artefacts	*'It was almost like I'd actually done a professional job with the windmills in Cumbria.'* *'So like you can give someone a book and say, 'Write an essay on this', and you're not going to be attracted are you? But if you have got your own, it gives you much more of an interest because it's just something that, instantly like, is personalised to you.'*

participants. Indeed, it is the equality of the other users within the Second Life environment and the ability to interact and communicate equally with them that makes the geocollaboration productive. However, the participants highlighted that the interactions among avatars in Second Life went beyond simple descriptions and explanations and involved debate to identify issues and the evaluation of conflicting ideas and opinions in effective group decision making. This indicates that the communication affordances of Second Life are sufficient to support the complex decision-making processes that are fundamental to many geocollaboration tasks. Indeed, even though Second Life communication extends to little more than a logged chat screen (effectively an internet messenger), users became immersed in the discussions and appreciated the speed with which individual's chat responses were relayed to the collaborators. The ability to review and recap the logged chat also offered communication functionality that is missing in most face-to-face encounters, and this promoted effective communication through the ability to seek clarification of ideas and challenge situations in which collaborator's interpretations of other's opinions was incorrect.

While the Second Life environment was used as the basis for displaying spatial data, it was also exploited by the participants as a way of communicating geographical information through large format posters and in-world presentations that acted as an effective and efficient method of outlining ideas to other collaborators. This enhanced the flexibility with which information could be exchanged and offered central discussion points to direct group-based decision making. Indeed, the integration or the spatial data and these other in-world artefacts was recognized as creating a flexible environment for geocollaboration in which *'everything could be done in one place'*. For many, the ability to create one's own artifacts to be used by other collaborators added significantly to the extent to which they developed a sense of ownership over the collaborative exercise; something which almost certainly promoted overall engagement in the exercise as a whole.

In summary, the combination of effective communication, data and information integration in Second Life led to it being recognized as an effective environment for scenario-testing in which the impacts of decisions could be reviewed and new actions proposed as a result. Indeed, evidence from the participants exists to show that Second Life was used in the generation of ideas and options, the negotiation of different views, the justification and trialing of different choices, and their execution in the virtual environment. In other words, all of McGrath's (1984) components of social interaction and group working were supported and integrated with spatial datasets and information, resulting in a highly promising geocollaboration environment.

Factors limiting the geocollaboration experience

Factors limiting the geocollaborative experience are outlined in Table 20.6. In line with the previously-discussed issues that acted to restrict learner engagement, the learning curve was frequently identified as a factor that limited the geocollaboration experience of many. While this is perhaps expected, it was more surprising to find communication and the group – both themes recognized as promoting the geocollaboration – being raised as limiting factors too. The communication affordances of Second Life identified as being so positive were those associated with relatively simple text-based communication technology, and it was clear that participants were able to use these for effective collaboration. However, participants recognized the limitations of text-based communication and the absence of the more subtle aspects of communication that may be important in face-to-face interaction, but are largely absent in text-based chat. These include physical and emotional expressions which, although supported to some extent in Second Life, are difficult to use effectively and are quite restricted. The result is a communication experience that lacks the depth of a face-to-face encounter and can become a rather clinical experience (although there may be benefits in terms of the clarity of the communication that results). Communication was also limited by the need to type – something which could be distracting and, where several collaborators were communicating simultaneously, could result in disorganized and unstructured chat logs that were difficult to follow and led to misunderstandings and difficulties in appreciating the views of the participants. Perhaps the most surprising factor to emerge was the fact that Second Life made it easy to avoid communication altogether by simply not joining in the chat discussion; something that is much harder to do in a face-to-face setting.

There is little doubt that Second Life promotes a sense of group working through the existence of avatars in the virtual world, however it is possible for anyone to have a Second Life avatar and it is not always easy to gain detailed knowledge of the real person behind the avatar. As a result, some participants were wary of other avatars and identified the lack of knowledge about the people behind them leading to a lack of trust in their opinions. In

Table 20.6 Factors limiting the geocollaborative experience

Theme	Key factors	Example quotes
The Learning Curve	Steepness of learning curve	*'It's kind of like, well it is like make believe, but then it's just weird how you have got to go through a learning process to even start to use it. Sometimes it was quite difficult to use and you just like sit there for half-an-hour just not knowing what you are doing or just doing the wrong thing.'* *'It was quite difficult at first to actually learn how to use Second Life. It was quite difficult I thought, compared to normal computer games.'* *'I've got quite a lot of experience with computers, not with these sort of environments, but with games and things, and I found it really difficult to use.'*
The Group	Knowing one collaborators	*'You don't know who the audience is. It could be a politician, it could be an anti-wind farm protestor, so you can't tailor what you are saying to them. Yes, in a normal presentation situation, you would have a pretty good idea who your audience was and in Second Life you wouldn't necessarily.'*
	Trusting collaborator's knowledge	*'You could easily over or underestimate the level of understanding that whoever you are talking has [in Second Life], whereas when I am talking to you I know that you are hopefully going to understand what I am saying.'*
Interfacing and Navigating the Virtual World	Control complexity	*'I've only used it once, but I thought it was a bit hard to be honest because we just couldn't get the controls right.'* *'We had to spend a long time getting used to how to use the controls and I still wasn't that good when we did it for real'*
	Navigating	*'When we were sort of discussing where to put the wind turbines, it was really difficult to communicate to figure out where to locate them and I had to actually physically move around to the other side of the dock and talk to my group and then point on the screen to where the wind turbines should be.'* *'I found it really difficult to know exactly where I was on the model. I needed a sort of North arrow to say, this is North. I mean Second Life has got some sort of arrows, North arrows and things, but I found that really difficult to understand.'*
Communication	Physical/ emotional expression	*'Its difficult to express yourself through the emotions that you display, as in you shrug, you nod your head and so forth. Trying to combine the two can be sometimes difficult with typing at the same time.'*
	Typing skills	*'The keyboard and the mouse itself is a limitation.'* *'Obviously typing is a lot more difficult than speaking. I can type pretty well so I'm ok with that. If you weren't so good, it would be quite frustrating I would imagine to try and communicate that way.'*

Table 20.6 *(Continued)*

Theme	Key factors	Example quotes
Communication	Organisation	*'The biggest limitation is obviously the chat procedure log. Not being able to control who's talking when and stuff, so when you're typing extensive information, you can even miss what someone says.'*
		'It was really difficult to know what everyone had said and reply to everyone and I was trying to explain how I'd went about locating the turbines and all the sort of criteria we had went through and it kept interrupting me and then I got confused and stuff. So that was a bit of a disadvantage.'
	Communication avoidance	*'There were bits in which I dropped out in which when you didn't ask the questions I felt, this just isn't working, you know. I'm not, it sounds a bit selfish but I'm not getting the attention which this debate deserves, you know. I'm here to have a debate and it's not working.'*
		'It was easy to dodge questions if you didn't know the odd one where you just went moving on . . .'

a geocollaboration this clearly has significant importance in terms of one's ability to weigh up the views and information being presented by other collaborators; especially as Second Life promotes equality in terms of one's ability to access and influence the debate.

Criticisms of Second Life were also directed at more spatial aspects of the experience and, in particular, the difficulties associated with navigating the in-world environment. The lack of a clear spatial referencing system made it very difficult to discuss detailed spatial issues – especially where comparative locations were central to the collaboration. Indeed, in some cases participants resorted to a physical, face-to-face discussion when facing detailed spatial issues. This certainly serves to reinforce the lack of a proper spatial framework (both in terms of referencing and semantically) within Second Life which is arguably one of the biggest challenges facing its use for geocollaboration.

20.6 Conclusion

Technological futures

There is little doubt that geocollaboration has moved out of the realm of the specialist GIS professional and into the hands of the general public and that Web 2.0 technologies have been central in the development of technological platforms for geocollaboration. Consequently, some training in geocollaboration technologies should be at the heart of all higher education programmes in spatial science and with 3D virtual worlds existing at the forefront of geocollaboration technology, the use of platforms such as Second Life within the classroom should be seriously considered. However, change is on the horizon and it can be expected that the future of the web will almost certainly mirror developments in the 3D visualization capabilities of film, television and mobile device displays – arguably resulting in the emergence of Web 3.0. The impact of such developments on virtual worlds is potentially significant. To date, they have been dominated by a client-server architecture that has resulted in a plethora of proprietary graphics and physics engines with very differing protocols and varied capabilities for the import and display of geographical information. Consequently, the import of GIS data into different virtual worlds involves workflows that cannot be easily transferred between virtual world clients and the protocols that they use. Some open source attempts at rectifying this have emerged (e.g. OpenSim (www.opensimulator.org)) and these have made it possible to create unique, user specified 3D content via a multi-platform, multi-user 3D application server which supports the protocols of the majority of virtual world clients. However, such toolkits require a high degree of programming and software development skill sets that are not always within the capabilities of many GIS educators. Consequently, they are unlikely to become the mainstream approach for integrating GIS data in virtual worlds for educational purposes. Of greater potential are the emerging 3D graphics standards such as Mozilla's WebGL which are set to become supported as standard by many web browsers (including Google Chrome, Safari and Firefox). If widely adopted and supported by virtual worlds, these have the potential to replace the need for the installation of specific clients and to make the between-world transfer of 3D content simple. For those wishing to integrate their GIS data into the virtual worlds of the future, the technical challenge is likely to become one of format conversion into emerging Web 3.0 standards. As these standards become established and in wide-spread use throughout the web, it would seem reasonable to expect a significant proportion of GIS educators to adopt them. However, the built-in level of support for the specific data models and affordances of geographic information that will be provided by the new generation of 3D web standards remains to be seen and this will be a crucial determinant of their geocollaborative value.

Challenges for the GIS educator

It is clear from the experiments reported in this chapter that Second Life has genuine potential as a geocollaborative platform, but that several important barriers to its enthusiastic uptake by GIS educators and students alike exist. Most importantly, Second Life has not been designed with spatial data or the GIScientist in mind and this results in significant difficulties associated with integrating spatial data into the environment, as well as limits to the type and extent of the geocollaboration that can be undertaken. Each spatial data format will require its own specific conversion workflows before it is able to be integrated into Second Life, and there is little in the way of generic conversion tools available to assist the educator in this task. One approach is to utilize the fundamental geometric affordances of the basic prim to create simple, incremental spatial learning objects (e.g. spherical, cylindrical and conical prims to show the impact of map projection (DeMers, 2010)). However, real spatial data sets are seldom sufficiently geometrically constrained to apply this approach. Alternatively, one can utilize open-source 3D modeling tools for GIS data conversion and, whilst this increases the flexibility with which GIS content can be integrated into Second Life, it requires the educator to gain yet more specialist 3D content creation skills. Indeed, it is difficult to see how most GIS educators, for whom geocollaboration in Second Life is likely to represent only a very limited element of their teaching activities, would justify the upfront time requirements for 3D content creation. However, the particular geocollaboration affordances of Second Life make it an excellent platform for supporting networks of

GIS educators and students who develop, create and share content and learning materials freely with one another and, in so doing, reduce the individual burden of content creation.

The extensive learning curve is not solely a concern for the educator, and the student will almost certainly face a significant learning curve before any valuable geocollaboration can be experienced in Second Life. This means that learners must also be prepared to commit significant resourcing and time to the learning activities that make use of the technology. This commitment will be difficult to sustain where the relevance of the learning activities is unclear to the learners, or where the nature of the geocollaboration required of the learners is unsupported by the affordances of Second Life. Similarly, it will be difficult to justify unless it can be shown that the use of Second Life is central to a significant block of learning activities beyond its use in occasional, isolated learning activities. Therefore, both of these aspects should be at the forefront of any GIS educator's mind when considering using Second Life as a geocollaborative teaching and learning platform.

The present difficulties associated with GIS data integration in Second Life, the steepness of the learning curve associated with gaining proficiency of interaction in the virtual world for both educators and students, and the limited communication affordances makes it difficult to recommend Second Life as an essential geocollaborative learning environment in GIScience programmes. In the short term it is most likely to remain a niche educational tool. However, the emergence of Web 3.0 has the potential to embed virtual world technologies in the everyday online experience of the future, and this is likely to be paralleled by greater uptake of virtual worlds across educational institutions and programmes. It also has the potential to make the process of integrating GIS data in virtual worlds far easier and more flexible. Together, these factors have the potential to make geocollaboration in virtual worlds a far more common component of GIScience education programmes of the future.

Acknowledgment

This work was undertaken within the DELVE project, funded by a Joint Information Systems Committee (JISC) Teaching and Learning Innovation Grant. The authors are particularly grateful to Dan Weaver for his efforts in the development of the Second Life terrain used in the Second Life experiment.

References

Baban, S. M. J., and T. Parry (2001) Developing and applying a GIS-assisted approach to locating wind farms in the UK. *Renewable Energy*, 24: 59–71.

Bishop, I. D., and D. R., Miller (2007) Visual assessment of offshore wind turbines: The influence of distance, contrast, movement and social variables. *Renewable Energy*, 32: 814–831.

Boulos, M. N., M. Scotch, K.-H. Cheung, and D. Burden (2008) Web GIS in practice VI: A demo playlist of geo-mashups for public health noegeographers. *International Journal of Health Geographics*, 7(38): 16.

CIBGIIT (2003) *IT Roadmap to a Geospatial Future*. Washington DC: National Academies Press.

DeMers, M. (2010) Second Life as a surrogate for experiential learning. *International Journal of Virtual Personal Learning Environments*, 1(2): 17–31.

Dodge, M., S. Doyle, A. Smith, and S. Fleetwood (1998) Towards the virtual city: VR and internet GIS for urban planning. *Proceedings of Virtual Reality and GIS*. London: Birkbeck College, 22 May, 1998.

Doyle, S., M. Dodge, and A. Smith (1998) The potential of web-based mapping and virtual reality technologies for modelling urban environments. *Computers, Environment and Urban Systems*, 22: 137–155.

Eisnor, D. (2006) What is neogeography anyway? (Blog, Dated May 27, 2006), http://platial.typepad.com/news/2006/05/what_is_neogeog.html (accessed June 11, 2009).

Gelernter, D. (1991) *Mirror Worlds: The Day Software Puts the Universe in a Shoebox... How Will it Happen and What Will it Mean?* New York: Oxford University Press.

Goodchild, M. F. (2007) Citizens as voluntary sensors: Spatial data infrastructure in the world of Web 2.0. *International Journal of Spatial Infrastructures Research*, 2: 24–32.

Haklay, M., A. Singleton, and C. Parker (2008) Web mapping 2.0: The neogeography of the GeoWeb. *Geography Compass*, 2: 2011–2039.

Huang, S. T., M. N. Boulous, and R. P. Dellavalle (2008) Scientific discourse 2.0: Will your next poster session be in Second Life? *European Molecular Biology Organisation Reports*, 9: 496–499.

Huang, B., B. Jiang, and H. Li (2001) An integration of GIS, virtual reality and the internet for visualisation, analysis and exploration of spatial data. *International Journal of Geographical Information Science*, 15: 439–456.

Hudson-Smith, A., and A. Crooks (2008) The renaissance of geographic information: Neogeography, gaming and Second Life. *University College London Centre for Advanced Spatial Analysis Working Paper*, 142: 1 15.

Keller, J. M. (1997) Motivational design and multimedia: Beyond the novelty effect. *Strategic Human Resource Development Review*, 1: 188–203.

Keller, J. M., and K. Suzuki (2004) Learner motivation and e-learning design: A multinationally validated process. *Journal of Educational Media*, 29: 229–239.

Kingston, R., S. Carver, A. Evans, and I. Turton (2000) Web-based public participation geographical information systems: An aid to local environmental decision making. *Computers, Environment and Urban Systems*, 24: 109–125.

McGrath, J. E. (1984) *Groups: Interaction and Performance*. Englewood Cliffs, NJ: Prentice-Hall.

MacEachren, A. M. (2000) Cartography and GIS: Facilitating collaboration. *Progress in Human Geography*, 22: 445–456.

MacEachren, A. M. (2001) Cartography and GIS: Extending collaborative tools to support virtual teams, *Progress in Physical Geography*, 25: 431–444.

MacEachren, A. M., and I. Brewer (2004) Developing and conceptual framework for visually-enabled geocollaboration. *Journal of Geographical Information Science*, 18: 1–34.

Magnuz (2008) 3D map of Sweden. *Second Life SLURL: Badger 56,56,402*. (accessed November 12, 2009).

Minocha, S., and N. J. Mount (2009) *Design of Learning Spaces in 3D Multi-user Virtual Environments*. Joint Information Systems Committee (JISC) Report on the DELVE Project. London: JISC.

Mount, N. J., C. Chambers, D. Weaver, and G. Priestnall (2009) Learner immersion engagement in the 3D virtual world: Principles emerging from the DELVE project. *Italics*, 8: 1–15.

Mount, N. J., and D. Weaver (2009) Wading through Derwent Water: Taking terrains from the real world to Second Life. *Proceedings of 17th Annual Geographic Information Science UK (GISRUK) Conference, University of Durham*, pages 275–284.

ODPM (2004) *Planning Policy Statement 22: Renewable Energy*. Norwich: HMSO.

O'Reilly, T. (2005) What is the Web 2.0? Design patterns and business models for the next generation of software. http://oreilly.com/web2/archive/what-is-web-20.html (accessed June 11, 2009).

Olson, G. M., D. E. Atkins, R. Clauer, T. A. Finholt, F. Jahanian, T. L. Killeen, A. Prakash, and T. Weymouth (1998) The upper atmosphere research collaborator. *Interactions*, May/June: 48–55.

Plale, B., G. Eisenhauer, K. Schwan, J. Heiner, V. Martin, and J. Vetter (1998) From interactive applications to distributed laboratories. *IEEE Concurrency*, April/June: 78–90.

Ramachandra, T. V., and B. V. Shruthi (2005) Wind energy potential mapping in Karnataka, India, using GIS, *Energy Conversion and Management*, 46: 1561–1578.

Rinner, C., C. Keβler, and S. Andrulis (2008) The use of Web 2.0 concepts to support deliberation in spatial decision making. *Computers, Environments and Urban Systems*, 32: 386–395.

Ritzema, T., and B. Harris (2008) The use of Second Life in distance education. *Journal of Computing Sciences in Colleges*, 23(6): 110–116.

Ryan, R. M., C. S. Bigby, and A. Przybylski (2006) The motivational pull of video games: A self-determination theory approach. *Motivation and Emotion*, 30: 347–363.

Tobler, W. (1970) A computer movie simulating urban growth in the Detroit region. *Economic Geography*, 46: 234–240.

Wood, J., H. Wright, and K. Brodie (1997) Collaborative visualisation. *Proceedings IEEE Information Visualisation '97*, Phoenix, Arizona, pages 253–260.

Wood, J., J. Dykes, A. Slingsby, and K. Clarke (2007) Interactive visual exploration of a large spatio-temporal dataset: Reflections of a geovisualisation mashup. *IEEE Transactions on Visualization and Computer Graphics*, 13: 1176–1183.

Wheless, G. H., C. M. Lascara, A. Valle-Levinson, D. P. Brutzman, W. L. Hibbard, B. Paul, and W. Sherman (1996) The Chesapeake Bay virtual ecosystem: Initial results from the prototypical system. *International Journal of Supercomputer Applications and High Performance Computing*, 10: 199–210.

Yee, N. (2007) Motivations of play in online games. *Journal of CyberPsychology and Behaviour*, 9: 772–775.

21

From location-based services to location-based learning: challenges and opportunities for higher education

David M. Mountain

School of Informatics, City University, Northampton Square, London, UK

21.1 Introduction

This chapter will consider the impact the relatively new discipline of Location-Based Services (LBS) has had on the teaching of Geographic Information Science and Technology (GIS&T) in higher education. In line with the overarching ambition of this book, it will attempt to place historic activity in context, report on current developments, assess where these current trends may be taking us, and highlight best practice for the teaching and use of LBS.

Prior to offering some definitions of LBS provided from diverse perspectives, we offer a discussion about why it is important to teach LBS and what is challenging in its teaching. The impact of LBS is interesting because, as well as being a subject that falls within the GIS&T curriculum, they are also receiving attention as pedagogic tools in their own right as part of the move towards mobile and blended learning (Traxler, 2009). For this reason, this chapter will not only consider LBS as an object of study, but also how LBS can be used as a tool to *enable* study: specifically how LBS fit within the theory of mobile and blended learning, and the opportunities and challenges these new tools present for higher education.

Why teach LBS?

Perhaps the over-riding reason for teaching LBS is the potential number of users of this technology compared with Geographic Information Systems (GIS). LBS are increasingly familiar: much has been written about them in academic journals (Raper *et al.*, 2007a; Raper *et al.*, 2007b) and the mainstream media (Cellan-Jones, 2008; Keegan, 2010) over

Teaching Geographic Information Science and Technology in Higher Education, First Edition.
Edited by David J. Unwin, Kenneth E. Foote, Nicholas J. Tate and David DiBiase.

the past ten years. This initial research and development has now translated into the roll-out and commercial take-up of thousands of LBS for diverse mobile platforms including the iPhone, Google Android and Symbian. At the time of writing, the global number of LBS subscribers was on course to meet predictions of growth from 7 million to nearly 300 million between 2006 and 2011 (Zimmerman, 2008). GIS are characterized as sophisticated, complex and often costly systems designed for use by small numbers of professional users. In contrast, LBS tend to be more specific, lightweight, user-friendly and relatively low-cost applications developed for a non-expert audience (Virrantaus *et al.*, 2001). As such, LBS are a channel through which a global audience with limited technological skills and experience can interact with geographic data and functionality, whose computational complexity they need not understand so long as it provides them with pertinent information in response to their needs. Most people can understand the concept of a 'buddy tracking' LBS that alerts you when trusted friends are within five minutes walk away, but it takes GIS expertise to appreciate the sophistication of the spatial analysis required to develop such an application. The main barrier to use is access to a mobile device with sufficient processing power to run LBS applications. An analogous trend was seen in the early 2000s, when web-based mapping applications, built on application programming interfaces (API) from the likes of Google, Microsoft and Yahoo brought spatial data and limited GIS functionality to anyone with access to a computer and the internet. It is necessary for students of GIS&T to understand this new channel if they are to exploit its potential in terms of disseminating spatial information and tools, and by the analysis of datasets generated by its users.

As of 2011, relatively little specific to the *teaching* of LBS has been written in the academic literature. In attempting to define the current domain of GIS&T, the 2006 *Body of Knowledge* report (DiBiase *et al.*, 2006) mentions LBS only briefly, as an application of positioning systems and their potential for field data collection. When considering future trends the report acknowledges that LBS are creating a mass market for GIS&T, which may extend its scope beyond the current audience of specialist practitioners. The teaching of technologies related to LBS has received more attention, particularly positioning systems, and wireless communication networks (DiBiase *et al.*, 2006).

There are many interrelated issues that combine to make the teaching of LBS challenging. LBS covers diverse subject matter requiring the consideration of technological aspects such as hardware components, software development, economic impacts and business models, and the handling of geographic information as well as social issues, such as those related to ethical use and privacy. Communicating the central theme of LBS – retrieving information that is relevant to one's location – tends not to translate well to the classroom where it can be difficult to appreciate the impact of the local environment and context on information needs. A further challenge is that for the last decade, LBS have been a discipline in a state of rapid and constant change. The infrastructure upon which LBS are built has evolved almost beyond recognition over this period. This requires the instructor to identify enduring themes and analytical frameworks that will remain useful to students as implementations of LBS continue to develop.

What are LBS?

Definitions of LBS are diverse and evolving. The definition from Brimicombe and Li (2006), 'the delivery of data and information services where the content of those services is

Figure 21.1 Map-based LBS on three different platforms. From left: Android, iPhone, Windows Mobile

customized to the current or some projected location and context of the user', conveys the key theme, which is the *in situ* provision of information relevant to the local environment. Many definitions are keen to distinguish LBS as *mobile* services and highlight how they differ from desktop computing. These characteristics include being nomadic, ubiquitous and personal (Alexander, 2004), a unifying theme being that access to information is no longer tied to a static personal computer.

LBS still tend to evoke images of two-dimensional map representations for the display of spatial information on portable devices (see Figure 21.1). Examples include backdrop maps of your surroundings (*where am I?*) and displaying local information (*what's around me?*). However, increasingly the use of location is more subtle, with less emphasis on the map itself than the retrieval of geographically relevant information (Raper, 2007). This is achieved in various ways. For most local searches conducted on mobile devices, it is the content, rather than its explicit spatial representation, that people are interested in (see Figure 21.2). When requesting local news, a map is of less importance than the news story itself. LBS encourage an egocentric perspective of the world where information is arranged relative to your location (Mountain and Liarokapis, 2007). Many LBS have accounted for this by replacing the map with augmented reality (AR) interfaces. Rather than overlaying spatial information on a backdrop of map data, with AR the physical world itself is the backdrop, which is annotated by the screen of a mobile device. Location can also be seen as an important component of an individual's *context*, the information that can help to characterize their situation. Context considers not only the physical environment but also human factors such as the characteristics of an individual, the activities they are currently engaged in, and their social environment (Jiang and Yao, 2006).

a. Venues from FourSquare | b. Travel information from UKTravelOptions | c. Local News Feed from Garv Technologies

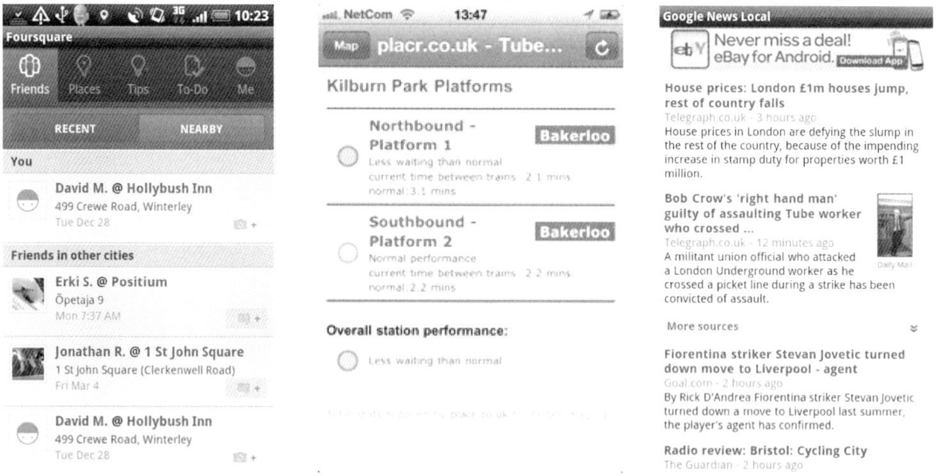

Figure 21.2 LBS serving local information

LBS perspectives

As a result of the large potential audience of LBS, and the diversity of the services that fall into this category, there are many approaches to teaching LBS. It is therefore useful to be able to analyse them from alternative perspectives fully to understand their scope. The following four perspectives are some of the most commonly adopted, either explicitly or implicitly, in LBS teaching and reflect the course learning outcomes and background experience of the cohort and teaching staff.

Technology infrastructure perspective

Steiniger *et al.* (2006) identify five technological components that comprise the infrastructure for LBS. A key pedagogic benefit of adopting a technology perspective, particularly as an introduction to the subject, is that it provides a more tangible description of the component parts that together comprise LBS. Different cohorts of students are likely to have different experiences of each of these five components and when considering the detail in which each component should be considered the instructor should be aware of the module aims and intended learning outcomes. The first component refers to the *mobile devices* with which users interact with LBS and their characteristics in terms of size, display, processing power, and storage. At one end of the spectrum the mobile device may be an entry-level mobile phone with a small screen and basic web browser. Increasing in sophistication, the device might be a 'smartphone' such as the Apple iPhone. Portable laptops, though bulky, may also be classed as mobile devices appropriate for LBS. The second component is the *communication network*, including cellular networks operated by mobile telecommunications companies and WiFi access points that provide a gateway to the internet (Benford *et al.*, 2005). The third component is the *position-determining technology* used to locate the mobile device. The widespread take-up of global navigation satellite systems (GNSS) such as the US Global Positioning System (GPS) in *sat-nav* devices and mobile phones has resulted in more people having a basic understanding of this component so that the curriculum can focus more on the limitations of satellite-based solutions and alternative positioning systems such as cellular and WiFi networks based on communications infrastructures. The fourth component is the *provider of the information* used by the LBS, which includes 'points of interest' databases and backdrop mapping data. The fifth and final component is the *location-based service providers* themselves, who develop application logic capable of integrating positioning information and geographic data, manage the flow of data over communication networks, and deliver the locally relevant information.

Services perspective

An alternative ontology sees LBS subdivided into discrete service categories, handling distinct functionality with which the LBS applications can be built. The Open Location Service (OLS) Standard (Open Geospatial Consortium, 2008) defines five categories with interfaces for requests to, and responses from, each service. This categorization provides an effective way to explore the types of questions LBS are designed to answer and how they work. If a course learning objective is software development, this might be the most appropriate perspective to adopt. The first of the five core services is the *directory* service, where the directory is a database (or other information repository) storing information about *features*, including their spatial location. Beyond the strict OLS definition of a feature as a place, product or service, increasingly these directories store information about intangible entities. Provided it can be associated with a spatial location, a feature can equally be a local shop or a 'blog' entry. These directory services can be searched using both non-spatial attributes and spatial criteria. The second service category is the *position gateway* service that provides a standard interface to the position-determining technology described above and returns a mobile device's current location. Third is the *location utility* service that can convert between textual descriptions of places, such as addresses or place names, and geographic positions, such as latitude and longitude. The gateway and location utility services are back-end services that an end user is unlikely to interact with directly, but which are essential in the overall architecture of LBS. Fourth are *presentation* services that render geographic information, including backdrop mapping and foreground features, such as a current location or points of interest, for viewing by an end user. The final service category in the OGC ontology is the *route* service. Given a start point, end point and optional preferences, this will calculate and return route geometry that can be used for navigation. Any implemented LBS will require interaction between some or all of these services to satisfy requests from end users. If you make a query to find out about nearby restaurants, a location gateway service will first establish your position, and then pass this to a directory service. The results from the directory service are then displayed over a backdrop map with your current location using the presentation service. Finally, the service may call upon the route service to provide you with navigation advice to get to this preferred destination. This perspective is appropriate for software developers and those interested in designing new services, who must be aware of the architecture of LBS.

Business perspective

Commercial aspects of LBS are the focus here, including business models, with emphasis on generating revenue. This provides an interesting analytical perspective for business students, who may wish to develop alternative models for how services are funded. For example, should it be by monthly subscription, or should it be free at the point of use but funded by advertisements? This can draw more broadly on theory from e- and mobile-commerce, but it is interesting to contrast the importance of local information with the allegedly 'spaceless' internet. One important theme here is the LBS value chain: the organizations required to implement LBS, and the services, infrastructure and data that they contribute (Department of Trade and Industry, 2004). This chain includes mobile operators, handset manufacturers, controllers of positioning systems, information content providers, regulatory bodies and application developers.

Social perspective

LBS also raise issues from social and political perspectives. Location-aware computing has become a microcosm of the wider privacy debate concerning storage, use and access to personal digital information related to individuals (Dobson and Fisher, 2003). By definition, LBS must know where you are, and this information is usually transferred beyond a user's handheld device over wireless and wired connections to remote servers. How long this information is stored, who should have access to it and for what purpose, raises social, political and legal questions that students of LBS need to consider. This perspective is well suited to those interested in information use and policy, whose motivation may be to ensure that developed services conform to legal requirements and voluntary codes of practice.

21.2 Teaching LBS as a subject

Learning theory for LBS

Three related themes that Gold *et al.* (1991) suggested should underpin learning in higher education remain relevant after nearly twenty years. First, *student-centered learning* encourages students to be proactive in the learning experience, take control of their learning materials and shifts the focus and pace to account for their own preferred learning style. In a world of digital learning resources and virtual learning environments, this is routine, but, as we will consider in Section 3 of this chapter, LBS can promote new ways of engaging with materials that are tailored to the individual. Second, *active learning* encourages students to develop beyond the role of passive receiver of knowledge, to knowledge explorers, and ultimately knowledge creators (Huggard and McGoldrick, 2005). Students are encouraged to try things, to be permitted to make mistakes and to learn from these (Race, 2006), a position summed up by Niels Bohr in the statement 'An expert is a man who has made all the mistakes which can be made, in a very narrow field'. In the following section, we will explore different approaches, suited to the lecture theatre, tutorial and computer laboratory, that encourage students into these explorer and creator roles. Finally, learning should be seen as an *open ended process*, providing knowledge and skills that will continue to develop beyond the duration of the formal teaching period. This section will consider how teaching LBS should highlight continuing self-reliance in keeping abreast with new developments, and placing these in the context of what has come before.

Teaching methods

Lectures

Lectures challenge the instructor to avoid a one way broadcast that naturally places students in the role of knowledge receivers. The lecture format is an opportunity to introduce the concept of LBS from one or more perspectives depending upon the background and experience of the cohort and the intended learning objectives. Structuring the session with tasks and opportunities for questions can shift the lecture to be more tutorial in style. The following exercise has been conducted at City University over several years, with different cohorts at both undergraduate and master's level, on GIS and Business Studies courses, and with class sizes varying between 10 and 100. Following an introduction to the topic, students are encouraged to brainstorm ideas for new LBS in small groups. The task requires groups to add a spatial component to an existing web service, or think of a completely new service. This requires no prior technological or development knowledge, but encourages students to think about what services would be desirable or useful in the first instance, and how they could go about implementing them in the second. Asking each group to elect a spokesperson as the first step and telling them that you will choose groups to speak at random is a good check to encourage students to focus on the task. Over the years some remarkably viable services have been proposed as a result of this exercise. With the advent of *Facebook*, spatial extensions to social networking sites tend to dominate, particularly friend-finders to encourage spontaneous meetings. Interesting location-aware mobile commerce applications have been suggested, such as a product locator that allows shoppers to locate clothes in their size in nearby stores, sparing them the trial of physically visiting every store. Personalized leisure applications are often mentioned, such as local surf services that build on wave prediction services to alert and direct you to waves that suit your style of surfing. Such group exercises promote the immediate application of received knowledge, consolidating understanding, and move the student rapidly into the role of knowledge creator.

Rapid and personalized feedback that reacts to an individual's ideas has been cited as one of the most important factors in allowing students to make sense and consolidate the knowledge that they are constructing for themselves (Race, 2006). This can be difficult for large cohorts, but a scalable approach is to ask one group to describe their idea, and then ask which other groups in the room had similar ideas. Any feedback you offer following this is clearly for all those groups, and not just the one that voiced their ideas.

Computer-based laboratory exercises

Computer exercises provide an opportunity for students to develop LBS for themselves, placing them in the role of knowledge creators. The appropriate level of sophistication for this development is dependent upon many factors, but, even for introductory courses for students with limited computing skills, meaningful practical exercises can be designed. The prior knowledge and skills of the cohort is perhaps the key limiting factor. Geography and GIS&T students will likely have limited computer programming skills, however LBS may equally be taught to computer scientists, information scientists, and those studying mobile computing or human-computer interaction, each of which will bring their own perspective and skills. Other important constraints to consider include the time to devote to the practical component of a module, and whether the work is group based or individual. For students with no development experience, or limited time, designing paper mock-up applications or simple prototypes that link between a series of predefined views can be an effective approach to defining and evaluating an application (Dix *et al.*, 2004).

For cohorts with basic development skills, students can be encouraged to make rapid progress by using existing API and development kits, such as Google's Android platform (Google, 2009). Providing code snippets that perform standard functions, such as getting a position or displaying a backdrop map, can further speed the development process.

Laboratory exercises have been developed at City University designed to provide hands on experience of the services perspective, described in Section 3, all of which are relevant to mobile computing. The first considers positioning methods, specifically implementations of the World Wide Web Consortium Geolocation API specification (2010). This API provides access to the geographic location of the *hosting device*, which may be a mobile phone, a laptop, or a desktop machine. This location can be determined using various technologies, from the highly accurate Global Positioning System (GPS) through WiFi positioning, and GSM cell ID, down to a lookup of the address associated with the device IP address at the coarsest spatial resolution. Increasingly, web browsers are implementing these location-aware features (Mozilla, 2010), allowing website developers to request the location of a client-device and to tailor their content on the basis of this. Various tutorials exist describing how to request the device location, such as the example from Mozilla for Firefox using javascript (Mozilla, 2010). A snippet of about ten lines of javascript in a simple HTML page will get the current latitude and longitude and display it as text. Asking the class to discuss how useful latitude and longitude are in isolation to the user, and what else might be required to build a useful LBS, leads to a discussion of further OGC web services, such as presentation (mapping) and directory (content) services, and leads on to subsequent laboratory exercises.

The second exercise combines the geolocation API built previously with a mapping service, such as the Google Maps API, to display the user's location over a backdrop map. Again, this can be implemented in a relatively short snippet of javascript code. By creating and evaluating this 'location over a map' web page using different devices, students get a tangible example of the accuracy associated with different positioning-determining technology. A desktop machine, without GPS or WiFi sensors, tends to provide a result only accurate to the nearest town: the databases matching IP address to locations in the physical world are coarse resolution. A WiFi-enabled laptop tends to be more accurate since the range of WiFi access points is typically tens of meters. Providing it has a clear view of the sky, a GPS enabled 'smartphone' is likely to give the most accurate results.

The final exercise we use demonstrates directory services, and how georeferenced content can be retrieved to provide a 'Where's my nearest' web service. Many owners of georeferenced content have started to provide access to their data repositories via online APIs, such as Yell, and Scoot. Adding a few more code snippets produces an application that displays a user's location and nearby points of interest over backdrop maps. Once implemented, the web application can be used as a prompt to discuss how such content can be generated, georeferenced and maintained. It can also be used to explore the information needs of LBS users asking in what scenarios the implemented web service is useful, what other information would be useful in different scenarios, and how could this be made available to the user.

There are several examples of laboratory-based development projects in the literature. Huggard and McGoldrick (2005) report on five years' experience of delivering an LBS module to final year undergraduate computer science students. Following a series of lectures and tutorials that provide context and encourage initial exploration of the topic, students are required to implement a novel prototype LBS. This assignment also encouraged some inspired and novel ideas for LBS, including location-based *flash mob* applications to coordinate group protest, and games that link the physical world with Tolkien's *Middle Earth*. Their evaluations suggest that this task gave both creative freedom and control over the project, encouraging a sense of ownership and critical thinking that placed the student firmly in the role of knowledge creator. Moreover, the assessed submissions were unique to each group, which actively discouraged plagiarism.

Small group tutorials

Tutorials offer an opportunity to discuss some of the issues surrounding LBS. One of the overriding social issues to consider is privacy. LBS have the capacity to track a user's location, either explicitly or implicitly, and build up over time a pattern of their movement. For some LBS, such as lone worker applications, their sole purpose is actively to track and report an individual's location and most LBS will transfer a user's position over the mobile network in order to query spatial databases where the transaction may be recorded (Benford *et al.*, 2005). The transfer and storage of this information can compromise an individual's privacy, and has proven to be an emotive and contentious issue. Small discussion groups are a good place to raise these issues, discussing how long information should be stored for, who should have access to it, and opportunities for misuse in the context of national legislation (such as the UK Freedom of Information and Data Protection Acts) and voluntary codes of practice (such as the UK Code of Practice for the use of passive location services). The misuse of these data in particular makes for a stimulating debate, which can take place at a number of levels. Dobson and Fisher (2003) cite misuse by individuals with the example of a dominating partner forcing their spouse to carry a tracking device. At an organizational level, privacy concerns can be related to the ownership of data recording your current and historical position, whether or not this could be sold to advertisers and the potential for location-aware *spam*. At an institutional or governmental level privacy can be discussed in the context of under what circumstances it is permissible for law enforcement authorities to access individual track logs, and the potential for misuse of national security legislation (Huggard and McGoldrick, 2005).

Outdoor field work

Fieldwork provides a better opportunity to demonstrate LBS than the classroom. In this context 'outdoors' need not mean a dedicated residential fieldtrip of the type so beloved of academic geographers, since simply getting students outside with mobile phone and GPS can be instructive. Alternative LBS can be compared with physical surroundings, prompting questions about the completeness and currency of the directory service used by the LBS. Asking students about their information needs while on the move and trying to satisfy those in an outdoor setting can be an effective approach to identifying the limitations of current services and opportunities for new ones.

21.3 LBS as a pedagogic tool

Mobile and blended learning

The field of LBS has a dual role to play in GIS&T teaching. As we have seen, it belongs on the curriculum as a subject in its own right, but it can also be used as a pedagogic tool

(Madge and O'Connor, 2004). Education evolves, taking advantage of new technology and responding to social and cultural changes in how people work, learn and access information (Guy, 2009). The internet and information era has had a huge impact on teaching and learning, for example the use of virtual learning environments (VLE) and a significant body of researchers is convinced that mobile computing will trigger the next paradigm shift. As a pedagogic tool, LBS fall under the umbrella of mobile and blended learning (Traxler, 2009), an emerging field exploring how mobile platforms are changing the pedagogic environment (Alexander, 2004). The focus here is the potential for migrating learning away from traditional environments such as the classroom or computer terminals, which are referred to variously as *tethered* (Salmon, 2007), *sedentary* (Alexander, 2004) or *static* (Mountain *et al.*, 2009).

Mobile learning strives to be more than using a mobile device as the interface to a virtual learning environment (Traxler, 2009), it also refers to the mobility of learners and constant accessibility of resources. Sharples *et al.* (2005) went beyond mobility in their distinction between mobile and static learning, stating that mobile learning should be a constructive and social process, which is both personalized to the learner and situated in its environment. Technology is diverse and may vary in sophistication from text messaging to augmented reality interfaces (Guy, 2009), but it is secondary to the definition. More important are the opportunities for learning to take place at new times, in new, possibly remote, places and to reconsider in light of this what we consider learning to be. Another distinction identified by Traxler (2009) is that with static e-learning the learner's attention is dedicated solely to this task with limited engagement with the physical world, whereas mobile learning is woven around a host of daily tasks, places and situations.

The impact of mobile technology on learning mirrors trends in the workplace where there has been a blurring of the traditional spatial and temporal boundaries. The opportunity to be in constant contact via mobile communications has led to more fluid schedules, with less emphasis on 'clocking in and out' at a specific place and time, and less need for staff to co-locate. Likewise VLEs, and now mobiles, have removed the need for lecturer and class to co-locate. Mobile learning theory must consider some of the broader issues that arise from this trend.

Traxler (2009) argues that it is difficult to formulate a theory of mobile learning since the field is volatile, where learning techniques that take advantage of emerging technology may lead to frequent rapid shifts in the definition of what it is. He suggests that there are three choices if such theory is to be developed. First, theory can be adapted from conventional e-learning, which raises concerns about how transferable this theory may be. If mobile learning is characterized primarily as an alternative interface through which a learner may interact with a VLE, then this is a valid approach to take. His second approach is to consider the mobile learning environment to be distinct and develop theory from new. Researchers focusing on learning opportunities afforded by personal networks and communities tend to adopt this approach, in particular the idea that with ubiquitous access to vast information resources, it is now less important to retain information than to know how to find relevant information where and when it is required. His third approach suggests that mobile learning theory attempts to subscribe to and apply more general abstract learning theory, for example where the focus is on the changing culture in which students learn, within which mobile technology is seen as just another tool (Traxler, 2009). Wali *et al.* (2008) develop this idea, suggesting that much of the dominant mobile learning theory is technocentric, with too much focus on the generic mobile human-computer interaction issues associated with

Table 21.1 Convergence between learning and technology (after Sharples *et al.*, 2005)

New Learning	New Technology
Personalised	Personal
Learner-centred	User-centred
Situated	Mobile
Collaborative	Networked
Ubiquitous	Ubiquitous
Lifelong	Durable

small screens and keyboards that has resulted in emerging theory being vulnerable to new technological developments. Shifting away from the technology towards a learner-centered perspective may focus on the enduring themes of mobile learning (Kukulska-Hulme *et al.*, 2009), such as social practice and new conceptualizations of learning, how to foster collaboration, and how mobile technology can advance the constructivist agenda.

Sharples *et al.* (2005) highlight convergent trends that are occurring simultaneously in education and technology. As summarized in Table 21.1, learning is striving to be personalized, ubiquitous, collaborative and situated, allowing students to share knowledge and solve problems regardless of the time or where they are, taking into account their personal needs and context. Mobile computing can satisfy many of these needs, for example, using networking technologies and portable devices for virtual co-location of learning materials and colleagues, and utilizing the location-aware capacity of mobile technology to encourage situated learning.

Location-based learning

Given the trend towards mobile learning, how can the location-aware property of mobile devices be exploited in teaching and learning? An important aspect of mobile learning is *situated learning* (Naismith *et al.*, 2004), which considers the relationship between learning and the physical and social context in which it occurs. Location-aware technology offers the opportunity to adapt the learning experience to the physical infrastructure, ambient information and social context, such as the presence of others (Kadirire, 2009). LBS have received attention as a pedagogic tool not only within GIS&T, but from other disciplines where learning materials can be linked to real locations. The learning materials associated with some subjects are relatively location insensitive, but others possess strong links to the physical world. Geography and history are prime examples, but the link between information and location exists in environmental science, engineering, architecture, sociology, politics, and many other disciplines. This presents both an opportunity and a challenge which has a historical precedent in the teaching of GIS. Forer and Unwin (1999) distinguished between GIS as an *object* of study, as opposed to a tool to *enable* study. The parallel for LBS is clear. As an object of study, LBS belongs on the curriculum of GIS&T and mobile computing courses. As a tool to enable study, LBS can be used across a much more diverse range of disciplines. Academics and teaching staff in those disciplines may be less proficient in their technical skills than those on GIS&T and mobile computing courses. They are far less likely

Figure 21.3 Situated learning where learners explore ambient information linked to the physical world at times and places of their choosing

to have come into contact with spatial data, or have received any training in using space as referencing framework. These are skills that must be mastered if LBS are to be used effectively as pedagogic tools outside the domains in which they are currently taught.

The association of information with physical space and real world objects can be seen as creating an *ambient intelligence* (Kukulska-Hulme *et al.*, 2009). This provides an opportunity to access learning materials in the context in which they are immediately relevant (Figure 21.3), which is seen as one of the main benefits of mobile learning (Benford *et al.*, 2005). Armstrong and Bennett (2005) argue the need for understanding *processes*, which may operate across space and through time. Furthermore these processes can occur at a range of spatial (from highly localized to global) and temporal scales (from seconds to millennia). Such processes are often taught in the classroom using abstract examples that require a significant cognitive leap on the part of the learner to understand how they manifest themselves in the physical world. When considering LBS from the perspective of ambient intelligence, as individuals move through physical space they can be simultaneously exploring an information space. Any location therefore could potentially act as an impromptu site of learning and provide tangible examples of abstract concepts. Several projects and systems have explored the possibility of using LBS and ambient intelligence as pedagogic tools.

The *mobile computing in geographic education* (*MoGeo*) framework described by Armstrong and Bennett (2005) illustrates the limitations of abstracted descriptions of context in the classroom and the potential for LBS. When teaching fluvial processes, an instructor might state that at the peak of a flood, the Iowa River had a discharge of 26,000 ft^3/s, but this is a figure that is hard to visualize or understand. When LBS is applied to education,

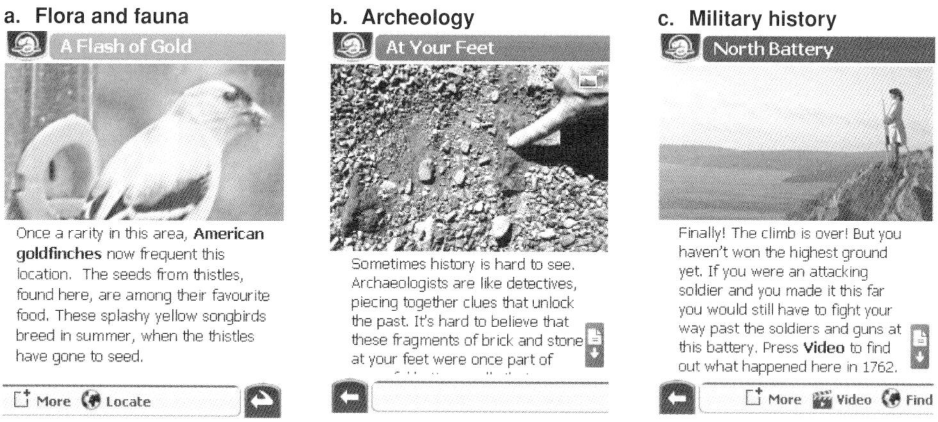

Figure 21.4 Alternative information spaces overlaid on physical space for Signal Hill, Canada

there is the potential for students to visit the river and access live discharge data from web feeds, which can act as a baseline by which to gauge this magnitude. Georeferenced multimedia data, such as photographs, aerial imagery, videos, and even anecdotal evidence from observers, provides the ambient intelligence that can be retrieved and explored by students *in situ*. Such interaction provides concrete examples of the abstract process, consolidating individual learning.

Camineo (2011), a spin out of the *WebPark* project (Mountain and MacFarlane, 2007), provides a platform for designing and implementing location-aware mobile guides aimed predominantly at outdoor recreational areas such as national parks. The guide provides access to information traditionally disseminated via guide books and notice boards. Since the information is spatially referenced, local searches will return information relevant to the searcher's surroundings. The guide has the capacity to 'push' information to a mobile device when users come within 'geofenced' buffers around its location. Since pushed information can be both overwhelming and distracting, there is a need to filter down from all nearby information objects to those of interest to the individual carrying the handheld device. The *Camineo* system allows different themes to be selected, so that two individuals following the same route may be exploring different information spaces comprised of different learning materials. One may be pushed information on the theme of flora and fauna, another the military history associated with the places that they pass (Figure 21.4).

Providing alternative experiences for different individuals based on their preferences presents a model of where teaching and learning may be headed in the future (Sprake, 2009). Since location can be seen as just one part of an individual's *context* (Jiang and Yao, 2006), future location-based pedagogic tools should aspire to account for this as part of a student's unique learning experience. Some important aspects of context are identified in the *MoGeo* framework (Armstrong and Bennett, 2005). The first is the classes on which a student is enrolled. By associating learning materials with different classes, each student may retrieve only that information relevant to them. This has a strong parallel with the *Camineo* example, where georeferenced content is organized by theme. A second aspect of context in education is an individual's progress through a course. Future location-based pedagogic tools can account for this by using the conditional release of materials, for example pushing current materials, allowing searching of previous materials, but hiding future materials, all based on an individual's progression.

The use of location-aware mobile devices and volume of spatially referenced information are both increasing rapidly, and there is the potential for a profound change in approaches to lifelong learning. Less than 1 per cent of the learning activities are conducted as part of formal, credit-based education (Sharples, 2000). The vast unstructured document collections on the internet have done much to promote life-long learning, but have required learners to develop new skills in clearly expressing their information needs, and in rapidly assessing the relevance of retrieved results. Location-aware computing offers a new way of filtering these collections with new opportunities for life-long learning. The information spaces we explore as we move through the physical world need not be small, bespoke collections in a single database or VLE, but can be any georeferenced resource on the web. The creation of such georeferenced resources was once the preserve of geographic experts. LBS provides opportunities for creation, as well as consumption, of such resources by a much wider audience (Goodchild, 2009). These emerging resources mean that location-aware learning need not be confined to discrete sessions, but can be infrequent and serendipitous. Given key themes that are of interest to the learner, their movement through the physical world may be the trigger for pushing information that is relevant to them. For example, a history enthusiast may define a trigger search phrase 'The English Civil War' that can remain active for many months. As they visit new locations as part of their daily life, they may be pushed georeferenced information from web resources that match this query as they come into proximity with it. This provides a radically new approach to materials. Traditional teaching methods have tended to encourage a linear perspective of knowledge acquisition. Location-aware computing offers an innovative approach to defining alternative routes through learning materials.

21.4 Conclusions

This chapter has presented two distinct but overlapping themes, the first related to the teaching of LBS as a subject in higher education, the second exploring the application of LBS as pedagogic tools, and the associated opportunities for teaching and learning. LBS demands a place on the GIS&T curriculum because it is becoming one of the primary channels through which people interact with geographic data, technology and functionality. LBS are mainstream services built for the consumer market: compared with GIS there are few barriers to using them. Distinct alternative perspectives have emerged in the teaching of LBS, some focusing on the driving technology, others on software architectures or the associated social or business issues. When deciding which perspectives to highlight, consideration should be given to the course learning objectives, and the level, skills and background of the student cohort. Consideration should also be given to the choice of teaching method. LBS is often characterized as a 'techie' subject best taught in the computer laboratory. However, the privacy debate associated with LBS may best be explored in small group tutorials and a better sense of context for local search achieved by encouraging engagement outside of the classroom.

LBS also has a role as a pedagogic tool, as part of mobile and blended learning, where the use of mobile platforms allows the *untethering* of education, promoting learning in places and at times of the learners choosing, breaking out from traditional teaching environments.

As a tool for learning, the use of LBS is not confined to the GIS&T curriculum. One pitfall to identify and avoid is that in this context, technology is the enabler, not the *raison d'être*. The instructor should focus on the opportunities for learning afforded by mobile devices and LBS: the learner's experience can become more personalized, learner-centered, situated, collaborative and ubiquitous as part of lifelong learning. The technology of the day develops rapidly and, rather than any one technological implementation, these approaches to learning should lead developments. The key opportunity for LBS is the relationship between learning and the physical and social context in which it occurs. Learning materials can be linked with locations: an information space overlaid on the physical world. This emerging infrastructure of ambient intelligence allows learners to filter information not only by relevant theme, but by proximity. This filtering by location and theme allows two individuals to explore different, personalized information spaces while navigating the same physical space. Instructors from all disciplines should consider how this resource can be applied to their field, and how their students would benefit from using LBS to explore this ambient information layer.

References

Alexander, B. (2004) Going nomadic: mobile learning in higher education. *EDUCAUSE Review*, 5(33): 29–35.

Armstrong, M. P., and D. A. Bennett (2005) A manifesto on mobile computing in geographic education. *Professional Geographer*, 57(4): 506–515.

Benford, S., A. Ramsden, G. Roussos, J. Traxler, and J. Trinder (2005) Future location-based experiences. http://www.jisc.ac.uk/media/documents/techwatch/jisctsw_05_01.pdf (accessed March 7, 2011).

Brimicombe, A., and Y. Li (2006) Mobile space-time envelopes for location-based services. *Transactions in GIS*, 10(1): 5–23.

Camineo (2011) *Digital Guide for Tourism*. http://www.camineo.com (accessed March 7, 2011).

Cellan-Jones, R. (2008) Location, location, location. http://www.bbc.co.uk/blogs/technology/2008/07/location_locationlocation.html (accessed March 7, 2011).

Department of Trade and Industry (2004) *Location-based services: understanding the Japanese experience*. http://www.oti.globalwatchonline.com/online_pdfs/36246MR.pdf (accessed October 5, 2009).

DiBiase, D., M. DeMers, A. Johnson, K. Kemp, A. Luck, B. Plewe, and E. Wentz (eds) (2006) *The Geographic Information Science and Technology Body of Knowledge*. Washington, DC: Association of American Geographers and University Consortium for Geographic Information Science.

Dix, A., J. E. Finlay, G. D. Abowd, and R. Beale (2004) *Human-Computer Interaction*. Harlow: Prentice Hall.

Dobson, J. E., and P. F. Fisher (2003) Geoslavery: Society must contemplate a new form of slavery, characterized by location control. *IEEE Technology and Society Magazine*, 22(1): 47–52.

Forer, P., and D. Unwin (1999) Enabling progress in GIS and education, IN: P. A. Longley, M. F. Goodchild, D. J. Maguire, and D. W. Rhind (eds) *Geographical Information Systems, Vol. 2: Management issues and applications*. New York: John Wiley & Sons, pages 747–756.

Gold, J. R., A. Jenkins, R. Lee, J. Monk, J. Riley, I. Shepherd, and D. Unwin (1991) *Teaching Geography in Higher Education A Manual of Good Practice*. Oxford: Blackwell.

Goodchild, M. (2009) Neogeography and the nature of geographic expertise. *Journal of Location Based Services*, 3(2): 82–96.

Google (2009) Google projects for android. http://code.google.com/android (accessed October 5, 2009).

Guy, R. (ed) (2009) *Evolution of Mobile Teaching and Learning.* Santa Rosa: Informing Science.

Huggard, M., and C. McGoldrick (2005) Practical positioning projects: Location Based Services in the laboratory, *35th ASEE/IEEE Frontiers in Education Conference.* Indianapolis : Institute of Electrical and Electronics Engineers, Session S3F. October 19–22, 2005. http://www.scss.tcd.ie/publications/tech-reports/reports.09/TCD-CS-2009-22.pdf (accessed August 14, 2011).

Jiang, B., and X. Yao (2006) Location-based services and GIS in perspective. *Computers Environment and Urban Systems,* 30(6): 712–725.

Keegan, V. (2010) The next challenges for mobile phones: Find me and tell me who I am. *The Guardian,* 14 January, 2010.

Kadirire, J. (2009) Mobile learning demystified, IN: R. Guy (ed) *Evolution of mobile teaching and learning.* Santa Rosa: Informing Science Press, pages 15–56.

Kukulska-Hulme, A., M. Sharples, M. Milrad, I. Arnedillo-Sanchez, and G. Vavoula (2009) Innovation in mobile learning: A European perspective. *International Journal of Mobile and Blended Learning,* 1(1): 13–35.

Madge, C., and H. O. Connor (2004) Online methods in geography educational research. *Journal of Geography in Higher Education,* 28(1): 143–152.

Mountain, D. M., and F. Liarokapis (2007) Mixed reality (MR) interfaces for mobile information systems. *ASLIB Proceedings,* 59(4/5): 422–436.

Mountain, D. M., and A. Macfarlane (2007) Geographic information retrieval in a mobile environment: evaluating the needs of mobile individuals. *Journal of Information Science,* 33(5): 515–530.

Mountain, D. M., H. I. Myrhaug, and A. Goker (2009) Mobile search, IN: A. Goker, M. Graham, and J. Davies (eds) *Information Retrieval.* London: Wiley, pages 103–130.

Mozilla (2010) *Using geolocation.* https://developer.mozilla.org/En/Using_geolocation (accessed March 7, 2011).

Naismith, L., P. Lonsdale, G. Vavoula, and M. Sharples (2004) *Literature Review in Mobile Technologies and Learning Technology.* NESTA FutureLab. http://www.futurelab.org.uk/resources/documents/lit_reviews/Mobile_Review.pdf (accessed March 7, 2011).

Open Geospatial Consortium (2008) Location service (OpenLS). Open Geospatial Consortium, http://www.opengeospatial.org/standards/ols (accessed September 20, 2011).

Race, P. (2006) *The Lecturer's Toolkit: A Practical Guide to Learning, Teaching & Assessment.* London: Routledge.

Raper, J. (2007) Geographic relevance. *Journal of Documentation,* 63(6): 836–852.

Raper, J., G. Gartner, H. Karimi, and C. Rizos (2007a) A critical evaluation of location based services and their potential. *Journal of Location Based Services,* 1(1): 5–45.

Raper, J., G. Gartner, H. Karimi, and C. Rizos (2007b) Applications of location based services: a selected review. *Journal of Location Based Services,* 1(2): 89–111.

Salmon, G. (2007) Notes for university teachers of the 21st Century, IN: S. Walker, M. Ryan, and R. Teed (eds) *Designing for E-learning Conference.* Greenwich, UK, University of Greenwich, pages 10–13.

Sharples, M. (2000) The design of personal mobile technologies for lifelong learning. *Computers and Education,* 34(3–4): 177–193.

Sharples, M., J. Taylor, and G. Vavoula (2005) Towards a theory of mobile learning. *Proceedings of MLearn 2005 conference,* Cape Town, H. van der Merwe and T. Brown (eds). http://www.mlearn.org.za/CD/papers/Sharples-%20Theory%20of%20Mobile.pdf (accessed August 14, 2011).

Sprake, J. (2009) Designing participant-generated context into guided tours. *International Journal of Mobile and Blended Learning,* 1(2): 19–38.

Steiniger, S., M. Neu, and A. Edwardes (2006) *CartouCHe – Lecture Notes on LBS: Foundations of Location Based Services.* Zurich, University of Zurich. http://www.e-cartouche.ch/ (accessed March 7, 2011).

Traxler, J. (2009) Learning in a mobile age. *International Journal of Mobile and Blended Learning,* 1(1): 1–12.

Virrantaus, K., J. Markkula, A. Garmash, and Y. V. Terziyan (2001) Developing GIS-supported Location-Based Services. *Proceedings of WGIS'2001 – First International Workshop on Web Geographical Information Systems*, Kyoto, Japan, pages 423–432.

Wali, E., N. Winters, and M. Oliver (2008) Maintaining, changing and crossing contexts: an activity theoretic reinterpretation of mobile learning. *Research in Learning Technology*, 16(1): 41–57.

World Wide Web Consortium (2010) *Geolocation API Specification*. http://dev.w3.org/geo/api/ (accessed March 7, 2011).

Zimmermann, A. (2008) *Dataquest Insight: Location-Based Services Subscriber and Revenue Forecast, 2006–2011*. Stamford, USA: Gartner Inc.

22

GIS is dead, long live GIS&T: an educational commentary on the opening of Pandora's box

Nicholas J. Tate

Department of Geography, University of Leicester, Leicester, UK

22.1 Introduction

The preceding four chapters have been concerned with what we might loosely describe as *digital worlds* technologies, which it can be argued are part of GIS&T. In addition there has been a particular focus on learning and teaching, specifically in the context of GIS&T. Echoing the distinction that has been made in reference to GIS technology being employed both to teach *about* and to teach *with* (Sui, 1995), a similar approach could be adopted for each of these technologies, where the pedagogic focus could be directed *inward* (teach about) or *outward* (teaching with). Indeed, this distinction is used explicitly in the chapter on LBS by Mountain. Furthermore, a number of possibly dialectical relationships can be suggested between technologies such as these (or computing more generally) and traditional GIS on the one hand, as well as that between such technologies and education (Sharples *et al.*, 2007). The first of these – the nature of and relationship between GIS and its constituent digital world technologies is the focus of the first part of this chapter. The second part is concerned more directly with pedagogy and explores how these digital world technologies might enable educational affordances of use for learning and teaching.

22.2 Digital worlds

Mark Weiser of the Xerox Palo Alto Research Centre coined the term *ubiquitous computing* in 1988 (Want, 2009) and in a series of publications in the early 1990s, of which Weiser

Teaching Geographic Information Science and Technology in Higher Education, First Edition.
Edited by David J. Unwin, Kenneth E. Foote, Nicholas J. Tate and David DiBiase.

(1991) is probably best known, he predicted a time when computers would effectively disappear, including the oft-cited quote (Weiser, 1991, 94):

> The most profound technologies are those that disappear. They weave themselves into the fabric of everyday life until they are indistinguishable from it.

This is frequently used in the context of the march of progress of computer technology. However, Weiser saw deeper. That some technologies disappear can be explained by familiarity accrued through *learning* (Weiser, 1991, 94):

> Such a disappearance is a fundamental consequence not of technology but of human psychology. Whenever people learn something sufficiently well they cease to be aware of it.

Weiser articulated a vision of a time when computational technology in the form of ubiquitous (or pervasive) computing would occupy a similar position to other invisible technologies (for example writing) that are learned and taken for granted to such a degree that they would blend into the background of everyday reality. Proponents of what became known as *ubicomp* such as Want (2009) have argued that this defines the present third age of computing characterized by many computers (e.g., smart phone, e-book reader, digital camera) per person. Want (2009) contrasts this with the first age, defined as the era of the Mainframe (many people to one computer), and the second age defined as the era of the Personal Computer (one person per computer).

What also makes Weiser (1991) particularly of interest is that establishing *location* was critical to fulfil the ubicomp vision: computers would need to know where they are. Today, location is recognized as a key element of ubicomp (e.g., Ley, 2007; Varshavsky and Patel, 2009) and this is reflected in the development of the field of *location aware computing* where location is determined via location technologies such as GNSS-GPS and RFID (Hazas *et al.*, 2004). On geographic scales this makes mobile GIS and LBS possible.

At the end of the first decade of the twenty-first century, we live in a world increasingly characterized by *geo-everything* technologies enabling the location/tagging of real and virtual people as well as objects relative to the earth (Johnson *et al.*, 2009). Historically, the mapping of such data has been in the hands of the specialist GIS professional via GIS&T. The advent of web-based mapping and the manipulation of geospatial data using the web-based map *mashup* has also become significant by empowering almost anyone with web access and a computer to participate in this process.

Batty *et al.* (2010) reported that as of February 2010 there were an estimated 2153 map mashups (listed at http://www.programmableweb.com/tag/mapping). By March 2011 this had increased to 2459. At the time of writing Japan is in the process of dealing with an earthquake crisis and associated nuclear emergency. One example of map mashup has become associated with such events: the *crisis-map mashup* for disaster-response (Liu and Palen, 2010). The Harvard Center for GIS Analysis set up a specific portal http://cegrp.cga.harvard.edu/japan/ for geo-spatial resources related to the incidents in Japan. This went live at 12 noon on 11 March 2011 (Berman, *personal communication* 2011) and as of 14 March 2011 comprised a mix of 25 websites and mashed-up web services to associated geospatial data resources. Contexts ranged from the anticipated maps of media resources, seismic activity, maps directly aiding disaster-response (e.g., maps of shelters and food supplies) to maps of those involved in the disaster (e.g., *tweets* and their locations in the vicinity of the earthquake). By 16 March 2011 this had grown to 29 with the addition of maps of radiation. An example of a mashup related to the crisis is that developed by Paul Nicholls at http://www.japanquakemap.com/. This makes use of USGS seismic data and Google Map's API to allow the mapping of the recent Japanese earthquakes over various recent time periods. This went live on 16 March and had accumulated over 100,000 hits in 12 hours (Nicholls, *personal communication*). The various examples referred to above comprise not only a mix of maps and data derived/produced from official sources (e.g., Japan's post-earthquake imagery from the Japanese Geospatial Information Authority) and universities, but also maps mashed up by volunteers using both official and unofficial data. It is the latter – creation of content by non-specialists – that is consistent with a Web 2.0 philosophy, and represents a significant change in the model by which geospatial data is created and used.

Whether this is Web 2.0 or more precisely *where* 2.0 (coined in the context of the O'Reilly-initiated conference series), the distinction between producer/GIS professional and user/non-specialist is becoming increasingly blurred as evidenced by the mix of producers/users and official/unofficial data sources in the above example. In contrast to a history of map provision from the top-down by agencies, *OpenStreetMap* is frequently cited as an example of a free and editable map product created bottom-up by users (Haklay and Weber, 2008). Neologisms such as *produsers* (Coleman *et al.*, 2009) or *prosumers* (Hodgins, 2010 after Toffler) are being used to describe the role of the non-specialist user/citizen in this map creation process. This has led to the development of what is variously termed *volunteered geographical information* (Goodchild, 2007), the concept of *citizens as sensors* (Goodchild, 2007), *crowdsourcing* (Hudson-Smith *et al.*, 2009b) and the development of *neogeography* (Goodchild, 2009, Hudson-Smith *et al.*, 2009a), all of which reference the widespread creation and use of web-based geospatial resources from the bottom-up by non-specialists.

The advent of web-based neogeography has been interpreted by some as heralding the democratization of GIS (Goodchild, 2008); a new revolution in GIS recognizing the current analytical limitations of Web 2.0 mapping (Batty *et al.*, 2010); a possible candidate for putting the people into a new type of GIS (Miller, 2007) and an opportunity to lever value-added geospatial services via applications such as ArcGIS.com (Dangermond, 2010). However, this has led also to concerns about the GIS discipline along the lines of 'whither GIS in a where 2.0 world?' This concern has been reflected in discussion in reviews and editorials of the consequences and implications of the emergence of neogeography for professional GIS (e.g., Sui, 2005, 2008a; Crampton, 2009) – what Dan Sui calls 'a new Geography without geographers' (Sui, 2008a, 5).

3D virtual worlds – getting inside the computer

Weiser (1991) considered virtual reality (VR) to be in diametric opposition to his vision of ubicomp. Twenty years later, with the advent of technologies such as Web 2.0/where 2.0, the convergence of these with mobile computer technologies as well as the broadening of VR to refer to a great variety of 'realities' with differing approaches to interaction, whether this distinction still holds is moot. VR can take many forms. Perhaps Hilary McLellan put it most succinctly: 'virtual reality systems enable users to become participants in artificial spaces created by the computer' (McLellan, 1996, 464). Usually these spaces are 3D. This encompasses not only the traditional fully- or semi-immersive VR but also what we can term 'through the window' VR where the degree of interaction is mediated via a standard desktop computer screen and controls (Fisher and Unwin, 2002), and the interactive 'virtual worlds' of the web. It is the latter which has led to a resurgence of interest in VR (Dalgarno and Lee, 2010) and it is this form of VR which has become most widely accessible.

There are various classifications of such web-based virtual worlds. For example de Freitas (2008) distinguishes virtual worlds based on the context of use into *role play worlds* (e.g., World of Warcraft); *social worlds* (e.g., Second Life); *working worlds* (e.g., Open Wonderland); *training worlds* (e.g., SAIC's On-line Interactive Virtual Environment) and *mirror worlds* (e.g., Google Earth). In contrast Bainbridge (2007) reserved the term virtual world for those virtual environments allowing people to be directly represented (usually as avatars) and interact. Warf and Sui (2010) and Sui (2008b) provide a similar distinction between *mirror worlds* on the one hand and *virtual worlds* on the other (the latter used as a catch-all for non-mirror world virtual environments).

Mirror worlds are more immediately relevant to GIS&T. Synonymous with virtual globes, these are defined as 'digital representations of non-virtual reality' (Warf and Sui, 2010, 202) or 'enhanced virtual models or replications of the physical world' (de Freitas, 2008, 15). Examples of virtual globe software include Google Earth, NASA's World Wind, Esri's ARC Explorer and Skyline Globe. These offer extensive navigation and visualization of a 3D model of the world and can incorporate similar data mashups to the 2D mashups (e.g., Tiede and Lang, 2010; Wood *et al.*, 2007). There are clear links to the concept of a Digital Earth (Aurambout *et al.*, 2008; Aurambout and Pettit, 2008; Butler, 2006; Craglia *et al.*, 2008; Foresman, 2008) however although a virtual globe may in some ways act as a Digital Earth, the scope of Al Gore's original Digital Earth vision was much broader (Riedl, 2007; Goodchild, 2008).

A virtual world called Second Life (SL) (http://secondlife.com) is a product of Linden Labs and was released in 2003 (Educause, 2008). By 2011 there were some 22 million registrations with just over 1.3 million having logged-in over the previous 60 days. It has received particular attention in both the media and academy as an example of what we may term a multi-user virtual environment. It is a web-based social world (de Freitas, 2008) that encourages communication and social interaction: in short a *social virtual world* (SVW) (Devisch, 2008; Hendaoui *et al.*, 2008). Central to SL is the use of an *avatar* to represent the individual, and the ability via this avatar not only to interact with other avatars within the 3D environment – termed *co-presence* (Delgarno and Lee, 2010), and to configure the 3D environment itself. In addition, SL operates an internal economy as well as an internal object format known as the 'prim' (Thompson, 2011).

By virtue of the social component, SL clearly lends itself to serious contexts that involve collaboration such as group decision making. For example Gordon and Manosevitch (2010) report a case study of using SL for community planning in Boston (Mass.). By importing geographical data into SL the environment can also act as an opportunity to visualize and collaborate in the context of such data within virtual space (Hudson-Smith and Crooks, 2009; Batty *et al.*, 2010). Faced with a disembodied mirror worlds such as Google Earth, and an embodied world such as SL, it is not surprising that a virtual space containing *both* co-present avatars along with a suitable representation of geography has been envisaged, variously termed a *virtual geographic environment* VGE (Gong *et al.*, 2009) or *second earth* (Roush, 2007; Sui, 2008b).

Mobile technologies – getting the computer outside

Closer to Weiser's (1991) vision of ubicomp is the development of mobile computing, and in the context of GIS&T the specific development of location based services (LBS). The definition of LBS is varied (Jiang and Yao, 2006) but it usually involves the interaction of elements of GIS, positioning and web technologies in the context of a mobile user (Steiniger *et al.*, 2006). Brimicombe and Li (2009, 2) suggest:

> Location-based services (LBS) are the delivery of data and information services where the content of those services is tailored to the current or some projected location and context of a mobile user.

The development and uptake of mobile technology has been significant: particularly the mobile (or cell) phone. Sharples and Roschelle (2010, 4) have noted:

> Now, almost every adult and adolescent child in industrially developed countries owns a multimedia communicator with more computing power than guided the first landings on the moon.

What has become known as Moore's law (which after Thompson, 2006 is the observation that capacity and performance of integrated circuits doubles every two years), is particularly relevant to mobile devices as it enables more functionality. Indeed, Roy Want has noted that mobile phones (smart phones in particular) are the closest to a ubiquitous computer and are likely to become the platform of choice for future software development (Want, 2009). This is supported by the statistics of market growth: in 2010 the GSA estimated that over the period 2005–2009 global shipments of GNSS-enabled mobile phones increased from 60 million to over 280 million units per year, with shipments likely to reach almost one billion units per year by 2020 (GSA, 2010). Not-surprisingly LBS are a particularly significant element of the positioning/GNSS market, with the GSA suggesting that by 2020 the LBS industry will be worth some €96 billion globally (and second only to in-vehicle road systems in terms of importance). The vision of the mobile phone providing both an interface to, and augmentation of the 'mirror world' identified above for not only people but objects as well to deliver augmented reality is very compelling (Smart *et al.*, 2007).

22.3 Learning with digital worlds

As Mark Weiser realized, in the context of profound/disappearing technology, the technology was only part of the equation: but learning is the key. It is probably fair to say that both the development of learning applications and learning theory relevant to digital worlds technology has lagged behind the development of the technology. However, there are numerous and increasing examples of learning and teaching applications appearing which harness both virtual world and mobile technologies, covering both the contexts of teaching *with* and teaching *about*.

The development of technology has required the development of specific theories related to learning in these environments (Sharples and Roschelle, 2010) in both *mobile*- or '*m-learning*' (Evans, 2008) and *virtual*- or '*v-learning*' (de Freitas, 2008) contexts along similar lines to *e-learning*. Underpinning much of the pedagogic use of digital worlds technology and theory construction is the participation metaphor of learning (Sfard, 1998) whereby learning is derived from dialogic *collaborative* activities. This contrasts with the more traditional acquisition metaphor of learning by which learning is a monologic and solitary activity (Sfard, 1998; Paavola *et al.*, 2004; Paavola and Hakkarainen, 2005). There are also clear links to various constructivist theories of learning as well as links to community of practice theory (Lave and Wegner, 1991) by which people in the workplace learn informally rooted in practice (Duguid, 2005).

As Lynn Songer notes in her chapter, navigating a 3D virtual world on a desktop computer has an element of 'wow'. But aside from conveying information, the obvious question is does such technology encourage learning – more specifically does a virtual 3D environment encourage more learning than a 2D or non-virtual environment? We are primarily concerned with the outward facing use of the technology here, where the subject of the learning is not the technology itself. In a general sense there is increasing recognition of cognitive affordances of video games and similar technologies for learning (for example Johnson, 2006 and Gee, 2007). In addition there is also evidence of widespread adoption of 3D virtual worlds in a variety of contexts throughout the academy. However, beyond the hype, Dalgarno and Lee (2010) suggest that in the face of little conclusive research the answer to this question is still equivocal. They have suggested that the characteristics of *representational fidelity* (degree of realism encompassing visual, tactile, auditory and kinaesthetic elements) and the opportunities for *learner interaction* as being the key and unique characteristics of 3D virtual worlds used for learning (which they term 3D virtual learning environments – VLE). Building on this and from an extensive review of current research they suggested five *potential* learning 'affordances' that such environments offer (Table 22.1).

Of course, not all of these are relevant to all learning contexts in which virtual worlds technologies are deployed. However, it ought to be possible to associate some of these affordances to 'typical' uses of the virtual world's technology within the realm of GIS&T, and to use such a mapping to compare the relative merits of such technologies for learning. As different technologies and different learning contexts are involved this mapping is not straightforward. Two are relatively easy to deal with: (3) might be considered to apply to some degree regardless of the technology/context (which we will effectively ignore for this purpose of this discussion) and (5) might be expected to apply to those environments enabling a degree of social co-presence. The remaining three are more difficult. Superficially in a GIS&T context (1) should apply in those instances where geographic information is

Table 22.1 Virtual world learning affordances (modified from Dalgarno and Lee, 2010; used with permission)

1: To facilitate learning tasks that lead to the development of enhanced **spatial knowledge representation** of the explored domain
2: To facilitate **experiential learning** tasks that would be impractical or impossible to undertake in the real world
3: To facilitate learning tasks that lead to increased intrinsic motivation and **engagement**
4: To facilitate learning tasks that lead to improved transfer of knowledge and skills to real situations through **contextualization of learning**
5: To facilitate tasks that lead to richer and/or more **effective collaborative** learning than is possible with 2D alternatives

being worked with in any realm, from a virtual city to a cave system and (2) is clearly appropriate in the context of elements of field teaching relating to real world environments which are inherently dangerous or impractical to access, as well as those environments in which direct experimentation is difficult or impossible. Affordance (4) is more difficult, but perhaps augments (1) and (2) by situating learning in a simulation of a real geographical environment. Hitherto the discussion has been strongly orientated towards outwards facing teaching (or learning) *with* GIS&T. In terms of teaching (or learning) *about* GIS&T one might anticipate that the same affordances apply i.e., (1), (2) and (4) apply but the nature and emphasis of each changes with the knowledge domain becoming more abstract (Dalgarno and Lee, 2010) and hooks to the real world becoming of lesser importance.

Section chapters

In the discussion that follows, each of the five affordances noted above is associated with the technology described in the first three chapters of the section using a scale in which × denotes negative association and √ is used to indicate positive association (and √√ indicates strong and √√√ stronger positive association).

Chapter 18: Virtual Geographic Environments – Gary Priestnall, Claire Jarvis, Andy Burton, Martin Smith, Nick Mount

1	2	3	4	5
√√√	√√√	(√)	√√√	×

Gary Priestnall *et al.* provide a comprehensive overview of what they term *virtual geographic environment* (VGE) technologies, although their conceptualization of this domain contrasts with the definition above in covering all VR environments used in geography. The focus of the subsequent discussion is primarily on the use of semi- and fully-immersive VR technologies which are closest to a 'traditional' VR system with dedicated hardware and software systems and specialist software to enable the creation of visually highly realistic environments. However, mirror worlds (specifically virtual globes), along with SL, the use of games engines and augmented reality (AR) are also discussed.

Priestnall and colleagues detail the specific work flow steps required to build a VGE from the ground up including addressing source data (via for example digital photogrammetry and LiDAR) as well as required levels of detail and specific object creation (for example specific types of vegetation). Although affordance (1) – enhanced spatial knowledge representation of the specific domain may be significant in a semi- or fully immersive VGE, as the environments are constrained in geographical scope and extent and allow limited access, freedom to explore is more difficult. What is lost in terms of width of view is gained in terms of depth of view and greater representational fidelity. The ability to set up specific experiential tasks and contextualise learning (affordances 2 and 4) related to specific domains are significantly enhanced and would be strongly associated with this technology. The use of an immersive VR model in the context of understanding GNSS satellite availability (Li *et al.*, 2010) is a specific teaching and learning example.

One element which comes through from all the chapters considering virtual words, is the considerable scope and versatility of the technology such that both teaching *with* and teaching *about* contexts can be facilitated. As Priestnall *et al.* note:

> VGEs can be used to facilitate teaching regarding GI principles [...] visualising the unseen [...] and to reveal mathematical form in a visual manner to those without a background in mathematical notation (Priestnall *et al.*, page 281–282).

A further issue which arises from the chapter is the question of which technology to adopt in a given pedagogic situation. With an increasing number of options to choose from this is a difficult choice. Priestnall *et al.* suggest a four phase sequential approach which combines both technological and pedagogic issues. Phase one encompasses task and audience definition followed by design issues (phase 2), technical implementation (phase 3) and then evaluation (phase 4). Here, the fitness-for-use of the technology is directly tied to specific learning outcomes which guide the desired degree of immersion and level of visual realism required:

> Thus, for example, stylised vegetation may not be appropriate for biogeographical learning where careful rendering of vegetation assemblages may be important, but where the emphasis is on landscape form the simpler photo-drape over a more detailed terrain model may be highly effective (Priestnall *et al.*, page 283)

What is certain is that beyond descriptions of the technology and example applications, the interweaving of the technology within a curricular context is essential.

Chapter 19: Using Web-based GIS and Virtual Globes in Undergraduate Education – Lynn Songer

1	2	3	4	5
√√	√(√)	(√)	√	×

In the chapter by Lynn Songer the use of both web-based GIS, and the virtual globes of mirror worlds are explored in the context of undergraduate geography education with a number of examples described. Certain issues which emerge from this chapter echo comments made both elsewhere (Patterson, 2007; Schulz *et al.*, 2008) and elsewhere in this book on the utility of virtual globes to support *spatial thinking*. The value of spatial thinking is clearly of relevance more broadly in geography but equally so in other disciplines (see Chapter 16 in this volume). However, there are clear limitations in using virtual globes where both the options for spatial analysis and the representational fidelity are limited, and there is no co-presence. Although mashups add significant functionality, this clearly limits affordance (1) relative to the more immersive systems described by Priestnall *et al.* and the lack of virtual co-presence excludes affordance (5).

In spite of some of the limitations, Songer identifies that one important characteristic of virtual globe technology is that it allows students the freedom to *explore* their virtual environments. Depending on whether the teaching and learning focus is using GIS&T *with* or *about*, this might range from exploring spatial/geographic elements of the geographic domain under consideration to exploring the fundamental spatial model and concepts underlying the virtual globe model (Goodchild, 2008). Affordances (1) and (2) are clearly relevant in an exploration context and also possibly (4) where students have the ability to think spatially. The ability to explore not only offers opportunities for students to become autonomous active learners but has implications for the teacher in scaffolding learning where a role change from information gate-keeper to learning facilitator/choreographer (de Freitas and Neumann, 2010) may be appropriate. This highlights a particular tension in that allowing students the freedom to explore may be contrary to the need to direct the learning process. Notwithstanding such a concern, de Freitas and Neumann (2010) have suggested that an appropriate model of learning within virtual worlds more generally might be directly related to exploration. Building on the experiential learning cycle of Kolb (1984) they have suggested an *exploratory model of learning* by which students gain experience directly from the exploration in, and reflection upon, virtual environments.

Songer also notes two significant advantages of virtual globe technology in contrast to more complex off the self GIS systems: it is relatively no/low cost simple technology with a shallower learning curve. Certainly these overcome well-known obstacles to using GIS technology in learning and teaching, and in this sense the rationale for use is similar to many of those which motivate free and open source software development. This contrasts with the semi/fully immersive VR and AR environments discussed by Priestnall *et al.* (Chapter 18) which are of significantly greater cost/complexity and the Second Life environment (Mount and Priestnall, this volume, DeMers, 2010) with significantly greater complexity and steeper learning curves.

Chapter 20: Trying to Build a Wind Farm in a National Park: Experiences of a Geocollaboration Experiment in Second Life – Nick Mount and Gary Preistnall

1	2	3	4	5
√	√(√)	(√)	√√	√√√

In the third chapter in Section 4 for this book, Nick Mount and Gary Priestnall use Second Life (SL) in a specific teaching and learning exercise: the decision making process around locating a wind-farm in the English Lake District. In addition to establishing a work flow for the inclusion of spatial data into SL (where the message is that this is not easy but can be done for a limited set of representations including surfaces) they include and assess a variety of student responses collected from post-exercise evaluation. From these responses it is clear that as anticipated the limited representational fidelity of real geographic environments in SL is a clear constraint. However, the clear advantage of SL (and in contrast to the examples above) is that co-presence allows for synchronous communication and collaboration within the virtual world. In other words, SL offers scope for *significant learner interaction*: the other main learning advantage of 3D environments suggested by Dalgarno and Lee (2010). By interacting with geospatial data this also enables another form of collaboration termed *geocollaboration* (after MacEachren and Brewer, 2004). Although students responded favourably about the ability to interact, an unanticipated finding was that students also found the environment to be in some ways limiting, and that virtual interaction in some respects was a poor substitute for face-to-face interaction. Of the technologies considered in Section 4, SL is the only one to allow co-presence, and therefore scores highly on affordance (5) allowing learner interaction. Given the current emphasis placed on collaboration/interaction in learning and teaching theory, this particular affordance of SL clearly explains the use of this 3D virtual environment in teaching. But although the environment is configurable, the limitations in terms of spatial data inclusion and representational fidelity must limit affordance (1) in comparison to the other technologies considered.

Away from the immediate context of virtual worlds, the use of mobile and LBS technology in the contexts of both teaching *with* and teaching *about* is the subject of the chapter by Mountain. Sharples *et al.* (2009) have observed that until recently the assumption has often been made that with the use of technology, most learning occurs inside the lecture theatre or practical laboratory. Mountain typifies the view from the mobile learning community and argues that:

> the use of mobile platforms allows the untethering of education, promoting learning in places and at times of the learners choosing, breaking out from traditional teaching environments (Mountain, page 340).

This provides an extremely persuasive view of a dimension of education which will surely grow in significance – the deployment of mobile technologies enabling collaborative learning in informal and situated environments. A good illustration of this is provided by Clough (2010) who examines the potential for informal collaborative learning using Web 2.0 and GPS in the context of *geocaching*. Echoing comments made above about versatility, Mountain provides a number of examples of the different learning and teaching contexts in which LBS can be used, ranging from the inevitable field teaching to small group discussions about privacy.

The m-learning context provides a final example of the development of bespoke learning theory. Sharples *et al.* (2007) have stressed the importance of conversation and communication in learning and have proposed a model of m-learning based on Engström's (1987) activity model. Here, Sharples *et al.* (2007) envisage learning as a framework which comprises not only an explicit tool role for technology, mediating between subject (the user) and object (the information), but also an explicit *semiotic* element where learning resources act as semiotic signs to mediate between subject (learner) and object (knowledge and skills). Although the theory has been developed and used in a mobile learning context, this model may well be appropriate for more generic adoption in the context of any of the technologies making up GIS&T.

22.4 Conclusion

The contention made here, is that the dialectical relationship between GIS and its constituent and related technologies, and the convergence of those technologies has meant that it is becoming increasingly difficult to identify where the boundaries of GIS start/stop and other technologies begin. In the context of Weiser's (1991) view of profound yet disappearing technology, elements of GIS have clearly converged with other digital worlds technologies (including the web) both inside and outside of the computer. Whether we call the result the first steps towards GIS/2 (Miller, 2006), a VGE (in the sense of Gong *et al.*, 2009) or the emergent *metaverse* (Smart *et al.*, 2007) the scope of GIS has expanded considerably.

Whether or not this means GIS as we knew it is dead, it most certainly means that GIS&T is a very different beast.

The result of this transformation has profound implications for education and the separate dialectical relationship between education and technology in the context of GIS&T raises all sorts of fundamental questions. These range from questions about the identity of learners: what, where, and how are they learning, as well as similar questions related to teachers; who they are, what, where and how are they teaching. A consequence of democratization may be a loosening of control which may be difficult to manage in an education context. A consequence of ubicomp may be a societal *technological unconscious* (Thrift, 2004). The combination of Moore's law, the rise of neogeography coupled with the emergence of new theories and practice of learning with technology firmly opens Pandora's box in terms of delivering GIS&T education.

Acknowledgements

The University of Leicester is acknowledged for granting a sabbatical to the author which enabled the writing of this chapter.

References

Aurambout, J.-P., and C. J. Pettit (2008) Digital globes: Gates to the digital Earth, IN: M. Ehlers, K. Behnke, F.-W. Gerstengarbe, F. Hillen, L. Koppers, L. Stroink, and J. Wächter (eds) *Digital Earth Summit on Geoinformatics 2008: Tools for Global Change Research*. Heidelberg: Wichmann, pages 233–238.

Aurambout, J.-P., C. J. Pettit, and H. Lewis (2008) Virtual globes, the next GIS, IN: C. Pettit, W. Cartwright, I. Bishop, K. Lowell, D. Pullar, and D. Duncan (eds) *Landscape Analysis and Visualisation: Spatial Models for Natural Resource Management and Planning*. Berlin: Springer, pages 509–532.

Bainbridge, W. S. (2007) The scientific research potential of virtual worlds. *Science*, 317: 472–476.

Batty, M., A. Hudson-Smith, R. Milton, and A. Crooks (2010) Map mashups, Web 2.0 and the GIS revolution. *Annals of GIS*, 16(1): 1–13.

Brimicombe, A. J., and C. Li (2009) *Location-Based Services and Geo-Information Engineering*. Chichester: Wiley.

Butler, D. (2008) The web-wide world. *Nature*, 439: 776–778.

Clough, G. (2010) Geolearners: Location-based informal learning with mobile and social technologies. *IEEE Transactions on Learning Technologies*, 3(1): 33–44.

Coleman, D. J., Y. Georgiadou, and J. Labonte (2009) Volunteered geographic information: The nature and motivation of produsers. *International Journal of Spatial Data Infrastructures Research*, 4: 332–358.

Craglia, M., M. F. Goodchild, A. Annoni, G. Camara, M. Gould, W. Kuhn, D. Mark, I. Masser, D. Maguire, S. Liang, and E. Parsons (2008) Next-generation digital Earth. *International Journal of Spatial Data Infrastructures Research*, 3: 146–167.

Crampton, J. (2009) Cartography: Maps 2.0. *Progress in Human Geography*, 33(1): 91–100.

Dangermond, J. (2010) GIS on the Web: A Big Step. http://where2conf.com/where2010/public/schedule/detail/12406 (accessed March 2, 2011).

Dalgarno, B., and M. J. W. Lee (2010) What are the learning affordances of 3-D virtual environments? *British Journal of Educational Technology*, 41(1): 10–32.

de Freitas, S. (2008) *Serious Virtual Worlds: A Scoping Study.* Bristol: Joint Information Systems Committee.

de Freitas, S., and T. Neumann (2009) The use of 'exploratory learning' for supporting immersive learning in virtual environments. *Computers & Education*, 52: 343–352.

Devisch, O. (2008) Should planners start playing computer games? Arguments from SimCity and Second Life. *Planning Theory & Practice*, 9(2): 209–226.

DeMers, M. (2010) Second Life as a surrogate for experiential learning. *International Journal of Virtual and Personal Learning Environments*, 1(2): 17–31.

Duguid, P. (2005) The art of knowing: Social and tacit dimensions of knowledge and the limits of the community of practice. *The Information Society*, 21: 109–118.

EduCause (2008) *7 Things You Should Know About Second Life.* http://net.educause.edu/ir/library/pdf/ELI7038.pdf (accessed March 2, 2011).

Engström, Y. (1987) *Learning by Expanding: An Activity Theoretical Approach to Developmental Research.* Helsinki: Orienta Konsultit.

Evans, C. (2008) The effectiveness of m-learning in the form of podcast revision lectures in higher education. *Computers & Education*, 50: 491–498.

Fisher, P. F., and D. J. Unwin (eds) (2002) *Virtual Reality in Geography.* New York: Taylor & Francis.

Foresman, T. W. (2008) Evolution and implementation of the Digital Earth vision, technology and society. *International Journal of Digital Earth*, 1(1): 4–16.

Gee, J. P. (2007) *What Video Games Have to Teach Us About Learning and Literacy.* New York: Palgrave MacMillan.

Goodchild, M. F. (2007) Citizens as sensors: The world of volunteered geography. *GeoJournal*, 69(4): 211–221.

Goodchild, M. F. (2008) What does Google Earth mean for the social sciences? IN: M. Dodge, M. McDerby, and M. Turner (eds) *Geographic Visualization: Concepts, Tools and Applications*, Chichester: Wiley, pages 11–23.

Goodchild, M. F. (2009) Neogeography and the nature of geographic expertise. *Journal of Location Based Services*, 3(2): 82–96.

Gong, J., W. Li, Y. Li, and M. Huang (2009) Networked collaborative virtual geographic environments: Design and implementation, IN: H. Lin, and M. Batty (eds) *Virtual Geographic Environments.* Beijing: Science Press.

Gordon, E., and E. Manosevitch (2010) Augmented deliberation: Merging physical and virtual interaction to engage communities in urban planning. *New Media & Society*, 13(1): 75–95.

GSA (2010) *GNSS Market Report*. Issue 1 October 2010. http://www.gsa.europa.eu/go/download the gsa gnss market-report (accessed March 11, 2011).

Haklay, M., and P. Weber (2008) OpenStreetMap: User-generated street maps. *Pervasive Computing*, 7(4): 12–18.

Hazas, M., J. Scott, and J. Krumm (2004) Location-aware computing comes of age. *IEEE Computer*, 37(2): 95–97.

Hendaoui, A., M. Limayem, and C. W. Thompson (2008) 3D social virtual worlds: Research issues and challenges. *IEEE Internet Computing*, 12(1): 88–92.

Hodgins, W. (2010) What shapes the future of learning? The third industrial revolution, IN: N. Paine, and E. Masie (eds) *The Masie Center's Learning Perspectives 2010.* Saratoga Springs, NY: The MASIE Center and The Learning CONSORTIUM, pages 119–131.

Hudson-Smith, A., M. Batty, A. Crooks, and R. Milton (2009b) Mapping for the masses accessing Web 2.0 through crowdsourcing. *Social Science Computer Review*, 27(4): 524–538.

Hudson-Smith, A., and A. Crooks (2009) The renaissance of geographic information: Neogeography, gaming and virtual environments, IN: H. Lin, and M. Batty (eds) *Virtual Geographic Environments.* Beijing: Science Press, pages 25–36.

Hudson-Smith, A., A. Crooks, M. Gibin, R. Milton, and M. Batty (2009a) Neogeography and Web 2.0: Concepts, tools and applications. *Journal of Location Based Services*, 3(2): 118–145.

Jiang, B., and X. Yao (2006) Location-based services and GIS in perspective. *Computers, Environment and Urban Systems*, 30: 712–725.

Johnson, S. (2006) *Everything Bad is Good for You*. London: Penguin.

Johnson, L., A. Levine, and R. Smith (2009) *The 2009 Horizon Report*. Austin, TX: The New Media Consortium.

Kolb, D. A. (1984) *Experiential Learning*. Englewood Cliffs, NJ: Prentice Hall.

Lave, J., and E. Wenger (1991) *Situated Learning: Legitimate Peripheral Participation*. Cambridge: Cambridge University Press.

Ley, D. (2007) Ubiquitous computing, IN: *British Educational Communications and Technology Agency Emerging Technologies for Learning: Volume 2*. Coventry: Becta, pages 64–79.

Li, J., C. H. Jarvis, and C. Brunsdon (2010) The use of immersive real-time 3D computer graphics for visualisation of dilution of precision (DOP) in virtual environments. *International Journal of Geographical Information Science*, 24: 591–605.

Liu, S. B., and L. Palen (2010) The new cartographers: Crisis maps mashups and the emergence of neogeographic practice. *Cartography and Geographic Information Science*, 37(1): 60–90.

MacEachren, A. M., and I. Brewer (2004) Developing a conceptual framework for visually enabled geocollaboration. *International Journal of Geographical Information Science*, 18(1): 1–34.

McLellan, H. (1996) Virtual realities, IN: D. H. Jonassen (ed) *Handbook of Research for Educational Communications and Technology*. New York: Macmillan, pages 461–497.

Miller, C. C. (2006) A beast in the field: The Google Maps mashup as GIS/2. *Cartographica*, 41(3): 187–199.

Patterson, T. C. (2007) Google Earth as a (not just) geography education tool. *Journal of Geography*, 106: 145–152.

Paavola, S., and K. Hakkarainen (2005) The knowledge creation metaphor – an emergent epistemological approach to learning. *Science and Education*, 14: 535–557.

Paavola, S., L. Lipponen, and K. Hakkarainen (2004) Models of innovative knowledge communities and three metaphors of learning. *Review of Educational Research*, 74(4): 557–576.

Riedl, A. (2007) Digital globes, IN: W. Cartwright, M. P. Peterson, and G. Gartner (eds) *Multimedia Cartography*, 2nd ed., Berlin/Heidelberg: Springer Verlag, pages 255–266.

Roush, W. (2007) Second Earth. *Technology Review*, 110(4): 39–48.

Schultz, R. B., J. J. Kerski, and T. C. Patterson (2008) The use of virtual globes as a spatial teaching tool with suggestions for metadata standards. *Journal of Geography*, 107: 27–34.

Sfard, A. (1998) On two metaphors for learning and the dangers of choosing just one. *Educational Researcher*, 27(2): 4–13.

Sharples, M., I. Arnedillo-Sánchez, M. Milrad, and G. Vavoula (2009) Mobile learning: Small devices, big issues, IN: N. Balacheff, S. Ludvigsen, T. de Jong, A. Lazonder, and S. Barnes (eds) *Technology-Enhanced Learning: Principles and Products*. Heidelberg: Springer, pages 233–249.

Sharples, M., and J. Roschelle (2010) Guest editorial: Special issue on mobile and ubiquitous technologies for learning. *IEEE Transactions on Learning Technologies*, 3(1): 4–5.

Sharples, M., J. Taylor, and G. Vavoula (2007) A theory of learning for the mobile age, IN: R. Andrews, and C. Haythornthwaite (eds) *The Sage Handbook of E-learning Research*. London: Sage, pages 221–247.

Smart, J., J. Cascio, and J. Paffendorf (2007) *Metaverse Roadmap 2007: Pathways to the 3D Web*. http://www.metaverseroadmap.org/MetaverseRoadmapOverview.pdf (accessed March 1, 2011).

Sui, D. (1995) A pedagogic framework to link GIS to the intellectual core of geography. *Journal of Geography*, 94(6): 578–591.

Sui, D. (2005) Will ubicomp make GIS invisible? *Computers, Environment and Urban Systems*, 29(1): 361–367.

Sui, D. (2008a) The wikification of GIS and its consequences: Or Angelina Jolie's new tattoo and the future of GIS. *Computers, Environment and Urban Systems*, 32(1): 1–5.

Sui, D. (2008b) Geography and GIS enter a Second Life. *Geoworld*, September 2008: 16–17.

Steiniger, S., M. Neun, and A. Edwardes (2006) *Foundations of Location Based Services.* CartouCHe1 – Lecture Notes on LBS, V. 1.0 http://citeseerx.ist.psu.edu/viewdoc/download?doi=10.1.1.94.1844&rep=rep1&type=pdf (accessed March 4, 2011).

Tiede, D., and S. Lang (2010) Analytical 3D views and virtual globes – scientific results in a familiar spatial context, *ISPRS Journal of Photogrammetry and Remote Sensing*, 65(3): 300–307.

Thompson, C. W. (2011) Next generation virtual worlds: Architecture, status and directions. *IEEE Internet Computing*, 15(1): 60–65.

Thompson, S. E., and S. Parthasarathy (2006) Moore's law: The future of Si microelectronics. *Materials Today*, 9(6): 20–25.

Thrift, N. (2004) Remembering the technological unconscious by foregrounding knowledge of position. *Environment and Planning D: Society and Space*, 22: 175–190.

Varshavsky, A., and S. Patel (2009) Location in ubiquitous computing, IN: J. Krumm (ed) *Ubiquitous Computing Fundamentals.* Boca Raton: CRC Press, pages 286–319.

Want, R. (2009) An introduction to ubiquitous computing, IN: J. Krumm (ed) *Ubiquitous Computing Fundamentals.* Boca Raton: CRC Press, pages 1–36.

Warf, B., and D. Sui (2010) From GIS to neogeography: Ontological implications and theories of truth. *Annals of GIS*, 16(4): 197–209.

Weiser, M. (1991) The computer for the 21st century. *Scientific American*, 265: 94–104.

Wood, J., J. Dykes, A. Slingsby, and K. Clarke (2007) Interactive visual exploration of a large spatio-temporal data set: Reflections on a geovisualization mashup. *IEEE Transactions on Visualization and Computer Graphics*, 13(6): 1176–1183.

Section V

Distance and e-learning

23

Media and communications systems in cartographic education

William Cartwright

School of Mathematical and Geospatial Sciences, RMIT University, Melbourne, Victoria, AUSTRALIA

23.1 Introduction

Classroom-based geospatial teaching and learning have traditionally depended upon paper artifacts, such as books, maps, atlases and charts. For schools in remote areas of Australia, where teachers supported entire communities and teaching resources were far removed from learning locations, students relied on supplementary materials delivered via correspondence school mail, School of the Air two-way radio, telephones, video and television, videotex (using Viatel in Australia), computers and satellite systems. Now teaching and learning is supported through materials delivered on CD-ROM, on computers and via the World Wide Web.

Education in Australia has always used new media and innovative methodologies for teaching existing subject matter differently and to explore how new topics can be included in curricula. Increased public access to computers and networks has led to an awareness that resources like discrete multimedia products and their distributed counterparts on the World Wide Web have revolutionized the way in which information is both accessed and used. Australian educators embraced the use of interactive multimedia when it became available on modest computer platforms and accessible via large-scale communications resources (Bearman, Kidd and Cesnik, 1995). These resources have proven useful for students throughout Australia, not only for students in remote areas of the country.

This section of the chapter provides a general overview of the technologies that have been used to provide education in rural and outback areas, so as to build a picture of the foundations of innovative use of technology to support educational delivery. It also showcases how Australian educators have embraced technology as a means of providing quality education that would otherwise be impossible to deliver. These technologies might have been as simple as using the mail to facilitate the movement of teaching materials

Teaching Geographic Information Science and Technology in Higher Education, First Edition.
Edited by David J. Unwin, Kenneth E. Foote, Nicholas J. Tate and David DiBiase.

between teachers and their (virtual) classroom participants, or more technology-enhanced means like the phonograph, television and the web, bringing the world to small, isolated classrooms. This history of the trials and methods that have been employed pre-New Media and pre-web provides a background to the later parts of this chapter, which address the use of contemporary communications technology.

23.2 Communications systems and the delivery of distance education

As noted previously, there has always been a demand for distance education in Australia. For example, at the turn of the century in Queensland, itinerant teachers visited hundreds of families on several occasions throughout a school year to provide educational support that supplemented the home schooling children received (Fowler, 1987). Over time, communications systems, initially the post, and later radio and satellite communications, were used to provide educational support. This section of the chapter briefly outlines some of these communications and technological applications.

Correspondence schools

In the State of Queensland, a state with a very low population in primarily rural areas, the Primary Correspondence School was fully developed in 1922. At its peak in World War II there were over 7000 children in the system. In 1986 there were still 1700 children enrolled (up to grade 7) (Fowler, 1987). From a Brisbane base, where all materials were produced and marked, the mail system was used to post educational resources, either weekly (grades 1 to 3) or fortnightly. Some of these printed materials were supplemented with audiotapes when this became feasible. A 'Home Tutor', usually a parent or governess, was also part of the education 'chain'. This person organized and supervised the course (Fowler, 1987). In Australia primary school students study the rudimentary elements of map use. Several excellent printed teaching materials have been produced, like the *Jacaranda Primary Atlas 3E*, which won the Australian Geography Teachers Association (AGTA) 2008 Award – Primary School Resources (Wiley Australia, 2008) and its complementary teachers package – *Jacaranda Primary Atlas 3E Teacher Resource Kit* and Making Maps (Kelly et al., 1992),

Correspondence courses fitted the need to provide of tertiary educational courses for individuals employed in the mapping industry, which often required extended periods in the field or operated in rural areas where no suitable course were otherwise available. So-called 'Technical Colleges' developed courses in close collaboration with industry. Technical Colleges delivered notes, practical assignments and examinations via the Australian Government's Post Master General's department (PMG) (now Australia Post).

In the State of Victoria the Melbourne Technical College (renamed the Royal Melbourne Institute of Technology in 1960, and later RMIT University (Murray-Smith and Dare, 1987)) developed and delivered a course in 'Survey Draughting' for cartographic draughtsmen and survey assistants. An independent school – the Melbourne Technical College Correspondence School was established to develop and deliver correspondence courses (Murray-Smith and Dare, 1987). Comprehensive notes and assignments were posted to students, who undertook the prescribed tasks and mailed the completed work back to their correspondence teacher. The cover (Figure 23.1) and contents (Figure 23.2) of typical documentation for such a course is provided below.

Melbourne Technical College

124 LATROBE ST., MELBOURNE

CORRESPONDENCE SCHOOL

ENGINEERING SURVEYING A

PAPERS 4 - 6

Prepared by

ARNOLD WILSON

Licensed Surveyor Vic.
Authorised Surveyor Tas.
M.V.I.S. & M.T.I.S.

1694

Figure 23.1 Melbourne Technical College Correspondence School – Engineering Surveying A – Papers 4–6 (Wilson, ca. 1971)

ENGINEERING SURVEYING A
PAPER 4 (CONTD.)

TRADING A SPORTS GROUND

Take spot levels on the grid system say not more than 50 feet intervals. Write out your levels on graph paper, each inch representing 50 feet. For diagrammatic purposes make a latum reduction from all levels so that you are dealing with levels reading between 0 and 10 (assuming here that there is no more than 10′ difference of levels in the arena.

To prepare a modified ISOMETRIC profile showing the actual grades in both directions along the grid lines proceed as follows. Let us say that grid level at A is 4.00. On both the Northing and Easting lines through A (on the grid) mark off 4.00 on scale 10′ to an inch, and project the points to intersect at L. Similarly plot the points M, N and O, all fall on lines at 45° inclined to the grid datum. Connect these points by lines running northerly and southerly and easterly and westerly.

This will give you approximate profiles in both directions from which you can visualise the 3 dimensions–length, breadth and depth. Make use of colored cotton threads to get an idea of ultimate grading along each line.

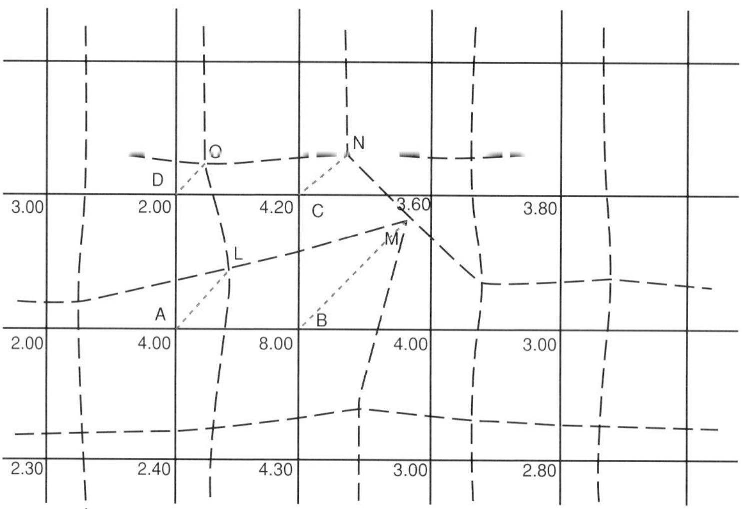

SKETCH FOR GRADING SPORTS GROUND

Figure 23.2 Melbourne Technical College Correspondence School – Engineering Surveying A – Papers 4–6: assignment detail. (Wilson, ca. 1971)

Students were given specific deadlines for submission of assignments, which was specified within the notes (Figure 23.3). The written notes allowed students to pace themselves, as most students undertook the course on a part-time basis.

The Australian State of Victoria was not the only state to provide correspondence courses and similar courses were provided in New South Wales.

School of the Air

The School of the Air (SOTA) started in Cloncurry, Queensland in 1960. It was designed to supplement Correspondence School materials by providing oral elements (Fowler, 1987).

NOTE TO CORRESPONDENCE STUDENTS

You must complete the assignments at the end of each unit and send them progressively to the College of External Studies for correction and comment.

Each unit will occupy approximately two weeks of your time. Your first test will occur after Unit 8.

Figure 23.3 Melbourne Technical College Correspondence School – Engineering Surveying A – Papers 4–6: 'Note to Correspondence Students. (Wilson, ca. 1971)

The School of the Air was the means by which many Australians in remote areas received their education. For example, the Cairns School of the Air offered radio lessons to children enrolled with the Primary Correspondence School in Brisbane and used transceivers. The radio frequency used was that of the Royal Flying Doctor Service. Western Australia had five School of the Air centres (serving 135 students) and one South Australian SOTA was based at Port Augusta.

UHF radio was also used to connect schools and children in classrooms. The Victorian Country Education Project trials used two-way UHF radio to provide one-way and two-way communication. This was later replaced by microwave connections. In 1985 the Western Australia Education Department experimented with satellite communications plus JF radio for 'downloads' and a combination of HF radio, telephone or satellite for return paths (Fowler, 1987).

Telephones

Telephones have long been used to connect students to teachers and students to students for class interaction. In 1983-84 South Australian correspondence schools used loudspeaker telephones for two-way communication. This was especially successful for practical music teaching and languages (Fowler, 1987). Later, with the general access to university-supplied computer systems via modems this method became redundant.

Video and videoconferencing

In the early 1980s the Australian Federal Government introduced its Federal Government Loan Video Scheme, which involved 142 children. The scheme used the ABC's School's Broadcasts and enabled State Education Departments to copy and make available off-air videos. This was supplemented with local (State) education video resources. Education Departments also loaned VCRs to families on an as-needed basis. Late night broadcasting (and subsequent video recording) of lectures was used by the Western Australian College of Advanced Education for its Bed program (Fowler, 1987). Once feasible, videoconferencing was used to facilitate lecture delivery, mainly among multiple campuses.

Video is still used to complement printed matter in courses offered to students in remote areas of Australia. Correspondence education, now called 'distance education' in Australia, still operates, but printed materials are now supplemented with video and discrete computer materials. The University of Southern Queensland's Distance and e-Learning Centre offers degree courses in Spatial Science and Spatial Science Technology (University of Southern Queensland, 2009). The University uses mail to distribute study books, selected readings from journals, newspapers or books, audio tapes and/or video tapes, computer managed learning, booklets, CD ROMs or floppy disks, assignment information, and exercises, tutorial questions and answers.

Videotex/teletext – viatel

Viatel, a public videotext service provided by Telecom, Australia, and based on the UK's *Prestel* service (Inglis, 1984), began operation in 1985. The system enabled the user to retrieve information from databases and to carry out transactions on them, including ordering products and services, banking and sending messages. The basic computer equipment was supplied by G.E.C. but software packages to enable access to Viatel were Australian designs, as was an improved gateway protocol concept. (Australian Academy of Technological Sciences and Engineering, 1988). Viatel was used by the New South Wales Correspondence School, the Education Department of Western Australia (which used the BBC Acorn computer and had the ability to emulate *Prestel* information) and the Education Department of South Australia (Hosie, 1985). Access to courses provided by distance education providers using *Viatel* was facilitated via a service called the Mentor 'Coursefinder', established by Inglis, at RMIT in Melbourne (Atkinson, 1987).

Computers, cables and modems

In the relatively early days of using computers in education Lundin (no date) noted that there were then at least five different types of computer services relevant to education and training, namely:

- electronic mail (email);
- bulletin boards;
- database/library access;
- computer-text conferencing; and
- file transfer

While an effective teaching application as a stand-alone device, the computer facilitated a better approach to distance education when materials were able to be delivered using CD-ROM and then subsequently connected by telephone cables and later, communications systems. These applications are considered later in this chapter.

Atkinson (1986) discussed the use of computer bulletin boards for distance education students. These became popular when home computers became affordable and also when the attitudes of telephone network authorities to data communicators changed in the late 1970s – early 1980s (Atkinson, 1986). This method of student access to educational resources was of great interest in the mid-1980s, when it became apparent that all students needed to be connected was a home computer, communications software, a modem and a telephone. Local area networks (LAN) were also established for on-campus access. In 1985 the Western Australia Schools Computing Branch used electronic mail between a number of its country schools using microcomputers connected via telephone modems. It was used to provide a means to build a student community and to facilitate submission of assignments and essays (Fowler, 1987). The system improved markedly when file transfer ability was added to bulletin board services (Atkinson, 1986). In 1986 there were over 100 bulletin boards in Australia (Withers, 1986, cited in Atkinson, 1986), and, by the end of the 1980s, there were over 180 computer bulletin boards operating in Australia (Withers, 1987, cited in Atkinson, 1987).

Satellites

Queensland Education used satellites as an alternative to radios for remote education provision and as an adjunct to the audio resources it provided by radio. A so-called Homestead and Community Broadcast Satellite Service (HACBSS) enabled remote users with an appropriate satellite dish to receive television programs from the Australian Broadcasting Commission (ABC) and commercial programs as well. Off-peak broadcasting time was used to transmit educational programs for recording and for replay during school hours. AUSSAT, the Australian communications satellite, was used for this purpose (Fowler, 1987). A combination of Apple IIe computers, satellite-delivered videos and a fortnightly two-way video link provided a suite of materials and teaching methods which, in combination, gave students a rich multi-medium experience.

23.3 New Media educational packages

Wikipedia (2010) defines New Media as '. . . a term meant to encompass the emergence of digital, computerized, or networked information and communication technologies in the later part of the 20th century'. The forms of New Media used for education includes Hypertext, digital documents and images (still and moving), sound and interactive products. Modes of New Media delivery included CD-ROM initially, and later the web to provide timely and comprehensive map-enhanced learning suites. Educational products designed to help teach the basics of geography and mapping have long been available. Many were produced as paper products, others as mixed media packages containing printed materials, photographic slides and audiotapes. In time the publishing platform migrated to CD-ROM and then to the web. This section of the chapter provides information about the use of New Media in its widest context. It begins by looking at some of the early trials with New Media formats like microformats and multiple media and then moves to analogue and optical applications, and finally to the web.

Microfiche

Microfiche was entrained as a medium for the storage of maps during the late 1970s and early 1980s. Exploratory work was carried out by a research team led by Massey (Massey, Poliness and O'Shea, 1985) which produced an atlas depicting the socio-economic structure of Australia on thirteen monochrome microfiche. Each microfiche featured one of Australia's capital cities, states and territories. This revolutionary product was jointly produced between the Geography Department at the University of Melbourne and Latrobe Comgraphics. It was used to support the teaching of geography to secondary students in Australia who, prior to the production of this atlas, had no comprehensive and inexpensive mapping resources to study. The microfiche medium was accepted and used in the teaching courses related to geography.

Mixed media

The *Bold Atlas of Australia* was produced for visually impaired students in Australia studying geography at the primary school level. It contained a printed book containing simple maps and an accompanying audiocassette that described the maps. Teachers of sighted children also sought out the *Bold Atlas* for classroom instruction because the simplicity of its innovative map designs made it easier to teach the rudiments of the geography of Australia. Classifications depicted on thematic maps were kept simple, limited to a maximum of just four classes, and the symbology too was straightforward. Pages from the atlas are shown in Figure 23.4.

A number of other mixed media slide-tape applications were developed for use in geography courses for Australian senior high school students and first-year university students. Audio Visual resources, a Melbourne-based company that specialized in slide-tape resources for education commissioned a number of cartography slide-tape packages, developed by Cartwright (1987a-c). Topics included:

- Map production;
- Map reading; and
- Map Types.

RMIT University also produced a slide-tape package in-house that covered map printing (Cartwright, 1983).

Multiple media, later known as multimedia or rich media, made it easier to produce teaching products than the methods used earlier for Roberts' and Cartwright's slide-tape packages. The inclusion of many media items, controlled interactively, revolutionalized how media was accessed. Tertiary educators adopted multimedia enthusiastically, which encouraged the development of more innovative products.

Videodisc

Videodiscs were created for geography education, for example to teach map-reading skills (Carstensen and Cox, 1988). The medium was not used in teaching *per se* in Australia, but it was used at RMIT to illustrate how the technology could be employed to provide rich media representations of geographical artifacts. RMIT Library purchased the *Domesday* videodisc package in 1986 for £4000, at the time a considerable sum of money for a teaching resource. The videodisc package provided an interactive picture of Britain in the 1980s (Atkins,

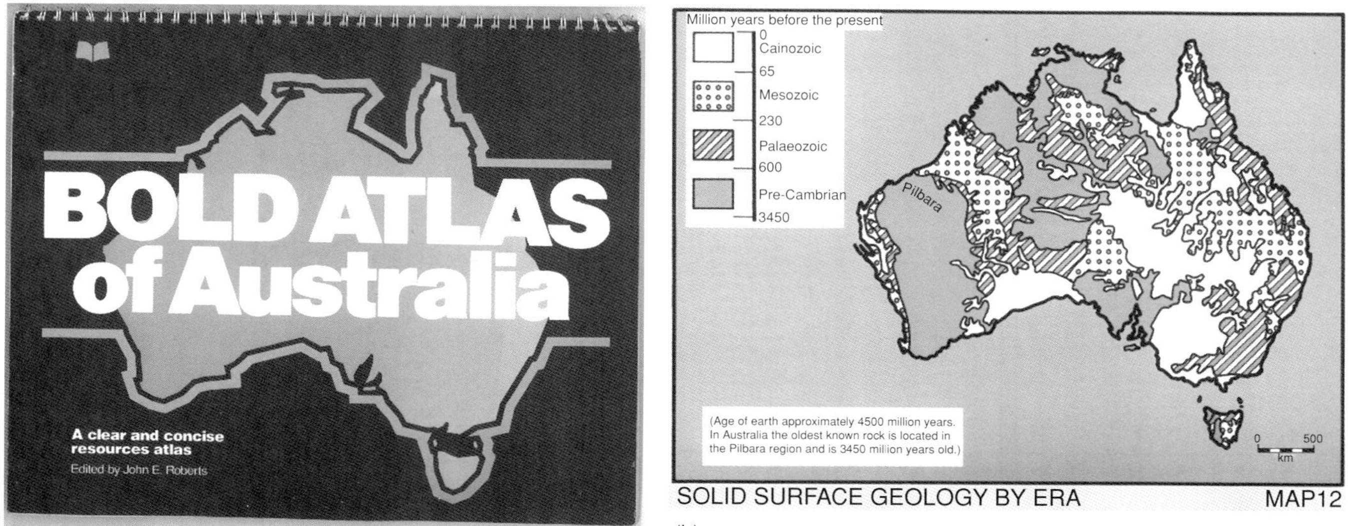

(a) (b)

Figure 23.4 **a.** Bold Atlas of Australia. **b.** Typical page. From The Bold Atlas of Australia edited by John Roberts. Reproduced by permission of Scholastic Australia Pty Limited

1986, Goddard and Armstrong, 1986). It was jointly produced by the British Broadcasting Commission (BBC), Acorn Computers and Philips to commemorate the 900th anniversary of William the Conqueror's tally book.

CD-ROM

The publishing world embraced CD-ROM as an alternative to paper. Book abstracts, bibliographic references and encyclopedias, were published on this optical storage media. Of the many educational packages developed, two example products from Australia are *Geoskills*, a CD-ROM package developed at the School of Human and Environmental Studies the University of New England (Jones, 2000) and *Understanding Map Projections*, a multimedia learning package developed at RMIT University (Cartwright, 1999). A National Teaching Development Grant from the Australian Federal Committee funded the development of *Understanding Map Projections* for University Teaching and Staff Development (CUTSD), 1998. It was designed as an introductory teaching package for tertiary students and senior secondary school students studying geography. As stipulated by funding guidelines, the package was made available free for use in Australian universities and senior high schools.

Understanding Map Projections was designed to teach the concepts of map projections using visual methods to depict properties usually specified using mathematical formulae. It allows students to view graphic portrayals of various map projections and to explore how the continental shapes distort as projection variables are altered. The concept behind project development was that, after using the package, students who previously had little or no knowledge about map projections should have a much greater appreciation of map projections and their use to depict the earth (3 dimensional) on a plane (2 dimensional) paper map or computer monitor. Students would also be better able to identify particular projections and choose the appropriate one to use for the display of particular parts of the earth and different mapping themes.

The package was developed as a modular product to provide teaching support resources, rather than offering a stand-alone course. Its design was based on the assumption that instructors would build their own teaching activities around the package and use the various elements to construct applications that suit a student group using the package. This was to ensure that it could be used at both early tertiary levels and advanced post-primary levels. At RMIT University the package replaced some of the elements previously taught by traditional methods, i.e. lecture and practical exercises.

The package was designed around four modules, which are presented at the 'entry' page (Figure 23.5).

The 'overview' section provides a general introduction to the package, the comprehensive Glossary of Terms and information about the production team. The Glossary contains a generous coverage and explanation of the terms used throughout the package). The 'History' section contains a history of map projections and associated developments using text which provides a treatise on the history of map projections (Figure 23.6) and interactive timelines, covering four periods of the development of map projections – from simple graphical depictions to more accurate mathematically derived constructions (Figure 23.7).

The third element of the package is the 'Projections' section. Users are able to explore the elements of the three types of map projections: cylindrical, azimuthal and conical.

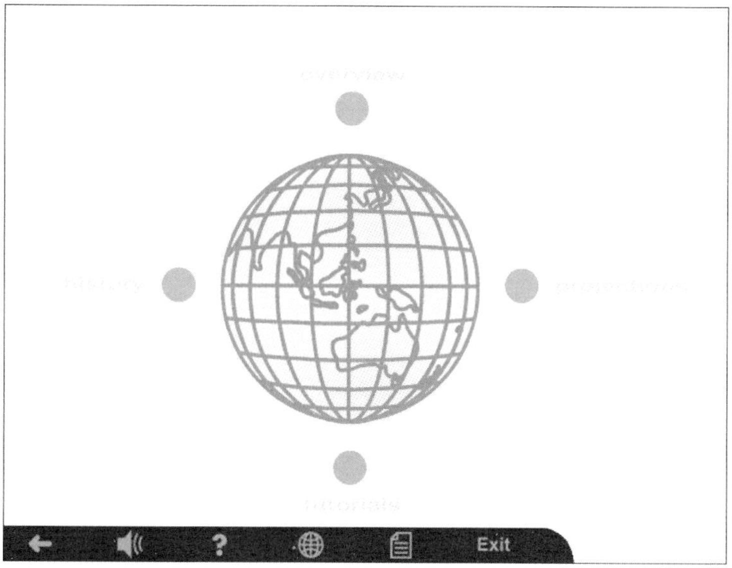

Figure 23.5 'Entry' page of the multimedia package *Understanding Map Projections*

Two 'special' map projection cases are also expanded upon. In the example illustrated in Figure 23.8, a map as a virtual cylinder is wrapped around the globe to illustrate the basic principles of the projection.

Finally, tutorials included in the package allow students to explore the attributes of map projections and discover how to choose one to minimize distortions in area, distance or azimuth. A tutorial based on selecting desired attributes of map projections is shown in Figure 23.9.

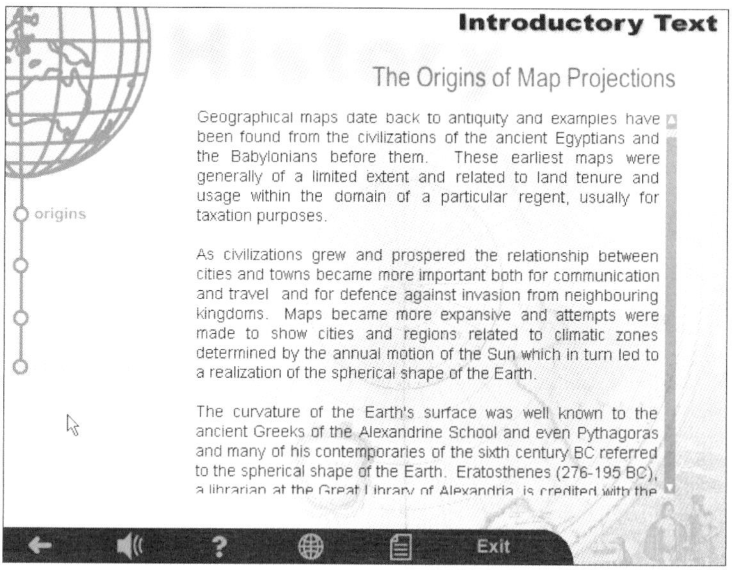

Figure 23.6 Text describing the Origins of Map Projections

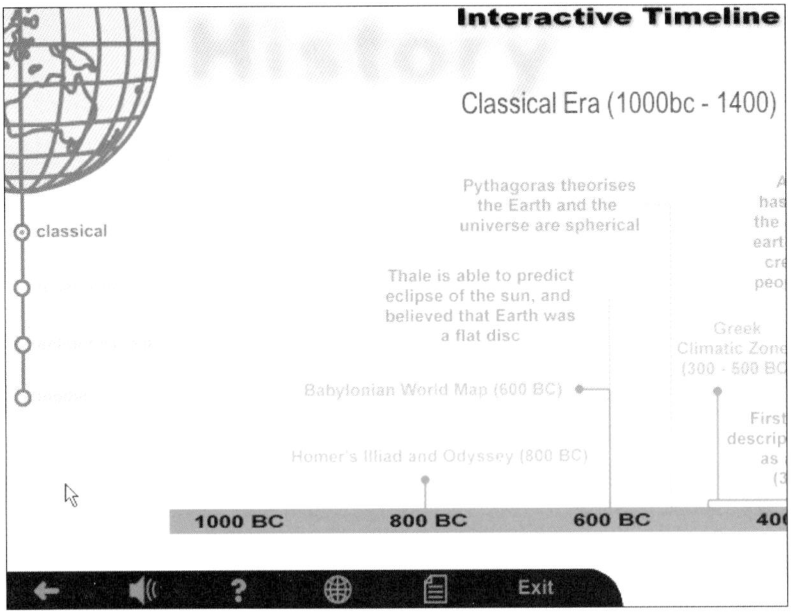

Figure 23.7 Interactive Timeline

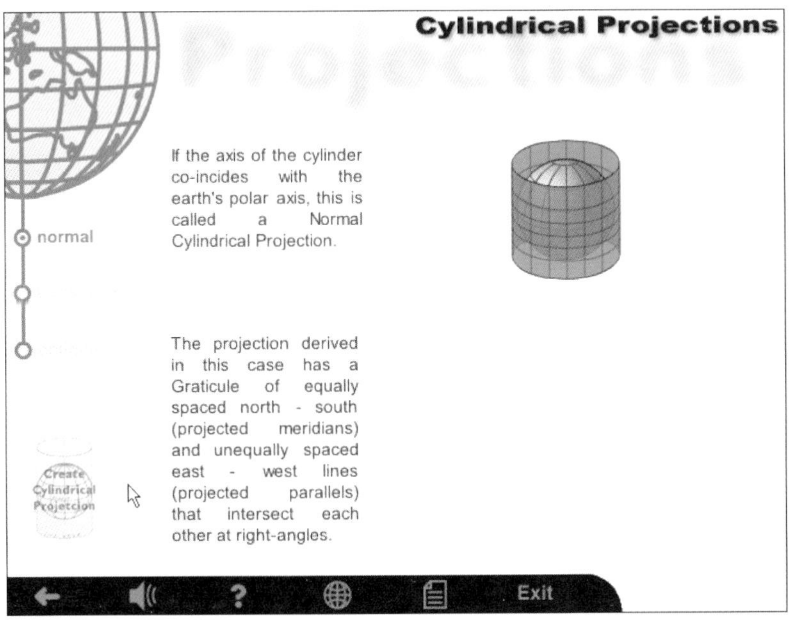

Figure 23.8 Simulation of the Cylindrical Projection basics

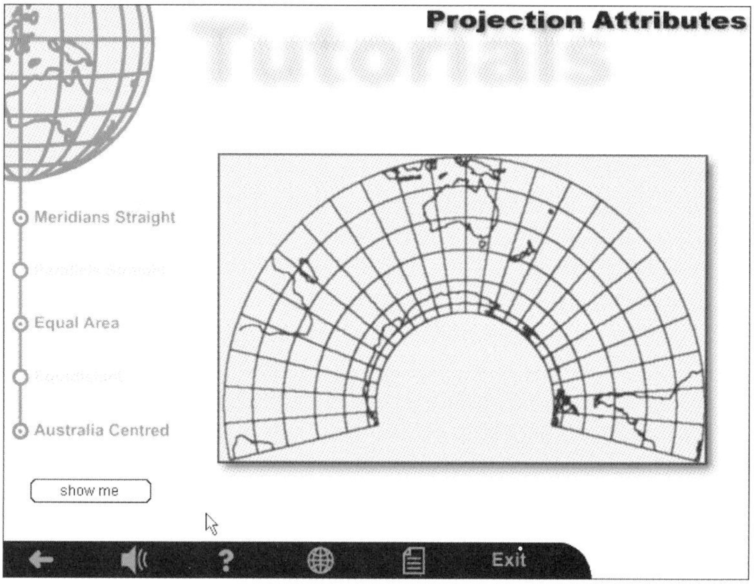

Figure 23.9 Projection attribute selection and resultant map projection example

The CD-ROM package can be mailed to students in remote areas of Australia, where broadband internet access is restricted, or unavailable altogether. In this context, independence from internet access is distinct advantages of the CD-ROM medium.

Online

Beginning in the early to mid-1990s, the World Wide Web provided a new delivery method for new media educational resources. There exist a number of universities that deliver English language courses worldwide using the internet. For example, the British Open University has been a provider of courses to many disciplines, including geography, for many years. But, perhaps the biggest impact was the announcement of the Massachusetts Institute of Technology's OpenCourseWare initiative on 4 April 2001. In his announcement MIT President Charles M. Vest said, 'The idea behind MIT OpenCourseWare (MIT OCW) is to make MIT course materials that are used in the teaching of almost all undergraduate and graduate subjects available on the web, free of charge, to any user anywhere in the world' (MIT, 2001).

Now, many universities, for example Birkbeck College (University of London), City University, and Penn State University, provide web-delivered courses for the spatial sciences globally – if not for free. Birkbeck College developed an on-line MSc in Geographical Information Science by Distance Learning (Harris, 2003). The program was conducted entirely online, providing students with an alternative to face-to-face lectures and tutorials. London's City University Department of Information Science introduced their online master's program in 2000. It also can be completed entirely online, although a face-to-face program is also offered (City University, 2000). Penn State's *World Campus* provides online Master of GIS and Postbaccalaureate Certificates in GIS and Geospatial Intelligence (Directions

Magazine, 2009). These courses allow students to undertake study when and where they are best able to participate.

The web is now considered a viable means to conduct tertiary courses and course materials throughout Australia, including resources for cartographic education. For some Australians a web-based course may be the only course available. Examples of the many web-delivered courses and course components for the geospatial sciences include Curtin University of Technology's School of Spatial Sciences. Curtin offers three courses via the web: the Graduate Certificate in Geographic Information Science, the Graduate Diploma in Geographic Information Science and the Master of Science (Geospatial Science) – Geographic Information Science Stream (Curtin University, 2011). The Faculty of Engineering and Surveying at the University of Southern Queensland offers certificate, diploma, and degree programs in Geographic Information Systems via distance education, including online learning resources (University of Southern Queensland, 2009). And RMIT University provides program components via the Internet, supported by tutors who are engaged to support students via email (Shortis and Cartwright, 2000). Further online provision of courses in the geospatial sciences is viewed as an essential area for future development in Australia.

RMIT University's online modules are facilitated through its Distributed Learning System (DLS). The DLS uses the Blackboard learning management system as the backbone for online course provision. A typical DLS course support page is shown in Figure 23.10.

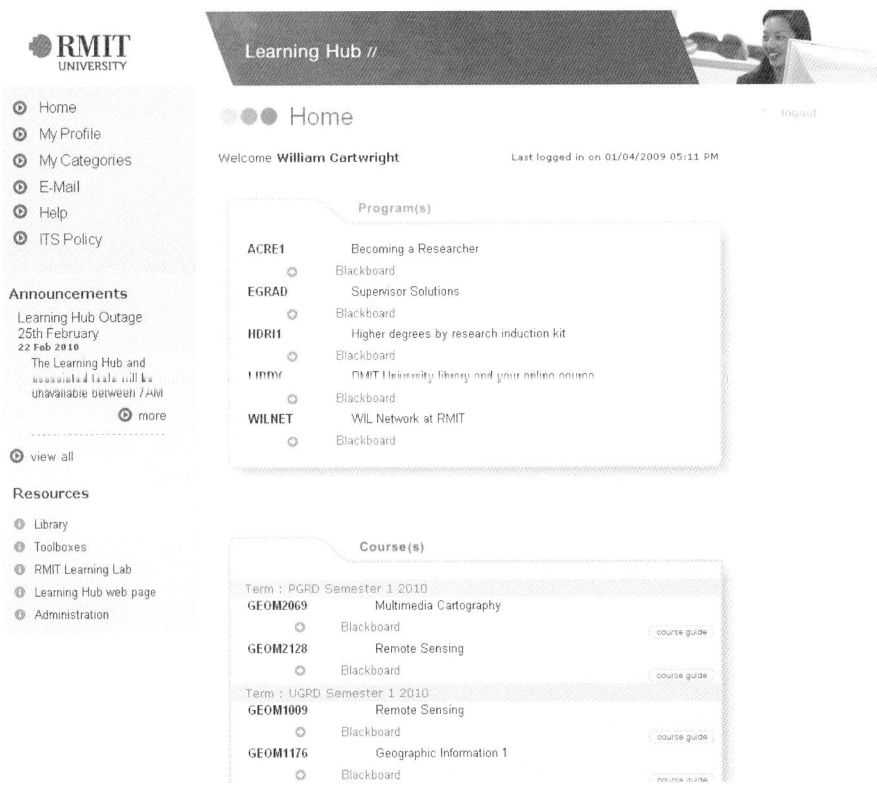

Figure 23.10 RMIT's use of Blackboard to facilitate on-line delivery through a University-wide system – Distributed Learning System (DLS)

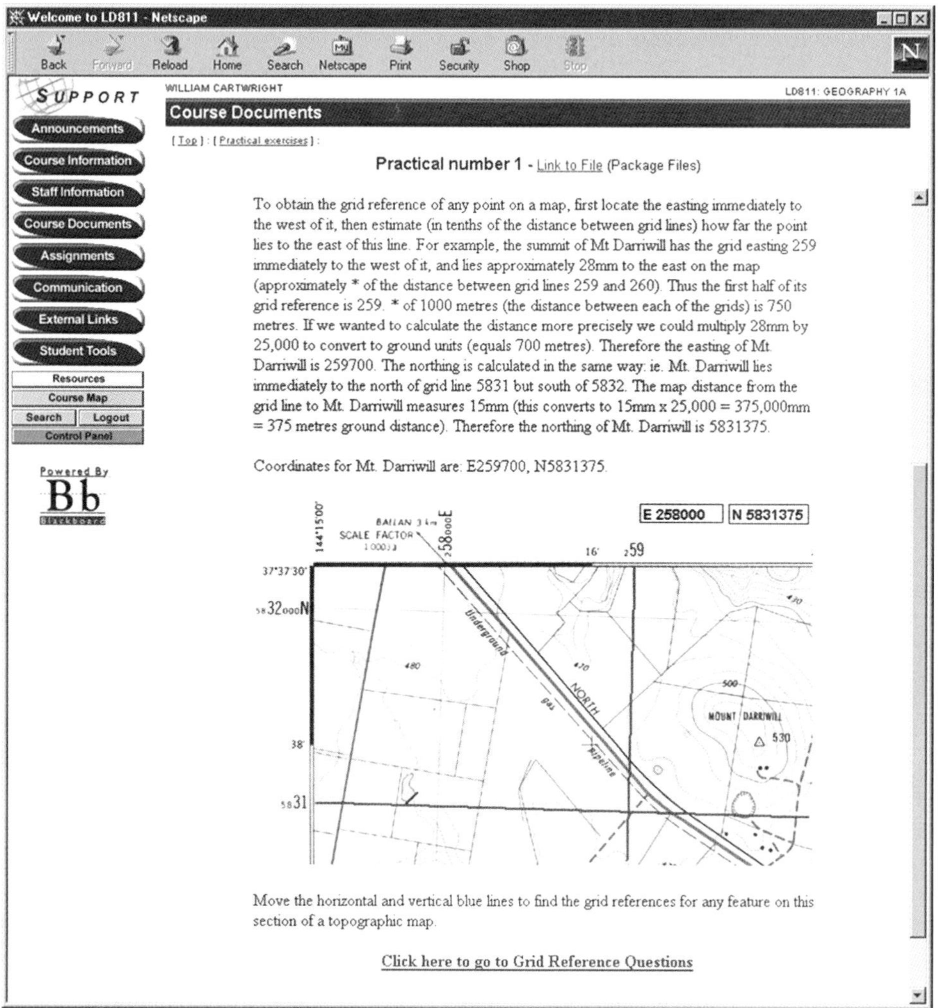

Figure 23.11 RMIT's use of Blackboard to facilitate on-line delivery through a university-wide system

Typical course delivery includes course notes, background reading, tutorials and practicals. Figure 23.11 shows a practical component of the course 'An Introduction to Topographic mapping'.

Online modules in geography delivered via RMIT's Blackboard system also include practical examples that are enhanced by links to supplementary internet-delivered packages. Colin Arrowsmith of RMIT has developed a rich resource for students studying geography at the University (Arrowsmith et al., 2005). Arrowsmith's courseware – 'A Virtual Geomorphic Atlas' (Arrowsmith and Cartwright, 2008) is facilitated via the DLS. Components of the course link to practical examples delivered using Google Earth. Students 'move' through the virtual landscape in Google Earth (Figure 23.12) and links to tutorials and practical exercises (Figure 23.13) are made from the Google Earth application.

Figure 23.12 Google Earth components of 'A Virtual Geomorphic Atlas'. (Image courtesy of Colin Arrowsmith, RMIT University)

Figure 23.13 Linked tutorial from 'A Virtual Geomorphic Atlas'. (Image courtesy of Colin Arrowsmith, RMIT University)

Figure 23.14 'Metro Map' graphic interface to the CD-ROM package for the course 'Multimedia Cartography and Geographical Visualization'

Hybrid

Like mixed media, distributed electronic media can facilitate the delivery of multi-modal teaching resources. The author developed a hybrid package for students studying the course 'Multimedia Cartography and Geographical Visualization 1', part of the bachelor degree in Cartography and Geographical Visualization at RMIT University. An interactive CD-ROM was developed as a discrete resource, which contained lectures, tutorials and practicals. To facilitate an interesting entry into the course and a way in to additional elements the CD-ROM package was presented through a graphic interface resembling a 'metro map'. As shown in Figure 23.14, the lecture, tutorial and practical components of each of the four course modules can be accessed via hot spots on the metro map.

A mouse click linked to other interactive packages containing tutorials (see example Figure 23.15) and practicals (see example Figure 23.16).

The course also has a presence on the RMIT University Distributed learning System. Students can access course materials via this university website (Figure 23.17). The site provides standard Blackboard features.

The website was further developed to contain all of the tutorials and practical components contained on the CD-ROM (Figures 23.18 and 23.19).

The hybrid package allows for students to access materials interactively without the need to have internet access or to access materials through the university's Blackboard system when they have internet access.

These latter sections of the chapter have outlined how New Media and communications systems have been employed by educators in the geospatial sciences to both enhance and facilitate efficient communication of educational materials. The examples have focused on the Australian geospatial sciences educational community, so as to provide an insight into the historical underpinnings of distance education using various media and communication

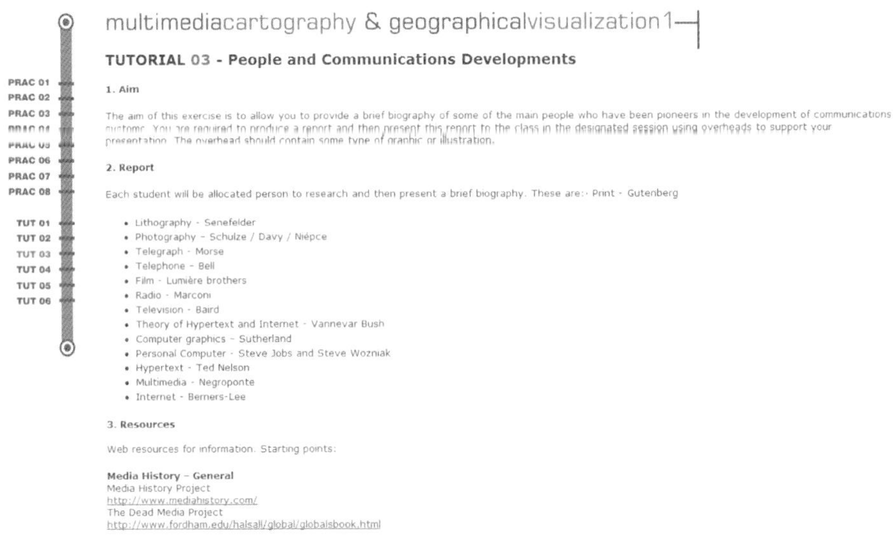

Figure 23.15 Tutorial from the CD-ROM package for the course 'Multimedia Cartography and Geographical Visualization'

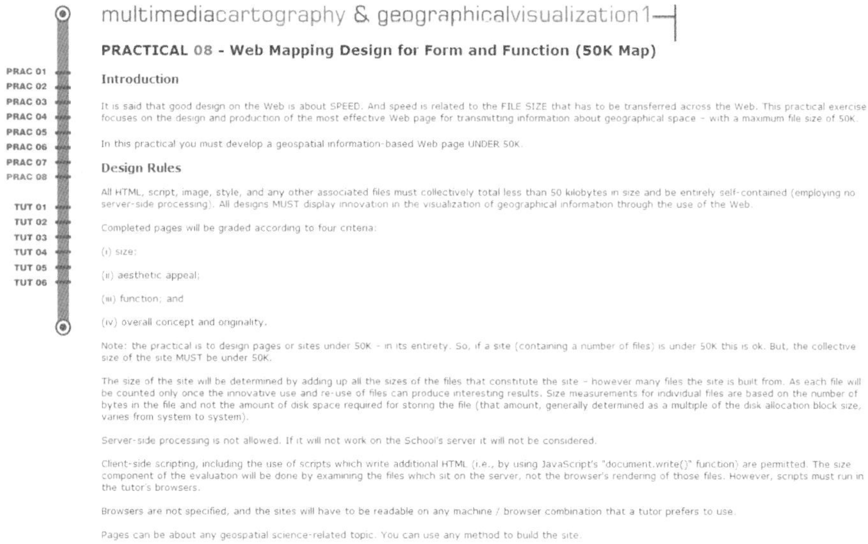

Figure 23.16 Practical from the CD-ROM package for the course 'Multimedia Cartography and Geographical Visualization'

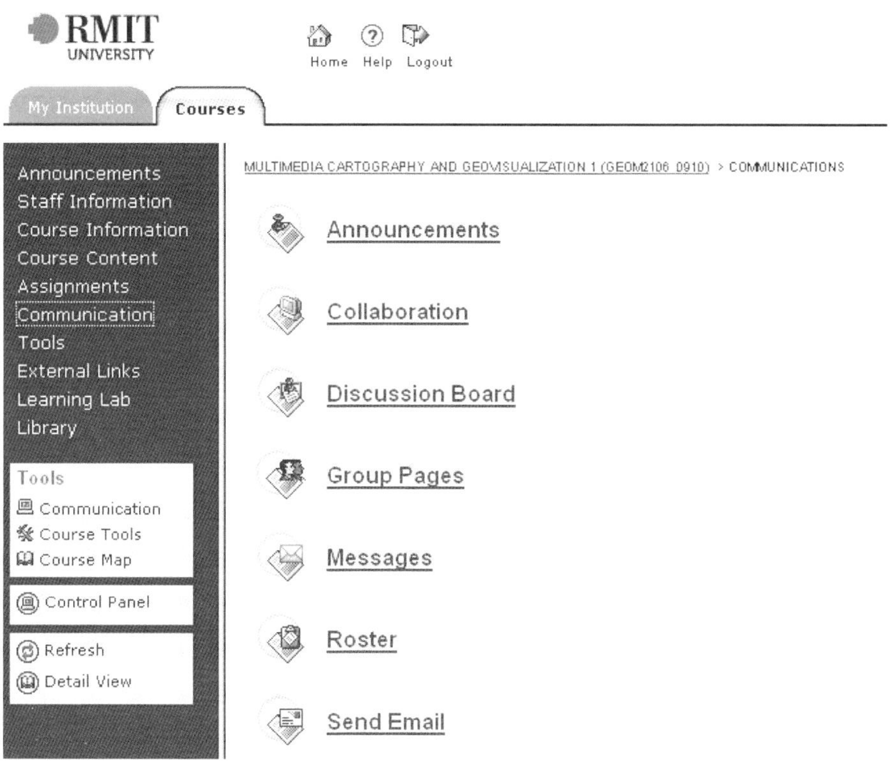

Figure 23.17 RMIT University Blackboard interface for the course 'Multimedia Cartography and Geographical Visualization'

Figure 23.18 'Multimedia Cartography and Geographical Visualization' – web-delivered tutorial on 'People and communications development'

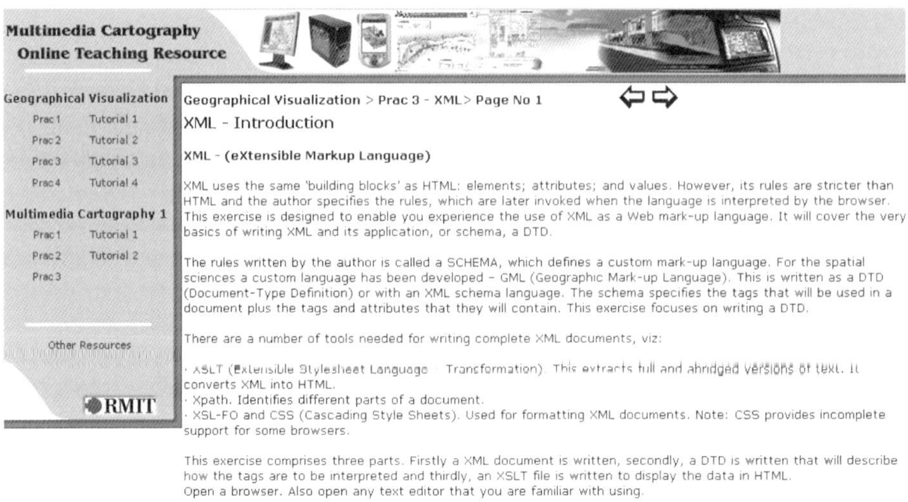

Figure 23.19 'Multimedia Cartography and Geographical Visualization' – web-delivered practical on 'People and communications development'

systems to better provision Australian students with appropriate course content. Initially, the focus on such course development was focused on students who lived in remote regions and needed to be provided with courses using methods that exploited the available media and communications systems. Later, discrete electronic media was used, followed by the internet and the World Wide Web. Innovative courseware continues to be developed to support face-to-face teaching courses. Also, demand for the provision of courses persists in remote parts of the country. Contemporary media and communications systems are still being employed to ensure the delivery of quality education courses.

23.4 Conclusion

Using New Media in educational courses has always been part of Australian education. The use of contemporary electronic media and communications systems is now commonplace. Materials can be produced much more efficiently than in the past, and delivered to students across the country and internationally using readily available and media technologies. Courses can be delivered entirely online, or New Media can be included in distance education offerings for students in remote areas, where internet access is non-existent or erratic at best, using discrete media like CD-ROMs or DVDs. Interestingly, the main impetus for New Media innovations in Australian geographic and cartographic education has been and continues to be the human and physical geographies of Australia itself.

References

Arrowsmith, C., A. Counihan, and D. McGreevy (2005) Development of a multi-scaled virtual field trip for the teaching and learning of geospatial science. *International Journal of Education and Development using Information and Communication Technology*, 1: 3. http://ijedict.dec.uwi.edu/viewissue.php?id=5 (accessed February 13, 2011).

Arrowsmith, C., and W. E. Cartwright (2008) A virtual geomorphic atlas for undergraduate teaching of geomorphology. *Proceedings of the SSI Queensland Conference*, Brisbane, Queensland: Spatial Sciences Institute, Queensland, July 17–19, 2008.

Atkins, S. (1986) The Domesday project. *Media in Education & Development*, 18(3): 110–113.

Atkinson, R. (1987) Development priorities for computer communications in distance education. Paper presented at ASPESA 8th Biennial Forum, University of New England, Armidale, NSW, 27–31 July 1987. http://www.roger-atkinson.id.au/pubs/confs/aspesa87.html (accessed February 15, 2011).

Atkinson, R. (1986) Computer bulletin boards for distance education students. Conference paper, Ed Tech '86, Perth, Western Australia, 2–5 December. http://www.roger-atkinson.id.au/pubs/confs/edtech86atk.html (accessed February 15, 2011).

Australian Academy of Technological Sciences and Engineering (1988). *Technology in Australia 1788-1988*. http://www.austehc.unimelb.edu.au/tia/572.html#2355 (accessed February 13, 2011).

Bearman, M., M. Kidd, and B. Cesnik (1995) HIV hypermedia project: Evaluation and update. *Proceedings of ASCILITE'95*, Melbourne: Australian Society for Computers in Learning in Tertiary Education, pages 16–23.

Cartensen, L.W., and A. B. Cox (1987) The map-environment interface: Surrogate travel by videodisc. *Proceedings of AutoCarto 8*, Baltimore, Maryland: ASPRS/ACSM, page 594.

Cartwright, W. E. (1983) *Maps and Printing: Adaptation of Technology* (Slide/Tape). RMIT, Department of Surveying.

Cartwright, W. E. (1987a) *Map Production* (Slide/Tape). Melbourne: Audio Visual Resources.

Cartwright, W. E. (1987b) *Map Reading* (Slide/Tape). Melbourne: Audio Visual Resources.

Cartwright, W. E. (1987c) *Map Types* (Slide/Tape). Melbourne: Audio Visual Resources.

Cartwright, W. E. (1999) Understanding map projections. *UniServe Science News*, 14: 24–27.

Curtin University (2011) Spatial sciences – Online distance learning. http://spatial.curtin.edu.au/distance/ (accessed February 13, 2011).

Directions Magazine (2009) Penn State online GIS program earns top national honor, October 23, 2009. http://www.directionsmag.com/press.releases/?duty=Show&id=38959 (accessed February 13, 2011).

Fowler, B. (1987) Aussat and all that! Reaching the Australian outback. *Australian Journal of Educational Technology*, 3: 2, 119–128. http://www.ascilite.org.au/ajet/ajet3/fowler.html (accessed February 13, 2011).

Goddard, J. B., and P. Armstrong (1986) The 1986 Domesday project. *Transactions of the Institute of British Geographers*, 11(3): 290–295.

Harris, R. (2003) Building a GIScience community in cyberspace: Reflections on GIScOnline. *Journal of Geography in Higher Education*, 27(3): 279–295.

Hilsberg, T. (1999) The Internet and higher education in Asia. Paper presented at Blackboard Learning '99 Conference, Washington, DC, USA, December 5–7.

Hosie, P. (1985) Promises, promises – Viatel and education. *Australian Journal of Educational Technology*, 1: 1: 39–46. http://www.ascilite.org.au/ajet/ajet1/hosie.html (accessed February 12, 2011).

Inglis, A. (1984) British Telecom introduces Prestel micro-computing. *Educational Technology Newsletter*, June: 1–2.

Jones, A. (2000) Geoskills – The development of a computer-aided learning package for spatial data. *GIS User*, 40: 32–33.

Kelly, A., W. Cartwright, D. Fraser, B. Williams, and R. Schnitzer (1992) *Making Maps*. Melbourne: Bookshelf Publishing Australia Pty. Limited.

Lundin, R. (n.d.) *Flexible Delivery of Continuing Professional Education: Models, Issues and Trends.* http://www.usask.ca/dlc/Context.html (accessed February 13, 2011).

Massey, J., J. Poliness, and B. O'Shea (1985) Mapping the socio-economic structure of Australia – A microfiche approach. *The Globe*, 23: 32–38, Melbourne: The Australian Map Circle.

Massachusetts Institute of Technology (2001) MIT to make nearly all course materials available free in the World Wide Web. http://web.mit.edu/newsoffice/nr/2001/ocw.html (accessed February 13, 2011).

Murray-Smith, S., and A. J. Dare (1987) *The Tech: A Centenary History of the Royal Melbourne Institute of Technology*. South Yarra: Hyland House.

New South Wales Department of Technical and Further Education College of External Studies (197?) *Geodesy Units 1–10*, Survey Certificate, Stage 3.

Roberts, J. E. (1983) *Bold Atlas of Australia*. Sydney: Ashton Scholastic.

Shortis, M., and W. Cartwright (2000) An overview of curriculum transformation and multimedia delivery. *Proceedings of the Trans-Tasman Surveying Conference, Queenstown*, New Zealand, August.

University of Southern Queensland (2009) Faculty of Engineering and Surveying – Geographic Information Systems (GIS). http://www.usq.edu.au/engsurv/areas/gis (accessed February 13, 2011).

Wikipedia (2010) New media. http://en.wikipedia.org/wiki/New_Media (accessed February 27, 2010).

Wiley Australia (2008) The Jacaranda Primary Atlas third edition wins an Australian Geography Teachers Association 2008 award! *Wiley Australia News*, September. http://www.johnwiley.com.au/html/latestnews/latestnews_10.html (accessed February 12, 2011).

Wilson, A. (1971) Engineering Surveying A Papers 4–6, Melbourne, Australia: Melbourne Technical College Correspondence School.

Withers, S. (1986) On-line information. *Australian Personal Computer*, 7(8): 20–23.

Withers, S. (1987) Communication. *Australian Personal Computer*, 8(7): 235–239.

24

UNIGIS – networked learning over a distance

Josef Strobl

University of Salzburg, Center for Geoinformatics (Z_GIS), Salzburg, AUSTRIA

24.1 Introduction

One area in which GIS&T educators have innovated is in distance learning using a variety of delivery methods and approaches and often in collaborative ventures. As they have discovered, it is one thing to develop materials in a collaborative venture, but quite another to deliver them into differing educational systems. In this, the UNIGIS experience is particularly interesting. In the early 1990s, universities from the UK, Netherlands and Austria joined forces to offer distance learning programs in what back then was simply termed 'Geographic Information Systems'. Although UNIGIS started using a hardcopy correspondence format, the emerging internet immediately was used for all communications, and later served as a base for content management and integrated learning platforms. Today the UNIGIS network of universities (www.unigis.net) includes partners on all continents, ranging from the original Manchester-Amsterdam-Salzburg cluster in Europe, to the Americas, Southern Africa (Breetzke, 2007), an emerging presence in Australia, and to a rapidly growing network in Asia. These partner universities are now complemented by study centres in many additional locations. These are operated by an academic (or other) institution in a defined region and are under the academic and organizational responsibility of one of the UNIGIS partner institutions. Students tutored at local study centres are registered at the host (partner) institution which awards the final qualifications. Such has been the development of the network that it is now possible to register for a UNIGIS program in at least twenty-five institutions worldwide and its programs admit close to 1000 new students each year.

This idea of online distance learning in GIS&T was of course not unique to the UNIGIS consortium, but was also taken up by a range of academic institutions and by some publishing houses. Few succeeded in the longer term, with exceptions being professional courses in the Penn State's *World Campus*, *GIScOnline* from Birkbeck, University of London, and a long history of successful programs hosted from Kingston University, also in London.

Teaching Geographic Information Science and Technology in Higher Education, First Edition.
Edited by David J. Unwin, Kenneth E. Foote, Nicholas J. Tate and David DiBiase.

Other initiatives, like the *Virtual Geography Department* (Solem, 2000) or the *Swiss Virtual Campus* (see Chapter 26) contributed numerous lessons-to-be-learned.

This chapter is not intended as a how-to set of guidelines, and certainly not as a record and assessment of all developments in nearly two decades of work. Rather, it is presented as a case study that illustrates some of the more important considerations that have emerged from both successes and failures in the UNIGIS network. Although we review issues related to the development of the network, the character of our students, our approach to the curriculum and so on, of particular interest is likely what we have to say about the organizational framework itself and the way that it supports a variety of business models.

24.2 Networks

UNIGIS has a very wide geographical spread of partner institutions, and this is only made possible by a single core principle. Although standards, quality and expertise provided by UNIGIS qualification are expected to be 'global', access, assistance and communications (including language of instruction) are localized as much as possible. This 'local studies for global qualifications' approach requires a well organized network of independent institutions and could not be replicated by the strictly centralized approach adopted by other distance learning providers in the field.

The sustainability of UNIGIS as a widely distributed 'federated' and highly flexible approach to GIS&T education in very different locations is clearly linked to its character as a network of partners. Such a network has high resilience and can adjust better to local variations in things such as curriculum, legal idiosyncrasies and fee levels (lower fees in one market have to be supported by lower cost through localization of resources). At the same time such a network can take advantage of common ground while retaining independence when it is needed. In UNIGIS, central management is vested in the *UNIGIS International Association* (UIA), which is a foundation incorporated in the Netherlands with participating academic institutions as members. It is governed by the association board with an annual general assembly at one of the member sites. However, under what is actually a very thin layer of network coordination by UIA, there is a growing body of bilateral and multilateral agreements and cooperative action.

24.3 Target audience

Distance learning actually is not only characterized by its mode of delivery, but most importantly it is also defined by the particular target student audience (see Strobl, 2008). Most UNIGIS students are currently active as GIS&T practitioners, and a majority hold previous degrees and has substantial professional experience. All think that GIS&T is an expanding area offering increasing future professional opportunities and want either to establish themselves fully in this field, or to improve their career options by means of a formal qualification and acquisition of a more structured knowledge base. Interestingly, this target group of ambitious, upwardly mobile and highly motivated professionals turns out to be quite similar in many different countries, societies and environments. 'Average' UNIGIS students are in their thirties, have collected professional experience after a first undergraduate cycle of studies, are starting their families and thus are 'settling down'. At the same time they are recognizing the value of a geospatial approach in their professions, looking at stabilizing and enhancing their careers and wanting to make sure they stay at the ' top of their game' in the coming years.

In the current age of 'Lifelong Learning', this kind of continuing education increasingly tends to be the rule rather then the exception. As the speed of many (technological and other) developments seems to accelerate, 'updating and enhancing' one's qualifications is frequently required, and professional careers as well as personal lives cannot always be put on hold to 'go back to school'. In-service continuing education when holding down a job and tending to family and other social obligations implies not only temporal constraints, but also geographical limitations. Distance learning offers the requisite flexibility in scheduling and place of learning. This is illustrated by the fact that in German speaking regions the geographical distribution of students does not exhibit a significant distance decay from Salzburg, but rather is determined by clusters in the general population.

Even though GIS&T is an ICT-oriented discipline, a broad majority of students do not consider themselves to be ICT professionals. First, they regard themselves as belonging to some original domain or discipline such as Planning, Resource Management or Ecology and, typically, in any intake of students no one discipline will have a share of higher than 15 per cent. For them, GIS&T is essentially an enabling technology, a methodology that allows them to do a better job in these domains. This clearly demonstrates the position of GIS&T as a *transversal* discipline, offering geospatial concepts, thinking (National Research Council, 2006), methods, and spatially explicit communication methods (like 'mapping') to many different disciplines (see Figure 24.1). In short, recognizing the value of GIS&T from personal experience, 'having a problem and looking for a solution' distinguishes the typical UNIGIS student from undergraduates obtaining a qualification and then starting to look for an application domain.

Although UNIGIS was originally conceived as a postgraduate program aiming at master's (MSc) qualifications, it became clear early on that a significant subset of candidates were not aiming at an(other) academic degree, but rather at a recognized qualification in geospatial methodology and technologies that could be obtained in a shorter timeframe than a traditional master's. Such a 'professional diploma/certificate', branded as 'UNIGIS

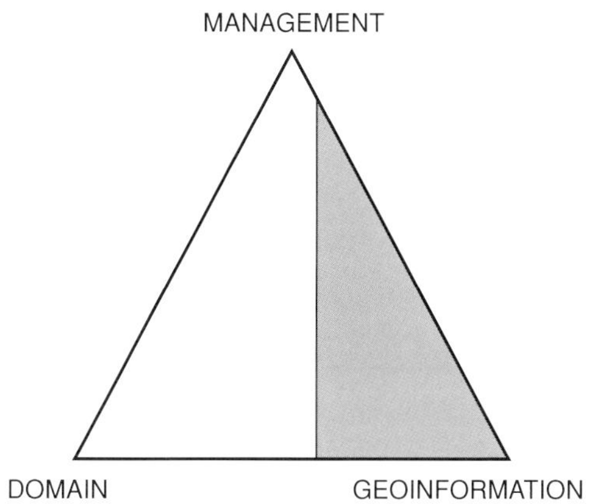

Figure 24.1 Qualification emphasis offered by UNIGIS with major weight on geospatial concepts and methods, a management perspective, and building upon (application) domain knowledge

professional', is offered in some regions either as a first step towards an MSc (e.g. in the UK), or as a separate, parallel study program (e.g. in the German language program). In the latter case admission requirements are somewhat lower than for the MSc option, although in practice most applicants already hold academic degrees and are looking for an add-on qualification in GIS&T.

While the above mentioned target groups are surprisingly similar across most UNIGIS partner institutions worldwide, a somewhat different type of student is emerging in some Asian countries with a focus on select partners in UNIGIS India. Typical applicants there have completed a first degree and immediately afterwards try to enhance their employability and career prospects with an additional, postgraduate qualification. This in turn means that these students study full-time, and distance learning does not individualize their learning contexts in the same ways. Indeed, sometimes classroom settings are used to facilitate online access and tutoring.

Given these additional target audiences, it is not a trivial task to translate these needs into common curricula and academic delivery frameworks and it is to these that attention now expands. Although practitioners from various application domains continue to pursue GIS&T add-on qualifications, student numbers from technical (or integrative – like Geography) disciplines looking towards GI to start their careers are increasing.

24.4 Curricula

In any field curricula serve several important and partially overlapping purposes, such as defining objectives and outcomes, communicating the focus of study programs, defining expected qualifications for both students and employers, and facilitating the comparison of alternative study options. In a new discipline like GIS&T it is at the same time both harder and easier to define curricula when compared to well-established disciplines. It is harder, because work has to start from a beginning in which the range of options and potential contents can seem overwhelming. It is easier because there is less 'baggage' in the form of traditions and 'always-have-been-important' topics and it is possible to have a more forward-looking focus on outcomes and less on the identity of a discipline. Starting with the *NCGIA Core Curriculum* and more recently culminating in the *UCGIS Body of Knowledge* (DiBiase *et al.*, 2006), the GIS&T community has worked hard to build its identity through a common understanding of what graduates are expected to understand and able to do. In each of these (and many other) curriculum development initiatives some fundamental questions have been explored:

- Are the required ICT qualifications included, or considered prerequisites? Some curricula and textbooks continue to start from the very old fashioned 'bits-and-bytes' discussion (Toppen, 1992).
- Should we have to teach students in some user-discipline context? Particularly in continuing education, students know quite well *what* they intend to do with their GIS&T, they want to learn *how* to do it and to do it well!
- How should we incorporate still rapidly evolving areas of spatial technologies, such as data acquisition through imaging, LiDAR, GNSS, SensorWebs, UWB and other indoor positioning and tracking technologies? Are these constantly updated parts of a

curriculum, or rather instantiations of generic concepts related to how we measure the 'what-where-when' of phenomena?
- With what emphasis do we include indispensable topics that can legitimately be considered independent disciplines? Examples of this, from a very long list, include remote sensing, surveying and cartography.

It follows that, as in many applied, technology-oriented curricula, there is a 'demand conflict' between concerns of immediately employability and longer term conceptual and generic knowledge transcending technology generations.

The specific challenge to postgraduate, continuing education curricula like UNIGIS arises from the fact of a very diverse prior exposure to explicit spatial thinking, geospatial concepts and, for example, 'mapping practice' in the target audience. As an admission criterion we keep considering the idea of 'spatial disciplines', that is, disciplines with a mapping tradition such as Forestry, Planning, Transportation, Geography and Geology, whose graduates are more likely intuitively to relate to approaches to spatial analysis as compared to, for example, a Computer Science graduate, who will bring different strengths. In one attempt to balance starting positions, UNIGIS Salzburg developed two 'propaedeutical' learning modules, *ICT for Novices* and *Geography for non-Geographers*, teaching basics of ICT and spatial thinking and concepts respectively. Experimentally, these modules were prescribed for 'borderline' admission candidates and helped with final admission decisions, motivating some candidates to retract their applications, and providing others with a smoother start.

Essentially, the key issue of prerequisites is by and large neglected in most GIS&T curriculum development efforts. This issue is of course more critical the later in life a curriculum is aiming, and is obviously most relevant when students join from very different experiential backgrounds, academic disciplines, cultures, and application domains. Most likely this variety cannot be dealt with adequately through strict prerequisites and remedial learning, but rather flexible implementation of curricula with supportive tutoring and instruction as discussed in the next section of this chapter.

Another closely related issue is giving credit for prior learning. Candidates with a variety of backgrounds bring with them formal qualifications and academic credit invariably overlapping with at least some of the curriculum requirements. Curricula focusing less on the process and more on outcomes and what-students-can-do can deal with requests for such credit for prior learning more easily by telling the students to 'just do the assignments and demonstrate your competences'!

As shown in Figure 24.2 UNIGIS maintains a common core curriculum to cross-reference programs at various member institutions and to serve as a common guideline. Beyond this core, partners are free to define and offer elective subjects allowing students to emphasize themes and objectives of particular interest. Maintaining a balance between compulsory core subjects and electives is an interesting issue, but currently is around fifty percent core, twenty five percent electives and twenty five percent thesis.

In several academic UNIGIS partner institutions, curricula for geoinformatics programs as well as the qualification of suitable teachers, development of materials and distance learning *per se* were entirely new, even though highly qualified academic departments already existed. In this, European Union (EU)-supported curriculum development initiatives under the *Tempus* framework also turned out to be helpful in Poland, Croatia and Central Asia while at the same time offering participating UNIGIS sites from other countries the chance

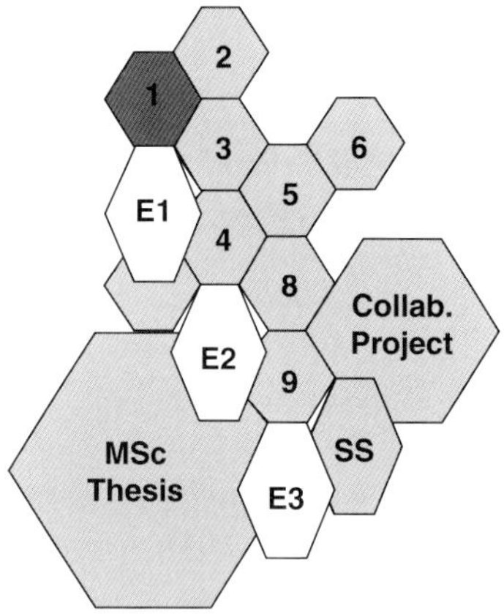

Figure 24.2 UNIGIS (University of Salzburg) Curriculum Schema with compulsory modules (1-9), electives (E1-3), summer school (SS), collaborative projects and master's thesis

to reflect on their practice, often initiating substantial change in all institutions participating in the project.

One example for a recent curriculum debate was centered on the applications/geoinformatics issue: are graduates expected to be primarily competent application developers, or should they first understand the methodology of application logics in their domains? Parts of the industry are clearly demanding 'computer scientists with a spatial touch', while the broad range of application domains requires some development skills, but more of a focus on 'what to do' instead of 'how to implement'. These divergent demands can hardly be covered in any single program, so UNIGIS is experimenting with blocks of electives focused on application development, spatial database design, or related qualifications to respond to the demand for advanced ICT skills without changing course away from an applied program supporting the full spectrum of application disciplines.

In Europe, curricula are expected to be 'Bologna-conformant', based on the European Credit Transfer System (ECTS) measuring student effort to achieve defined outcomes. Twenty-five to thirty hours are rated as 1 ECTS, and a full master's program typically requires 120 ECTS over four semesters. ECTS, though, is not well defined for continuing education, part-time studies, overlap with professional practice and competence-oriented qualifications. Implementing curricula usually follows national or institutional regulations governing the academic institutions awarding a qualification, therefore necessarily leading to differences in frameworks. This is particularly noticeable in interdisciplinary and international cooperative programs like UNIGIS, where a strong common denominator and divergent local constraints have to be balanced.

24.5 Organizational framework

The academic norm is to take an agreed curriculum and implement it in a traditional environment of classes and classrooms, lectures and laboratories followed by examinations over a number of semesters. All universities have a clearly defined *modus operandi* of how successfully to make this happen, but when UNIGIS was launched, none of the departments involved had any experience of how to manage and organize an academic program not tied to classrooms, not timetabled by hour and semester, and not involving regular face-to-face contacts between students and instructors. Today a *modus operandi* has evolved across different UNIGIS partner sites that groups admitted students into cohorts or *intakes*, synchronizes their progress through obligatory *modules* (= classes), with materials provided through online learning management systems, and with written assignments uploaded and then graded by instructors.

It is interesting to note that in virtually all partner universities successful participation in the UNIGIS network was a bottom-up initiative led by GIS&T faculty and that the infrastructural and organizational role of the university apparatus varies widely. Although some functions, such as learning management systems, student databases and quality management, are run by the university, in a majority of instances a parallel infrastructure is run by the UNIGIS unit itself (and sometimes shared across partner sites) to take care of all the specific requirements that arise from distance learning in a continuing education context.

24.6 Business models

The overall business model is the most decisive factor for successfully organizing quality postgraduate programs by distance learning, but once again in practice the key factors are quite similar across the UNIGIS partners. In them, the development and running of a new distance learning study program represented a very substantial investment, but neither the initial investment nor the subsequent operation are fully incorporated into the overall university financial structure. Even though national fee systems vary widely, UNIGIS typically is implemented as a separate financial operation in each university with a mandate for full cost recovery including institutional overheads. A number of scale effects (cf. DiBiase and Rademacher, 2005) mean that, to keep study fees at an acceptable level, development and other costs have to be shared and spread over multiple institutions.

This latter point was a key incentive for UNIGIS to develop into an international network. The effort of development and delivery, the breadth and depth of required faculty expertise across many fields (Foote, 1999) and infrastructures for learning would have required impossibly high student numbers to be financially viable. Such large numbers immediately run counter to important objectives like personalized student contact through tutoring and supervision and building of knowledge in tightly connected student communities. Moreover, in many instances the market area of a single institution as defined by language, region and target group(s) would not have supported such high student numbers on a sustained basis. The development of a network of institutions, where cost, expertise and ideas are shared means that new partners now, roughly 20 years after the original inception of UNIGIS, face a very low startup barrier, and can offer a program they most likely otherwise would not be able to sustain on their own.

That said, organizational and business models at partner universities differ widely. In some cases operations are tightly integrated into the university structures, reflected as one cost/revenue cycle operating completely within the institution's business systems. In more cases, though, UNIGIS is run more like a project within the academic rules of the institution, but all or most faculty are employed or contracted directly as UNIGIS staff and all expenses have to be borne directly from revenues.

Overall, distance learning requires new lines of organizational thinking, reaching across and beyond the traditional boundaries of institutions. As proposed in the Networked Organization of Distance Education (NODE) concept and illustrated in Figure 24.3, online, or

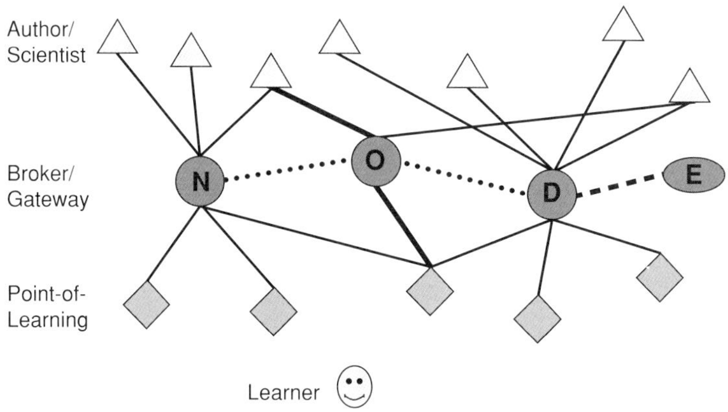

Figure 24.3 NODE – Networked Organization of Distance Education

physical points-of-learning provide student access to institutional (university?) frameworks that in turn connects learners with authors and facilitators of knowledge. In order for it to be sustainable each step in such an educational value chain needs to be properly reflected in the business model.

As outlined above, the scaling and sharing of efforts between and among UNIGIS sites on average by far exceeds the within-university resourcing, resulting frequently in more of an intra-UNIGIS dependency than individual university integration. Therefore it is not surprising that UNIGIS partner sites now consider themselves a *Virtual University of Geographic Information Science* reaching across many 'real' universities.

24.7 Online learning didactics

Initially, none of the founders and early actors in UNIGIS had any serious experience in distance education. Starting out with a correspondence course certainly does not provide much experience relevant to today's online environments. While many strategies and operational steps had to be developed from experience, the guiding principle actually did not change, and it is tightly coupled to the particular target group at the core of the UNIGIS initiative.

Although the UNIGIS originators knew nothing of it, the network developed at the same time as the ideas of situated learning and communities of practice (CoP) (Lave and Wenger, 1991). A CoP is loosely defined as a '... group of people who share an interest, a craft, and/or a profession ... and then proceed towards ... through the process of sharing information and experiences with the group that the members learn from each other, and have an opportunity to develop themselves personally and professionally'. Due to the characteristics of a mature target audience and the UNIGIS focus on continuing education, students went from being a group of practitioners through a community of learners to a community of practice. The latter is manifested in the *Club UNIGIS* alumni association that some graduates consider the most important of the outcomes from their studies, as it offers an effective professional network that some alumni use as a key foundation for their professional activities. As so often in advanced continuing education, academics need to recognize that our role is less and less that of teacher, but more and more that of facilitator and moderator. Even in a new intake of UNIGIS students, most of them are already GIS&T professionals, and can share a treasure trove of skills, experience and domain expertise by far exceeding what the academic faculty can offer.

A common language of concepts and an environment conducive to learning are important first steps towards facilitating students learning from each other and with each other. The concept of the network, which at the beginning of this chapter was focused on institutions, actually might be even more important for the learning process among mature, motivated students. This is why UNIGIS courses start with face-to-face workshops, designed to entice students to establish small learning groups. Even though follow-up communication will be mostly by electronic means, this initial learning in study groups might at first sound counter-intuitive in an online distance learning course, but it actually is what keeps individuals motivated, secures high completion rates and already well thought-out questions to instructors. These workshops are the first step towards the development of communities of practicing professionals within *Club UNIGIS*.

Obviously, acknowledging learning as a social activity is the foundation when developing practical guidelines for online didactics. The use of social web technologies facilitating group interaction in a learning environment has already been discussed in Strobl (2007). Since then, a variety of new technologies have offered themselves for consideration. Emphasizing peer interaction as a foundation for learning certainly does not take away responsibilities from teachers. Working with a given number of students in an online rather than a residential class turns out to be more work for the instructor, not less – a factor which by now is widely recognized and, for example, discussed in Beaudoin (1990) from the perspective of the resulting lower teacher-to-student ratios. The need for a full time tutor who is a first point of contact for students has been established from experience and positive feedback over many years. Instructors at sites with less domain expertise receive second-level backstopping by module authors and recognized experts.

Some of the key lessons we have learned about on line distance education are:

- Active learning starts from a constructivist position, but in addition offers the advantage of providing feedback to students, an insight into 'what I can do'.
- Providing feedback may well be the most important factor in sustaining motivation. Whether through quizzes, a successful result from an analysis, or from personal interaction, learners need regular and frequent feedback on their work.
- A need for quick turn around help, particularly when working with sometimes obstinate and idiosyncratic technologies (We are of course not talking about GIS&T software here!). Learners can get quickly frustrated with very minor glitches.
- Providing a clear sense of purpose through examples, case studies and applications from real world domains.
- Establishing the link to professional practice through the use of professional software, which enables students immediately to see the mutual benefit and overlap between what they are learning and their professional job.
- Collaborative projects/assignments offer a break from individualized work, focus on teamwork as an important facet of professional practice, strengthen a learning community, and often lead to impressive results.

This list of lessons learned could be extended, but should make one well-known conclusion very clear: an academic programme is much more than textbook content and an exam. The overall social environment, the process of learning in all its variants, and the consideration of each student as an individual – all have to be taken into account in a more explicit and conscious way for distance learning programs.

24.8 Quality assurance

Quality assurance and management (QA) are acknowledged to be indispensable and critical elements for successfully sustaining academic programs. In the context of a distributed, multi-partner distance learning arrangement there are some factors to be considered beyond the standard institutional and external accreditation and quality monitoring procedures. Overall, the entire network 'benefited greatly from the initial QA process and practices based on the mature UK system' (Petch, 2010, see http://www.qaa.ac.uk/) but with the expansion of the network to cover new regions, themes, contents and actors, QA cannot now be considered fully satisfactory. In fact, distance learning poses several particular challenges peculiar to itself. Some, such as assuring the identity of a student actually having submitted an assignment, are relatively easily resolved, but others, such as rating the professional relevance of a learning outcome in a local market, are not so simple and some, such as ensuring the direct compatibility of level of graduates in opposite parts of the globe, are virtually impossible. QA is one of the core tasks for UIA and network management. While institutions will and mostly do take responsibility for assuring the quality of their own qualification awards, securing the integrity of the brand and level of UNIGIS qualifications worldwide is as difficult as it is essential.

As Jim Petch (*personal communication* 2010), one of the UNIGIS originators, has commented: 'Distance learning and e-learning are but first steps in the intelligent industrialization of education. Still there are lots of lessons to learn. But this is where the main frontier for education lies. And QA is what will enable it to be intelligent and not just a haphazard by-product of technology.' He goes on: 'End-to-end QA for e-learning is crucial for using e-learning as agent of change – and a key issue for enabling next generation teaching and learning. A *sine qua non* for sustained development of distance education!'

Practical measures for quality assurance in UNIGIS vary greatly, but for the most part rely on established methods whose overall aim is to produce an ongoing, high level QA culture across a diverse and heterogeneous network:

- Joint curriculum development.
- Benchmarking of curricula against standard industry curricula.
- Cross-site training workshops for UNIGIS faculty.
- Mutual/peer review of learning materials.
- Sample review of student assessments, a process that is not always trivial considering the multitude of languages involved.
- Sample co-review of theses.
- Masters thesis competition, and submitting theses to external competitions which provides an incentive for high performance.
- Publication of papers under external review.

Again, this list can be extended but it serves as an illustration of the many practical steps required to establish a thorough QA environment; this is of course a process that by definition will never be finished.

24.9 Conclusions: successes and challenges

Why has the UNIGIS distance learning program in GIS&T proved to be so successful for over two decades (see Strobl and Car, 2009)? In retrospect it is possible to identify factors that have enabled the establishment of the program globally and that have very little to do either with the technology used for delivery of the courses or, for that matter, the detail of the curricula on offer. Possibly the most important has been the development of trusting relationships among partner sites, between students and instructors, and among fellow students. UNIGIS would never have emerged if some of the key actors had not successfully worked together before. In turn this reflects a shared set of aspirations and objectives (Molendijk and Scholten, 2005) and a focus on the social contexts of learning. The flexibility and adaptability of a bottom up approach and an extensive network has enabled the sharing of resources and innovations, and co-operation with the GIS&T industry has enabled the program to stay close to both market needs and student expectations.

This is not to say that there are no remaining challenges associated with sustaining the growth of the network and the evolution of the technology. As UNIGIS has grown, so it has become more diverse, leading to greater possible differentiation in institutional and individual agendas and making the maintenance of an appropriate quality assurance regime increasingly difficult. At the same time there is an increasing need for curriculum maintenance and reinvention to respond to increasingly rapid changes in the technological environment in which we are situated.

Acknowledgement

James Petch has contributed significantly to the content of this chapter both as the founder of UNIGIS, and by sharing specific thoughts and comments on some of the themes and issues discussed. Thanks are also due to university leaderships for 'letting this happen' in the early stages and trusting an unproven concept.

References

Beaudoin, M. (1990) The instructor's changing role in distance education. *The American Journal of Distance Education*, 4(2): 21–29.

Breetzke, G. (2007) A critique of distance learning as an educational tool for GIS in South Africa. *Journal of Geography in Higher Education*, 31(1): 197–209.

DiBiase, D., M. DeMers, A. Johnson, K. Kemp, A. Luck, B. Plewe, and E. Wentz (eds) (2006) *The Geographic Information Science and Technology Body of Knowledge*. Washington, DC: Association of American Geographers and University Consortium for Geographic Information Science.

DiBiase, D., and H. Rademacher (2005) Scaling up: How increasing enrollments affect faculty and students in an asynchronous online course in geographic information science. *Journal of Geography in Higher Education*, 29(1): 141–160.

Foote, K. E. (1999) Building disciplinary collaborations on the World Wide Web: Strategies and barriers. *Journal of Geography*, 98(3): 108–117.

Lave, J., and E. Wenger (1991) *Situated Learning: Legitimate Peripheral Participation*, Cambridge: Cambridge University Press.

Molendijk, M., and H. J. Scholten (2005) From local heroes toward global communicators: The experiences of the UNIGIS network in educating GIS professionals worldwide. Nuffic Conference *'A Changing Landscape,'* The Hague, 23–25 May http://www.nuffic.nl/pdf/os/em/molendij.pdf (accessed August 15, 2011).

National Research Council (2006) *Learning to Think Spatially: GIS as a Support System in the K-12 Curriculum.* Washington, DC: National Academies Press.

Solem, M. N. (2000) The virtual geography department: Assessing an agent of change in geography education. *Journal of Geography in Higher Education*, 24(3): 353–364.

Strobl, J. (2007) Geographic learning in social web environments, IN: S. Catling and L. Taylor (eds) *Changing Geographies: Innovative Curricula.* Oxford, UK: IGU Commission for Geographical Education and Oxford Brookes University, pages 327–332.

Strobl, J. (2008) Digital earth brainware. A framework for education and qualification requirements, IN: J. Schiewe and U. Michel (eds) *Geoinformatics Paves the Highway to Digital Earth.* Osnabrück, Germany: Institut für Geoinformatik und Fernerkundung, Universität Osnabrück (gi-reports@igf), pages 134–138.

Strobl, J., and A. Car (2009) Continuing professional education via distance learning – success factors and challenges: A case study based on the worldwide UNIGIS network. *Österreichische Zeitschrift für Vermessung & Geoinformation*, 1: 109–112.

Toppen, F. (1992) GIS education in the Netherlands: A bit of everything and everything about a bit? *Cartographica*, 28(3): 1–9.

25
The Esri Virtual Campus

Nick Frunzi
Esri, Redlands, California, USA

25.1 History

The Virtual Campus is Esri's e-learning solution, providing a platform for both synchronous and asynchronous online courseware. Launched in July 1997, the Virtual Campus has supported an average of 10,000 students per month, totaling over half a million students to date.

Historically, Esri has offered instructor-led classroom-based training since 1988. Originally the classroom educational experience began as a ten-day class that was offered to Esri customers at their workplaces, or onsite at Esri's offices throughout the US and at Esri's International Distributor offices worldwide. In the early 1990s, class offerings were reduced to two, three or five days, allowing students to be away from their primary jobs for shorter periods of time.

The evolution of the web presented Esri with a new channel to reach customers. A corporate vision emerged to build a campus on the web. Long before the advent of web-based virtual reality, Esri decided to create a place on the web for a community of students to participate and share in the growth of GIS. The Virtual Campus was designed to provide asynchronous courses, allowing students to select from a range of classes and consume at their own pace. A virtual library was envisioned to become the largest GIS reference on the web. Online discussion forums allowed students to form communities, either by interest, industry, or perhaps a proctored class. Lastly, a student could customize the look of the Virtual Campus choosing from four web skins and color schemes, to make it their place, their campus.

As the Virtual Campus evolved, Esri realized that the ability to customize the look of the Campus was not something of high value to students. Esri also found that the vision of building professional collaborative web communities was a premature idea. Examining today's landscape of social media on the web, this might seem unimaginable, but in the late 1990s, there wasn't enough critical mass to make these online collaborative communities succeed. Only early adopters of discussion forum technologies and those with the idea of building communities on the web were successful in creating virtual communities. This

Teaching Geographic Information Science and Technology in Higher Education, First Edition.
Edited by David J. Unwin, Kenneth E. Foote, Nicholas J. Tate and David DiBiase.

was not a mainstream collaboration and communication methodology at that time. In fact, Esri experienced the opposite of the cliché, 'Build it and they will come'. While students enthusiastically embraced the Virtual Campus, they did not embrace the community aspects. This was also due to the transient nature of the students; they used the Virtual Campus for training, not networking. Subsequently, discussion forums, as well as the self-customized Virtual Campus, were eliminated in 2000.

25.2 Structure

The type of learning offered on the Virtual Campus is asynchronous distance learning, meaning the courses are self-paced, available to be taken anytime from anywhere a student has an internet connection. There are no instructors involved in the student's on-demand learning experience. Originally, the courses on the Virtual Campus were developed in this paradigm because education delivered in a distance context with instructor involvement was precluded by technical issues, including the lack of bandwidth available to students. This led to a 'lower-tech' solution provided by asynchronous distance learning.

The material was designed to be modular. Courses were initially comprised as modules of individual lessons (Figure 25.1). Each module took approximately three hours for students to complete. The lessons within the modules provided GIS concepts, examples and exercises, including text, descriptive graphics and software screen captures – an important asset for the hands-on exercises. Each module included an exam, allowing students to self-assess their understanding and application of the material. Overall course lengths varied between four and six modules, with the first module being free, allowing students unfamiliar with e-learning to 'try before you buy.' Required for the course exercises, the data was made available for download. In addition, students could download ninety-day evaluation versions of the software if they did not want to purchase a production license or perform the course exercises using their production system.

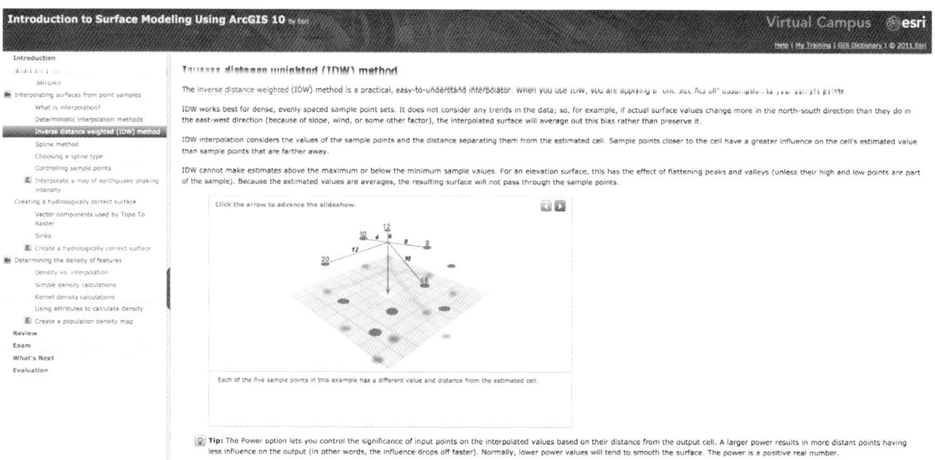

Figure 25.1 Example of modular design used to group lessons

In 2001, Esri added a technical support option to the Virtual Campus. Prior to this point, Esri had always provided *ad hoc* support to students as needed. However, usage of the Virtual Campus had grown so much that dedicated support staff was deemed necessary. Indeed, the added support increased student success and satisfaction and is a service Esri continues to offer to this day.

There are several ways to access courses on the Virtual Campus. Classes can be purchased individually or through one of three subscriptions or Annual User Licenses. A Standard Annual User License allows an organization to buy courses in bulk at a 5–50 percent discount scaling on volume. This Annual User License only includes courseware owned and authored by Esri. The Premium Annual User License allows an organization to buy courses in bulk at a 10–60 percent discount, scaling on volume. This license includes all courseware on the Virtual Campus, including any third-party authored courses. The final license is called an 'Educator Annual User License,' affording the same features as the Premium License, but offered only to educational institutions, libraries and museums. Discounts for this license scale from 40–65 percent.

A major change in Esri's licensing model took place in 1999. The recurring feedback from faculty at universities and colleges was that although they were very happy to use the Virtual Campus in their courses and incorporate it into their curricula, they were greatly concerned about long-term funding to support this adoption. By example, if their school administrators decided to cut the Virtual Campus License from the school budget, the faculty would no longer be able to use Esri's materials, thus disrupting their courses and subsequently their curriculum. In order to remove this barrier, Esri added unlimited use of the Virtual Campus as a feature of each of its University Site Licenses.

The University Site License program is one that allows educational institutions to have as many copies of Esri software products as it wishes. By adding the Virtual Campus to this existing software licensing program, a school no longer had to pay any additional fee to use the Virtual Campus. Removing this hindrance allowed extensive adoption of the Virtual Campus in GIS/GIScience curricula. At the time of this change in 1999, 40 percent of the students that were active on the Virtual Campus each month were academic students, not professionals. That number has ballooned over the past ten years. By August 2009, four-fifths of the students on the Virtual Campus were academic students. In 1999, eleven educational departments took advantage of this program and by 2005 over 700 educational departments were using the Virtual Campus actively as part of their curriculum. Over the ten years 1999–2009, the retail value of the Virtual Campus courses taken by students is estimated at over $50 million. This is an impressive contribution to encourage the growth of GIS and the use of e-learning throughout higher education. Esri considers this essential to furthering the availability of geospatial education.

25.3 Curriculum

The initial structure of the Virtual Campus curriculum was centered on three main areas. Esri software training provided students an opportunity to learn the multiple software applications. GIScience courses provided an understanding of GIS and its application in solving problems. Domain-specific training authored by third-party subject matter experts in various fields provided a focus on technology within an educational context. In fact, Esri partnered with several of these subject matter experts in designing the curriculum.

In 2001, Esri partnered with the authors of *Geographic Information Systems and Science* (GIS&S) to produce a hybrid e-learning course offering, entitled *Turning Data into Information Using ArcGIS 8*. The course was designed to complement the GIS&S book and reinforce its concepts through exercises and examples. Although the book was not a requirement for course completion, when combined, the two provided a well-rounded pairing of academic and practical instruction. When the book was updated to a second edition in 2004, the course was updated to use ArcGIS 9. Since its introduction, over 5100 students have taken this e-learning course, and as such, Esri views the collaboration with the authors as a success, and is dedicated to continuing the relationship with the author team. The course will soon be updated to complement the forthcoming third edition of the book.

Another area of quiet success has been the Virtual Campus Library. In 1999, Esri acquired a GIS bibliography from Duane Marble, now professor emeritus at Ohio State University, which was republished as the core of the Virtual Campus Library. At the time of acquisition, the bibliography contained around 13,000 items. Through continued acquisition and item entry, the library now has over 110,000 records and is considered the largest bibliography of GIS and GIScience freely available on the web.

25.4 Instructional delivery models

While the Virtual Campus was very successful from its inception, in 2001, Esri decided it was finally time to offer a synchronous e-learning product, known as Live Training Seminars. A Live Training Seminar (LTS) is one hour of free synchronous content featuring an instructor lecturing with PowerPoint presentations and software demonstrations through the Virtual Campus. Interactivity is achieved via a virtual chat, where the students pose questions. At intervals throughout the hour, the instructor will answer, via voice, the questions that were posed via chat. An assistant in the room collects the questions for the instructor. Currently, Esri provides LTSs approximately every six weeks, three times during a given day at 9:00am, 11:00am and 3:00pm PST. There are over 2100 attendees on average for each day Esri offers the sessions. This model uses a streaming media server and the only requirement from the student's perspective is a Windows Media plug-in on their computer and Virtual Campus access. This model became a quick success and as such Esri developed a dedicated studio space for LTS production.

Esri also built derivative works from the live events. A Training Seminar (TS) is an asynchronous version of an LTS. Esri digitally records each LTS, selects the best presentation of the three and conducts a small amount of post-processing. The TS is then uploaded to the Virtual Campus to be freely viewed by anyone at anytime. This model of educational offering has grown beyond being simply a by-product of an LTS. Esri now creates recorded seminars to teach important topics that do not necessarily need the added value of a live instructor to interact with and answer questions.

A second derivative work was the Web Workshop (WW), implemented between 2001 and 2005. Web Workshops differed from the LTS and TS by two important characteristics. The Web Workshop was a 'for fee' educational offering derived from a TS. In this case, Esri created an exercise using the software to allow the student to not only hear the lecture and see the PowerPoint slides and demonstrations, but to actually download the data and use the software hands on to accomplish similar tasks as outlined in the TS. These Web Workshops became very popular and useful in an on-demand learning environment, allowing short focused topics and the ability for the student to reinforce learning through practice. Over time Esri also wanted to produce shorter courses in the original format, so the WW model was eliminated and replaced with one-module courses, removing confusion and simplifying Esri's educational offerings.

In 2004, Esri added another synchronous offering: the Instructor-led Virtual Classroom (ILV). In this model, Esri wanted to expand the success of the LTS and WW into a larger event by offering a longer period of instruction. There was concern that the current delivery methods would not keep a student's attention for an extended time period. To mitigate this concern, training was offered in three hour sessions on Tuesday, Wednesday and Thursday (mornings or afternoons) of a given week. This allowed Esri to manage any technical issues with the students on the Monday before the class began. The ILV instructional model was similar to the LTS in that there was a live instructor presenting PowerPoint slides and demos. Esri used the existing LTS studio and media streaming equipment. Different than the LTS, the ILV offered software exercises. Using a Citrix virtual environment, students were able to take training without having any impact on their production environments. Importantly, it allowed the instructor to 'shadow' the student's work if they needed assistance, by viewing their progress through Citrix and even taking over control of the machine as necessary to provide further instruction or assistance. ILVs also introduced phone conferencing in the virtual classrooms, allowing voice, instead of virtual text chat, so the students would have the ability to speak to each other and the instructor.

Initially, Esri used the ILV model for very specific and somewhat specialized topics, which proved moderately successful, though very expensive to provide. The required support staff and the complexity of the solution required changes. In addition, Esri was only able to provide these classes from the Esri headquarters in Redlands, California. Esri has instructors in nine other offices throughout the United States and, since they did not have any access to the studio, these instructors were unable to teach in this format. To solve this problem and move the training out of the LTS studio, Esri incorporated the use of Microsoft's Live Meeting and more recently Adobe's Connect products. These technologies allow Esri to have any instructor, anywhere in the country teach a class to students anywhere in the world, and not to be tied to the facility in Redlands.

The year 2005 brought the introduction of Esri's use of the Sharable Content Object Reference Model (SCORM). The SCORM specification is promoted by the US Department of Defense as a standard e-learning specification allowing content to be easily packaged and consumed by Learning Management Systems (http://www.adlnet.gov/Technologies/scorm/default.aspx). SCORM allows Esri to take courses out of the Virtual Campus and give them to customers, mostly in the US Federal Government, for consumption in their own LMSs. The biggest lesson learned from this effort was that the SCORM standard was designed under a different educational philosophy than Esri's courses. SCORM assumes that you want to have full linear control over the learning experience. SCORM can force learners to visit every page of a course in a specific order before taking a test, and then restrict them from revisiting pages or retaking the test. Esri has a more fluid, non-linear philosophy; less concerned about the order the student visits, or consumes content, as long as they pass the test. Esri also allows students to take the test as many times as they wish, and re-read the material as often as they like.

In 2007, looking to meet another need for on-demand training, Esri began producing the Instructional Series Podcast as a new e-learning option. These free mp3 files contain approximately 10 minutes of instruction about a single topic with no supporting, i.e. visual, materials. With a minimum of resources Esri can produce a new podcast every few weeks using a laptop, the Audacity recording software and a microphone. Although there is a small amount of post-production, these offerings are very quick and inexpensive to produce. The podcasts have been exceptionally positive with an average of over 65,000 downloads per year.

Assessing the most successful training on the Virtual Campus, a number of consistent themes emerge. Courses that include 'Tips and Tricks' are very popular. This type of training is often seen to fill the gaps in a student's knowledge by helping them be more productive. Topics with very technical information provided in a simple instructional way are also met with great satisfaction. Classes with exercises focusing on real-world problems allow students to finish the class and immediately apply what they learned in their work or study. Classes with a comprehensive discussion of a topic, mixed with asynchronous delivery, allow self-directed learners to consume, absorb and implement at their own pace, something not seen in synchronous online or in classroom events. Short, just-in-time training options like the LTS, TS, WW, one-module courses and podcasts are very popular, not only due to low or no cost, but primarily because they provide the education that the student needs, when they need it.

25.5 Challenges to success

Looking holistically at the Virtual Campus, Esri has also had to acknowledge and address shortcomings. As mentioned earlier, discussion forums were eliminated and in fact, have not returned. There are still internal dialogues at Esri as to whether a replacement community does in fact need to be built, and if so, the inherent value of a community based around the Virtual Campus itself rather than specific areas such as universities, industry associations or professional groups which currently exist on the web.

A limitation that Esri has experienced for a number of years was the inability to offer extensive classes in our server and enterprise-level technologies as part of our e-learning offerings. This training was provided through LTS, TS and podcast, but while these methodologies are very useful, Esri was unable to provide web classes that would allow the student to use the software during exercise scenarios. When taking a web course with exercises, students are allowed to download trial versions of the software to use with the class. Downloading and installation is not an issue with ArcGIS Desktop, Esri's GIS software for desktop and laptop PC's. Installation of ArcGIS Desktop is as straightforward as installing any PC-based software. Installing a product like ArcGIS Server however, requires extensive permissions as well as configuration and tuning skills. This added complexity as well as the inability to install Esri's exercise datasets into production server environments has prevented Esri from providing in-depth server and enterprise-level training through e-learning.

Third-party course authoring has had inconsistent results. Often Esri worked with subject matter experts who were excellent classroom instructors, but had a difficult time creating materials for asynchronous consumption by students. Another issue has been the timeliness of delivery. In all cases, and not unexpectedly, third party authors had other primary means of employment and developing a course for the Virtual Campus was a secondary priority. This often caused issues where course development started using a specific version of Esri's software but was not completed until after a new software version release, which meant a course had to be rewritten or at least updated before it was published, adding to time and costs.

A derivative of the third-party work was translation of some of the courses, specifically into Spanish, Portuguese and German. The Spanish and Portuguese offerings were encouraged and produced by Esri's distributors. The German course was authored by Josef Strobl at the Universität Salzburg and was used in classes that he taught. This was a very successful course due to Dr Strobl's dedication and involvement as well as the specific purpose of the course design. Conversely, the Spanish and Portuguese classes, while properly translated, have not been used in any appreciable volume.

Success has almost always been attached to focus. From 2000 through 2007, Esri had an excellent relationship with Penn State and its World Campus through the efforts of David DiBiase. Beginning in 2000, Esri featured a Penn State banner on the Virtual Campus home page. In 2002 Penn State authored two courses, *Penn State Edition: Cartographic Design* and *Penn State Edition: Learning ArcGIS I (for ArcView 8.1, ArcEditor 8.1, and ArcInfo 8.1)*. These were successfully offered for several years. Today the Virtual Campus hosts David DiBiase's *Understanding Geographic Data* course which is the Virtual Campus edition of Penn State's *The Nature of Geographic Information*, part of Penn State's online Certificate Program in GIS.

Course completion rates have been an ongoing issue. Being that the original course design was asynchronous and lacked much interactivity, averaging 24 or more hours of instruction, course completion rates tended to be below 36 per cent. In many cases there was just too much material to keep the learner continually engaged. Another factor in this statistic was that, in academic settings, instructors would select individual modules in different courses as needed to support their curriculum. This is possible since the entire catalog was available to them through the University Site Licence. This need points to what works very successfully today, and the future direction of the Virtual Campus: shorter engagements for specific topics.

A continuing challenge for Esri has been the Esri Course Catalog search feature. As of this writing, there are 218 items in the course catalogue. As Esri transitions to shorter, more focused web courses, the obvious by-product would be a catalogue containing more classes than in the past, when the majority of the training on the Virtual Campus was four or six-module courses. This fragmentation of material is very good for the student in an on-demand context, but can result in a frustrating search for the appropriate course. As a result of this limitation, Esri has started a project to overhaul and replace how searches are performed on the Virtual Campus, with the intent of making the catalogue much easier to navigate.

25.6 Forging a new path

Looking forward, with the success of the ILV model, and more of society being familiar and comfortable with e-learning, Esri has recently embarked on offering full-day synchronous instruction via the web. After a few technical difficulties, the first test offering was met with complete success and was enjoyed by both students and the instructor. Esri took the feedback from the event and offered another equally successful event a few weeks later. While the sample size of 38 students is not comprehensive from a statistical sampling perspective, it provided enough data to move forward. Esri will continue to expand the program with the objective that it can deliver most, if not all, of the brick and mortar classroom curriculum in a synchronous e-learning context. This will allow Esri to extend the reach of its current classroom-based instructor-led offerings, by being able to leverage the e-learning platform to deliver synchronous training to a student's desk, regardless of location. From a technological perspective, Esri has moved off the Citrix platform for software exercise work and is now using a hosted platform, allowing students to access virtual machines in a hosted environment using Esri's software and course data.

Taking the experiences that Esri has learned over the past 12 years, observing the use of the asynchronous and synchronous e-learning on the Virtual Campus, as well as what has been seen in the classroom, Esri is in the process of completely redesigning how it provides asynchronous e-learning. The intent is to move away from the linear learning model with an abundance of supporting material, to a student-centered, just-in-time delivery methodology for the entire Virtual Campus, not simply the LTS, TS, and podcasts that are currently offered. Students will be able to access very specific topics when they need to learn about a finite function or task. They can consume a broader collection of topics in any order they choose. Within the courses, there will be a dramatic reduction of text, an (optional) addition of voice, and more interactivity. This shift will allow students to quickly apply what they learn. The main method to accomplish this will be the extensive use of exercises, allowing the student to spend more time actively practicing, rather than reading. This modularization and exercise focus is also intended to achieve a long-standing educational goal: the reuse of materials both in an e-learning and classroom context. Since Esri will now be offering its classroom education on the web, it makes sense to use the same exercise materials for classroom and online training. At a minimum, the target is extensive repurpose, if complete reuse is not an option. Finally, the use of virtual machines and a cloud-based solution will also allow Esri to teach server and enterprise-based training courses in a web-based delivery model.

Given the current success of podcasts, Esri is assessing video podcasting, where again, students with a portable media player can not only listen to, but view instructional materials. The design will be similar to that of existing podcasts, but extended into video: 10 minutes of instruction on a specific topic, limited words visually presented, and more voice and demonstration activity. The video podcasts would be accessed in two ways. First, on a computer connected to the Virtual Campus, which allows Esri to provide a high-resolution solution for those without bandwidth or screen limitations. Alternatively, students can download the video podcast (in either mpeg or another format) to a portable media player and consume the material from the device. A future expansion of this type of delivery might also be to cellular devices, allowing Esri to provide geospatial education to areas in developing nations where there is limited internet infrastructure, but a robust cellular availability.

As previously discussed, discovering smaller educational items in the catalogue will be more dependent on search. In addition to improving the existing search feature, Esri will develop a needs assessment tool wherein students can enter information about their level of skills and knowledge and what they wish to achieve. The search tool can then lead the students to a collection, or perhaps even a single learning item to satisfy their educational need.

In comparison to the age of the web, Esri's Virtual Campus has been providing geospatial e-learning for a long time. Esri's Virtual Campus is an integral component of the future of geospatial education in academia, as well as in business. As such, Esri continues to keep ahead of new developments in e-learning technologies and pedagogy and intends to continue to evolve the platform to meet the needs of current and future generations of students and business professionals alike.

26

Delivering GIScience education via blended learning: the GITTA experience

Robert Weibel[1], Patrick Lüscher[1], Monika Niederhuber[2], Thomas Grossmann[2] and Susanne Bleisch[3]

[1] *Department of Geography, University of Zurich, Zurich, Switzerland*
[2] *ETH Zurich, Institute of Terrestrial Ecosystems, Zurich, Switzerland*
[3] *FHNW University of Applied Sciences Northwestern Switzerland, Institute of Geomatics Engineering, Muttenz, Switzerland*

26.1 Introduction

Over the past decade, programs to fund the development of e-learning content in higher education institutions have been implemented in various countries. Examples of such programs include an initiative called JORUM (JORUM, 2011) funded by the Joint Information Systems Committee (JISC) in Britain (JISC, 2011); the program 'New Media in Education' funded by BMBF, the Federal Ministry of Education and Research in Germany; and the Swiss Virtual Campus (SVC) program in Switzerland (SVC, 2011). Common to all these programs was the objective to promote the use of new media and information technology in teaching and learning, and the creation of synergies through reuse of teaching materials. Typically, projects funded under these programs were required to consist of several networked partner groups, preferably at different higher education institutions. Over the years, the above initiatives have funded hundreds of e-learning projects across many academic disciplines. The Swiss Virtual Campus program alone funded more than 80 projects in the years 2000 to 2008, in a small country like Switzerland (SVC, 2011).

GITTA (Geographic Information Technology Training Alliance; GITTA, 2011) is a project to develop e-learning content that has been funded by the Swiss Virtual Campus programme between 2001 and 2004, with additional funding for content maintenance between 2004 and 2008. It was originally initiated by ten institutes of seven Swiss universities, federal institutes of technology and universities of applied sciences, respectively. These groups formed a consortium, the so-called *GITTA Consortium*. The key aim was to create a pool of lessons

Teaching Geographic Information Science and Technology in Higher Education, First Edition.
Edited by David J. Unwin, Kenneth E. Foote, Nicholas J. Tate and David DiBiase.

for teaching GIscience and technology (GIS&T) to which every partner would contribute specific lessons in their own field of specialization and expertise, and which would then allow these partners to build their own, tailored courses from the lesson pool. With the implementation of the lessons in different fields of GIS&T the goal was to cover comprehensively a broad range of topics and use this to supplement traditional teaching in a blended learning setting.

This chapter describes, from the perspective and experience of GITTA, different options of integrating e-learning lessons into the curriculum at various partner institutions, and of blending e-learning with more traditional forms of teaching to achieve blended learning (Graham, 2005). Earlier papers have dealt in detail with other issues that we consider relevant in e-learning, such as a platform-independent, interoperable technical infrastructure (Fisler and Bleisch, 2006); open educational resources (OER) and issues of copyright and licensing (Fisler and Weibel, 2006); as well as strategies for ensuring the sustainability of e-learning content (Weibel *et al.*, 2009).

The chapter is structured as follows. Section 26.2 provides an overview of the content that is offered by GITTA. Section 26.3 explains the pedagogical model that is used in GITTA and gives hints on the technical implementation of this model. Section 26.4 then represents the main part of the chapter, presenting and discussing three examples of the integration of GITTA content in GIS&T courses at three different institutions, illustrating the breadth of different options that can be chosen for content integration to achieve blended learning. Finally, Section 26.5 wraps up the chapter by conclusions and an outlook.

26.2 An overview of GITTA

GITTA covers a wide range of topics in GIS&T. More than 40 lessons exist today, grouped into thematic modules, as illustrated in Figure 26.1. These modules focus on conveying theoretical and technical concepts. Six *theory modules* exist, organized into a basic and an intermediate level, depending on the level of complexity of the concepts presented:

1. *GI-Systems:* On the basic level this module offers an initial introduction to GIS&T. It shows what a GIS is and what it can do, using various application examples. It also reviews the typical functionality of a GIS and provides information about existing commercial products and their areas of application. On the intermediate level two lessons exist on spatial partitioning and indexing, as well as on spatial data compression.
2. *Data Capture:* The Data Capture module introduces students to sources of geographic data and to the broad process of digital capture and data pre-processing. It focuses on selected methodologies for capturing relevant geographic data and is structured into the following content: overview of sources, primary sources, derived sources and metadata quality.
3. *Database Systems:* This module introduces the concepts and architectures associated with databases. Specific terms are discussed and the characteristics of the database approach in GIS&T are compared to other systems. The fundamentals of database architecture are followed by an explanation of conceptual and logical data modeling and an introduction to the Structured Query Language (SQL).
4. *Spatial Modeling:* This module presents the main concepts – projections, location and topology – that drive the modeling of the geographic space. It gives an overview of

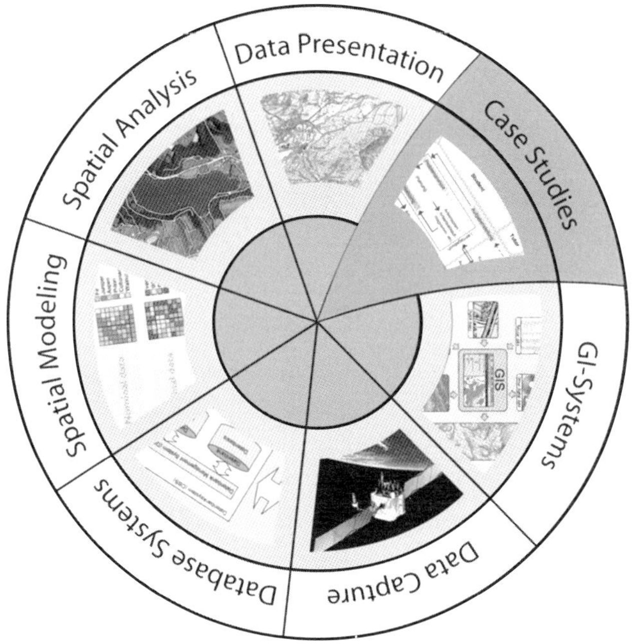

Figure 26.1 Thematic scope and organization of GITTA content. Theory modules (white) and case studies (dark)

spatial modeling processes, using regularly and irregularly distributed observation units (field vs. entities, raster vs. vector models). Furthermore the properties of spatial information are analysed from the perspective of measurement scale and origin (measured, derived, or interpreted).

5. *Spatial Analysis:* The main goal of this module is to provide the technical knowledge required for the development of analytical GIS applications. Lessons include fundamentals and techniques for dealing with continuous and discrete spatial variables, spatial queries, terrain analysis, suitability analysis and accessibility analysis. These topics are dealt with on the basic level and later refined at the intermediate level.
6. *Data Presentation:* The basic module describes the history and use of maps, while introducing the different map types available. Its emphasis is on topographical cartography and the components required for map development, including graphical design and legibility rules. A further focus is placed on cartographic generalisation concepts, procedures, and methods. In the lessons of the intermediate level, thematic map design as well as mapping with a GIS and cartographic software is discussed.

Every theory module contains between four and 13 lessons, encompassing basic and intermediate level lessons. Every lesson accounts for two to three hours of learning time for the student. As will be explained in more detail in the following sections, the lessons not the modules form the main building blocks for developing actual courses based on GITTA materials. The role of modules is thus typically restricted to assigning lessons to broad thematic categories.

Besides the theory modules, five *case studies* have been created in which students are confronted with realistic sample cases that are relevant to the practice of GIS&T. In these case studies, the practical problem solving skills of the students are trained (Niederhuber *et al.*, 2005). In each case study, a description of the assignment and original data material is provided. The student is then expected to develop solutions for solving the assignment using GIS tools. For example, the assignment of one case study consists in modeling the potential habitat of given animal species for a study area, using methods of GIS-based multi-criteria analysis. It usually takes the student around 30 hours to solve such a case study, roughly equivalent to a small term project. In a typical pedagogical setting, the student is expected to solve the assignment largely independently, but a teaching assistant will be available to provide assistance either through electronic communication in a learning management system (LMS) or face-to-face.

In addition to the above content of the original project, the GITTA website also offers access to 15 lessons that were generated in the CartouCHe project (CartouCHe, 2011). CartouCHe was another e-learning project funded by the SVC, focusing on cartographic themes, including multimedia cartography and location-based services, respectively.

26.3 Pedagogical approach

As mentioned above, the original project consortium consisted of ten groups at seven different universities in Switzerland, spread across several disciplines. The main motivation was to exploit synergies in teaching across GIS&T-related disciplines and extend the scope and depth of GIS&T courses within the partner institutes. By sharing of materials, the intention was to facilitate the introduction of new elements and topics into existing courses, and in some institutes even the first-time establishment of GIS&T courses. Over time, nearly 40 authors and contributors created the lessons and case studies described above. Obviously, when such a large number of authors are involved, measures must be taken to ensure that complexity can be managed. First, the overall design of the project had to be highly modular, allowing multiple authors to develop content concurrently, yet largely independently, in different languages. Second, pedagogical frameworks had to be found for lessons and case studies, respectively, that could serve as guidelines for authors.

We believe that in GITTA we found a pragmatic and simple, yet effective solution. In order to meet the first of the above requirements, the content was organized as a *pool* of lessons and case studies rather than as a closed, static course that follows a fixed order. Thus, the *lessons* and *case studies* form the key elements used to build individual courses by integration. The choice and order of lessons used for a particular course then depends on the needs of the teacher who assembles lessons to build his/her own course. To embed lessons or case studies into a particular course or curriculum, all that is needed is a link to the required lesson or the import of an IMS or SCORM content package (IMS, 2011; SCORM, 2011) into a learning management system (LMS) such as Blackboard, Moodle, or OLAT. Sample scenarios of the integration of GITTA content to build blended learning courses are provided in the following section.

In order to satisfy the second requirement regarding the pedagogical design, a common pedagogical framework was defined for lessons. Each *lesson* consists of different *units*, which are based on an adapted version of the e-learning model called ECLASS, originally developed

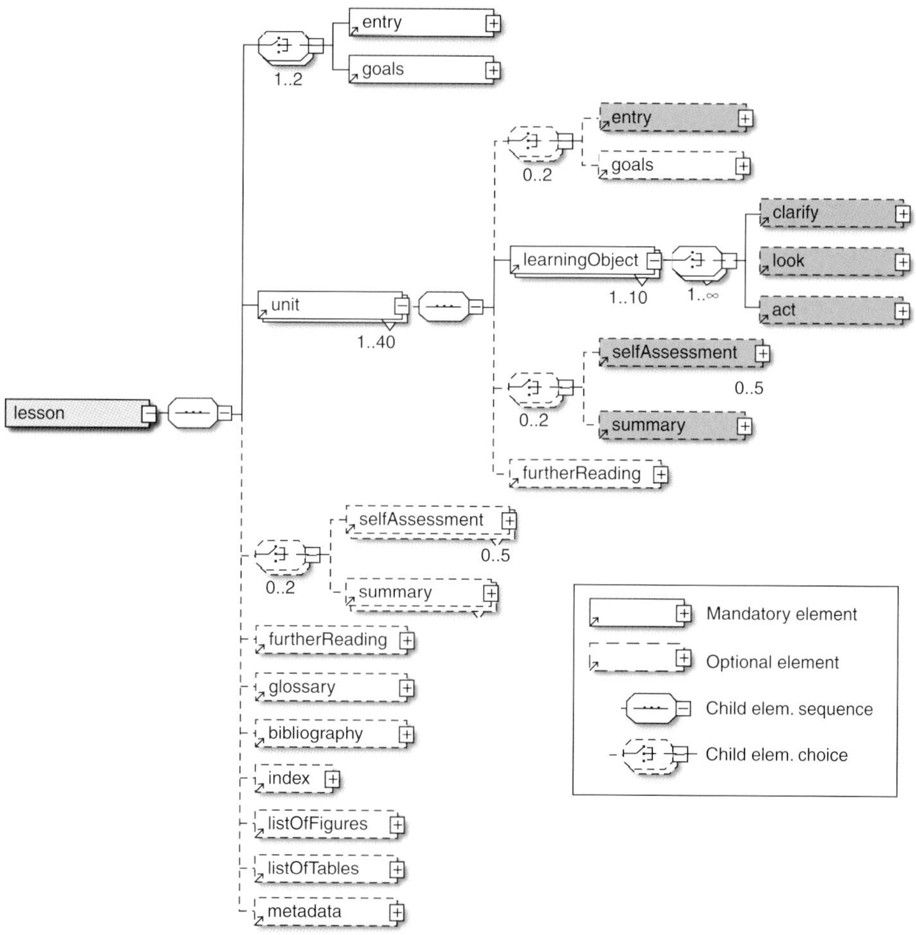

Figure 26.2 Structure of a lesson

by Gerson (2000). This straightforward model is based on practical experience by Gerson. It builds mainly on the constructivist learning paradigm (Jonassen and Land, 2000), and integrates elements of the design of learning sequences (Horton, 2000). Of the various possibilities for learning sequences reviewed by Horton (2000), this model convinced by its simplicity and flexibility, criteria that are important when dealing with many authors in a consortium.

In our adaptation of Gerson's pedagogical model, the acronym ECLASS expands to {Entry, Clarify, Look, Act, Self-Assessment, Summary} and provides a pedagogical structure that contains crucial learning elements (see also Figure 26.2):

- *Entry* gives students an overview of what can be learned, including the learning objectives for each unit or lesson.
- *Clarify* forms the bridge between the general concept (scenario or context) and the specific details taught.

- *Look* provides the opportunity to review examples of what is being taught in the lesson or unit.
- *Act* lets students practice key concepts. Ideally, this is accomplished by providing interactive media like Flash or SVG animations.
- *Self-Assessment* lets students evaluate themselves what they have learned, for instance through interactive multiple-choice tests with instant feedback.
- *Summary* recapitulates what has been taught in the lesson or unit.

In comparison to Gerson's version of the ECLASS model, we have replaced the original Share element with the Summary part. In our experience, the sharing of concepts learned in a lesson among students can either be accommodated in the Act element of an ECLASS learning sequence or, more importantly, it can be taken care of by other instruments of blended learning (e.g. discussion sessions in the classroom or on a discussion forum, or assignments in a case study).

The original ECLASS model was further extended by additional elements such as glossary, bibliography and metadata that allow creating a self-contained e-learning course. Furthermore the three elements Clarify, Look and Act together form a so-called *'learning object'*. Clarify, Look and Act elements can be used in any order or even multiple times within a learning object. This allows an author to start with an example (Look), followed by the theory related to the example (Clarify) or the other way round. Even the uncommon but sometimes successful approach is possible in which the student starts with a short exercise (Act) and only after having tried out some solutions reads the theoretical background (Clarify) and sees some real world examples (Look). The ECLASS model is on the one hand flexible enough to represent different learning scenarios (e.g. those described in Horton, 2000) and meets the requirements of the creators of the e-learning lessons. On the other hand it ensures that the content complies with the defined pedagogical guidelines.

The above pedagogical guidelines have been mapped to an XML framework that allows an author to check whether the pedagogical model has been used correctly. This framework is called eLML, short for 'eLesson Markup Language'. Fisler and Bleisch (2006) describe the structure and the features of eLML in detail and compare it to other XML-based frameworks for use in e-learning. eLML was originally developed within the GITTA consortium and released as an open source project on SourceForge.net in 2004. Since 2006 it is hosted and maintained by the Multimedia and E-Learning Services (MELS) of the University of Zurich. In the meantime new features and tools were added, such as a web based WYSIWIG editor named Firedocs. eLML is now used internationally and successfully by e-learning projects in a variety of disciplines (eLML, 2011).

Figure 26.3 shows a screenshot from a GITTA lesson. The comparison with the elements of Figure 26.2 reveals that these elements are also readily visible to the student. The navigation pane shows the hierarchical structure of the lesson (depicted by a round dot), units (shown by triangles), and learning objects (using squares). Also, the other elements such as glossary, bibliography etc. are visible. The figure further shows the integration of a Flash interaction.

As Niederhuber *et al.* (2005) describe, GITTA *case studies* use their own pedagogical model, which is different from the ECLASS model, but is nevertheless implemented in eLML. Case studies usually start with an Entry element followed by two units where Clarify elements are used first to describe the assignment of the case study and second to explain the general instructions and conditions.

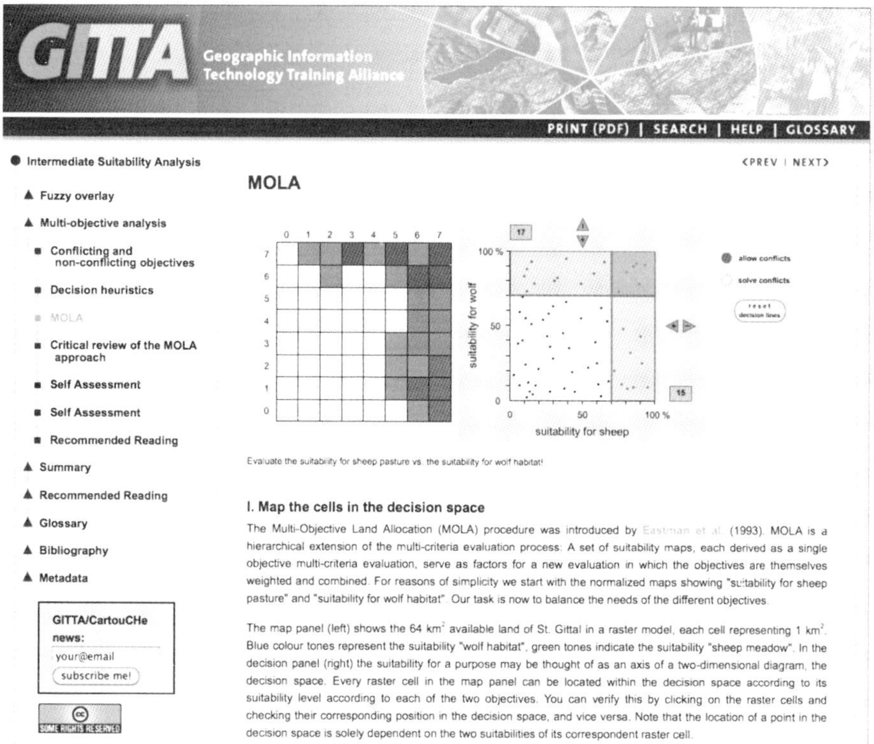

Figure 26.3 Screenshot of a GITTA lesson

26.4 Achieving blended learning with GITTA

As mentioned in the introduction, GITTA was funded by the Swiss Virtual Campus (SVC) program, a national initiative to develop e-learning within Swiss universities. The SVC sought to encourage development of materials with a broad-based appeal by stipulating that any project must have a minimum of three separate universities as partners. Very importantly, the SVC also required that partners guarantee that developed materials be integrated into existing curricula. The GITTA project, among the groups of the project consortium, created a lively discourse about issues of pedagogy, didactics and curricula, as well as about different teaching and learning cultures in the different language regions, schools and disciplines involved. Owing to the heterogeneity and the diverse requirements of the partners involved, GITTA content was not developed as a single, monolithic course, but as a pool of lessons (cf. §26.3) which form the building blocks for integration into tailor-made courses. In the remainder of this section, we will present and discuss three examples of the integration of GITTA lessons and case studies into particular courses at three partner institutions. Of course, more alternatives for integration exist, including pure e-learning, which would however only make sense in a distance learning context.

Table 26.1 Structure of the course 'Introduction to spatial analysis using GIS'. Shading indicates use of GITTA lessons and case study

Week	Topic	Lecture / Tutorial	Practicals
1	Course introduction	L	-
2	GIS basics	L	PR1
3	Conceptual data models and data structures	L	PR1
4	Operations on entities	L	PR2
5	GITTA tutorial: Spatial Queries	T	PR2
6	Operations on fields	L	PR3
7	GITTA tutorial: Terrain Analysis	T	PR3
8	Applications of GIScience I	L	PR4
9	Applications of GIScience II	L	PR4
10	GITTA tutorial: Suitability	T	PR5: GITTA case study
11	GIS in Switzerland	L	PR5
12	GITTA tutorial: Accessibility	T	PR5
13	-	-	PR5
14	Wrap-up session	L	-

Example #1: replacement of conventional lectures, revision through tutorials

The first example describes the pedagogical structure of the course *Introduction to spatial analysis using GIS*, which encompasses six credits in ECTS (European Credit Transfer and Accumulation System) points. The course is offered to second year bachelor students in geography at the University of Zurich. About 120 students attend the course every year.

Table 26.1 shows the course structure. The course is made up of a lecture-style part where theoretical background is given, and a practical exercise part. Four of the lectures are conducted as blended learning sessions using GITTA lessons (shaded grey in Table 26.1) and a case study forms the basis for the longest of the practicals. The remaining lectures and practicals take place in traditional fashion.

For each GITTA lesson a two-stage process is pursued:

1. The first stage constitutes the preparation and acquisition of the learning content. The students receive an email with a link to the corresponding GITTA lesson. The email also contains instructions in preparation of the tutorial session, that is, instructions on selected 'Act' and 'Self-Assessment' elements that should be solved before the actual tutorial takes place. Equipped with this information, the students then work through the GITTA lesson independently.
2. In the second stage a revision of the lesson content is made in a tutorial session, with mandatory attendance for all students. The tutorial replaces the traditional lecture and takes 45 minutes (instead of 2 × 45 minutes of the lecture). The full class of usually around 120 students is split into six parallel tutorial groups of about 20 students that are directed by qualified tutors.

Tutorial sessions aim to consolidate the student's knowledge of the subject matter by various active tasks. The sessions are structured as follows: They start with a brief Q&A session to clarify points that might have remained unclear to students during the self-study stage. In the remainder of the tutorial session, individual tasks are reviewed (e.g. 'Act' and 'Self-Assessment' elements that were assigned in the instructions) and group exercises are carried out. The following three examples should illustrate these kinds of tasks:

- The lesson 'Terrain Analysis' contains an exercise for manually calculating the visibility of two points in a given TIN-based terrain model. Based on this exercise, fundamental aspects of visibility analysis are again discussed.
- In the tutorial for the GITTA lesson 'Accessibility', the students receive a worksheet with tasks such as drawing buffers for certain geometries, constructing Thiessen polygons for a given distribution of points manually, and carrying out distance transformations for points in a raster environment. The pedagogical motivation of these exercises is to understand mathematically what is usually done at the push of a button. This enables the students to understand and critically appraise the results produced by the computer.
- The tutorial session for the lesson 'Suitability' consists of a (conceptual) habitat modelling exercise that is conducted in small groups of four to five students (splitting the tutorial group into four to five small groups). Each small group receives a summary of the habitat requirements for a specific species. Within a timeframe of 20 minutes, they have to specify the elements of a habitat model based on these habitat requirements, including the identification of the data that are needed (e.g. terrain data, vegetation etc.), and the GIS operations that are required to process the basic data into the information required by the habitat model. Each group then briefly presents their solution to the tutorial class, and discusses it with the plenary. Alongside a revision of the content of the lesson, the students learn to defend their work in front of a plenary and challenge solutions presented by their colleagues.

The following observations were made: The students are usually well prepared and participate actively in the tutorials. Both the students and the lecturers enjoy this interactive form of teaching. The tutorials allow for a closer contact with the students (compared to the traditional lectures in front of 120 students). For the students, there is extra effort involved (self-study of the GITTA lesson *plus* tutorial session), but the consolidation of the subject matter is more profound and in the end-of-term exam the questions relating to the GITTA lessons are usually answered better than others, suggesting a positive effect of blended learning. For the teachers, the extra effort involved in terms of preparation is minimal. However, more instructors are needed: Three instead of one, as two sets of three parallel tutorial sessions are held per GITTA lesson (to accommodate the six tutorial groups).

Besides the theory-oriented lessons, GITTA material is also used for the fifth practical of the course, which consists of the GITTA case study 'Habitat analysis in the Swiss National Park' (see Table 26.1, right-most column). The subject of the case study links to the previously held GITTA tutorial on 'Suitability' and the theory is applied in a realistic setting, using commercially available GIS software.

Example #2: e-learning as supplement, revision by wiki-based summaries

The Department of Environmental Sciences at ETH Zurich offers the course 'Introduction into Geographic Information Science and Technology', worth five ECTS. The number of participants varies between 30 and 60 students. GITTA lessons are used to revise and

Table 26.2 Overview of lecture topics and supplementary GITTA lessons. Arrows indicate deadlines for wiki entries

Week	Topic	Self-study (GITTA lessons)
1	Course introduction	
2	GIS basics	
3	Spatial models I: Vector	
4	Spatial models II: Raster & TIN	
5	Data modeling & management	- Introduction to Database Systems ↙ - Database system concepts and architecture
6	Data capture	
7	Spatial analysis I: Vector	- Spatial Queries ↙ - Accessibility
8	Spatial analysis II: Raster	- Terrain Analysis ↙ - Suitability Analysis
9	GIS in practice I	
10	GIS in practice II	- Cartographic Data Representation - Presentation and visualisation Needs ↙ - Layout design settings / graphical semiology
11		
12	Exam	

improve knowledge of concepts and techniques that were introduced in traditional lectures of the course.

The course contains a theory track and an exercise track. In the following, we focus on the theory track only. The learning content is taught in ten interactively designed face-to-face lectures, taking 90 (2 × 45) minutes each. For selected topics (see Table 26.2), self-study of GITTA lessons serves to revise and deepen the theory that was introduced in the lectures.

At the end of lectures supplemented by e-learning, students are given a short introduction to the corresponding GITTA lessons (Table 26.2). In order to revise the theory covered by the GITTA lessons, the students are requested to work on specific assignments and publish their results as a wiki page. A particular assignment may include several learning objectives. These were designed based on the cognitive levels of learning of the taxonomy by Bloom (1956). This taxonomy organizes the learning process into six levels of increasing cognitive skills: knowledge, comprehension, application, analysis, synthesis and evaluation. Bloom's taxonomy was later reworked (Anderson and Krathwohl, 2001), with the main differences being that verbs were used to denote the cognitive level, and that the positions of synthesis and evaluation were swapped.

By way of example, Table 26.3 shows an extract of the learning objectives and the topics for wiki summaries that were defined for the GITTA lesson 'Introduction to Database Systems'. The levels of learning according to Bloom are shown in brackets for each task. Creating short summaries as wiki entries, the students obtain an overview of the characteristics of the database approach as well as its advantages and disadvantages. By explaining these characteristics by means of examples, the students not only reproduce mere fact knowledge, but have to put their knowledge into a different context. Additionally, the competence to construct examples of application scenarios is demanded. Thus, the students learn to select the appropriate approach for data organization and storage within GIS.

Table 26.3 Learning objectives and wiki topics for the GITTA lesson 'Introduction to Database Systems'

Learning objectives	Wiki topics
1. You know and understand the characteristics of database systems and are able to explain these characteristics using an example (levels 1 & 2: knowledge & comprehension)	Briefly outline the characteristics of the database approach and give an example for each characteristic.
2. You know the advantages and disadvantages of database systems. For a specific GIS task, you are able to decide whether to use a file based or a database approach (levels 2 & 4: comprehension & analysis).	Discuss the advantages and disadvantages of the file based vs. the database approach. Briefly describe two scenarios for both of them for a selected GIS task.

Experience (Autumn semester 2009) has shown that the students created wikis accurately and used them intensively for preparing the final exam, but didn't amend or improve them. This turned out to be unsatisfactory, since the teachers themselves do not comment or correct the content of wikis (a fact that was communicated clearly at the beginning of the course). In order to discover errors concerning the content of wikis, one solution is to give the students the opportunity to present and discuss their wikis at the end of the following lecture.

Example #3: e-learning course using virtual collaboration

The third example presents the course 'Cartography', which is part of the module *Geoinformatics V* of the BSc in Geomatics at the FHNW University of Applied Sciences Northwestern Switzerland. The course encompasses 1 1/3 ECTS, equivalent to approximately 40 hours of work, spread over four weeks. Some 20 students usually take part in this course. Table 26.4 gives an overview of the course.

The course starts with a tutorial in the first week where students are introduced to the setup of the course and to the GITTA materials. The course is designed to be completed in self-study using the three GITTA lessons listed in Table 26.4. The first three weeks are intended for studying the material. Additionally, there is a lecture on (geo)information visualization in the third week, since GITTA does not cover this topic. All of the materials and tools (e.g. timetable, forum) are provided via a learning management system (Moodle).

All three GITTA lessons contain exercises. For this particular course, four of them were selected for mandatory completion. The students have to publish their result in the corresponding forum by the end of each week. By way of example Figure 26.4 shows the exercise 'Map Brainstorming' of the lesson 'Cartographic Data Representation'. The students are requested to discuss the content and structure of a thematic map online in groups of two to three students. Figure 26.5 illustrates as a second example the exercise 'Map Critique' from the lesson 'Layout Design'. This exercise is completed by the students individually.

For each of the exercises the students also have to deliver at least one peer feedback, which has to be published as soon as possible, but at the latest by the end of week 4 (cf. Table 26.4).

Table 26.4 Overview of the course 'Cartography'

Week	Tutorial	Self-study (GITTA)	Practicals
1	Introduction: - Cartography course - GITTA lessons	- Cartographic data representation	PR1: Map Brainstorming
2		- Layout design settings / graphical semiology - Generalization of map data	PR2: Map Critique
3	Lecture: Introduction to (geo)information visualization Exercise: Diagrams in Excel		PR3: Generalization
4			PR4: InfoVis Critique Peer feedbacks for all practicals due

Additionally, the students receive feedback from the course tutor for the InfoVis practical (PR4), as well as on the other practicals when requested.

The following observations have been made with this course: The students appreciate the well-structured e-learning lessons. The peer feedbacks proved to be of value, since the students engaged more deeply in the materials when they had to comment on the work of fellow students as well. Occasionally the students needed a reminder to complete all

Map Brainstorming

Gather in a group of 2 to 3 students for an online whiteboard discussion.
For the discussion, please keep the following aspects in mind:

- First, make an agreement on who is uploading the minutes of your discussion into your workspace.
- Take a minute to get a first overview of the assignment.
- Then start to share and discuss your ideas with each other. Allow at least 10 minutes for this communication part. Keep to the objective of the assignment!

Your task is to make a map of the population density in Europe for the year 2002. This map will be included in a schoolbook for 9-10 year old children. Find a collective answer to the following questions:

- What exactly do you want to be shown on this map, and by which means?
- What are the sequential steps you will follow?
- Optional: Draw a very simple preliminary sketch of your map.

Your course instructor will give you further information about the presentation of the exercise and times/deadlines.

Figure 26.4 Example of a mandatory exercise: 'Map Brainstorming' (GITTA, 2011)

Map Critique
Objective: Exercise your sensitivity to Layout Design.

Task
Find a map or a map related illustration in a newspaper, leaflet, book, internet, etc. Prepare a well-designed written discussion taking the following into account: Present your discussion on 2-3 A4/letter size pages together with a digital version of your map example. The text should be clear, to the point and without spelling errors including the following information: Name, Date, Map reference/Source as well as the time taken to complete the exercise. Your course leader will decide on the language to use (e.g. German/French/Italian/English). The questions below may be considered as a thread for your discussion:

- Is the scale well chosen? Propose a better solution.
- Are the map elements well positioned? How could the positioning be improved?
- Is the colour composition done well? How could it be improved?
- Is the typography appropriate? Any reasons for improvement?
- Have the legibility rules been respected?
- Can you find aspects where the layout might be improved?

Presentation
Submit your results (mail or printed) by the deadline set by the course instructor.

Figure 26.5 Example of a mandatory exercise 'Map Critique' (GITTA, 2011)

feedbacks. It was also realized that it is important for the tutors to be present on the electronic forums as well and make additional comments about the exercises. Generally, such comments have a motivating effect and stimulate more extensive online discussions. However, this is a very time-consuming task that should not be underestimated.

26.5 Discussion and conclusions

In this chapter, we have focused on experiences made with the integration of e-learning materials from the GITTA project into courses at participating partner institutions. On the basis of three examples taken from three different courses on different topics of GIScience at three different participating universities, we have shown that GITTA lessons and case studies can be integrated in very different ways. In terms of course integration, the great strength of GITTA is its flexibility and the many possibilities that it offers for integration and re-combination of content, as well as the capability of tuning to the requirements of a particular course and curriculum. The modular structure and content also fosters course integration in a blended learning mode, rather than trying to rely on a 'ready-made', pure e-learning course. Conversely, experience and student feedback has also shown that measures of accompanying the e-learning part are essential to the success of a blended learning course. Blended learning isn't simply about interspersing a few e-learning lessons into a conventional course. We have shown different examples of accompanying measures such as face-to-face revision sessions, face-to-face or online group discussions, online exercises, or short summary presentations by students as effective ways of consolidating the theory conveyed in e-learning lessons.

When the GITTA project started ten years ago, it united groups from several different disciplines contributing to GIScience, from different language regions, and from different types of higher education institutions in Switzerland. Hence, the organization, content and pedagogical approach of the project had to be modular and flexible. GITTA consists of a pool of lessons for theory teaching and case studies for practical exercises. The individual lessons and case studies all follow a common pedagogical approach and structure, but how exactly these materials are integrated into actual courses is not predefined, except for the general aim of using materials preferably in blended learning mode. Taking the pool-of-lessons approach rather than developing a pre-cast, monolithic course structure bears the danger of heterogeneity and lack of continuity when several lessons are integrated to form an actual course. On the other hand, this approach provides maximum flexibility to adapt to the needs of the particular curriculum that hosts the course. Furthermore, GITTA content is commonly used in a blended learning mode, hence the teacher has the possibility to provide the 'thread' between individual lessons. Responding to requests by teachers who use GITTA materials, we are currently even considering unbundling lessons into smaller building blocks. In many situations of blended learning, it would be preferable not to use entire lessons, but instead have access directly to individual units of a lesson, or even 'goodies' such as particular Flash animations or illustrations.

The ECLASS model that is used for GITTA lessons and units provides a straightforward structure that contains all the necessary elements for content development guided by learning objectives, and offers the flexibility that is needed for content development in a multidisciplinary consortium of authors. Basically, it's simple but it works very well for our purposes. However, despite the fact that self-tests are included in GITTA lessons, they do not incorporate online assessments that could be used for evaluating and grading the students' performance in a course. Also, as a consequence of not using a particular learning management system, progress monitoring of students is not possible. These two points may be seen as drawbacks of our strategy that builds on modularity. However, since it is possible to import GITTA content into LMS through SCORM and IMS archive formats, both points could be built into a particular course at the level of the LMS, using the corresponding functions of the LMS, if so desired.

Compared to other e-learning projects GITTA could be blamed for taking a rather technocentric approach. This can, however, be largely explained by the history of the project. Remember that GITTA was, from the outset, a networked project between several partners from different disciplines and universities. Thus, the main criteria were portability and extensibility of content, and that in turn called for the use of standards. To achieve maximum portability and extensibility, it was clear that we had to use XML. However, at the time when we started in 2001, there was simply no LMS available that would allow integrating XML sources and that wouldn't impose its rigid structure. Hence, we had to get more technical than we originally had wished to, which in turn led to the development of eLML. Although it's not an academic or pedagogical achievement, the eLML framework is certainly one of the contributions that we are proud of, as it is now used by a range of other e-learning projects worldwide (eLML, 2011).

Sustainability is a major issue in e-learning projects. While research projects usually have a fixed duration and can be concluded once the final reports and publications have been generated, e-learning projects are basically only successful if they persist indefinitely. Unfortunately, that also implies that content has to be continuously updated, particularly in an area such as GIScience, where the half-life of knowledge is relatively short.

As mentioned in the introduction to this chapter, several countries have known programmes that supported e-learning projects in higher education, particularly in the early 2000s. Many of these projects are now dysfunctional or have disappeared completely, while GITTA is still alive after ten years.

In an earlier paper (Weibel *et al.*, 2009) we have identified four reasons that we believe have enabled GITTA to persist so long: (1) modular content design; (2) flexible, platform-independent technical framework; (3) open content strategy; and (4) solid organizational structure. The first two points have been discussed in some detail above and need no further consideration. A fifth point should be added: stable funding. Swiss universities are in the fortunate situation of enjoying relatively generous and stable base funding, meaning that small updates and extensions as well as the maintenance of the technical infrastructure can be covered from the budgets of the participating Swiss partner institutions. However, beyond that, we believe that the third and fourth point also contribute to the sustainability of GITTA in a major way. Very early on, the GITTA partners decided to make the content freely available under a Creative Commons license (Fisler and Weibel, 2006). Through a simple registration mechanism on the GITTA website (GITTA, 2011), the content can be downloaded as SCORM or IMS content packages and integrated into an LMS, or it can be accessed directly from the project web server (GITTA, 2011). Since the end of 2006, the GITTA project is managed by the GITTA Association, which provides a simple and efficient organizational structure that is open to accept new partners. This organization does not only coordinate the project, but it is also responsible for technical support and provides help with the course integration of GITTA materials. In the summer of 2010, a survey was conducted among the subscribers of the GITTA Newsletter. Besides documenting the widespread, worldwide usage of GITTA content, this survey also provided valuable feedback on the further development of content. Among others, requests for translations to other languages were made, and new lessons or modifications of existing lessons proposed. Very importantly, more than forty respondents indicated that they would be willing to help with the further development of GITTA content, as translators, reviewers of existing lessons, authors of new or extended lessons, or developers of animations. This pool of potential volunteers can be used to recruit new authors and contributors, thus exploiting the benefits of the open content strategy and contributing to extending the sustainability of the overall project.

Acknowledgements

The authors of this chapter represent a much larger group of individuals who have contributed to the GITTA project as authors, designers, instructors, testers, and in other important ways. We gratefully acknowledge these contributions and hope that we have adequately reflected the spirit of the project.

References

Anderson, L. W., and D. R. Krathwohl (eds) (2001) *A Taxonomy for Learning, Teaching, and Assessing: A revision of Bloom's taxonomy of educational objectives*, New York: Longman.

Bloom, B., M. Englehart, E. Furst, W. Hill, and D. Krathwohl (1956) *Taxonomy of Educational Objectives: The classification of educational goals. Handbook I: Cognitive domain*, New York: David McKay, Longmans.

CartouCHe (2011) *Cartography for Higher Education in Switzerland.* http://www.e-cartouche.ch (accessed March 11, 2011).

eLML (2011) eLesson Markup Language (eLML) homepage. http://www.elml.org (accessed March 11, 2011).

Fisler, J., and S. Bleisch (2006) eLML, the eLesson markup language: Developing sustainable e-learning content using an open source XML framework. *WEBIST 2006 - International Conference on Web Information Systems and Technologies*, 11–13 April 2006. Setubal, Portugal.

Fisler, J., and R. Weibel (2006) *GITTA:* Open content material for GIS education. *EUGISES 2006 Conference*, 7–10 September 2006, Krakow, Poland.

Gerson, S. M. (2000) E-CLASS: Creating a guide to online course development for distance learning faculty. *Online Journal of Distance Learning Administration*, 3(4). http://www.westga.edu/~distance/ojdla/winter34/gerson34.html (accessed March 11, 2011).

GITTA (2011) Geographic Information Technology Training Alliance (GITTA) homepage. http://www.gitta.info (accessed March 11, 2011).

Graham, C. R. (2005) Blended learning systems: Definition, current trends, and future directions, IN: C. J. Bonk, and C. R. Graham (eds) *Handbook of Blended Learning: Global perspectives, Local Designs.* San Francisco, CA: Pfeiffer, pages 3–21.

Horton, W. (2000) *Designing Web-Based Training.* Chichester, England: John Wiley & Sons.

IMS (2011) Content packaging specification, Version 1.1.4. IMS Global Learning Consortium, Inc. http://www.imsglobal.org/content/packaging (accessed March 11, 2011).

JISC (2011) Joint Information Systems Committee (JISC) homepage. http://www.jisc.ac.uk (accessed March 11, 2011).

Jonassen, D. H., and S. M. Land (2000) *Theoretical Foundations of Learning Environments.* Hillsdale, NJ: Lawrence Erlbaum.

JORUM (2011) Jorum homepage. http://www.jorum.ac.uk (accessed March 11, 2011).

Niederhuber, M., H. R. Heinimann, and B. Hebel (2005) e-Learning basierte Fallstudien zur akademischen Ausbildung in der Geoinformatik: Methodisches Konzept, Umsetzung und Erfahrungen, IN: J. M. Haake, U. Lucke, and D. Tavangarian (eds) DeLFI 2005: 3. Deutsche e-Learning Fachtagung Informatik, *Lecture Notes in Informatics*, 66: 34–45.

SCORM (2011) Sharable Content Object Reference Model. Advanced Distributed Learning. http://www.adlnet.gov/Technologies/scorm/default.aspx (accessed March 11, 2011).

SVC (2011) Swiss Virtual Campus (SVC) homepage. http://www.virtualcampus.ch (accessed March 11, 2011).

Weibel, R., S. Bleisch, S. Nebiker, J. Fisler, T. Grossmann, M. Niederhuber, and L. Hurni (2009) Achieving sustainability in e-Learning for GIScience – Experiences from the GITTA Project. *Geomatica*, 63(2): 109–118.

27

GIS&T in the open educational resources movement

David DiBiase

John A. Dutton e-Education Institute, Penn State University, University Park, Pennsylvania, USA

27.1 Introduction

This chapter positions higher education in geographic information science and technology (GIS&T) in relation to the open educational resources (OER) movement. After defining OER and the movement it denotes, I compare several initiatives designed to promote free sharing of GIS&T-related educational resources and, in one special case, free access to graduate education. The chapter concludes with an argument that the most important potential benefit of OER is open education itself, and that the proprietary interests that keep education closed should be resisted.

GIS and geography educators were quick to recognize the potential of the internet and World Wide Web (web) to discover, acquire and share digital educational resources. Ritter (1994) published one of the earliest papers in the geospatial community that announced *Mosaic*, the first graphical interface web browser, and promoted awareness of the web's potential. Thoen (1994, 47) captured the excitement of discovery in his observation that 'to say that the Internet is huge is an understatement. It contains literally *gigabytes* of programs, data sets, source code, online news services and thousands of discussion groups . . .' [emphasis added].

Among others, Bishop and colleagues (1992, 1995) promoted the use of data and software available via the internet and web to enrich teaching and learning in remote sensing of the environment, environmental conservation, and geomorphology. Crampton's (1995) overview of 'Cartographic resources on the World Wide Web' profiled several of the earliest web-based educational resources for GIS&T, including Plewe's TIGER Mapping Service, Foote's Virtual Geography Department (see below), and Crampton's own Virtual Bosnia Field Trip. Mirroring the proliferation of the internet itself, use of web-based resources for GIS&T education has become commonplace since then. Today educators routinely use free online resources maintained by public agencies, private individuals and businesses. A few examples include the tools provided by the US National Geodetic Survey

Teaching Geographic Information Science and Technology in Higher Education, First Edition.
Edited by David J. Unwin, Kenneth E. Foote, Nicholas J. Tate and David DiBiase.

(2010), David Rumsey's Map Collection (Rumsey, 2010), and the Esri Mapping Center (Esri, n.d.).

Following the leads of the early innovators, numerous GIS&T educators have themselves become contributors of online resources. Volunteered resources by geospatial educators support teaching and learning of topics spanning most of the ten knowledge areas identified in the *GIS&T Body of Knowledge* (DiBiase *et al.*, 2006), including *Cartography and Visualization* (e.g. Brewer and Harrower, 2009), *Spatial Analysis* (e.g. de Smith, Goodchild and Longley, 2011), *Geospatial Data Properties* (e.g. Snyder and Voxland, n.d.) and *GIS&T and Society* (DiBiase, Wright and Harvey, n.d.). Beyond these and other important contributions, however, formal OER projects tend to result from coordinated organizational strategies rather than *ad hoc* individual efforts.

Of the various definitions of OER the Organization for Economic Cooperation and Development's (OECD, 2007, 30) may be the most widely cited:

> open educational resources are digitised materials offered freely and openly for educators, students and self-learners to use and reuse for teaching, learning and research.

Under this definition OER includes:

> Learning content: Full courses, courseware, content modules, learning objects, collections and journals.

and

> Tools: Software to supported the development and use, reuse and delivery of learning content, including searching and organization of content, content and learning management systems, content development tools, and online learning communities.

Readers who have shared their educational resources informally for years – via unrestricted websites, for instance – may wonder, 'how is 'OER' different from what I already do?' Formal OER projects are distinctive in at least four respects (Table 27.1). First, truly open resources are not only freely available for use, they are also licensed for legal re-use by teachers, learners and anyone else, ideally using standard rather than idiosyncratic license agreements. Second, formal OER projects make it easier to re-use resources by providing them in standard formats that can be imported into learning management systems or content management systems, or in environments that facilitate collaborative authorship. Third, like open source software projects, formal OER initiatives are characterized by active developer and user communities. And fourth, successful OER projects provide incentives for resource providers to maintain and expand high-quality content. Granted, few OER projects embody all these characteristics. However, projects that incorporate even some offer clear advantages over isolated personal initiatives.

Table 27.1 Characteristics of formal OER projects

1. Standard licenses allow legal use and re-use.
2. Standard digital formats that facilitate re-use and/or collaboration.
3. Active communities of authors and users.
4. Incentives for sustained participation.

27.2 The OER movement

I use the word 'movement' here in its sense of a group of people who share a common ideology and who try together to achieve certain general goals (WordNet, 2009). It seems to me that the common ideology shared by OER proponents are the beliefs that education ennobles humankind, and that education is at its best when learners are encouraged to construct knowledge actively, often by 'remixing' elements of knowledge and expression produced by predecessors (Jenkins, 2006; Lessig, 2008). More than ideology, these shared beliefs may constitute the 'moral ideal' that is one of the defining characteristics of the education profession (Davis, 2002).

The founders of the OER movement were inspired by the success of certain free and open source software (FOSS) projects – particularly the Linux operating system – in synergizing the efforts of many independent volunteer developers. As Raymond (2001, 16) stresses, 'the most important feature of Linux . . . was not technical but sociological.' 'Who would have thought,' Raymond (2001, 21) asks, 'that a world-class operating system could coalesce as if by magic out of part-time hacking by several thousand developers scattered all over the planet, connected only by the tenuous strands of the Internet?'

Founders and thought leaders gather annually at an Open Education conference, which in its sixth year (2009) attracted over 200 on-site participants and many more online followers. Among the most influential leaders is David Wiley of Brigham Young University. While still a PhD student at Utah State University in 1998, Wiley coined the term 'open content' and created an early license agreement that promoted content sharing while preserving authors' copyright (Wiley, 2006; Smith, 2009).

Wiley provides evidence of the scope and momentum of the OER movement in a report to the Organization for Economic Cooperation and Development, in which he estimates that more than 2,500 open-access courses are available from over 200 universities (Wiley, 2007). Nearly all of these have appeared within the past ten years, and the proliferation of open courseware appears to continue unabated. Many of these institutions' OER offerings can be searched and accessed through the OpenCourseWare Consortium and the OER Commons among others.

27.3 Sustainability of OER initiatives

The OER movement captured the attention of educators everywhere in 2001 when the Massachusetts Institute of Technology with much fanfare announced its OpenCourseWare Initiative (Vest, 2006). With substantial philanthropic support and industry partnerships, MIT set out to make educational resources used in all its classes freely available worldwide under the recently developed Creative Commons license. By September 2009, MIT's OpenCourseWare initiative listed 1,900 'courses' (Massachusetts Institute of Technology, 2010). Links to courseware are organized by the academic department, and individual resources may be located by keyword search. A sample search on 'GIS' in 2009 yielded 333 results, sorted by relevance. The first ten results included six HTML pages of lecture notes and reading lists (some with links to further resources) and four PDF files consisting of exported presentation slides, assignments, or discussion notes.

Considering how plain many of them appear to be, it's easy to underestimate the impact of MIT's open educational resources. In 2009 I had the chance to ask Chuck Vest, who was

MIT's president when the OCW initiative was conceived and announced, how he responds to the many sceptical observers who have dismissed the initiative as 'hype.' Rather than resort to website traffic counts or other statistics, Vest described how OCW resources had been used by the Bahá'í Institute for Higher Education to create an 'underground university' that counteracts the Iranian government's denial of higher education opportunities to Iranian Bahá'ís. How many of us produce educational resources that have such an impact?

As Wiley (2007) points out, however, the MIT example is unique, and because of its high cost and reliance on philanthropic support, probably unsustainable. In 2007 the OCW initiative employed 29 people and had an average annual budget of $4.3 million. While acknowledging MIT's success in attracting foundation support and vendor partnerships, Wiley (2007, 8) concludes that there is 'very little chance that any other institution will be able to replicate the MIT model.'

Other higher education institutions have launched OER initiatives, but none so far has embodied a sustainability plan of the sort that Wiley characterizes as 'OCW 2.0' (Wiley, 2009a). His own alma mater, Utah State University, offers open courseware associated with 80 different courses (Utah State University, n.d.). Utah State's relatively modest OCW project employed just a full-time director and some student assistants and cost only about $125K per year to operate through 30 June 2009. Then, however, the director was laid off due to budget constraints after support from the Hewlett Foundation and state legislature was exhausted. Wiley called the dismissal 'heartbreaking' (Parry, 2009).

Rice University has shown that it is possible to grow a substantial OER initiative with minimal centralized university support. In September 2009 Rice's Connexions project (Rice University, n.d.) listed 14,838 reusable modules in 796 collections. Like OER Commons, Connexions is a 'referatory' that provides links to resources served locally by contributing institutions. Authors contribute resources voluntarily from institutions around the world with little or no backing. Wiley (2007) points out that Rice's success at sustaining its decentralized OER initiative reflects the presence of an influential champion. For the same reasons Wiley's own departure from Utah State may have undermined the sustainability of that project.

In the following section I review a cross-section of major OER initiatives related specifically to GIS&T with reference to the four characteristics outlined in Table 27.1 above.

27.4 GIS&T in OER

The Geographer's Craft and Virtual Geography Department projects

Ken Foote was among the first to organize a web-based collection of open resources for GIS education beginning with the Geographer's Craft project in 1992 (Foote, 1997). This was a year-long course that used an active-learning, problem-solving approach to introduce geographic research techniques, all built around hypermedia, web-based course materials. By 1996, with funding from two NSF grants, Foote and his students had created one of the first comprehensive, online bodies of educational resources in geography, including 14 units on key topics in GIScience. Foote found that within months of units going online, file downloads from outside the university far exceeded those made by his students at the University of Texas. The files were being used across all internet domains (.edu, .com, .mil) and from internet addresses worldwide. The resources continue to be widely used, and those written by Peter Dana on map projections, coordinate systems, GPS and geodetic datums are cited widely in digital and paper reference materials and still top lists of internet search results on those topics.

The widespread use of the Geographer's Craft resources suggested that a similar sharing of materials might be possible if other faculty were willing to contribute. From 1996 through 1999 his National Science Foundation-funded 'Virtual Geography Department' attracted over 100 contributors whose interests and expertise spanned the discipline. Foote's stated objective – 'to develop a web-based clearinghouse for high quality curricular materials and laboratory modules that can be used by students and faculty all over the world' (1999, 113) – typified later OER projects. However, his broader goal was to exploit the web to promote and sustain 'intradisciplinary collaboration' (Foote, 1999, 108). To this end, the project emphasized workshops in which educators worked together to learn web publishing skills and pedagogical strategies for using web-based resources in higher education.

Foote's 1999 article 'Building disciplinary collaborations on the World Wide Web' compares several kindred projects – including the Virtual Geography Department – in regard to project goals and the strengths and weaknesses of strategies adopted to achieve them. Project sustainability was a key concern, as was the oft-cited lack of incentives for sustained voluntary faculty contributions. As Foote seemed to expect, several of the high-profile initiatives he compared were soon abandoned (e.g. the Core Curriculum in GIScience, successor to the NCGIA's Core Curriculum in GIS project) or stopped short of fulfilling their potential as OER clearinghouses (e.g. the Alexandria Digital Library).

The Virtual Geography Department itself still exists (Foote, 2004), but its contents are dated. For example, as of 31 August 2009, only five of the 34 courses linked from the Virtual Department's 'Geographic Information Science' resource page offer open and up-to-date syllabi and laboratory exercises, and most of those provide required exercise data only to registered on-campus students. Rights to re-use resources vary. One syllabus even states that 'use of these materials by other instructors in their courses is expressly forbidden without my written permission.' Most resources are provided as HTML documents, word processing or PDF documents (Table 27.2).

Project reviewer Michael Solem (2000, 353) concluded that the Virtual Geography Department succeeded in 'diffusing innovative practice in geography by training faculty members in Web pedagogy and online curriculum development', despite the fact that 'some participants failed to follow through with new online materials after the conclusion of the workshops . . .' (Solem, 2000, 363).

Anderson (2009) describes several types of business models that include provision of 'free' goods. Of these, the Virtual Geography Department typifies a 'non-monetary market.' The primary incentive for voluntary contributions in a non-monetary market is the enhanced reputation that accrues to authors and/or institutions from the widespread distribution and

Table 27.2 OER characteristics of the Virtual Geography Department

1. Standard licenses allow legal use and re-use?	No – mixed
2. Standard digital formats that facilitate re-use?	Mostly HTML, word processing and PDF documents with some standardized metadata descriptions
3. Active communities of authors and users?	Not sustained
4. Incentives for sustained participation?	No

use of their works. As Foote himself observed, however, that incentive is inadequate for most academic geographers since such contributions are rarely included among the criteria by which university faculty members are awarded promotion and tenure. Foote (*personal communication*, 2009) also notes that:

> for most faculty, sharing teaching materials – putting them out in public – is a foreign and uncomfortable experience. Though they do this with their research writings, they are far more hesitant to do the same with their teaching materials.

For these and other reasons, sustainability has proven as elusive for the Virtual Geography Department as for most of the other projects that Foote compared in 1999. One exception is the UNIGIS project.

UNIGIS International network

Of the projects he compared in 1999, Foote was most optimistic about the UNIGIS. Founded in 1990 and expanded by educators in the UK, Austria and the Netherlands, UNIGIS began as a print-based correspondence course, then migrated to web-based distance learning in the late 1990s. Ten years later, despite many organizational and technological changes, UNIGIS International continues to thrive. Ten universities in Europe, Africa, South and North America operate nodes. UNIGIS students register in and earn postbaccalaureate certificates, diplomas and masters degrees from one of the participating universities, but may earn credits for modules offered by several different institutions. Partners share curricula and educational resources, including revisions and translations. They also share marketing and administrative costs (Molendijk and Sholten, 2005). Foote observed that 'formal collaborations that have permanent staff and means of funding, such as the UNIGIS project, may offer a more viable, long-term model for developing collaborations' (Foote, 1999, 114).

However, UNIGIS is not an OER project. Educational resources created and shared by consortium members are not open to others except fee-paying students. Indeed, the essence of the relationship between members is an exclusive license agreement that governs access to educational resources copyrighted by the consortium. Therefore the original question Foote posed in 1999 remains unanswered: Is it possible to create a sustainable OER project for GIS?

Penn State 'World Campus'

In North America, UNIGIS nodes compete for students with several universities that offer distance education in GIS&T, including Penn State University. Penn State's online GIS certificate and master's degree programs attract about 1500 enrollments annually from about 400 students who register through the university's online 'World Campus' (Penn State, 2011). Key to Penn State's success is a university policy that rewards entrepreneurialism by returning a large share of tuition revenue to academic units who create and sustain online programs. In the fiscal year 2008-09, for instance, the share of tuition revenue returned to the Penn State program was $2.2M, much of which supported salaries of the fifteen full-time-equivalent instructors and support staff.

As of September 2009, 14 of the program's 26 online courses were at least partly available as open educational resources. The open 'courseware modules' consist mainly of HTML pages and associated graphics that are served through an open source content management system (Drupal). This is paired with a password-protected learning management system (ANGEL) in which select materials and communications are shared only with registered students who pay tuition and earn academic credit. The courseware is contributed voluntarily by faculty members and is licensed for non-commercial re-use through a standard Creative Commons share-alike version 3.0 license. Users are invited to submit comments and requests to faculty authors through Penn State's College of Earth and Mineral Sciences' Open Educational Resources initiative (Penn State College of EMS, n.d.).

The costs of maintaining these open resources (which are also used in classes by fee-paying students) are charged to the programs' operating budget, along with faculty salaries and related expenses. Since 2008 the Penn State program's marketing strategy has included open access to select courseware. The rationale for this strategy is based on the expectation that in an increasingly competitive higher education marketplace, adult learners will choose to register with an institution whose educational resources are open-access and of superior quality. Feedback from one student suggests how the strategy works:

> The ability to access course information . . . was critical in my decision to choose Penn State over other distance education providers. Distance education was new to me and I had some concerns regarding quality and value. When I discovered the wealth of well-presented information provided for GEOG 482 and other courses in Penn State's GIS program, I immediately felt an increased level of comfort with the quality of education I would be receiving (Foster, *personal communication* 2009).

The Penn State program exemplifies the type of business model Anderson (2009) calls 'freemium.' In this 'most common' strategy, online businesses give away a free good to many users but earn revenue from a relative few who are willing to pay for additional features. As Anderson (2009, 185) observes,

> . . . a college education is more than lectures and readings. Tuition buys direct proximity to ask questions, share ideas, and solicit feedback from academics . . . for universities, free content is marketing.

Time will tell if Penn State's OER strategy is sustainable. From Foote's (1999) perspective a weakness may be that the approach is motivated primarily by competitiveness, not cooperation.

In regard to the distinguishing characteristics of formal OER initiatives outlined above, the Penn State approach to open education embodies three of the four characteristics: its resources are licensed for legal re-use; it provides access to a community of authors as well as a collection of resources; and it provides incentives for contributors (whose salaries depend wholly or partly on the quality of their work products and the success of the marketing strategy) (Table 27.3). The Penn State initiative falls short in regard to technical interoperability, however, since it fails to provide resources in standardized exchange formats like IMS Content Packages (IMS Global Learning Consortium) and SCORM (shareable content object reference model). From a technical perspective the most

Table 27.3 OER characteristics of the Penn State World Campus program

1. Standard licenses allow legal use and re-use?	Yes
2. Standard digital formats that facilitate re-use?	No – Mostly HTML, word processing and PDF documents
3. Active communities of authors and users?	Yes
4. Incentives for sustained participation?	Yes

ambitious collection of open educational resources in GIS&T may be the GITTA project (Geographic Information Technology Training Alliance, 2006).

GITTA project

GITTA is a joint project of 10 groups at 7 Swiss universities and federal institutes of technology that created 6 multi-lingual online modules to supplement classroom-based GIS&T education. Established in 2001 with support from the Swiss federal government, the GITTA project was one of 50 contributors to the Swiss Virtual Campus (2009) which promotes online and blended learning in Swiss higher education institutions. By September 2009 the Swiss Virtual Campus listed 82 courseware projects and promised 30 more to come.

The 6 GITTA modules available in 2009 consisted of over 40 lessons (23 English, 12 German and 5 French) plus 8 case studies (6 German and 2 French). Lessons included 6 to 13 HTML pages of text and graphics (including some Flash and SVG) plus quizzes and questions, bibliographies, glossaries and metadata. The modules are freely available to anyone who subscribes to the project newsletter, and are licensed for use and re-use through a Creative Commons Attribution-Noncommercial-Share Alike 2.5 Generic license. In addition to HTML pages and printer-friendly PDF files, the modules are provided as standards-compliant IMS and SCORM archives that can be imported to commercial and open-source learning management systems such as Blackboard and Moodle.

In sum, the GITTA project embodies at least 3 of the 4 distinguishing characteristics of formal OER initiatives: its resources are freely available and licensed for legal re-use; it provides access to a community of authors; and it facilitates re-use by providing resources in standard interoperable formats (Table 27.4). Strengths include the sophisticated technical and pedagogical frameworks within which its modules were designed. A formal sustainability plan is in place (Weibel *et al.*, 2009), though it's unclear that the non-monetary incentives to courseware authors will succeed in sustaining their participation. To support continuing development of lessons and modules after its five-year grant, the GITTA project formed an

Table 27.4 OER characteristics of the GITTA project

1. Standard licenses allow legal use and re-use?	Yes
2. Standard digital formats that facilitate re-use?	Yes
3. Active communities of authors and users?	Yes
4. Incentives for sustained participation?	Maybe

association of dues-paying members in 2006 (Grossman, Weibel and Fisler, 2008). Since dues are modest, and since benefits to dues-paying members appear to be not much greater than those enjoyed by users who access its resources for free, GITTA's business model more resembles a 'non-monetary market' like the Virtual Geography Department than a 'freemium' strategy like Penn State's. Ken Foote would approve of the fact that one of the Swiss Virtual Campus' stated objectives is to 'strengthen collaboration among universities' (Swiss Virtual Campus, 2009). It remains to be seen if the project will prove to be more sustainable over the long term than the Virtual Geography Department, which was founded with similar goals. In the short term, a €25K MedidaPrix prize awarded to the project in 2008 is sure to help.

LUMA-GIS

Perhaps the most formidable sustainability challenge in GIS&T higher education is the online master's degree program offered by the University of Lund in Sweden. The Lund University master's in geographical information systems (LUMA-GIS) is free – students admitted to the program pay zero tuition. Not surprisingly, the Lund program is popular – as of September 2009, 766 students had been admitted, with 1,789 more enrolling in individual courses. The 2,555 total active students participate online from 91 countries (Mårtensson, *personal communication*, 2009).

Lund began developing online courses in 1999. Development accelerated in 2001 when it and nine partner institutions gained support from the European Commission's Leonardo da Vinci program for vocational education and training (Onstein and Mårtensson, 2004). In 2004 Lund established a complete eleven-course online master's degree, which includes a final thesis project defended in person. Students are welcome to study at their own pace, part-time or full-time. Although student demographics vary widely, the typical Lund online student is single, male, over 30 years of age, works full-time, and studies from home (Mårtensson, Pilesjö and Galland, 2007). Five years after the master's program was established, only five students had defended theses and earned degrees. Mårtensson (2009) speculates that this low completion rate is due to the program's 'relatively low priority' in adult students' busy lives.

Given the willingness of students to pay substantial tuition and fees for online master's degrees at other institutions, why does Lund give away its degree and its faculty members' time and expertise? One explanation is that higher education is tuition-free (or nearly so) in many European countries – including Sweden – where taxpayer support for public higher education is significantly greater than in the US. However, this explanation fails to account for the number of students that the Lund program serves. Mårtensson (2009) reports that the financial support the Lund program receives from the Swedish national government is really only sufficient to support the staff and facilities needed to supervise about 50 graduate students. He and his colleagues accept many more because they're committed to 'capacity building of GIS in developing countries.' Besides meeting this need, the primary incentive for faculty is to 'place Lund University on the map.' For these reasons the LUMA-GIS program exemplifies the 'non-monetary market' business model.

The LUMA-GIS program is not an OER project. Its courseware is available only to registered students through a password-protected course management system. The program is pertinent to this discussion, however, insofar as it is motivated by the same 'moral ideal' that inspires OER advocates and projects. Also relevant is the sustainability that LUMA-GIS has demonstrated to date.

Discussion

Like LUMA-GIS and the Virtual Geography Department, the GITTA project embodies a 'non-monetary market' business model. Unlike LUMA-GIS, GITTA is an OER project. To succeed where the Virtual Geography Department and most of its contemporaries could not in sustaining an active developer community, the GITTA project needs to deliver added value to its dues-paying member organizations and to deploy dues income in ways that incentivise participation by authors. In the US, where public support is inadequate to offer free university education, entrepreneurial institutions may create mechanisms for deploying tuition revenue in ways that provide incentives to authors. Where this happens, as at Penn State, the 'freemium' business model may be a viable means for sustaining OER initiatives.

Ironically, OER may prove to be more sustainable where taxpayer support for higher education is least (in the US, for example), since competition for tuition-paying students in such places provides a justification for OER as a marketing strategy. The justification follows from the expectation that in an increasingly competitive higher education market, rational adult students will choose providers whose courseware is open for inspection and is of the highest quality. But regardless of an institution's level of taxpayer support or competitive position, how can it hurt to share educational resources with others who can't afford to pay or who don't need a degree?

27.5 GIS&T journals in OER

Open-access publishing

The need for open educational resources in GIS&T education may be most acute in the arena of scholarly publishing. Obviously teachers and learners in higher education – particularly in graduate education – need ready access to original source materials like academic journals. As subscription costs increase, however, research libraries are forced to be more and more selective about the titles they provide their patrons. Following the concentration of ownership of journal titles by a relatively few for-profit publishers (including Elsevier, Candover and Cinvenn, Thompson and Wiley) (Munroe, 2007), the cost of journal subscriptions has increased far beyond the rate of inflation in recent years. For example, Edlin and Rubinfeld (2004) observe that overall prices of for-profit journals are now as much as 500 percent higher than non-profit journals.

A 2008 survey of 45 academic libraries (an international sample of two-year and four-year colleges, research universities and small hospitals) concludes that 'journal publishers have been able to continuously increase prices because they control peer review and this control of peer review has not been challenged by academics themselves' (Primary Research Group, 2008, 28). About a quarter of survey respondents believe that open-access publishing is slowing the rate of increase, while nearly half of others believe it will eventually have some effect.

Open-access publishing in GIS&T

In 2009 the Directory of Open Access Journals (Lund University Libraries, 2011) listed 4,355 open-access scholarly journals (perhaps five percent of all scholarly journals). The list included three journals whose keywords include 'GIS,' twelve journals concerned with 'cartography,' and 44 with 'geography.' However, only two open-access journals were included among the 46 leading geographic information science (GIScience) journals identified by Caron *et al.* (2008) – the *URISA Journal* and *Mappemonde.* (The open *Journal of Spatial Information Science* hadn't appeared yet.) Caron and colleagues' study addressed the absence of a comparative analysis of research publications in the relatively young and ill-defined GIScience field. They combined a Delphi study of '40 international experts' and a quantitative comparison of journal citation rates (specifically, JCR impact factors) to identify and rank leading periodicals. The *URISA Journal* provides an instructive example.

In 1998–99 Harlan Onsrud offered to serve as editor on the condition that the URISA board of directors agreed to publish an open-access version of the journal. Onsrud, was concerned about escalating costs of academic journal subscriptions, and about scholars' responsibility to 'maximize dissemination of our works and our readership' (Onsrud, *personal communication* 2009). At the time, URISA president Joseph Ferreira stated that 'while commercial publishers best make progress through exclusivity and control, the *URISA Journal* editors believe that science and new knowledge is best advanced through an intellectual environment of openness and freedom' (URISA, 1999). Ferreira's position thus presages his MIT faculty colleagues' recommendation 2001. As of September 2009 there were 192 articles in 39 issues of the *URISA Journal* freely available for use and re-use (URISA, 2009), making this one of the richest open educational resources collections in the GIS&T field. A shortcoming is that URISA's license limits re-use of digital articles to URISA members.

Besides the obvious benefits to educators and students, what benefits accrue to URISA as an organization, and to authors who contribute research articles (Table 27.5)? Certainly the journal's reputation has not suffered. The *URISA Journal* is ranked 14th in relative importance among 46 GIScience periodicals in Caron and colleagues' 2008 analysis. Nor has open access hurt the organization financially. According to URISA Executive Director Wendy Nelson, both membership in the organization and library subscriptions have been stable since 2000 (Nelson, *personal communication* 2009). And the journal's editor Jochen Albrecht (*personal communication* 2009) confirmed that submission rates haven't been affected either. Authors who contribute manuscripts to the *URISA Journal* apparently see neither advantage nor disadvantage in open-access publishing (Albrecht, 2009). This impression is consistent with the equivocal findings of bibliometricians who have attempted to document such advantages.

Table 27.5 OER characteristics of the *URISA Journal*

1. Standard licenses allow legal use and re-use?	Non-standard; re-use rights limited to URISA members
2. Standard digital formats that facilitate re-use?	N/A – PDF files
3. Active communities of authors and users?	Yes
4. Incentives for sustained participation?	Yes

Do authors benefit from open-access publishing?

It's reasonable to assume that authors would prefer to publish in open-access journals if they knew that their work would be more widely read and cited. Craig *et al.* (2007, 4) observe that several 'early studies have shown correlation between free online availability . . . and higher citation counts.' Antleman (2004), Subler (2004) and Eysenbach (2006) are among those who provide evidence that open-access publishing 'provably increases the visibility and impact' of authors' work (Subler, 2004, 8).

However, while acknowledging the association between citation rates and open access, critics like Craig and colleagues warn against inferring causality since confounding factors are usually not taken into account in such studies. For example, a 'selection bias' suggests that authors who tend to be more frequently cited also tend to make their articles freely available (Moed, 2006). Furthermore, it's well known that the generality of apparent citation effects is limited due to the culturally specific nature of scholarly publishing and citation behaviors across disciplines.

It's hard to say, therefore, if publishing in open-access journals is beneficial for individual authors. At the same time, however, there is no evidence that open-access publishing has been detrimental to one of the first GIS&T professional associations that attempted it (URISA).

27.6 Justifying OER initiatives in GIS&T

Financial considerations aside, why should professional associations like URISA make their publications freely available? Why should higher education institutions and their faculty members give away their educational resources? One reason is the conviction that sharing such resources freely is the 'right' thing to do. One participant in the 2009 Open Education conference reported that participants discussed OER as a 'moral imperative' (Camplese, 2009). Can OER be justified on ethical grounds?

For a moral imperative to exist, one or both of two conditions must exist: either (a) people have a right to free educational resources, or (b) educators are duty-bound to provide them. In fact, neither is the case.

In regard to rights, Article 26 of the United Nations' Universal Declaration on Human Rights (United Nations, 1948) does state that 'Everyone has the right to education' and that 'education shall be free, at least in the elementary and fundamental stages.' However, the declaration goes on to state that 'higher education shall be equally accessible to all on the basis of merit.' In other words, the declaration recognizes the right of higher education institutions to be selective. If institutions have a right to choose which students gain access to its human resources (faculty), then it follows that institutions also have the right to restrict access to educational resources. OER is therefore not a right that higher education institutions are bound to honour.

What about our duties as educators and editors? At a minimum, these are codified in institutional statements of professional ethics like Penn State's (1996). This policy states that faculty members' primary responsibilities are to 'seek and to state the truth as they see it' and to preserve, protect and defend academic freedom. In regards to professors' obligations to society, the policy does state that they are obliged to 'promote conditions of free inquiry . . .' This could be taken to mean that faculty members are duty-bound to publish only in open-access journals and to share all educational resources freely under

Creative Commons licenses. Unfortunately, that interpretation is contradicted by common practice. No faculty member at Penn State or elsewhere would pass up an opportunity to be published in *Science*, for example, on the grounds that it is a breach of professional ethics to publish in a proprietary, limited-access journal.

Therefore, neither in fact nor in practice are educators in higher education institutions bound by rights or by duty to participate in OER initiatives.

27.7 Sustaining OER initiatives in GIS&T

The foregoing is not to suggest that 'opening' educational resources is a bad idea. Like other proponents I believe that sharing resources freely comes close to what philosopher of professions Michael Davis (2002) calls the 'moral ideal' of the education profession. My point is that if OER is not justifiable solely on ethical grounds, the case must be made that it can be a sound business strategy. Unfortunately there is as yet no evidence available to support that claim. Although OER has a relatively long history in GIS&T, the *URISA Journal* may be the field's only sustained formal OER project. And those closest to that project have no evidence of advantages or disadvantages accruing to contributors, users or the organization.

Recent developments are encouraging, however. In Europe and other places where taxpayer support for higher education keeps tuition low, non-monetary markets like the GITTA project may prove sustainable if contributors perceive sufficient value in enhanced reputation, increased collaboration and the satisfaction of participating in a 'gift culture.' Where tuitions are high, as in the US, entrepreneurial institutions like Penn State may succeed creating what Wiley (2009a) calls 'OCW 2.0' – a 'new generation of OpenCourseWare projects . . . built around sustainability plans.'

> [Such] second generation projects [could be] integrated with distance education offerings, where the public can use and reuse course materials for free (just like first generation OCWs) with the added option of paying to take the courses online for credit (Wiley, 2009a).

It's also possible that the international UNIGIS distance learning network could recognize the potential of an OER 'freemium' to expand markets and goodwill. Foote's optimism about UNIGIS may still be justified.

27.8 Opening education

The importance of OER has less to do with the resources themselves than with the learning they foster. One way open resources foster learning is by enabling knowledge construction by students who 'remix' existing works. Examples include the online map 'mash-ups' that students create using borrowed code and the open application programming interfaces (APIs) of mapping engines like Google Maps. Another way is to engage students in conversations that are not restricted artificially to registered students and their faculty instructors.

David Wiley demonstrated the potential of 'open learning' when he welcomed any interested person to participate in an online graduate seminar on open education, without requiring that external students register or pay tuition to his university. In addition to the eight students who registered for the class at his university, over 50 more individuals worldwide discovered the class and participated in discussions and assignments. Wiley (2009b) reported that the length and thoughtfulness of his registered students' writing assignments were noticeably improved when they were challenged to share their work openly with an audience not limited to classmates and teachers.

I would prefer to conduct my own graduate class – Professionalism in GIS&T – in the same open manner, but am prevented from doing so by copyright restrictions on key scholarly journal articles. Instead, while my discussions with students are cloistered in password-protected learning environments in which closed publications can be shared legally, discussions on similar topics take place in public blogs without consideration of the literature that might elevate such discussions from the level of personal opinion to that of public scholarship. For example, as my autumn 2009 class privately discussed the nature of the emergent GIS profession, a widely-read blogger stated – without the slightest supporting evidence – that 'GIS is on its way out as a profession . . .' (Meltz, 2009). I did challenge my students to add their thoughts to the blog, and did post a comment myself. However, restrictions posed by copyright holders prevented us from sharing literature that could have enriched that discussion. We were free to share references, of course, but access to the referenced works is restricted to the privileged few who enjoy access to a research library.

The ultimate objective of the OER movement, therefore, is not just free resources, but freeing education itself from the cloistered, inauthentic learning environments that characterize too much of higher education in general and GIS&T education in particular.

27.9 Conclusion

GIS&T educators routinely use free online resources, and many educators freely share resources of their own. Individual OER initiatives tend not to be sustainable, however, because they lack incentives for ongoing maintenance and user support. Programmatic and institutional OER initiatives are more likely to be sustained if they are designed to advance an organization's strategic plan. Examples of sustained OER initiatives remain rare, however, because 'free' business models have only recently been articulated and are not yet widely trusted. The best example in GIS&T higher education may be the open *URISA Journal,* but even in that case benefits to contributors and to the organization are not yet clearly evident.

The most important benefit of OER is the possibility of opening education itself – to enable students to become more active participants in more authentic learning environments. That benefit may not be realized until and unless senior scholars – whose reputations and promotions have already been secured – contribute more manuscripts, peer reviews and editorial services to open-access publications. Formal programmatic and institutional OER initiatives are needed because the obstacles to open education – especially the proprietary control over access to knowledge – are themselves institutionalized in disciplinary cultures that willingly donate labour to the businesses that own and profit from peer-review processes.

Acknowledgements

Ulrik Mårtensson, Kenneth Foote, Robert Weibel, Josef Strobl and Khursro Kidwai contributed comments that improved this chapter. Remaining shortcomings are solely my responsibility. An earlier version of this chapter appeared as an article in *Cartographic Perspectives* 64. It is 'remixed' here with the kind permission of the North American Cartographic Information Society.

References

Anderson, C. (2009) *Free.* New York: Hyperion.

Antleman, K. (2004) Do open-access articles have a greater research impact? *College & Research Libraries*, 65: 372–382. http://www.lib.ncsu.edu/staff/kaantelm/do_open_access_CRL.pdf (accessed August 27, 2009).

Bishop, M. P., R. M. Hubbard, R. L. Ward, M. S. Brinkley, and T. K. Moore (1992) Computer network resources for physical geography instruction, *Journal of Geography*, 92: 102–109.

Bishop, M. P., J. F. Shroder, and T. K. Moore (1995) Integration of computer technology and interactive learning in geographic education. *Journal of Geography in Higher Education*, 19(1): 97–109.

Brandalize, M. C. B., and A. F. B. Antunes (2009) The pedagogical experience in teaching cartography and GIS using public data and software. *Proceedings of the 24th International Cartographic Conference.* Santiago, Chile, 15–21 November.

Brewer, C., and M. Harrower (2009) *ColorBrewer 2.0.* http://colorbrewer.org (accessed January 9, 2011).

Camplese, C. (2009) OpenEd 2009 recap. (Blog) http://www.colecamplese.com/2009/08/opened-2009-recap (accessed August 25, 2009).

Caron, C., S. Roche, D. Goyer, and A. Jaton (2008) GIScience journals ranking and evaluation: An international Delphi study. *Transactions in GIS*, 12(3): 293–321.

Craig, I. D., A. M. Plume, M. E. McVeigh, J. Pringle, and M. Amin (2007) Do open access articles have greater citation impact? *Journal of Informetrics* 1(3): 239–248.

Crampton, J. (1995) Cartography resources on the World Wide Web. *Cartographic Perspectives*, 22(Fall): 3–11.

Davis, M. (2002) *Profession, Code, and Ethics.* Burlington, VT: Ashgate.

de Smith, M., M. Goodchild, and P. Longley (2011) Voronoi/Thiessen polygons. http://www.spatialanalysisonline.com/output/html/VoronoiThiessenpolygons.html (accessed January 9, 2011).

DiBiase, D., M. DeMers, A. Johnson, K. Kemp, A. Luck, B. Plewe, and E. Wentz (eds) (2006) *The Geographic Information Science and Technology Body of Knowledge.* Washington, DC: Association of American Geographers and University Consortium for Geographic Information Science.

DiBiase, D., D. Wright, and F. Harvey (n.d.) Ethics education for geospatial professionals. http://gisprofessionalethics.org (accessed January 9, 2011).

Edlin, A. S., and D. L. Rubinfeld (2004) Exclusion or efficient pricing – the big deal bundling of academic journals. *Antitrust Law Journal*, 72: 119–158.

Esri (n.d.) Esri Mapping Center. http://mappingcenter.esri.com/ (accessed January 9, 2011).

Eysenbach, G. (2006) Citation advantage of open access articles. *PLoS Biology* 4(5): 692–698. http://www.plosbiology.org/article/info:doi/10.1371/journal.pbio.0040157 (accessed August 27, 2009).

Foote, K. E. (1997) The Geographer's Craft: Teaching GIS in the web. *Transactions in GIS*, 2(2): 137–150.

Foote, K. E. (1999) Building disciplinary collaborations on the World Wide Web: Strategies and barriers. *Journal of Geography*, 98(3): 108–117.

Foote, K. E. (2001) The Geographer's Craft. http://www.colorado.edu/geography/gcraft/contents.html (accessed January 9, 2011).

Foote, K. E. (2004) The Virtual Geography Department. http://www.colorado.edu/geography/virtdept/contents.html (accessed January 9, 2011).

Geographic Information Technology Training Alliance (2006) GITTA. http://www.gitta.info/website/en/html/index.html (accessed January 9, 2011).

Grossmann, T., R. Weibel, and J. Fisler (2008) Sustainability of e-learning projects – the GITTA approach. *Proceedings of the 2008 Congress of the International Society for Photogrammetry and Remote Sensing*, Beijing, China, 3–11 July. http://www.isprs.org/congresses/beijing2008/proceedings/6a_pdf/2_WG-VI-2/03.pdf (accessed September 3, 2009).

Jenkins, H. (2006) *Convergence Culture*. New York: NYU Press.

Lessig, L. (2008) *Remix*. New York: Penguin Press.

Lund University Libraries (2011) Directory of Open Access Journals. http://www.doaj.org (accessed January 9, 2011).

Mårtensson, U., P. Pilesjö, and L. Galland (2007) A survey of dropouts from GIS distance learning courses. 10th AGILE International Conference of Geographic Information Science. Aalborg University, Denmark, 8–11 May.

Massachusetts Institute of Technology (2010) MIT *OpenCourseWare*. http://ocw.mit.edu (accessed January 9, 2011).

Meltz, D. (2009) GIS is dead – long live GIS. (Blog) http://donmeltz.com/blog/index.php/2009/10/25/gis-is-dead-long-live-gis (acccessed November 13, 2009).

Moed, H. F. (2006) The effect of 'open access' upon citation impact: An analysis of ArXiv's condensed matter section. http://arxiv.org/abs/cs.DL/0611060 (accessed August 27, 2009).

Molendijk, M., and H. J. Scholten (2005) From local heroes toward global communicators: The experiences of the UNIGIS network in educating GIS professionals worldwide. Nuffic Conference 'A Changing Landscape,' The Hague, 23–25 May. http://www.nuffic.nl/pdf/os/em/molendij.pdf (accessed October 18, 2009).

Munroe, M. H. (2007) *The academic publishing industry: A story of merger and acquisition*. http://www.ulib.niu.edu/publishers/ (accessed August 27, 2009).

National Geodetic Survey (2010) NGS geodetic tool kit. http://www.ngs.noaa.gov/TOOLS/ (accessed January 9, 2011).

OER Commons (2007) http://www.oercommons.org/ (accessed January 9, 2011).

OpenCourseWare Consortium (n.d.) http://www.ocwconsortium.org (accessed January 9, 2011).

Onstein, E., and U. Mårtensson (2004) The e-GIS project: European level developments of flexible learning models with geographical information science (GIS) for vocational training. Fourth European GIS Education Seminar, Villach, Austria, 2–5 September.

Organization for Economic Cooperation and Development (2007) *Giving Knowledge for Free: The Emergence of Open Educational Resources*. http://tinyURL.com/62hjx6 (accessed July 24, 2009).

Parry, M. (2009) Utah State U.'s OpenCourseWare closes because of budget woes. *Chronicle of Higher Education* (3 September) http://chronicle.com/blogPost/Utah-State-Us-OpenCourseWare/7913/?sid=wc&utm_source=wc&utm_medium=en (accessed September 7, 2009).

Penn State University (1996) Policy AD-47: General standards of professional ethics. http://guru.psu.edu/policies/AD47.html= (accessed March 3, 2009).

Penn State University (2011) World Campus. http://www.worldcampus.psu.edu/ (accessed January 9, 2011).

Penn State College of Earth and Mineral Sciences (n.d.) Open educational resources initiative. http://open.ems.psu.edu (accessed January 9, 2011).

Primary Research Group Inc. (2008) *Survey of Academic & Research Library Journal Purchasing Practices*. New York: Primary Research Group.

Raymond, E. S. (2001) *The Cathedral and the Bazaar*. Sebastopol CA: O'Reilly Media.

Rice University (n.d.) *Connexions*. http://cnx.org/ (accessed January 9, 2011).

Ritter, M. E. (1994) Mosaic and the World Wide Web. *Cartographic Perspectives*, 19(Fall): 44–47.

Rumsey, D. (2009) David Rumsey Map Collection. http://www.davidrumsey.com (accessed January 9, 2011).

Smith, M. S. (2009) Opening education. *Science*, 323: 89–93.

Snyder, J., and P. Voxland (n.d.) Interactive Album of Map Projections. http://projections.mgis.psu.edu/ (accessed January 9, 2011).

Solem, M. N. (2000) The virtual geography department: Assessing an agent of change in geography education. *Journal of Geography in Higher Education*, 24(3): 353–364.

Subler, P. (2004) Open access overview. http://www.earlham.edu/~peters/fos/overview.htm (accessed August 3, 2009).

Swiss Virtual Campus (2009) http://virtualcampus.ch (accessed January 9, 2011).

Thoen, B. (1994) Access the electronic highway for a world of data. *GIS World*, 7(2): 46–49.

Unwin, D. J., M. J. Blakemore, P. Dale, R. G. Healey, M. Jackson, D. J. Maguire, D. Martin, H. Mounsey, and J. Willis (1990) A syllabus for teaching geographical information systems. *International Journal of Geographical Information Systems*, 4(4): 457–465.

UNIGIS International (n.d.) http://www.unigis.net (accessed January 9, 2011).

United Nations (1948) *Universal declaration of human rights*. http://www.un.org/en/documents/udhr/ (accessed January 9, 2011).

URISA (2009) Journal Electronic Archives. http://www.urisa.org/PSS_journal_archives (accessed January 9, 2011).

Utah State University (n.d.) Utah State OpenCourseWare. http://ocw.usu.edu/ (accessed January 9, 2011).

Urban and Regional Information Systems Association (1999) URISA Journal breaks new ground. Press release. Des Plaines, IL: URISA.

Vest, C. (2006) Open content and the emerging global meta-university. *Educause Review* May/June: 18–30.

Weibel, R., S. Bleish, S. Nebiker, J. Fisler, T. Grossmann, T. Niederhuber, C. Collet, and L. Hurni (2009) Achieving more sustainable e-learning programs for GIScience. *Geomatica*, 63(2): 109–118.

Wiley, D. (2007) On the sustainability of open educational resources initiatives in higher education. Organization for Economic Cooperation and Development (OECD), Centre for Educational Research and Innovation. http://www.oecd.org/dataoecd/33/9/38645447.pdf (accessed January 9, 2011).

Wiley, D. (2009a) The Future of OCW, and 'OCW 2.0', Iterating toward openness pragmatism over zeal. (Blog) http://opencontent.org/blog/archives/881 (accessed August 25, 2009).

Wiley, D. (2009b) Openness, disaggregation and the future of education. Keynote address at 2009 Teaching and Learning with Technology Symposium, Penn State University, April 28. http://www.youtube.com/watch?v=VcRctjvIeyQ (accessed August 3, 2011).

WordNet (2009) Princeton University. http://wordnetweb.princeton.edu (accessed January 9, 2011).

28

Experiences in 'e-' and 'distance-' learning: a personal account

David J. Unwin

School of Geography, Birkbeck College, University of London, London, UK

28.1 Preamble: precursors

Most of my academic career from, say, 1965 to retirement in 2002 was undertaken in a computing environment that changed rather more rapidly than my intellectual ability to make best use of it. As a young graduate student I used the then 'giant' (32k words) new IBM System 360 mainframe at University College London for what from a 45 years on perspective look like some laughably simple calculations in which I fitted trend surfaces to spatial data and computed crude nearest-neighbour statistics for some point patterns. My first academic job in 1967 was as Lecturer in Climatology at the University College of Wales at Aberystwyth, where a very tolerant Head of Department allowed me my head to develop computing as a formal part of the undergraduate curriculum. Aberystwyth was (and is) a very small town where everyone knows everyone else. In those more leisured days over often prolonged coffee breaks we were able to be really academic, talking across disciplines and without today's fear of not getting our next single authored research paper out on schedule. Importantly, many of these conversations were with computer scientists charged with providing a campus wide computing service based around a large (now 48k words!) mainframe machine. It was their early realization that, as well as providing a batch service for research computing, this machinery could provide a service to students by way of timesharing terminals and a throughput dominated by small jobs that was critical to what the geography department was able to do. Our teaching had two aims. The first was to ensure that all our students went out with some exposure to computing, and to this end we taught students simple programming using the FORTRAN language (Dawson and Unwin, 1976). Many made good use of this training and went out into the world to populate the UK's rapidly growing IT industry. The second was to maximize our use of computing as a means of what I used to think of as increasing the bandwidth of our teaching methods, for which we developed a series of simple teaching resources, such as a simple SYMAP-like

Teaching Geographic Information Science and Technology in Higher Education, First Edition.
Edited by David J. Unwin, Kenneth E. Foote, Nicholas J. Tate and David DiBiase.

line-printer choropleth mapping routine attached to a routine that enabled students to perform principal components analysis on a selection of variables from a common data bank and map the results (see Unwin, 1980).

In 1973 I moved to the University of Leicester into what at the time seemed like the computing Dark Ages, where exactly the same hardware ran just a batch service and arguments raged, notably with the physics department, about whether undergraduates could even be allowed to use the available computing resources. Mercifully, sanity prevailed (i.e. we won the argument) and a new machine, purchased with the blessing of the physicists because it had a 60-bit word length and therefore was assumed to be a good thing for numerical work, turned out to be a superb student computing machine that allowed time sharing and multi-access over a campus wide terminal network. An internet search suggests that this must have been around 1979/1980, and it led us to develop a series of e-learning resources.

In particular, I was massively influenced by the paper by John Silk (1979), who at that time was a colleague on the founding editorial board of *Journal of Geography in Higher Education* (JGHE), in which he described his use of simulation to teach spatial statistics and the difficult, but fundamental, concept of the sampling distribution. I copied this exercise, but also developed resources to teach decision making under uncertainty loosely based on Gould (1963) and a simple zero-dimensional surface energy balance routine using a modified version of Outcalt's (1971) version of Myrup (1969). The teaching context of this latter exercise is described at http://www2.glos.ac.uk/gdn/abstracts/a32.htm and in Unwin (1981). In these same years I was also heavily involved in national initiatives to develop computing in UK's primary and secondary schools through membership of the Steering Committee of the quaintly named *Geographical Association Program Exchange*, and had developed close contacts with the world of education by way of membership of the editorial boards of *Teaching Geography* and the fledgling *Journal of Geography in Higher Education* (JGHE).

These same 1980s also saw the rise of the micro-processor based 'personal computer' with machines like the CBM 'PET', APPLE (both 1977), Sinclair ZX80 (1980), BBC Model B (1980), IBM PC (1982), and Sinclair Spectrum (1982) rapidly being applied to teaching geography. My colleagues Vince Gardiner and David Maguire were quick to realize their potential both in allowing us to create computer drawn maps 'in house' (see Unwin, 1982) and to underpin a step change in allowing rapid analysis of student collected data during the typical (in UK) one-week student field class (Gardiner and Unwin, 1986). That so much of this was reported in early volumes of the *JGHE* will come as no surprise to anyone involved with the journal in those early days, when, if only to fill the available space, shortage of 'copy' and a reluctance on the part of colleagues to engage with education, meant that both editors and editorial board were often pressed into reporting their teaching. As an aside, and as one of the founding editorial board, I am delighted to see how what began as an in-house publication from what was then Oxford Polytechnic has become part of our pedagogic bedrock.

What become abundantly clear in all this was that, with sufficient imagination and good machine access, computers could help student learning over the entire range of what then was considered to be the content of a degree programme in geography. I can and will justify this by reference to constructivist notions of education and the critical ability of the machinery to allow students to make mistakes (i.e. 'experiment') without subsequent penalty and in the privacy of an interactive session.

28.2 The arrival of GIS&T and the creation of educational resources

Although some of these experiences had to do with climatology, quantitative geography and spatial analysis, very little was specifically about GIS&T as we now understand it (see for example the contents of Unwin and Dawson, 1985). Technologically this was still the era of 'mainframe/workstation GIS' (see Chapter 1), but it was inevitable that what self-evidently worked for teaching in other areas of the curriculum might also be used to teach GIS&T.

As ever, luck played a part in three ways. First, David Maguire and I were successful in an application to the UK's central funding agency to do with university computing (the Joint Information Systems Committee) to host a *Computers in Teaching Initiative Centre for Geography, Geology and Meteorology* (Unwin and Maguire, 1990), which provided resources to enable us to spread the good word about e-learning through the community of academic geographers in UK. As a not entirely irrelevant aside, secure funding also enabled us to attend the very first meetings of GIS&T educators organized by Duane Marble and to make contact with kindred spirits in USA that have remained to this day. Second, we were able to tap into bursaries from the *European Social Fund* (ESF) to underpin the development of a face-to-face masters course in GIS (Strachan and Langford, 1994) that initially relied quite heavily on reuse of the materials in the NCGIA *Core Curriculum* (see Chapter 4). Third, all this activity led in 1988 to our group being designated by the ESRC as one of a number of *Regional Research Laboratories*, also tasked in part at least to spread the good word about GIS&T.

In all these activities I was fortunate to count as colleagues some very talented and pedadagogically concerned colleagues, especially David Maguire, Alan Strachan, Mitchel Langford, Bill Hickin and Kamie Kitmitto, together with Jo Wood and Jason Dykes, two graduates of the first cohort to survive the master's programme. Our first grant-aided project to develop teaching materials started in 1987 and involved joint work with colleagues from the University's X-ray Astronomy group to develop a software tutorial system for digital image processing that we called *LIPS* (Leicester Image Processing Suite). Coded by Mitchel Langford in FORTRAN using GKS graphics and implemented on a VAX workstation driving some very expensive colour graphic terminals this very much belonged to the 'mainframe/workstation' phase recognized by Tate and Unwin (2009, see Chapter 1), but it was used successfully in teaching for a number of years and from it we learned a great deal about tutorial system design (see Langford and Unwin, 1993). The experience led to a series of 'how to do it' manuals designed for newcomers to GIS&T using the Arc/INFO, IDRISI and GRASS systems (Hickin *et al.*, 1992; Langford, 1993a and b; Langford and Wood, 1995; Strachan *et al.*, 1993a and b). Of these materials, *Getting Started in GIS* (Langford, 1993a), a series of worked exercises with data implemented in the DOS version of Eastman's IDRISI system, was by far the most popular and bits of it can still be found by diligent internet searches.

These GIS&T educational resources were alike in several respects:

- In their underpinning assumptions about what was required, style and content, they reflected the computing environment and educational imperatives of the time as outlined in Tate and Unwin (2009) and Chapter 1 of this volume, concerned to teach in some depth *about* GIS with a strong training imperative.

- They consisted of a mixture of 'analogue' materials (text, OHP foils) and digital (data disks) media distributed by the postal service.
- Their creation was the work of a substantial team of authors with a very strong group ethos and dynamic.
- Although the reproduction costs were charged, all the development work was underwritten by grant aid from a central agency concerned to promote GIS&T use in UK's universities.

I have no doubt that, 20 years on, the same group of people working today would have produced something rather different. First, as with the GITTA experience reported in their chapter in this volume by Robert Weibel and his colleagues in Switzerland, instead of teaching how to use a specified GIS, they would concentrate, in the way the GITTA materials do, on fundamentals of the underlying GIScience. Second, as with Esri's *Virtual Campus* described by Nick Frunzi in his chapter, delivery would be entirely via a website making use of the integration of media onto the single, sustainable delivery platform that the internet has provided. Whether or not they would have been distributed free of any charge seems to me to be moot. In his contribution, my co-editor David DiBiase presents a powerful argument against 'the proprietary interests that keep education closed' and in favour of educational resources being some flavour of what has become known as 'open', citing as exemplars projects such as those from MIT and the University of Lund (Sweden). Having spent many hours in discussion about e-learning with senior university decision making staff (see Section 28.4 below) and, indeed, having for a short period been one such, the Pro-Vice Master responsible for IT provision in my own college, I am not quite as convinced on this issue as my fellow editor.

28.3 Learning at a distance

In 1992 I moved to join Birkbeck College, University of London where the geography department had an international reputation as a centre of excellence in GIS&T based largely around the work of, amongst others, David Rhind, John Shepherd and Jonathan Raper. The department already had a successful programme of short courses aimed at professionals entering into work with GIS&T and participated in a face-to-face master's programme in GIS&T taught in collaboration with staff from University College London's Departments of Geography and Geomatic Engineering, but its undergraduate student roll was falling rapidly. It should be explained that almost all undergraduates at Birkbeck are mature students who take their degrees while they remain in work, so that almost all the teaching is by way of face-to-face evening classes, taken, quite literally 'on the way home' from a full-time job in central London. For various reasons the 1990s saw a decline in the numbers wishing to read for a first degree in geography. It seemed prudent to defend against probable staff cuts by 'levering' as best we could from the research expertise in GIS&T and broadening the geographic range of our offering.

The solution Jonathan Raper, Richard Kelly and I adopted with some modest financial assistance from the central college development fund was a wholly-internet delivered, two-year, master's level course in GIScience called *GIScOnline* (see Harris, 2003, Elsner, 2005) that ran for the first time in 1998. Nobody in the college or the university had done anything like this before so that we very much had to make things up as we went along. *GIScOnline* was not only a success, in terms of the primary, if 'internal', objective of sustaining our student numbers, developing it was also a massive learning experience for all of us.

Perhaps the major lessons learnt were:

- Relatively speaking, the irrelevance of advanced technology. Initially all we used was standard email, *AOL Instant Messenger* and HTML web pages. Times have changed and in its current version the course is implemented in a proprietary virtual learning environment (VLE), but such tools were simply not freely available to us in 1997/8. Use of a VLE glues the necessary components together, but it is by no means essential.
- In his detailed, evidence-based, account of the first five years development of the course Richard Harris (2003, 280) suggests that 'it is easy to overplay the extent to which distance learning is a 'different beast' from more traditional courses; the implied emphasis on geographical distance and separation can be particularly misleading'. He points out that students in their residences reading a textbook or writing an essay are in some sense just as 'distant' as those on the other side of the planet and that the same students in a lecture theatre are likewise at a distance from their teacher. The key issue for course design was to improve teacher/taught communication but also to lubricate student/student interactions. For the former, use was originally made of a simple but highly effective device whereby we asked every student to create a précis of that week's stimulus materials, to add some questions about these materials, and to return these via email to the relevant tutor. I doubt that many instructors in conventional face-to-face work have a 'conversation' with all their class at least once every week. For the latter, student to student interaction was facilitated by asynchronous messaging, a variety of small collaborative projects, and a 'virtual conference' based on teams of three to five students working through problems together (for details, see Harris, 2003). In retrospect, I am prepared to argue that the quality of the student learning experience was and remains at least as good as that experienced in a face-to-face offering. There is nothing 'special' or technically clever in any of this, it is simply good, concerned teaching. As the accounts in Section IV of this book show developments using more recent technologies can enhance these communication channels even further and I am sure that students who have grown up with *Facebook, Second Life* and other social media will have no problems in applying them to their learning activities.
- It is one thing to argue about the quality of the instruction and learning using the internet, but rightly and properly the university authorities initially expressed concern to maintain the integrity of the awards made, the need to respect the legal framework of education in countries other than the UK, and so on. Although they were not available at the time, the national guidelines on quality assurance of distance learning (QAA, 2010) would have provided an excellent check list. In practice we played with what English cricketers refer to as a 'straight bat'. First we had a formal unseen and supervised terminal examination at student suggested centres convenient for them that provided at least a half of the final assessment. Second, we maintained the usual UK system of having a suitably qualified external examiner to oversee the assessment and to make suggestions for the continuing development of the course. Scheduling 'unseen' examinations to take place over just about every time zone such that there is no possibility of collusion is an exercise in one aspect of the geography of the planet (the distribution of world population as a function of longitude) that I can recommend.

- It is necessary to be sensitive to cultural differences in both the curriculum and teaching methods in higher education and the diversity of student's prior experience, a point explored by Elsner (2005). A simple example will suffice. In the UK any student given a mark of 70% or over for coursework would be delighted and strive hard to keep up that standard whereas for many US based students such a mark would be equated with near failure.
- Finally, a totally unexpected gain was in the impact of *GIScOnline* on at least some of the teaching in the equivalent face-to-face offering. Working at a distance forces one carefully to articulate intended learning outcomes, provide little test exercises to engage the learner's attention, and so on. All of this was rapidly introduced into the face-to-face course with what I think was a major improvement in facilitating student learning. I note that my former colleagues at Birkbeck have taken things further along this road and now offer a menu of courses with different degrees of what we now refer to a 'blended learning' (see http://www.bbk.ac.uk/gisc/).

The experiences in GIS&T education at a distance in Australia reported by William Cartwright in their chapter underlines some of these same lessons. First, although the contexts differ, in both cases there is a primary motivation to provide education over as wide an area as possible using whatever technology fits the particular bill. Second, as Harris (2003) reminds us 'distance' can be interpreted in many ways and in some sense all education is at a distance. Any or all of the technologies Cartwright lists could profitably be used in almost any educational environment.

GIScOnline and the RMIT on-line courses represent just one possible model for provision of advanced education in GIS&T at a distance, but it is by no means the only one. For example, the approach adopted in *GIScOnline* was extremely staff resource intensive and would not easily 'upscale' or be portable into other academic contexts. Above all what these web based courses indicate is that there is a substantial latent demand, worldwide, for these kinds of offering.

28.4 UK eUniversities Worldwide Limited (UKeU)

Meeting this demand and up-scaling of e-learning was one of the main issues that the *UK eUniversities Worldwide Limited* (UKeU) was set up to address. The idea of a central agency to deliver UK degrees to a global audience over the internet emerged at the height of the dot-com bubble during 1999–2000. After very up-beat consultants' reports on the market, together with advice from e-learning specialists on how to go about it, the *UK Higher Education Funding Council* (HEFCE) adopted a public/private partnership business model, committed itself to an investment of some £62m, and the UKeU emerged. UKeU did not have its own degree awarding powers, the delivery of e-learning within UK wasn't encouraged, and some very ambitious recruitment targets were established, with delivery making use of a learning platform to be produced by a private business partner, *SUN Microsystems*. These constraints in turn meant that as quickly as possible the company had to have available a menu of first degree level learning programmes from UK universities of international repute mounted on an untested software platform that were of sufficient quality to attract substantial student numbers and with an emphasis on what the marketing experts assumed was a massive, emerging market in Asia.

Perhaps as a result of my previous work, I was offered a board-level post in UKeU as its Learning Programmes Director, and for two years ran a team charged to liaise with the universities to source such programs and oversee their development. Hindsight is of course the most accurate of the sciences but, in retrospect and given the constraints, failure was almost inevitable. In 2004 HEFCE decided to wind up the company, at a time when it was reported to have cost £50m yet had only 13 available courses for which only 900 students had been recruited. A series of reports have investigated the reasons for this failure from several perspectives, including a very critical *House of Commons, Education and Skills Committee* (2005a) report and government response (2005b) and a careful factual overview by an e-learning specialist involved in both the initial setting up of the project and as its Director for Special Projects (Bacsich, 2005). These documents should be essential reading for any institution contemplating major investment in e-learning at distance. From the government's point of view the main reasons for failure suggested were insufficient market research to determine levels of demand for online education leading to a 'supply-led' mentality, an inability to attract further private investment, and a far too heavy emphasis on the technology needed to establish and run the promised learning platform. From a pedagogic point of view the UK *eLearning Research Centre* (Conole *et al.*, 2005, page 1) instead suggest that UKeU's 'early demise sounds a warning note to all of use involved in e-learning' and focuses attention on a 'mismatch between those with more of a business oriented vision for UKeU and those more interested in the academic aspects and the potential educational innovation'. I do, of course, have my own views on all this, but the main lesson seems to me to be that, unlike the way UKeU was asked to develop, in e-learning it pays to learn in small, relatively unambitious but secure, steps much as the UNIGIS collaboration described by Josef Strobl in his chapter in this volume has been developed.

28.5 Virtual graduate seminars

Heading into retirement as UKeU was wound up, I was approached by the *World Wide Universities Network* (WUN), an association of (then) 14 research-led universities, to facilitate the setting up of what was to become its *Global GIS Academy*. The intention was to build on an active collaboration in teaching GIS&T involving sites at Penn State, Leeds and Southampton to create a virtual academy of people in the network interested in GIS&T. This involved a variety of initiatives, but a major activity was to organize and run a series of graduate and faculty level virtual seminars using an extended desk-top video and audio conferencing system that enabled participation from almost anywhere in the world with an appropriate computer, client software and broadband internet link. The seminars were open to anyone to 'attend' and were advertised to members of the US *University Consortium for Geographic Information Science* (UCGIS) and various research groups in UK. Attendance varied according to the popularity of the topic and/or status of the presenter but at maximum we had 60 client machines seen by perhaps 80 individuals located primarily in Norway, the Netherlands, England and the USA. These ran from 2005 through to 2009 when I decided that it was time for me to move on to other projects. The full list reflected a series of concerns in GIS&T, together with some that had a direct training motivation:

2005 Ethics for GI professionals (with David DiBiase, Penn State)
2006 Hot topics in GISc

2007 Spatio-temporal modelling in GIS
2007 Graduate Research in GISc
2007 Public Participation GIS (with T Nyerges, Washington State)
2008 Neogeography
2008 Spatial analysis in R
2009 Dynamic modelling in GIS environment (with S. Carver, Leeds)

Each series consisted of six presentations by well-known experts in the relevant field, not all of whom were members of the WUN network. Because all the media used were digital, it was possible to archive the summary slides used, a complete, re-playable video and audio recording of the presentation, and a log of the questions asked by the audience to create a useful resource that can be accessed via the internet re-used by instructors. In addition for some of the series we were able to pool and distribute relevant resources such as bibliographies, ideas for student follow up work, and suggested intended learning objectives. Although some members of the network made use of the available resources (notably those associated with the series on Public Participation), I remain disappointed that by and large people were unable to 'lever' good graduate level teaching using them.

Unlike running a face-to-face series in which recruiting speakers is often the main issue, there was no problem in persuading people to talk, the real difficulties arose as a consequence of everyone's unfamiliarity with the environment. In part this was a simple software training issue that could be handled by a brief training session ahead of each talk, but in part also there are a whole new set of opportunities and required protocols associated with working at distance. The huge advantage is the flexibility that such work allows. I was able, for example, to 'attend' one seminar when nominally on vacation from the comfort of the computer lounge of a cruise ship in the Mediterranean and on another occasion when traveling in excess of 100mph on a train from Edinburgh to London. At least one presenter gave a talk via WiFi from the comfort of a hotel room in upstate New York, and so on. The series has not been continued, but I note that in 2010 and following a broadly similar approach, the WUN network ran some 77 virtual seminars in a wide variety of topics. Given that the necessary technology is widely available, I see no reason why almost any group of academics or students cannot do something similar, and, of course, it opens the way to using the internet not only to share face-to-face teaching but also to archive this for viewing, as is now common with standard television. The issues that such a capability raises aren't technical but are managerial, social and economic and, as many academics seem wrongly to assume, they will not go away. If I were still in a position that enabled me to do it, I would attempt to link any graduate school with which I was involved with others in this same way.

28.6 2006 statistics.com: towards a future for GIS&T e-learning?

In 2003 David O'Sullivan and I published the first edition of a relatively basic GIS&T text with the title *Geographic Information Analysis* (O'Sullivan and Unwin, 2003). A year or so later I was approached by a company not known to me at the time (*statistics.com*) with the suggestion that I turn at least some of the content into a short course to join a menu of such standard format courses they offered and delivered wholly over the internet. Work for each run of a course involves four weekly lessons in each of which students visit a virtual learning environment (currently *Moodle*), read some stimulus material either in a set text or from downloaded files, participate in a threaded discussion with their tutor and/or fellow students, and complete an assignment that is graded. At the time of writing my first course *Spatial Statistics with GIS* has run nine times, with a further offering *Spatial Analysis Techniques in R* in its first year of operation. These two courses are a small section of a menu of over 100 that are available covering a wide spectrum of statistical and computational approaches and applications. Courses all have tutorial support, with the listed faculty drawn mostly from the USA (48 from 62), with India (6), UK (3), New Zealand (2), Belgium (2) and Bhutan (1) lagging behind.

As of early 2011, just over 220 people have taken my courses, and from an e-learning/GIS&T market perspective the interesting thing is that at least 43% of them have '.edu' or '.ac' email with a further 31% from a miscellany of '.org' affiliations. Several past students already have PhDs in some chosen non-geography discipline, with an evident demand from fellow academics in other disciplines who have a GIS available to do some geographical analysis for them, but lack the educational background to do this correctly. At the moment classic '.com' affiliations are in the minority (18%), but, as GIS&T technology becomes more and more a part of everyday life, I would not be at all surprised to see this proportion increase. Given that *statistics.com* has its home in Arlington VA, it's not surprising that some 76% of students to date are from the USA with an equal split between Europe and the rest of the world making up the rest. It is easy to over-interpret these data, but one conclusion I draw is that, as GIS&T has become a mainstream technology and has begun to become divorced from academic geography, so these raw numbers indicate a latent demand for short, tutor-led, vendor-independent education and training in basic concepts. Although around half of each cohort of students choose to submit assignments for grading, my impression is that, for the moment at least, some form of certification isn't important; what matters is the ability to get rapidly 'up to speed' in the essentials. Provisional credit for certain introductory courses at *statistics.com* can be obtained if needed by way of the *American Council on Education*, some courses can be packaged to create a certificated *Program in Advanced Statistical Analysis* and, although purists might object, I see no reason why courses should not also be amalgamated to create individually customized degree programmes. The result would be an 'eUniversity' built from the ground upwards making use of components that individually are offered to professionals already in the workplace 'just in time' and on a 'needs' basis. I can honestly put on record that leading active and concerned professionals through the intricacies of spatial statistical analysis has given me nothing but pure pleasure and in doing it I have learned a great deal not only about the specifics of GIS&T but also about public perceptions held by them of the more general value of academic geography.

Of course, our friends at Esri in Redlands may well argue that this is exactly what their *Virtual Campus* already does, and what concerns me is that the evident success of these commercial programmes must indicate that, with some exceptions, the traditional 'geography' academy is in some sense failing to meet a real societal need. Instead it hangs tenaciously onto a model of higher education, in which young students follow expensive, residential, largely face to face courses that are more-or-less prescribed for them, last for three years or so, and lead to a standard qualification that is assumed to be a good thing. For most of the planet this model of higher education has proved to be highly contingent in both time and space, a feature of education in the developed world that arose during the late nineteenth century that gained most acceptance for just the second half of the twentieth century. I would be the first to recognize that, in common with many others, I benefited in countless ways from the opportunities that this system offered but as I write in the UK what seem to me to be already excessive student university tuition fees are now being trebled to (up to) a maximum of £27,000 ($44,000 or so) for a three year course. Whether such a cost, additional to subsistence, learning materials, and the foregoing of three years income remains a sensible and gainful investment in a young person's future is to say the least moot. Perhaps a more sensible approach would be to recognize that the old model is no longer sustainable and is incapable of responding to the needs of our planet in the present century and instead replace it with a variety of forms of e-, distance- and blended-learning.

If I were starting out again as an educator in GIS&T, all I know for sure is that I'd look much more towards UNIGIS, *Campus World*, the *Virtual Campus* and *Statistics.com* for my models than I would towards the traditional diet.

Acknowledgement

I am grateful to Peter Bruce, President of s*tatistics.com* for access to, and permission to use, the student data for the course discussed in Section 28.6.

References

Bacsich, P. (2005) *The UKeU Reports: UKeU overview*. UK Higher Education Academy, Report 01. http://www.heacademy.ac.uk/assets/York/documents/resources/web0279_ukeu_reports_1_overview.doc (accessed August 15, 2011).

Conole, G., A. Carusi, and M. de Laat (2005) Learning *from the UKeU experience*. eLearning Research Centre, University of Southampton. http://www.elrc.ac.uk/download/publications/ICEpaper.pdf (accessed August 15, 2011).

Dawson, J., and D. J. Unwin (1976) *Computing for Geographers*. Newton Abbott: David and Charles.

Elsner, P. (2005) GIS teaching via distance learning experiences and lessons learned. *Planet*, 14: 28–30. GEES Subject Centre, The Higher Education Academy, UK.

Gardiner, V., and D. J. Unwin (1986) Computers and the field class. *Journal of Geography in Higher Education*, 10: 169–179.

Gould, P. R. (1963) Man against environment: A game theoretic framework. *Annals of the Association of American Geographers*, 53(3): 290–297.

Harris, R. (2003) Building a GIScience community in cyberspace: Reflections on GIScOnline. *Journal of Geography in Higher Education*, 27: 279–295.

Hickin, B. W., D. J. Maguire, and A. J. Strachan (1992) *Introduction to GIS: the ARC/INFO Method*. Leicester: Midlands Regional Research Laboratory, University of Leicester.

House of Commons, Education and Skills Committee (2005a) *UK eUniversity: Third Report of Session 2004–2005*, HC 489. London: HMSO. http://www.publications.parliament.uk/pa/cm200405/cmselect/cmeduski/205/205.pdf (accessed August 15, 2011).

House of Commons, Education and Skills Committee (2005b) *UK eUniversity: Government's Response to the Committee's Third Report of Session 2004–2005*, HC 489. London: HMSO. http://www.publications.parliament.uk/pa/cm200506/cmselect/cmeduski/489/489.pdf (accessed August 15, 2011).

Langford, M. (1993a) *Getting Started in GIS*. Sheffield, England: CVCP/USDU, University of Sheffield.

Langford, M. (1993b) *Moving On in GIS*. Sheffield, England: CVCP/USDU, University of Sheffield.

Langford, M., and D. J. Unwin (1993) The life and death of LIPS: Some lessons from the design and use of courseware for teaching digital image processing. *Institute of Electrical Engineers, Digest* 190: 1.1–1.4.

Langford, M., and J. Wood (1995) *GRASS Seeds: A Beginner's Tutorial.* Leicester, England: Academic Support for Spatial Information Systems, University of Leicester.

Myrup, L. O. (1969) A numerical model of the urban heat island. *Journal of Applied Meteorology,* 8: 908–916.

O'Sullivan, D., and D. Unwin (2003) *Geographic Information Analysis.* New York: John Wiley & Sons, Inc.

Outcalt, S. (1971) A numerical surface climate simulator. *Geographical Analysis,* 3: 379–392.

QAA, Quality Assurance Agency for Higher Education (2010) *Code of practice for the assurance of academic quality and standards in higher education. Section 2: Collaborative provision and flexible and distributed learning (including e-learning).* http://www.qaa.ac.uk/Publications/InformationAndGuidance/Pages/Code-of-practice-section-2.aspx (accessed August 16th 2011).

Silk, J. (1979) Use of classroom experiments and the computer to illustrate statistical concepts. *Journal of Geography in Higher Education,* 3(1): 13–25.

Strachan, A., and M. Langford (1994) Diploma/M.Sc. in GIS – the Leicester experience. *Proceedings EGIS'94,* Paris, France, pages 243–251. http://libraries.maine.edu/Spatial/gisweb/spatdb/egis/eg94092.html (accessed August 15, 2011).

Strachan, A. J., B. W. Hickin, D. J. Unwin, and K. Kitmitto (1993a) *Getting Started with PC ARC/INFO.* Sheffield, England: CVCP/USDU, University of Sheffield.

Strachan, A. J., B. W. Hickin, D. J. Unwin, and K. Kitmitto (1993b) *Getting Started with Workstation ARC/INFO.* Sheffield, England: CVCP/USDU, University of Sheffield.

Tate, N. J., and D. J. Unwin (2009) Teaching GIS&T. *Journal of Geography in Higher Education,* 33(Supplement 1): S1–S6.

Unwin, D. (1980) Make your practicals open-ended. *Journal of Geography in Higher Education,* 4(2): 39–42.

Unwin, D. (1981) Teaching a model-based climatology using energy balance simulation. *Journal of Geography in Higher Education,* 5(2): 133–138.

Unwin, D. (1982) Maps on micros. *Journal of Geography in Higher Education,* 6(2): 163–169.

Unwin, D. (1986) Editorial: The best thing since sliced bread. *Journal of Geography in Higher Education,* 10(1): 3–5.

Unwin, D., and P. Dale (1990) An educationalist's view of GIS: Some educational and sociological concerns. *Journal of Geography in Higher Education,* 14(2): 166–169.

Unwin, D. J., and J. Dawson (1985) *Computer Programming for Geographers.* London: Longman.

Unwin, D. J., and D. J. Maguire (1990) Developing the effective use of information technology in teaching and learning in geography: The Computers in Teaching Initiative Centre for Geography, *Journal of Geography in Higher Education,* 14(1): 77–82.

Conclusion

29
Ways forward for GIS&T education

David DiBiase[1], Kenneth E. Foote[2], Nicholas J. Tate[3] and David J. Unwin[4]

[1]*John A. Dutton e-Education Institute, Penn State University, University Park, Pennsylvania, USA*
[2]*Department of Geography, University of Colorado at Boulder, Boulder Colorado, USA*
[3]*Department of Geography, University of Leicester, Leicester, UK*
[4]*School of Geography, Birkbeck College, University of London, London, UK*

29.1 Introduction

The preceding chapters describe the rich heritage and current state of the art of GIS&T in higher education. In this concluding chapter we look ahead to challenges and opportunities that learners, educators, and higher education institutions will face in years to come.

29.2 Learner needs and aspirations

Many GIS&T educators share the conviction that everyone would be better off with basic spatial thinking skills and awareness of the roles and implications of geospatial technology in society. However, learners have different needs and expectations depending on their age, experience and educational goals. Notwithstanding the increasingly blurry distinction between professional and other users as discussed by Tate in Chapter 22, in general, GIS&T education must address three different kinds of students: (1) those who aspire to careers in the GIS&T field as well as current practitioners who seek continuing professional development, (2) those who are exposed to GIS&T in the context of being a facilitating technology in other disciplines, and (3) others whose professional or liberal educations could be enriched by awareness of GIS&T.

Education for GIS&T professionals

Consensus about the nature and scope of the global GIS&T enterprise has been elusive. Consequently, reliable estimates of the size and characteristics of the GIS&T workforce have been

Teaching Geographic Information Science and Technology in Higher Education, First Edition.
Edited by David J. Unwin, Kenneth E. Foote, Nicholas J. Tate and David DiBiase.

Table 29.1 Geospatial occupations defined by US Department of Labor Employment and Training Administration

Occupation	Estimated employment (2008)	Projected growth (2008-18)	Projected growth rate (2008-18)
Geospatial Information Scientists and Technologists*	209,000	72,600	Average (7-13%)
Geographic Information Systems Technicians*	209,000	72,600	Average (7-13%)
Remote Sensing Scientists and Technologists*	27,000	10,100	Average (7-13%)
Remote Sensing Technicians*	65,000	36,400	Average (7-13%)
Precision Agriculture Technicians*	65,000	36,400	Average (7-13%)
Geodetic Surveyors*	58,000	23,300	Faster than average (14-19%)
Surveyors	58,000	23,300	Faster than average (14-19%)
Surveying Technicians	77,000	29,400	Much faster than average (≥ 20%)
Mapping Technicians	77,000	29,400	Much faster than average (≥ 20%)
Cartographers and Photogrammetrists	12,000	6,400	Much faster than average (≥ 20%)

*New occupations established late 2009
Source: DOLETA (n. d.)

hard to come by until recently. For example, only since late 2009 has the US Department of Labor Employment and Training Administration (DOLETA) recognized a full complement of ten occupations that span the range of professional activities in GIS&T. Recently added to longstanding 'geospatial' occupations like surveyors and cartographers and photogrammetrists are newer occupations including geospatial information scientists and technologists, remote sensing scientists and technologists, and GIS and remote sensing technicians (Table 29.1). Geospatial software programmers and application developers are not included.

The employment estimates presented in Table 29.1 are not additive because some overlap. For instance, estimates for geospatial information scientists and technologists and geographic information systems Technicians are identical (209,000 and 72,600) because both are based upon U.S. Bureau of Labor Statistics estimates for a single occupation: 'computer specialists, all other.' However, even when all redundant estimates are eliminated, the preliminary estimates are still impressive: as many as 450,000 geospatial professionals employed in 2008, and up to 180,000 additional positions to be created by 2018 in the US alone (DOLETA, n. d.). Analysts also estimate that nearly 180,000 additional workers will be needed in the US by 2018. Estimates of the size of the geospatial workforce beyond the US are even harder to find, though Longley and colleagues reckoned that there were some two million GIS users worldwide in 2005 (Longley *et al.*, 2005). Though exact counts are unavailable, there is little doubt that the market of individuals who need preparation for careers in the GIS&T field, and for continuing professional development throughout their careers, is large and growing.

How are current and aspiring professionals to know what to look for in a GIS&T education or training program? Because no subject-specific accreditation exists for GIS&T programs in higher education (except for the few programs associated with schools of engineering, landscape architecture, and other state-regulated fields such as via the Royal Institution of Chartered Surveyors in the UK), credentials and curricula vary widely. Early curriculum development efforts described elsewhere in this book proceeded without the benefit of a comprehensive, industry-vetted specification of workforce needs.

A series of efforts over the past ten years has gradually brought workforce needs into focus. A pioneering effort by workforce development specialists at the University of Southern Mississippi in the US led to the original *Geospatial Technology Competency Model* (GTCM, Gaudet *et al.*, 2003). The *GIS&T Body of Knowledge* (DiBiase *et al.*, 2006) later fleshed out the breadth and depth of technical and analytical competencies that comprise the GIS&T field. These and related efforts in the US and Europe (e.g., Quality Assurance Agency for Higher Education, 2007; Solem *et al.*, 2008) culminated in 2010 in DOLETA's own *Geospatial Technology Competency Model* (DiBiase *et al.*, 2010). The new GTCM is an open, industry-validated specification of the general and industry-specific know-how needed for worker success in the three GIS&T industry sectors: positioning and data acquisition, analysis and modeling, and software and application development. The new GTCM provides an unprecedented opportunity for learners to self-assess their mastery of geospatial competencies, to identify gaps in their knowledge, and to choose among educational institutions and programs according to their alignment with industry needs and individual career trajectories.

Education for professionals in related fields

Many learners seek to master GIS&T sufficiently to use it effectively in a kindred field, rather than to enter one of the specialist occupations outlined in Table 29.1. To this end they are more likely to encounter the subject in the context of learning *with* GIS&T in one or a few applied classes, rather than *about* GIS&T in the context of a comprehensive degree program. To appreciate the scope of this educational need, consider the number and variety of industries to which Esri markets its software products (Esri, 2011). Education, one of Esri's largest markets, is increasing considered a 'horizontal market' that cuts across all of the other 'vertical markets,' each of which has its own particular educational requirements. Both Phoenix and Sinton (Chapters 2 and 3) discuss efforts to foster 'teaching with GIS&T' in new fields such as business and education, and in liberal arts institutions. In Chapter 16, Bodenhamer and Gregory provide a similar discussion in the context of the humanities. Recent major initiatives to promote GIS&T education and spatial thinking in kindred fields include the Center for Spatially Integrated Social Science's *Spatial Perspectives for Analysis in Curriculum Enhancement* (SPACE) program in the US, and the *Spatial Literacy in Teaching* (SPLINT) consortium in the UK.

General education

The majority of learners will never use geospatial technology professionally. However, they will certainly share in its benefits (and risks), and may contribute as 'produsers' or 'prosumers' of geospatial data products and services (see Tate, Chapter 22). Many students are unaware of the GIS&T field, and those who are may be thwarted from pursuing the subject by the prerequisite requirements associated with specialized educational opportunities tailored to the needs of aspiring professionals. Such students are harder to reach than those who are already aware that GIS&T is relevant to their lives and work. One way to engage them is informal GIS&T education like the *Geospatial Revolution* series (Penn State Public Broadcasting, 2010). Another effective way is to expand colleges' and universities' 'general education' curricula to include GIS&T.

General education denotes "the portion of an institution's curriculum that is required of all students—'the part of a university that is concerned most directly with the individual student's responsibility to society at large'" (Miller, 1988, 2). According to the Association of American Colleges and Universities (2011), nearly 80 per cent of US higher education institutions have adopted a broad set of learning outcomes that apply to all students. General education outcomes typically include 'information literacy,' a hallmark of GIS&T education. Tsou and Yanow (2010) describe several examples of GIS&T courses approved as general education offerings, and predict 'a significant increase of GIS&T courses available in general education in the coming years' (p. 53). Expanding access to GIS&T education by enlarging its footprint in general education curricula constitutes a grand challenge for the GIS&T education community.

GIS&T education for all: ethics and spatial thinking

'Ethical reasoning' is another of the learning outcomes commonly included among institutions' general education objectives. GIS&T professionals and others need to be aware of the ethical implications of society's increasing reliance on geospatial technologies. Professionals have special obligations that are embodied in codes of ethics and rules of conduct such as those promulgated by the GIS Certification Institute (GISCI, 2008), the Association of American Geographers (AAG, 2009), and other organizations.

In Chapter 14, DiBiase *et al.* discuss a variety of ethical issues associated with GIS&T. These include cross-cutting issues of privacy and confidentiality such as the tracking of the locations and movements of individuals or vehicles using dynamic or ground-based sensors and the use of high-resolution aerial and satellite imagery that allow identification of individuals or vehicles. Although these kinds of surveillance are often implemented for innocuous, practical reasons (though sometimes not), the results can be dangerous. Beyond the issue of surveillance and monitoring, data collected and collated by location in GIS (either by street or coordinate location) are used to create information mosaics – profile or propensity files – which can be intrusive. The widespread use of GIS in decision making also means that data may be misused or misrepresented, or that the inappropriate data or erroneous data may lead to harm. Equally important is how geospatial technologies, because of their expense and complexity, can privilege access to important data to the advantage of some nations, organizations, or individuals, and to the detriment of others.

Because the ethical implications of GIS&T have the potential to affect ordinary citizens as well as professionals, the GIS&T education community is duty-bound to incorporate formal ethics education in general education classes as well as in professional curricula. Though much has been written about the ethics of GIS, this book is the first to provide practical guidance about how moral reasoning skills can be effectively taught and learned using GIS&T case studies.

Spatial thinking

In their study of competencies valued by employers in the government, higher education, for-profit and not-for-profit sectors, Solem *et al.* (2008) found that 'spatial thinking' was the most frequently cited geographic skill area. Until recently, however, spatial learning has received little attention among learning scientists and education policy makers in comparison to the emphases placed on verbal and mathematical. The US National Research Council (NRC) report *Learning to Think Spatially* (2006) has ignited new interest in spatial thinking. According to the NRC, the power of spatial thinking is not limited to geography or environmental sciences. Instead it is fundamental to the ways in which people in many disciplines understand and think about many different processes and patterns. As the report states:

> Spatial thinking – one form of thinking – is based on a constructive amalgam of three elements: concepts of space, tools of representation, and processes of reasoning. It is the concept of space that makes spatial thinking a distinctive form of thinking. By understanding the meaning of space, we can use its properties as a vehicle for structuring problems, for finding answers, and for expressing solutions. By expressing relationships within spatial structures, we can perceive, remember, and analyze the static and, via transformations, the dynamic properties of objects and the relationships between objects. We can use representations in a variety of modes and media to describe, explain, and communicate about the structure, operation, and function of objects and their relationships. Spatial thinking is not restricted to any domain of knowledge, although it may be more characteristic of architecture, medicine, physics, and biology, for example, than of philosophy, business administration, linguistics, and comparative literature (National Research Council 2006, ix).

As Jo *et al.* note in Chapter 15 (page 227), their study of GIS textbooks and course syllabi indicate that the issue of spatial literacy and learning is still underdeveloped. They

> found little evidence of explicit instruction in, or opportunities for students to develop, spatial thinking skills . . . For those GIS instructors who intend to develop spatial thinking skills in their students, some texts offer more support than others, but by themselves are unlikely to produce students who possess effective spatial thinking skills. Thus, if spatial thinking is to be developed, it must be embedded in the classroom and laboratory.

The NRC report challenges the GIS&T education community to devise new ways to use GIS and related technologies to support the development of spatial thinking skills, and to demonstrate their effectiveness empirically. It also challenges educators to produce case studies that illuminate the roles of spatial thinking in geographical and environmental analysis (Sinton and Schultz, 2008; West, 2003).

29.3 Challenges and opportunities for educators

Pedagogy and andragogy

Historically, the graduate training of GIS&T educators in higher education seldom included much serious engagement with the scholarship of teaching and learning – what has come to be known as the 'science of learning' (Bransford *et al.* 2000). Neither did many early GIS&T educators have opportunities to consult with professional learning designers to help ensure the educational effectiveness of their teaching. Most new teachers simply emulated their mentors, learned on the job, and with the best of intentions went with what seemed to work. Even when peers were called upon to review teaching performance, they too often lacked any formal background in learning theory, research or design.

Times have changed, however. Accrediting bodies now expect higher education institutions to define learning outcomes in measurable terms, and to provide tangible evidence that outcomes have been achieved. In the UK for example, the majority of universities require inexperienced faculty/staff to obtain a postgraduate certificate in some aspect of higher education. The need to demonstrate educational effectiveness has led to a heightened interest in pedagogy among GIS&T educators. In Chapter 9, for example, Schultz describes how 'active' pedagogy can foster higher-order thinking by GIS&T students. Elsewhere, Summerby-Murray (2001) discusses use of problem-based inquiry based on constructivist principles in the context of students' GIS analyses of historic buildings. King (2008) demonstrates that problem-based learning can be conducted effectively even in an asynchronous online GIS&T class.

As these and other authors note, inquiry-based, active pedagogies are labor-intensive, and require educators 'to be comfortable with the notion that they cannot possibly know all things about GIS and must be willing to learn alongside their students' (Schultz, Chapter 9, page 141). Given the imperative of continuing professional development for GIS&T professionals, more and more GIS&T educators will be challenged to help learners whose professional experience and operational knowledge of geospatial technologies are greater than their own.

To address this challenge effectively, some educators will need to rethink the appropriateness of *pedagogy* – literally the art and science of teaching children – for experienced adult learners. By contrast, *andragogy* – the art and science of teaching adults – 'has gained wide acceptance as a set of assumptions designed to guide the development of programs for adults' (Rossman, 2000). One key assumption is the research-based generalization that 'as individuals mature, their need and capacity to be self-directing . . . increases steadily from infancy to preadolescence, and then increases rapidly during adolescence' (Knowles *et al.*, 2005, 62). Andragogy 'advocates learner control (at the very least, substantial input) over not only the objectives but also the learning strategies as well as evaluation procedures' (Rachal, 2002, 213). The need to adapt teaching approaches to accommodate the increasing diversity of students' age, experience, and goals poses a formidable challenge to GIS&T educators (DiBiase and Kidwai, 2010).

Curricula

Several chapters in this volume demonstrate how far the GIS&T education community has come in our understanding of what 'curriculum' means, and what curriculum design, development, and assessment entails. Over the past thirty years, our notion of curriculum has evolved from collections of course notes, to a Delphi-mediated consensus about topics, to multiple (undetermined) pathways through a comprehensive body of knowledge, to an ontology that provides the knowledge structures needed to create adaptive curricula that respond to individual learner needs. As it has broadened in concept, so too have some curriculum development processes expanded to include a broader array of stakeholders. For example, at one of our institutions (Penn State), one individual designed the sequence of courses comprising an online certificate program in GIS in 1997–98. By 2003–04, however, when that certificate program was expanded to a complete online masters degree program, an advisory board composed of external and internal experts designed the new curriculum by consensus.

In Chapter 4, Kemp provides a first-hand account of some of the pioneering efforts in GIS&T curriculum design. In Chapters 5 and 6, Prager and Foote discuss curricula as solutions to design problems constrained by intended learning outcomes and concerns about the scope and sequence of topics, all situated within particular educational contexts. Prager highlights use of the *GIS&T Body of Knowledge* (BoK) as a resource to help formalize learning outcomes.

More recently, as described above, the US Department of Labor has endorsed a new *Geospatial Technology Competency Model* (GTCM) that may prove to be more useful still for design and assessment of GIS&T curricula. In relation to the *BoK*, for example, the new GTCM offers several advantages, including:

1. Holistic – In addition to GIS&T-specific technical competencies, the new GTCM identifies a broad range of personal, academic, and workplace competencies needed for worker success.
2. Selective – the new GTCM highlights high-priority competencies, rather than attempting to catalogue a complete corpus of GIS&T knowledge. This is beneficial since selectivity and prioritization are defining aspects of curriculum design.
3. Revisable – at one-fifth the length of the *BoK*, the new GTCM will be easier to revise in response to rapidly evolving technologies, and readily adapted to contexts beyond the US.
4. Industry-validated – whereas the *BoK* is chiefly an academic work, senior representatives of the various GIS&T professions produced the new GTCM in collaboration with workforce analysts, and in response to comments received during a public review period.

In combination with occupation-specific job analyses (DOLETA, n.d.; Johnson, 2010), the new GTCM affords an unprecedented opportunity to assess the alignment of GIS&T curricula with workforce needs. The National Geospatial Technology Center (GeoTech Center) is currently developing an online self-assessment tool for use by colleges and universities that host GIS&T programs. The tool is intended to provide the benefits of an external advisory board without the associated expenses. However, because most of the academic departments that offer GIS&T courses and programs are not subject to discipline-specific accreditation, there is little incentive for faculty to voluntarily participate in what can be an arduous assessment process. This presents a challenge to academic leaders to inspire their faculties to approach assessment in a conscientious way, a challenge that will be especially difficult in parts of the world where assessment is not so well established – such as in the US.

Keeping up with technology and technology-enabled learning

As discussed in Section 4 of this volume, the scope of GIS&T has expanded considerably to encompass a wide range of technologies. GIS was originally defined as a system to capture, store, check, integrate, manipulate, analyse and display spatial data. Today this definition would seem to equally well describe a range of technologies that handle geospatial data. Hardware, for example, is becoming more specialized, as is evident in relatively recent emergence of LiDAR sensors and systems. Software includes the current crop of geobrowsers, as well as the growing variety of software systems to analyse and display earth observation data ranging from active systems such as LiDAR and SAR to passive systems such as digital photogrammetry. Even in the context of 'traditional' GIS, the proprietary software products supplied by commercial developers are now augmented by an ever-expanding set of open source software. A corresponding variety of user conferences, from the Esri International Users Conference to FOSS4G, also competes for users' attention. In short, the task of keeping up with advances in geospatial technology is more difficult than ever.

Meanwhile, increasing stringent security measures adopted by college and university information technology (IT) departments – partly in response to legislation concerning the privacy of student records – is making it ever more difficult for educators to provide authentic access to GIS&T software. By 'authentic access' we mean the ability to grant advanced students the administrator privileges they need to be able to install, configure and customize server-based, enterprise-class GIS software products. To reconcile the divergent needs of educators and of IT staff, software companies may need to provide hosted (a.k.a. 'cloud') server solutions that educators and students can access independently of university networks.

For GIS&T educators, keeping up with geospatial technologies is only half the problem. Equally challenging is the need to stay current with rapidly evolving learning technologies and environments, including mobile devices, virtual worlds, gaming, learning management and content management systems, and the bewildering array of media production tools and platforms (such as Jing, VoiceThread, and YouTube). Section 5 of this volume considers several of these, with particular emphasis on the potential affordances of mobile devices and virtual worlds ('m-learning' and 'v-learning'). One interesting unresolved question is whether mobile and immersive learning environments will converge or diverge; i.e., will m-learning better accommodate learners' busy lifestyles, while v-learning requires them to leave one life behind while immersing in a second?

To date, the ability to teach and learn at different times and places is perhaps the most consequential affordance of information technology for GIS&T education. In the US, at least, enrollments in online distance education are growing at a much faster rate than higher education enrollments overall (17 percent versus 1.2 percent). During the autumn 2008 semester, over 4.6 million US students took at least one online course. More than one out of four US higher education students now participate in at least one online course (Allen and Seaman, 2009).

Overwhelming evidence now exists that, when conducted properly, e-learning is at least as effective as face-to-face instruction (e.g., Bernard *et al.*, 2004; Sitzmann *et al.*, 2006; Means *et al.*, 2009). Also well known are the characteristics of effective online teaching, most important of which are frequent and meaningful interactions among students and instructors (e.g., Zhao *et al.*, 2005). Not all subjects are equally amenable to e-learning. For example, subjects that require student access to 'wet' labs, and those that require extensive field activities (such as land surveying) can be simulated online, but perhaps not in cost-effective ways. However, most of the competencies specified in DOLETA's new GTCM can readily be acquired online, especially via instructor-led distance education. Several examples of successful e-learning enterprises in the GIS&T realm are discussed in Section 5.

E-learning is certainly no substitute for the formative social experiences that younger learners need. However, it is surely becoming the preferred mode for the continuing professional development of adult learners, such as current GIS&T professionals and aspiring career changers. E-Learning therefore represents both a challenge and opportunity for the GIS&T education community, particularly in light of the potential to share students among institutions, as discussed below.

29.4 Challenges and opportunities for institutions

Institutional interoperability

Interoperability is a key topic in GIS&T. Authors of the *Strawman Report* that led ultimately to the *BoK* identified interoperability as a 'cross-cutting theme' (Marble *et al.*, 2003, 19). The format of the *BoK* itself was selected '[t]o ensure interoperability with corresponding documents in computer science, information science, project management, and other fields' (DiBiase *et al.*, 2006, 39). More recently, the new GTCM (DOLETA, 2010, 23) includes '[i]dentify[ing] the factors that affect the interoperability of geospatial software applications' among the 'critical work functions' required for the success of workers in the software programming and application development sector of the geospatial industry. The Open Geospatial Consortium (OGC, 2011), a voluntary consensus standards organization involving over 400 organizations, was founded to promote 'collaboration among and between OGC members to define, document, and implement open standards that solve geospatial interoperability problems.'

Sondheim *et al.* (1999, 348) point out that interoperability has a range of meanings. In its technical sense, the term refers to the ability to 'exchange data and to handle queries and other processing requests' by making use of 'a common understanding of the data and requests.' Kemp (Chapter 4, page 55) discusses an early attempt to envision 'a knowledge base of 'interoperable education objects' that could be snapped together in any number of ways to create personalized education opportunities.' The pragmatics of collaborative learning design meant to result in reusable and localizable learning objects are further considered by Durham *et al.* (2009), and by DiBiase in Chapter 27.

More generally, however, interoperability means 'the ability to move easily from one system to another' (Sondheim *et al.*, 1999, 347). Interoperable higher education institutions, therefore, are those that accommodate exchanges of students, student records, as well as educators, and that assure the compatibility of the credentials they confer. For example, articulation agreements (discussed by Johnson in Chapter 13) facilitate the transfer of academic credits from one institution to another, ensuring that credits count toward the academic credentials students seek.

An exemplary effort to increase institutional interoperability in higher education is the Bologna Process, which seeks to create a 'European Higher Education Area' that 'facilitate[s] mobility of students, graduates and higher education staff,' among other goals (Bologna Secretariat, 2010). By 2010, 47 countries (including many that are not in the EU such as Australia) have begun efforts to create compatible national qualifications frameworks leading to bachelors, masters and doctoral degrees that have comparable meaning and value throughout the higher education area, and which facilitate academic partnerships among institutions. A related effort called the Tuning Process (Tuning Process, n.d.) complements

Bologna's focus on educational *systems* (the purview of national governments in Europe) by aiming to reform educational *structures* – including subject-level learning outcomes and competences – which higher education faculty oversee. The Erasmus Mundus master of science program in geo-information science and earth observation for environmental modelling and management, a graduate degree program that combines the resources of five European universities, is one example of the synergy that institutional interoperability fosters (Erasmus Mundus MSc GEM program, n.d.). Unfortunately, no comparable effort to reform higher education systems and structures is yet underway in the US. One consequence of this is that relatively few articulation agreements are in place among GIS&T programs at US two-year colleges and four-year universities (Johnson, Chapter 13), and few programs participate in formal inter-institutional partnerships that allow students to take advantage of more than one institution's capabilities.

Barriers to institutional interoperability are social and cultural, but also technical and legal. In particular, students' ability to participate in educational offerings at more than one institution at a time (via e-learning, for example) may be stymied by the perceived need to share student records among incompatible information systems while assuring confidentiality as required by law. An important part of a technical solution to this problem is Shibboleth, an Internet2 middleware product that supports inter-institutional sharing of web resources subject to access controls (Internet2 Middleware Initiative, 2011). Leung *et al.* (2009) describe how students enrolled in Leeds University's (UK) master's program in geographic information systems participate in an online class on GIS programming and customization at Penn State University without creating redundant student records. Leeds students register for an independent study class at their home institution. The requirements of the independent study are to fulfill all requirements of the Penn State class. Shibboleth-enabled authentication allows Leeds students log into the Penn State learning management system *using their Leeds IDs and passwords*. At the end of each class term, Penn State instructors relay students' marks to the Leeds faculty member who supervises the independent study. The Leeds instructor translates the Penn State mark into a corresponding Leeds mark, accounting for cultural differences in how marks are awarded in the US and UK. A translation table is included in the revenue sharing agreement established between Leeds and Penn State (and the University of Southampton). Eighty per cent of tuition revenue that students' pay to their home institution is transferred to the partner institution that provides the online class. In a similar manner, Penn State's online masters program shares students with Southampton, which provides a popular online class on GIS for analysis of health. Boundary-bridging technologies like Shibboleth challenge education providers to develop deeply collaborative relationships that can reduce redundant content production and allow students to access a greater variety of educational opportunities within international networks of partner institutions.

Internationalization

The GIS&T enterprise is global in scope, as is the need for GIS&T education (Phoenix, 2004). GIS&T is often applied to problems that cross national boundaries, such as natural and man-made emergencies, environmental monitoring and management, and the spread of infectious disease. Despite this, Kemp and Frank (1996) note that GIS&T curricula are often designed locally, by individuals or small groups of educators who tend to emphasize their own particular interests and specialties.

Countering this tendency are several noteworthy efforts to internationalize GIS&T curricula, including the UNIGIS International consortium (Strobl, Chapter 24), the Erasmus Mundus masters program, the Association of American Geographers' online center for global geography education (Solem *et al.*, 2003), and the formalized sharing of students among e-learning programs at Penn State, Leeds University and the University of Southampton (Wright, Treves and Martin, 2009). Still, barriers to international collaboration remain high. Barriers include incommensurate degree and assessment requirements, uneven access to technology, diverse learning styles, differences in academic cultures and instructor preparation, lack of incentives for faculty collaboration, and the difficulty of executing inter-institutional articulation, intellectual property, and revenue sharing agreements.

Notwithstanding these obstacles, the potential of international and inter-institutional collaboration to enrich GIS&T education is compelling. Among a variety of strategies for managing the development and delivery of internationalized curricula, Solem *et al.* (2006, 155) emphasize the potential of internet-enabled distance learning to foster international collaboration in support of the lifelong learning needs of geospatial professionals:

> Distance education offers the possibility of sharing professional development options among partner institutions, each of which offers its own unique, and ideally complementary, learning opportunities.

The possibility of pooling online GIS&T courses led by educators at multiple institutions worldwide leads to an ambitious vision of a 'meta-university of GIS&T' capable of providing a uniquely diverse and truly adaptive curriculum:

> Adult learners ought to be able to earn degrees and certificates by successfully completing curricula that they design in consultation with qualified advisers, within a[n international] marketplace of distance learning providers that is coordinated by one or more educational brokerages such as the Worldwide Universities Network (WUN) (Solem *et al.*, 2006, 155).

Creating an online meta-university constitutes another grand challenge for GIS&T educators. Obstacles to such an achievement are formidable. International activities meant to increase institutional interoperability (discussed above) offer some encouragement. Meanwhile, however, international collaboration remains an intention of many senior administrators, but rarely leads to truly collaborative ventures (Dixon and Tammaro, 2003).

29.5 Conclusion: the GIS&T community of practice

Institutional barriers notwithstanding, one the most distinctive features of GIScience is the way educators have worked cooperatively across disciplinary and national boundaries to innovate and improve practice. Whilst such collaboration has frequently been formal in the form of organized consortia (The *SPACE* and *SPLINT* consortia in the US and UK, and the *DialogPLUS* project that involved both US and UK educators), much collaboration has been of a very informal variety. This sense of shared, collaborative mission is often now described as a 'community of practice' (CoP) or 'professional learning community.' CoPs are:

> . . . groups of people who share a concern, a set of problems, or a passion about a topic, and who deepen their knowledge and expertise in this area by interacting on an ongoing basis . . . These people don't necessarily work together every day, but they meet because they find value in their interactions. As they spend time together, they typically share information, insight, and advice. They help each other solve problems. They discuss their situations, their aspirations, and their needs (Wenger *et al.*, 2002, 4–5).

This is an apt description of much that has occurred in GIS&T education over the past twenty years (Unwin and Tate, 2008; Tate and Unwin, *in preparation*). It is true that academic life as a whole – the departments, disciplines and associations with which we are so familiar – fit within this broad definition. What's interesting about GIS&T education is that it cuts across many of these traditional disciplinary divisions. These are the situations in which Wenger *et al.* are most interested. Their concern is how to create, manage and use knowledge within organizations such as companies and businesses where there is no tradition of sharing and improving practice over extended periods of time, or in situations involving expertise which cuts across traditional disciplinary or professional lines – as does GIS&T. They argue that, in these settings, the ability of a business, organization, or community to learn and improve is increasingly essential to its success in the global knowledge economy.

But how can CoPs be sustained and expanded, and in what situations are they effective or ineffective? Much of the recent research on CoPs focuses on these questions (Cox and Richlin, 2004; Mullen, 2009; Samaras *et al.*, 2009; Stoll and Louis, 2007). Interestingly, many of the successful case studies in this literature focus on how CoPs can be used to help learners and educators cooperate across distances and institution types which would make other types of sharing difficult (Kimble *et al.*, 2008). In many ways, the findings of these studies reinforce the seven principles Wenger *et al.*, (2002, 49-64) outline for cultivating communities of practice by:

1. Designing for evolution by building on existing knowledge and networks rather than by reinventing what already exists.
2. Opening a dialog between inside and outside perspectives.
3. Inviting different levels of participation.
4. Developing both public and private community spaces for sharing and learning.
5. Focusing on value to the community and community members.
6. Combining familiarity with excitement so that routines are supported, but also varied to keep topics fresh and interesting.
7. Creating a rhythm for the community so that it has a sense of energy and movement to keep members involved.

This list aligns closely with points raised in Chapter 1 and elsewhere in this book about the dynamism of the GIS&T education community. Almost point for point, the rise of GIS&T has involved projects supporting these seven principles. This is not to say that every principle has been realized equally, or that the work of building a GIS&T education CoP is complete. Instead, we see this book as a means of rekindling excitement and re-energizing the community at a critical moment in its history. The first generation of innovators is now retired or retiring, so this is clearly a time to consider how the GIS&T CoP can be sustained into the future. But both the theory and practice of GIS&T is being transformed by the field's move into cyberspace. Perhaps the time is right to consider how best to sustain and expand the GIS&T education CoP by building upon past successes and lessons and rethinking goals and opportunities. In this sense, there is much that can be shared elsewhere in geography and other fields. Our CoP is a model worth emulating in terms of building and sustaining innovation in educational theory and practice.

We began this book by observing that the central challenge for educators is that GIS&T is changing so rapidly. Agility and openness to learning continuously – including from our students – is needed more than ever. Fortunately, none of us has to confront these challenges alone, so long as collaboration remains a defining feature of the GIS&T education community of practice.

References

Association of American Colleges and Universities (2011) *AAC&U Member Survey Findings.* http://www.aacu.org/membership/2009surveytalkingpoints.cfm (accessed March 20, 2011).

Allen, I. E., and J. Seaman (2010) *Learning on Demand: Online Education in the United States, 2009.* http://sloanconsortium.org/publications/survey/pdf/learningondemand.pdf (accessed March 23, 2011).

Association of American Geographers (2009) Statement on professional ethics. http://www.aag.org/cs/about_aag/governance/statement_of_professional_ethics (accessed March 20, 2011).

Bernard, R. M., P. C. Abrami, Y. Lou, E. Borokhovski, A. Wade, L. Wozney, P. A. Wallet, M. Fiset and B. Huang (2004) How does distance education compare with classroom instruction? A Meta-Analysis of the Empirical Literature. *Review of Educational Research Fall 2004,* 74(3): 379–439. http://rer.sagepub.com/cgi/content/abstract/74/3/379 (accessed March 23, 2011).

Bologna Secretariat (2010) About the Bologna process. http://www.ond.vlaanderen.be/hogeronderwijs/bologna/about/ (accessed March 24, 2011)

Bransford, J. D., A. L. Brown, and R. R. Cocking (eds) (2000) *How People Learn: Brain, Mind, Experience, and School.* Washington DC: National Academies Press.

Cox, M. D., and L. Richlin (eds) (2004) *Building Faculty Learning Communities. New Directions for Teaching and Learning,* 97. San Francisco: Jossey-Bass.

DiBiase, D., M. DeMers, A. Johnson, K. Kemp, A. T. Luck, B. Plewe and E. Wentz (eds) (2006) *Geographic Information Science and Technology Body of Knowledge.* Washington DC: Association of American Geographers and University Consortium for Geographic Information Science.

DiBiase, D., and Kidwai, K. (2010) Wasted on the Young? Comparing the performance and attitudes of postadolescent undergraduates and postbaccalaureate students in identical online classes. *Journal of Geography in Higher Education,* 34(3): 299–326.

DiBiase, D., T. Corbin, T. Fox, J. Francica, K. Green, J. Jackson, G. Jeffress, B. Jones, B. Jones, J. Mennis, K. Schuckman, C. Smith, and J. Van Sickle (2010) The new geospatial technology competency model: Bringing workforce needs into focus. *URISA Journal* 22(2): 55–72. http://www.urisa.org/files/URISA%20Journal%20Vol.22%20Issue%202.pdf (accessed March 4, 2011).

Dixon, P., and A. M. Tammaro (2003) Strengths and issues in implementing a collaborative inter-university course: the international master's in information studies by distance. *Education for Information,* 21(2–3): 85–96.

DOLETA (n.d.) *O*NET Online.* Washington DC: U.S. Department of Labor Employment and Training Administration. http://online.onetcenter.org/ (accessed March 20, 2011).

DOLETA (2010) *Geospatial Technology Competency Model.* Washington DC: U.S. Department of Labor Employment and Training Administration. http://www.careeronestop.org/competencymodel/pyramid.aspx?geo=Y (accessed March 24, 2010).

Durham, H., K. Arrell, and D. DiBiase (2009) Collaborative learning activity design: Learning about the Global Positioning System, In: P. Rees, L. Mackay, D. Martin and H. Durham (eds), *e-Learning for Geographers: Online Materials, Resources, and Repositories.* Hershey, PA: Information Science Reference, pages 38–51.

Erasmus Mundus (n.d.) MSc GEM program http://www.gem-msc.org/ (accessed March 24, 2011).

Esri (2011) Industries. http://www.esri.com/industries.html (accessed March 24, 2011).

Gaudet, C., H. Annulis, and J. Carr (2003) Building the geospatial workforce. *URISA Journal* 15(1): 21–30. http://www.urisa.org/files/Gaudetvol15no1.pdf (accessed March 20, 2011).

GISCI (2008) Code of ethics and rules of conduct. http://www.gisci.org/code_of_ethics.aspx (accessed March 20, 2011).

Internet2 Middleware Initiative (2011) Shibboleth. http://shibboleth.internet2.edu/ (accessed March 24, 2011).

Johnson, J. (2010) What GIS technicians do: A synthesis of DACUM job analyses. *URISA Journal* 22(2): 31–39. http://www.urisa.org/files/URISA%20Journal%20Vol.22%20Issue%202.pdf (accessed March 4, 2011).

Kemp, K. K., and A. U. Frank (1996) Toward consensus on a European GIS curriculum: The international post-graduate course on GIS. *International Journal of Geographical Information Systems*, 10(4): 477–497.

Kimble, C., P. Hildreth and I. Bourdon (2008). *Communities of Practice: Creating Learning Environments For Educators.* Charlotte NC: Information Age Publishing.

King, E. F. (2008) Can PBL-GIS work online? *Journal of Geography*, 107(2): 43–51.

Knowles, M. S., E. F. Holton, and R. A. Swanson (2005) *The Adult Learner*, 6th ed. Amsterdam: Elsevier.

Leung, S., D. Martin, R. Treves and O. Duke-Williams (2009) Exchanging e-Learning Materials, Modules, and Students, In: P. Rees, L. Mackay, D. Martin and H. Durham (eds), *e-Learning for Geographers: Online Materials, Resources, and Repositories.* Hershey, PA: Information Science Reference, pages 20–37.

Longley P. A., M. F. Goodchild, D. J. Maguire, and D. W. Rhind (2005) *Geographic information systems and science*, 2nd ed. Chichester: Wiley.

Marble, D. F., and members of the Model Curricula Task Force (2003) *Strawman Report: Model Curricula.* Alexandria, VA: University Consortium for Geographic Information Science.

Means, B., Y. Toyama, R. Murphy, M. Bakia, and K. Jones (2009) *Evaluation of Evidence-based Practices in Online Learning: A Meta-analysis and Review of Online Learning Studies.* Washington DC: U.S. Department of Education. http://www2.ed.gov/rschstat/eval/tech/evidence-based-practices/finalreport.pdf (accessed March 23, 2011).

Miller, G. (1988) *The Meaning of General Education: The Emergence of a Curriculum Paradigm.* New York: Teachers College Press Publications.

Mullen, C. A. (ed) (2009) *The Handbook of Leadership and Professional Learning Communities.* New York: St. Martin's Press.

National Research Council (2006) *Learning to Think Spatially.* Washington, DC: The National Academies Press.

Open Geospatial Consortium (2011) The OGC process: How the pieces fit together. http://www.opengeospatial.org/ogc/process (accessed March 24, 2011).

Penn State Public Broadcasting (2010) Geospatial Revolution. http://geospatialrevolution.psu.edu/ (accessed March 24, 2011).

Phoenix, M. (2004) Global GIS education. *GIM International*, February: 35–37.

Quality Assurance Agency for Higher Education (2007) *Subject Benchmark Statements: Geography.* http://www.qaa.ac.uk/Publications/InformationAndGuidance/Documents/geography.pdf (accessed August 15, 2011).

Rachal, J. R. (2002) Andragogy's detectives: A critique of the present and a proposal for the future. *Adult Education Quarterly*, 52(3): 210–227.

Rossman, M. H. (2000) Andragogy and distance education: together in the new millennium. *New Horizons in Adult Education*, 14(1). http://education.fiu.edu/newhorizons/journals/vol14n1.htm (accessed March 20, 2011).

Samaras, A. P., A. R. Freese, C. Kosnik, and C. Beck (eds) (2009) *Learning Communities in Practice*. Dordrecht, The Netherlands: Springer Press.

Sinton, D. S. and R. Schultz (2008) GIS and mapping applications for reasoning and critical thinking about the environment and society, IN: M. N. Solem, and K. E. Foote (eds), *Teaching College Geography: A Practical Guide for Graduate Students and Early Career Faculty*. Upper Saddle River, NJ: Prentice-Hall, pages 69–80.

Sitzmann, T., K. Kraiger, D. Stewart, and R. Wisher (2006) The comparative effectiveness of Web-based and classroom instruction: A meta-analysis. *Personnel Psychology*, 59: 623–664.

Solem, M., S. Bell, E. Fournier, C. Gillespie, M. Lewitsky, and H. Lockton (2003) Using the Internet to support international collaborations for global geography education, *Journal of Geography in Higher Education*, 27(3): 239–254.

Solem, M., L. Chalmers, D. DiBiase, K. Donert, and S. Hardwick (2006) Internationalizing professional development in geography through distance education. *Journal of Geography in Higher Education*, 30(1): 147–160.

Solem, M., I. Cheung, and B. Schlemper (2008) Skills in professional geography: An assessment of workforce needs and expectations. *Professional Geographer*, 60(3): 356–373.

Sondheim, M., K. Gardels and K. Beuhler (1999) GIS interoperability, IN: P. A. Longley, M. F. Goodchild, D. J. Maguire, and D. W. Rhind (eds), *Geographical Information Systems. Volume 1: Principles and Technical Issues*. New York: Wiley, pages 347–358.

Stoll, L., and K. S. Louis (eds) (2007) *Professional Learning Communities: Divergence, Depth and Dilemmas*. Maidenhead, UK: Open University Press.

Summerby-Murray, R. (2001) Analyzing heritage landscapes with historical GIS: Contribution from problem-based inquiry and constructivist pedagogy. *Journal of Geography in Higher Education*, 25(1): 37–52.

Tate, N. J., and D. W. Unwin (in preparation) *Collaboration in Giscience Education: Towards a Community of Practice*.

Tsou, M.-H., and K. Yanow (2010) Enhancing general education with geographic information science and spatial reasoning. *URISA Journal* 22(2): 45–54. http://www.urisa.org/files/URISA%20Journal%20Vol.22%20Issue%202.pdf (accessed March 4, 2011).

Tuning Process (n. d.) Tuning educational structures in Europe. http://tuning.unideusto.org/tuningeu/ (accessed March 24, 2011).

Unwin, D. J., and N. J. Tate (2008) Collaborating in GIScience education: Past trends, future possibilities. *Proceedings EUGISES2008*, September 11–14, Cirencester, UK. http://www.eugises.eu/proceedings2008/Unwintate_eugises_final.pdf (accessed March 10, 2011).

Wenger, E., R. McDermott, and W. M. Snyder (2002) *Cultivating Communities of Practice*. Cambridge, MA: Harvard Business Press.

West, B. A. (2003) Student attitudes and the impact of GIS on thinking skills and motivation. *The Journal of Geography*, 102(6): 267–274.

Wright, J., R. Treves, and D. Martin (2009) Challenges in the reuse of learning materials: Technical lessons from the delivery of an online GIS MSc module. *Journal of Geography in Higher Education*, 33: 78–87.

Zhao, Y., J. Lei, B. Y. C. Lai, and H. S. Tan (2005) What makes the difference? A practical analysis of research on the effectiveness of distance education. *Teachers College Record*, 107(8): 1836–1884.

Index

Teaching Geographic Information Science and Technology in Higher Education, First Edition.
Edited by David J. Unwin, Kenneth E. Foote, Nicholas J. Tate and David DiBiase.